Telecommunications Crash Course

Steven Shepard

Third Edition

McGraw Hill Education

New York Chicago San Francisco
Athens London Madrid
Mexico City Milan New Delhi
Singapore Sydney Toronto

Library of Congress Cataloging-in-Publication Data

Shepard, Steven.
[Telecom crash course]
Telecommunications crash course / Steven Shepard. —Third edition.
pages cm
Previous editions entitled: Telecom crash course.
Includes index.
p. cm.
ISBN 978-0-07-183266-3 (paperback)—ISBN 0-07-183266-1
(paperback) 1. Telecommunication. I. Title.
HE7631.S54 2014
384.068—dc23 2014011369

McGraw-Hill Education books are available at special quantity discounts to use as premiums and sales promotions or for use in corporate training programs. To contact a representative, please visit the Contact Us page at www.mhprofessional.com.

Telecommunications Crash Course, Third Edition

1 2 3 4 5 6 7 8 9 0 DOC/DOC 19 18 17 16 15 14

ISBN 978-0-07-183266-3
MHID 0-07-183266-1

The pages within this book were printed on acid-free paper.

Sponsoring Editor
Michael McCabe

Editorial Supervisor
Donna M. Martone

Production Supervisor
Lynn M. Messina

Acquisitions Coordinator
Bridget L. Thoreson

Project Manager
Vastavikta Sharma,
Cenveo® Publisher Services

Copy Editor
Ragini Pandey

Proofreader
Connie Blazewicz

Indexer
Cenveo Publisher Services

Art Director, Cover
Jeff Weeks

Composition
Cenveo Publisher Services

Meddle not in the affairs of wizards,
For you are crunchy, and good with ketchup.
—With apologies to J. R. R. Tolkien

This book is dedicated to my friend and colleague Joe Candido. Thanks for teaching me so much and for being there at all the right times. More adventures await—see you there.

Contents

Foreword

What a difference a decade makes.

The first edition of this book hit the streets in 2002, just after the calamitous telecom bubble had burst, volatilizing seven trillion dollars of market value into the digital atmosphere. Companies by the hundreds disappeared overnight and technology centers like Silicon Valley in California, Research Triangle Park in North Carolina, and Kanata in Ontario turned into ghost towns. Unemployed tech workers roamed the streets like zombies.

It was definitely not a good time to be in the technology industry. Everyone boxed up their logo-wear, fearing reprisals from people who had lost money in tech investments. Some hung around; many professionals left the sector completely, seeking the warmer climes of more stable industries.

Yet, despite the nuclear winter of the Tech Bubble I found myself dazzled by the technology that was available at the time and by the ongoing drama that defined the moribund telecom industry. I uploaded all of the chapters of the first manuscript of this book to my editor at McGraw-Hill using a 19.2 Kbps modem, which at the time seemed ridiculously fast. AOL, which had acquired Time-Warner, was a force to be reckoned with in the emerging digital media space. Nortel and Lucent were technology powerhouses that defined the industry. Mobility, which was just beginning to show its face, was still delivered using blocky phones that supported voice and voice only and that when used for more than about 15 minutes could heat a small house. The idea of national roaming, much less international, was something of a dream.

The Internet, such as it was, was still very much in its infancy. Nine years in the public domain, it was still pre-adolescent and had not yet become an unruly teenager. That would come later with the arrival of online porn, the browser wars, and low-cost digital compression.

Media meant television, radio, and VHS tapes. CDs were just beginning to show themselves as viable storage options. A large hard drive on a PC or laptop held 100 MB, far more than the average person would ever need. And the process of backing it up meant a handful of 3½ inch floppies or, if you were really sophisticated, an iOmega drive that connected via an EIA-232 connector. Later-stage machines had a built-in CD drive. The computer wars were well underway; Microsoft controlled the bulk of the market, and Apple was an up-and-coming computer manufacturer.

Information Technology—IT to its friends—was an enterprise function, a necessary and very expensive evil that corporations grudgingly paid for two reasons: one, because it supported business functions like accounting, finance, billing, operations, and maintenance; and two, because IT told them to. The IT organization behaved like a self-governing fiefdom, dictating its intentions and financial requirements to the rest of the corporation. Nobody questioned IT's decisions, motives, or spend because after all, we're talking about IT. *They know what they're doing, and we don't. So leave them be.*

Thirteen years and two editions later, things have changed—just a bit. In 2002, network processors—special computers that handled network traffic and which today we refer to as *routers*—began to appear in the network, and internet protocol (IP) started to take on an air of permanency in enterprise networks. Digitization was a much talked about topic, although cable companies were the ones who talked about it the most, not so much as a way to deliver a better video experience but rather as a way to compete more effectively with the telcos in their own game of voice and data transport. Meanwhile, the telcos decided that they should get into the content game, first as a joint venture with satellite providers, later as content providers across their own high-bandwidth optical networks.

As I look back on those heady days in preparation for writing this third edition of the *Telecom Crash Course,* I marvel at how much things have evolved in 10 or so years. Broadband really is—with wireless bandwidth routinely offering tens of megabits to mobile devices. No one refers to those devices as cell phones anymore, because the *phone* element is more of a functional footnote than it is a desired capability. The target market has become younger, more technically adept, and much more demanding. Telephone companies and their manufacturing partners are in dire straits as they struggle to redefine themselves in the face of declining relevance, ferocious competition and a perceived lack of vision. IT providers, which include hardware and software manufacturers, cloud providers, and data center operators, are finding themselves in the very uncomfortable place of no longer being in charge of their own destinies. Instead, IT has become the new purchasing department for technology, with control of that spend shifting to the business units.

And media companies? They have their own issues to deal with. Music companies once sold music; now they sell an experience that *involves* music. Movie companies once sold movies; now they sell a complete experiential surround, the center of which may be a movie but which may derive the bulk of its revenue not from ticket sales but from the sale of collateral products, character licensing and on-demand viewing. And to make matters worse, the traditional players in this space now find themselves having to anticipate the impact of the latest announcements from the real powerhouses in the media industry, the movers-and-shakers with names like Apple and Google, companies that understand one fundamental thing, a theme that will be woven through this entire book: *If you want to gain influence over a market, you must give up control of that market.* It's a simple yet profound observation, and to many, counter-intuitive. It is, however, true. Consider what Apple and Google have done: By giving up control of the content market (out of the hundreds of thousands of apps in the Apple Apps Store, Apple created only a handful, the rest created by third-party developers), they enjoy enormous influence over the platform market. And why does this matter? Because the platform— the device that is rarely more than a few inches from its owner—is the gateway to the experience enjoyed by the customer, and that is the single most important thing in this new world of telecom, media, and IT. By controlling the experience they essentially control the customer, creating stickiness, loyalty, and recurring revenue. But the revenue comes because of another observation that will be repeated throughout the book: *In this new world of technology enablement, companies in the game must accept the fact that they have two choices when it comes to revenue: They can have* some *of the money or* none *of the money.* Meaning? New competitive entities are emerging. The days of a single company being large enough and powerful enough to control a market are over; instead, the new competitive entity is an ecosystem of players made up of companies and individuals that contribute complementary products and services to the effort, resulting in a richer, more deliberate and customized experience delivered to the customer. Competition occurs between the ecosystems, not between individual companies.

Perhaps the most important message that this newest edition of *Telecommunications Crash Course* delivers is this: It simply isn't enough for telecom professionals to focus exclusively on telecom issues. Or for media professionals to focus solely on media. Or IT professionals to limit their focus to issues that solely affect the world of information technology. Why? Because the three domains are interlocked and interdependent. I don't care how good your network is, how universally deployed your broadband wireless service, or how pervasive your optical infrastructure; without good content to deliver over it, that network is nothing more than an exercise in infrastructure deployment. Similarly, the best content in the world does nothing for its developers without a good network over which to transport it to a paying customer and an IT infrastructure that makes it possible to store, analyze, track, and monetize the content. And the world's best data center is nothing more than a high-tech fortress, unless, of course, it is connected to a broadband transport network and houses a rich array of content.

As a result, the focus of the book has been expanded. It's still called the *Telecommunications Crash Course*, but it's now really a primer for the telecom, media, and technology sectors. Once you've taken the time to go through this book, you'll have a rich and complete understanding of the interplay among these three remarkable sectors and the co-dependent technologies they deliver.

Rest assured, the aftereffects of the bubble have largely subsided and the three sectors are growing and evolving and contributing nicely to the global economy. No longer are these sectors in the shadows: they now sit squarely in the bright light of a new reality, a reality that is powered by demand for always-on connectivity, an increasingly mobile lifestyle, a generation of millennials who depend on their mobile devices for staying in touch with their work, their family and their friends, and a blessing of content that can be accessed in a variety of ways—and monetized in a variety of ways. A new economy is being born, and we are lucky enough to be here as it happens.

Enjoy the book. As always, I thank you for your loyalty and friendship as readers and welcome your comments.

Steven Shepard
Williston, Vermont
January 2014

Steve@ShepardComm.com
Facebook: Telecom Crash Course

Introduction—and an Admonition

I began writing about technology in 1980. My first book came out that year; it was called *Commotion in the Ocean: A Technical Diving Manual.* At the time I was a professional SCUBA instructor and commercial diver living and working in California, and I had become annoyed by the lack of good teaching materials on the market for SCUBA instructors at the time. Most of the books still used old double-hose regulators, which haven't been in common use since Mike Nelson (AKA Lloyd Bridges) used them to foil bad guys on the 1950s TV show *Sea Hunt.* Those books were written by the original diving instructors, former Navy divers who believed that *real* divers had to be built like Tarzan, needed a 15-inch knife strapped to their thigh, and had to be able to hold their breath for three minutes while swimming at 100 ft. And they taught their classes and wrote their books that way. In other words, they taught and wrote as if they were teaching or writing for themselves, not the market of would-be divers who had an interest in the sport. So after trying unsuccessfully to teach divers using those inwardly focused books, I wrote my own manual. It was extremely popular because I wrote it in a conversational style, in plain language, and I directed it at new divers, not experienced "mossbacks." In other words, I focused outward, not inward.

There is a corollary here that we must keep in mind and which I will touch on at various places throughout this book. There are many books and training materials on the market today that purport to explain how technologies work, and they do so in excruciating detail. Similarly, many sales and marketing messages get delivered that are so over-the-top technical that they overwhelm the customer with techno-jargon and scare them away. In many cases, this happens because the authors of the books or training materials, or the sales professionals who are out there trying to move product, forget that their audience may not be as technical as they are, or may not be as enamored of technology as they are, or just don't care about the detailed inner-workings of the technology in question. An increasing number of sales encounters fall short because the person doing the selling or teaching the content or writing the book often forgets one simple but profoundly important thing: the technology itself isn't always the most important thing, but what it does, what it makes possible, what it enables, is *profoundly* important.

Please don't get me wrong: there are vast numbers of engineers and data architects and media specialists who care very much about the inner workings of the technologies that we're going to talk about in this book—and I'm glad they're out there. The work they do requires an intimate knowledge of the anatomical makeup of hardware and software, of muxes and servers, of protocols and standards, of data centers and central offices, and server rooms and security firewalls.

For those people there are hundreds of excellent resources on the market that cover every imaginable technology at breathtaking levels of detail. I have many of those books; I've even written some of them. And my job as a consulting analyst to the technology, media, and telecom (TMT) marketplace requires that I know that subject matter very well.

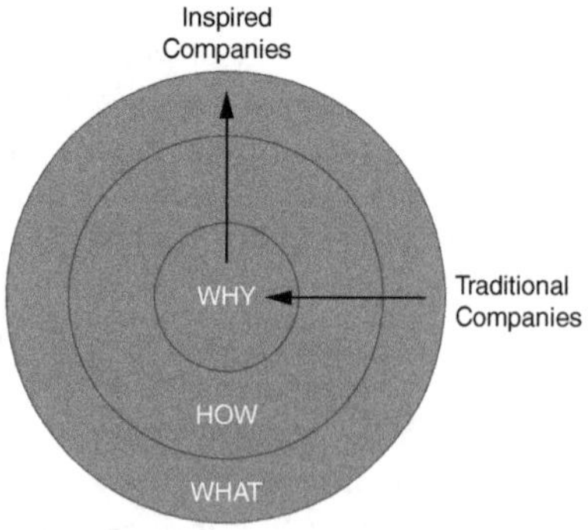

This book, however, is different. Instead of looking inward, it gazes outward, presenting technology in the context of the market, of the user, of the customer, not the people who design, build, maintain, troubleshoot, and sell it. Don't get me wrong—there's plenty of technology in this book, but it is presented to you with an overarching message: Why do I care about this? Why is this important to me, my company, my customer, my country, my family, the world? That's right: some of the technologies I will discuss in this book are literally changing the world in remarkable and profound ways.

Simon Sinek is a speaker and consultant to the sales and marketing world. In a TED talk he delivered in 2011[1] he explains the difference between "regular" companies and inspiring companies like Apple. He talks in the video about *the golden circle*, his explanation for why some companies stand head and shoulders above the competition, in spite of the fact that they make similar products. Traditional companies, he observes, begin the conversation with the customer at the periphery of the wheel—*what* they sell. They then progress to *how*

[1]Watch the YouTube video at *http://www.youtube.com/watch?v=d2SEPoQEgqA*.

the product works, is better, and the like, before finally ending with the *why*—as in, why the customer should buy the product.

Enlightened companies, on the other hand, begin in the middle: *why* we do what we do, followed by *how* we do it, followed in turn by a brief mention of the product, which is almost unnecessary for the customer to buy. At this point they will do business with you on the basis of who you are, not what you make.

Watch Simon's video. His message is powerful and aligns nicely with the overall message of this book. I will provide you with a solid education in technology, but more than that I will help you understand why it is so important from a social, economic, competitive, geopolitical—and of course—technical perspective, all of which are part of the *why* question.

So with that, let's get into it. Come take a trip with me.

African Journey

This is the African savannah. It is a hot summer day, mid-January, and the grass is shoulder high—a fact that prompted our driver to strongly suggest that we stay in the Land Rover (or "Landie," as the locals call them). There's no way of knowing what may be lurking in that grass, he explained; we agreed. It was tall enough to hide a rhino, and definitely high enough to cover a lion.

We are in Limpopo Province, about halfway between Polokwane and Tuli. The nameless dirt track is a mere suggestion of a road and is deeply rutted, causing the vehicle to sway precariously like a boat caught sideways in ocean swells. Ahead, a single bull elephant walks slowly down the road, oblivious to our presence. There is nothing else around: no buildings, no stores, not a single sign of people other than the occasional attempt at a fence, more to keep predators out than the goats it encloses in. Yet I note that I have five solid, unwavering bars of cellular signal strength on my mobile

We are on our way to an equally nameless village that is not far from here, a village where I have done some consulting work, believe it or not. A few years ago the local leaders of this village reached out to me and asked if I could help them bring communications technology to the area. I met with them the next time I visited South Africa, made some suggestions, and agreed to put them in touch with vendors I trusted who would help them select the few infrastructure elements they would need to get themselves connected. At the same time I also issued a warning: *Today, your children know the world as the distance they can comfortably walk from their village in a day. With communications technology they will become aware of a world that is much bigger and that will offer opportunities they cannot begin to have here. As a result they will grow up and most likely leave in search of employment in big cities. Are you prepared for that?* They were, they responded; they had already considered it, and while the departure of their children was not a pleasant prospect, the potential for better jobs and a brighter future overshadowed their misgivings. A year later the technology was installed and working.

But the story doesn't end there. Unintended consequences play a huge role in the developing world, and this place is no exception. This village and the villages that surround it are known for their crafts. Artisans in the region carve exquisite art from soapstone and hard African woods, and weave intricately patterned baskets from local grasses and, ironically, purloined telephone wire.

A handmade basket woven from telephone company wire.

The products created by the villagers here are stellar, but the market isn't. Most customers come into the village because they managed to get lost on the less-than-well-marked road to Kruger National Park, not because they are in search of local art work. But as they enter the village they slow down to look at the vendors who line the road, and many of them stop to make purchases. The problem is that many of the works of art that are for sale here are large—some of them far taller than I. This is not a problem, because a freight service sends a truck out every few weeks to collect the packages that have been prepared for delivery all over the world. Of course, "every few weeks" means that the customer will not see their purchase for a long time. By the time the packages are collected, transported to the coast or an airport, and shipped, two months may pass.

Of course, that was then. Today, a few years later, things have changed. As we approach the village I notice that the road is paved and much wider than the dirt road over which we've been driving for the last few hours. I still have five bars of service on my phone, but I also now have high-speed packet access (HSPA). And as we come around a curve in the road, I am gobsmacked by the sight of a large shipping facility, where before there was an open field. Large trucks are lined up to get into the place.

A local artist, selling his wares along the road.

Unintended consequences: (1) Bring Internet access and telephony to an area that hasn't had it before; (2) connect the regional schoolhouse to a broadband (WiMAX) wireless facility, (3) teach the kids how to use the Internet, and connect them to the global classroom; (4) Pay attention when the kids figure out how to put up a web site to sell local crafts, not to lost tourists, but to the world; and (5) sit back and enjoy a stable economy, a global education for your children, and growing regional prosperity. Welcome to the Internet economy. A combination of telecommunications access and transport, IT infrastructure, and the media elements displayed on the web site of this village quite literally put it on the map.

Want to see another example of the profound impact of TMT? Follow me, please, back to Johannesburg.

It is a hot, dusty African day. It's January, the peak of summer in the southern hemisphere, and the temperature hovers near 120 degrees. There is sparse shade, and what little I can find is occupied by troops of languid baboons—languid until attempts are made to get them to share. These animals have 2-inch incisors that are threateningly visible when they yawn—something they do a lot of in the shade—and which is an effective deterrent to my efforts to negotiate a small piece of the their space in the relative coolness of the tree's shadow.

I have stopped at the crossing of two dusty roads to take a closer look at technology in this distant corner of the world. I've spent a lot of time in Africa, Latin America, and Asia lately, and these places have opened their arms to me. For the first time in my writing career, I have found English lacking in its ability to express the profusion of feelings and sensations that these places create in me. I am enlightened, yet in the dark, about many things. I feel informed and aware, yet stricken with a sense of confusion and disconnection unlike anything I have ever felt in my travels to the far-flung corners of this planet. Just as I feel as if I am developing an insider's understanding of a place, something happens to knock it akimbo. South Africa, for example, has a global, sophisticated banking system, yet a large percentage of the country's population doesn't trust banks, preferring instead to keep their money *safely* hidden in their homes. Vast shanty towns spread scab-like over the landscape for endless kilometers. Yet these townships are girdled by a "commerce layer" so robust and vibrant that it defies description.

To my left is one of these townships, and approaching my car is a young resident of the place who looks at first glance like a mercenary soldier. His chest is covered by bandoliers that form a thick X, and were it not for his dazzling smile I might be worried. The bandoliers are not filled with bullets; they are cell phone chargers.

I have seen this fellow before; the last time we passed this way I stopped and talked to him. It is 8:30 in the morning, and I know that by noon he will have sold all of the charging cables and will instead be selling disposable razors, or papayas, or pecans, or carved African animals. By 2 p.m., he will move on to another product, perhaps flashlights, or kitchen utensils, or CDs; and between 5 and 8 p.m., prime commute time, he will sell the last of his inventory, which will most likely be newspapers, or cigarettes, or fresh melon juice, or various over-the-counter painkillers. Behind him his wife, hugely pregnant (with twins, I soon learn), tends a dung fire that burns in a 55-gal drum, over which she roasts fresh corn.

"So how are you, Richard?" I ask, remembering the name he has chosen to use because his Zulu name is too difficult for me to pronounce. "And how is your wife? You should be a father soon, I can see." He smiles, waves at his wife, who is 17. She smiles widely, brilliantly, and waves back. "I am fine sir, thank you. And yes, two weeks—the twins arrive in two weeks. It is a big day, sir, and we are ready. And where are you headed today?"

We chat briefly. The babies, who will be delivered at home by a traditional midwife, have a reasonable chance of a healthy birth provided there are no complications. Another elemental contrast: South Africa is the home of cardiac surgery pioneer Christiaan Barnard, who in 1967 replaced Louis Washkansky's failing heart with another, extending his life by 18 days and successfully completing the world's first heart transplant. South Africa is renowned for its modern medical capabilities. Yet recently, while walking on the beach in Eastern Cape Province with my wife, I met two Xhosa boys walking the same direction we were headed. Their bodies were painted face-to-feet with white mud because they were taking part in the ceremony that all Xhosa boys go through when they become men. Upon reaching 18 they leave their village or township and walk into the bush, where they build a small hut made of whatever they can find. There they will spend the next month, engaged in a ritual of purification. When they leave home they are given two buckets of white mud. One is to smear on their bodies, the sign of a Xhosa initiate who is in transition; the other is to eat. While they are in isolation the mud is the only thing they *will* eat, water the only thing they will drink. The mud gives a feeling of fullness. The ceremony culminates with ritual circumcision, performed by a village elder with traditional (read blunt, rusty, minimally sterilized) surgical tools, a vague comprehension of trauma management, and no anesthetic. Yet in a country with world-class healthcare, a shockingly large number of these young men die of shock or post-surgical infection.

It is no secret that Sub-Saharan Africa faces a health crisis of Armageddon-like proportions, a crisis the likes of which the world has not seen since the Middle Ages when plague and influenza burned through Europe. Not only do basic belief structures contribute to the problem, but so does the common practice of polygamy, not to mention a lackadaisical attitude toward prostitution. Truck drivers that travel north and south along the spine of Africa routinely sleep with different women each night, a practice that carries far less stigma in Africa than it does in the developed west. In the process they spread the virus more widely. Add to this the fact that many of the region's governments are disturbingly unstable—witness Zimbabwe's Robert Mugabe, who, in his zeal for singular power, has gutted his country through power seizures, convenient constitutional rewrites, and fear—and we have the formula for the greatest tragedy of our time.

Yet here I am, standing in front of a small township clinic. It is a small cinderblock building, about the size of a convenience store at home. I am here to meet the physician who runs the clinic. So far I'm not impressed; a hand-lettered sign taped to the inside of the window reads, disturbingly, "PAP SMEARS AND CIRCUMCISIONS WHILE YOU WAIT." Clearly they mean "no appointment required," but that's not what the sign says.

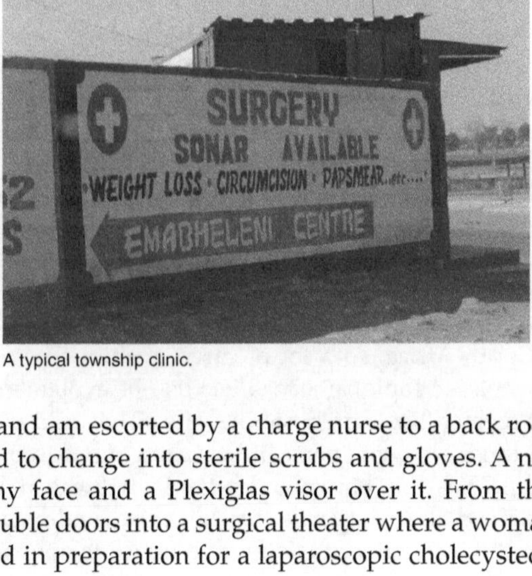

A typical township clinic.

I enter and am escorted by a charge nurse to a back room where I am directed to change into sterile scrubs and gloves. A nurse puts a mask on my face and a Plexiglas visor over it. From there I walk through double doors into a surgical theater where a woman has been anesthetized in preparation for a laparoscopic cholecystectomy—the removal of her gall bladder. The surgeons make four small abdominal incisions, into one of which they insert a tiny video camera. Surgical tools are inserted into the other incisions and the gall bladder is removed. The entire procedure takes 45 minutes, from open to close. The woman goes home that afternoon, groggy but feeling fine.

It might be important to note that the surgeons who performed this procedure were in Baltimore, Maryland, 8,000 miles away. They operated on the woman using a robotic surgical system that gave them a stereoscopic view of the sterile field and complete control over the tools they required to perform the procedure. The local healthcare professionals were there to change tools on the machine, monitor the status of the patient, deliver anesthesia, and close the incisions after completing the surgery.

But this book is not about healthcare; it is about technology. But think about the role that telecom, media, and technology play in this scenario. Suddenly, distance means nothing in the domain of healthcare. I wonder whether Frances Cairncross imagined this happening when she wrote "The Death of Distance" for the *Economist* Magazine in the late 1980s.

A shipping container that serves as a telephone company central office in sub-Saharan Africa.

Back to the crossroads. Sitting squarely in the open field across from me, the baboons and the shade is an international shipping container, the kind that are stacked on container ships. It is about 30-ft long and 10-ft high. A steady stream of people comes and goes, entering and leaving the container through the open doors on one end.

This container houses a switching office and customer service center for the cellular company that owns it. There are many of these containers scattered across the African landscape; they are cheaper than traditional brick and mortar structures. In fact, they cost nothing. While in Eastern Cape Province, I talked with the logistics manager of a large shipping company that moves cargo in and out of Durban, a major eastern port. Both Mercedes and BMW build their right-hand drive vehicles in South Africa, so a lot of cargo is shipped from Durban. "When the wireless people came to talk to us," he explained, "they had what at first seemed like a preposterous offer. They told us, "We need about 700 of those old containers, and we can't afford to pay you anything for them." We didn't quite understand what was in it for us, but as they explained their rationale, we began to understand. "We're going to use these containers as offices to house telephone company equipment. This equipment will serve to connect hundreds of cottage industries to the rest of the world, opening the world to products that today never see markets beyond Sub-Saharan Africa. Give us these containers and we'll make sure that you are the shipping company they use when the time comes to move product." Needless to say, we were quick to put the containers wherever they wanted them.

Stepping into the relative cool of the container, I'm charmed by what I see. The wall to my right has a row of cordless phones hanging in charging cradles. People who don't own phones can use these by buying a prepaid card. Halfway into the container is a counter, behind which a young girl takes orders for new service, collects phones brought in for repair, sells phones, recharges prepaid cards, and answers questions from customers. Behind her is a door with a cypherlock—the kind that has five silver buttons that must be pushed in a certain combination to open the door. This is important because the door opens into the switching equipment that provides wireless service to the local community. Charming is the fact that the door that protects the central office is a screen door. This is tropical Africa, after all, and temperatures in the summer can soar well above 130 degrees Fahrenheit. There are spinning ventilators on the roof of the container; the screen door ensures a steady flow of air for the equipment.

Across the street, cater-cornered from Richard, an old man and two young boys are setting up a rickety card table under a makeshift sunshade (a bed sheet) they have erected on four rough poles. On the card table they spread a collection of battered mobile phones of various vintages, all of which work. They then hang a sign on the front of the table that advertises available phone service for those who don't own a phone of their own. For a small fee, a person can make calls on one of the phones, paying by the minutes of use, carefully timed by one of the young boys using an equally battered stopwatch.

As we drive away from the intersection, hands wrapped around ice-cold Coca-Colas purchased from a roadside vendor taking shelter from the sun under a blue tarp, I shake my head as I try once again to get my arms—and my head—around what I see. Here is innovation at its best—succeeding because of strong beliefs about the future and the ability to shape it.

Africa offers hundreds of examples of the impact of the technologies we will cover in this book. It has become the place where mobile banking has become real, not because of the availability of technology but because of the fact that people don't trust banks. But they will trust a member of their village whom they know and who is willing to play the role of banker for everyone in the village, using secure cellular technology and mobile banking applications to ensure that everyone gets paid and has ready access to their funds.

Beyond Africa, of course, are equally compelling examples. A group of healthcare providers in southeast Asia uses mobile phones to text information to young women expecting their first children—what to eat to stay healthy, how to prevent infection, what to expect as your body changes in preparation for birth—and the infant mortality rate in the region plummets. A group of Bangladeshi women send text messages to illiterate members of society (but who have cell phones), and teach them to read. NGO Tostan's Community Empowerment Program saw similar results; after four months of sending SMS-based reading lessons to a large sample population, 73 percent were able to read the text messages they received,[2] an increase from nine percent at the beginning of the project.

The technologies that we will study in this book lie at the heart of global change. But they have just as much impact in the developed world as they do in developing countries. Telecom, IT, and media technologies are changing the way we work, live, play, educate ourselves, engage with each other, make decisions, influence politics, and do research. The impact of these technologies is spectacular. Consider the countless photographs[3] that emerged from the Arab Spring movement, thanking Facebook and Twitter for helping the people of the affected countries achieve what they are striving to do.

So with that, let's talk about the book.

Structure of the Book

This book consists of 11 chapters, as follows:

Chapter 1: The Changing Technoscape. This chapter sets the stage and introduces the argument that the telecom, IT, and media industries are converging and changing the rules of competitive engagement in exciting, dramatic, and irreversible ways.

Chapter 2: The Standards that Guide Us. Here we discuss the role of standards and include a brief discussion of the relevant standards that affect the IT and media sectors. We will also discuss basic network concepts and talk a bit about the history of the Internet as well as its evolving role in global communications.

Chapter 3: Data Communications Protocols. This chapter offers an overview of basic data communications concepts, mostly as a way to differentiate it from telecommunications. This chapter also dives somewhat deeply into the world of protocols, which are central to a clear understanding of the interplay among the three industry sectors.

Chapter 4: Telephony. In this chapter we talk about the world of telephony, which, despite all claims to the contrary, remains a vibrant and central force in the world of TMT. We discuss not only traditional telephony but VoIP as well, including the roles of SS7

[2] Tostan Web site, *http://www.tostan.org/area-of-impact/education.*

[3] *http://www.flickr.com/photos/international_center_for_transitional_justice/5413612600/*

and SIP and the emergent world of the IP PBX, sometimes called Broadband VoIP.

Chapter 5: The Byzantine World of Regulation. It may not be sexy, but it's important. Here we present an overview of regulation[4], its role, what it means, and the challenges regulators face as they attempt to regulate an ever-evolving industry. Painless, I promise.

Chapter 6: Premises Technologies. This chapter represents our foray into IT technologies beginning with an overview of the computer. We will also discuss premises technologies here, including local area networks (LANs), the growing role of gigabit ethernet, and a variety of access technologies including WiFi, ZigBee, FireWire, Thunderbolt, and USB.

Chapter 7: Content and Media. Here we discuss multimedia, video, and TV, as well as image coding schemes, digital content, the concept of Long Tail economics and its role in media. We will also discuss the central roles of smartphones and tablets.

Chapter 8: Access Technologies. This chapter is all about access technologies such as (on the fixed side) DSL and cable-based access technologies. We will also discuss the role of DOCSIS 3.0 and compare the role of the Cable Modem Termination System (CMTS) to that of the DSL Access Multiplexer (DSLAM). On the wireless side, we include the early history of radio before diving into the technologies themselves. We will discuss CDMA and GSM, HSPA, and LTE, as well as Bluetooth, RFID, and satellite solutions.

Chapter 9: Transport Technologies. In this chapter, we present brief overviews of such legacy technologies as frame relay and ATM, then provide a bit more detail about high-speed IP switching, optical networking, DWDM, channelized optics, and optical switching.

Chapter 10: The IP Takeover. As IP becomes the de facto standard for global networking, it is critical to have a good understanding of how it works and how it differs from the access and transport solutions that preceded it. Topics covered in this chapter include the inner workings of IP, IPv6, multiprotocol label switching (MPLS), and IP networking.

Chapter 11: The IT Mandate. This chapter builds on the material introduced in Chap. 4. This chapter offers discussions about the convergence of IT, telecom, and media; the expanding and evolving role of IT; the dramatic evolution of IT that has taken place in the last three years; cloud technologies; how data centers work; the increasingly important role of analytics; Big Data; security; the rise of Dumb Terminal 2.0; Bring Your Own Device (BYOD); and other

---

[4]Tostan Web site, *http://www.tostan.org/area-of-impact/education.*

topics that are shaping the technology landscape today. This chapter describes in significant detail the functional crossover that is occurring between the IT, telecom, and media sectors.

We conclude with some closing thoughts about the future of the TMT industry as a whole.

Intrigued? Good! Let's get into it.

Acknowledgments

One more time, with feeling.

I love to write—I often tell people that writing is something I am, not something I do—and as you can imagine, much of it is done alone, in my office or in a hotel room or on a plane, working in front of a computer screen (as I am doing now). Behind the screen, however, good writers stash a cadre of confidants upon whom they rely for editorial assistance, story guidance, and moral support. I am no different. Without those people I would produce much poorer words.

So my thanks go out to you all for all you do. Thank you so much for everything you have done to help me over the years. A special shout-out to Steve Chapman, my long-time editor at McGraw-Hill who approved the first edition of this book. Thanks, for everything, Steve.

In alphabetical order, I thank Phil Asmundson, Joe Candido, Anthony Contino, Phil Cashia, Steve Chapman, Linda Craenen, Bruce Degn, Lucie Dumas, Dave Heckman, Lisa Hoffmann, Linda Johnsen, Ron Laudner, Matt Leary, Dee Marcus, Roy Marcus, Gary Martin, Jacquie Martini, Mike McCabe, Dennis McCahill, Dennis McCooey, Paul McDonagh-Smith, Louis Morin, Jim Nason, Dick Pecor, Dan Pontefract, Glenn Ravdin, Gabriela Ruiz de la Flor, Kenn Sato, Mary Slaughter, Doug Standley, Dave Stubbs, Jay Tucker, Andrew Turner, Patricia Vaca, D Walton, Tim Washer, and Craig Wigginton.

I also thank my family. Sabine, you've put up with me for almost 38 years; thank you for all the insights, patience, friendship, and love. Cristina and Steve, thanks for teaching me, constantly, to be a better person. Mallory, thanks for bringing so much love to our family, and for giving us Ayla, our first grandchild, born 10 days ago as I write this. Children—and grandchildren—are the greatest gift imaginable.

Thanks to you all.

CHAPTER 1

The Changing Technoscape

This is a book about the remarkable interplay among the telecom, IT, and media sectors that defines the greater technology industry today. Never before have the interdependencies between these three sectors been so obvious, or so important.

There was a time, not all that long ago, when it was a simple exercise to assign companies to each of the three sectors. Telecom was the sovereign domain of telephone companies and thus played host to all of the usual suspects: Telefónica, Verizon, AT&T, TELMEX, British Telecom, TELUS, and so on. The technology domain hosted traditional IT players as well as software and hardware companies such as Cisco, Juniper, Alcatel-Lucent, Nortel, Ericsson, Siemens, Oracle, HP, Google, Yahoo!, etc. And the media sector was the home to such grand players as Sony, Disney, Vivendi, CBS, Universal Music, all of the Hollywood studios, and the New York Times to name a few.

But today, things aren't that simple. Look at the nested circles in Fig. 1.1. With the express understanding that the intersection area of the three circles is off-limits, where would you slot Apple into this equation? Clearly they are in the technology sector, given that they manufacture computers, tablets, iPhones, iPods, monitors, Apple TV devices, and software. But they are also one of the largest (if not *the* largest) distributor of music and videos in the world, and are responsible for redefining the music industry—and, perhaps saving it in the process. So clearly they are in the media sector as well. And thanks to the iPhone, one could reasonably argue that they play a key role in the telecom industry. So what *is* their industry? We could cheat and give it a name: "telemediology," since that incorporates elements of all three sectors, but that doesn't help us much. Instead of naming it, let's look at what the industry does and in the process explain the interdependencies between the three sectors.

As it turns out, each of the three sectors performs a unique and critical role in the new industry; a role that is clearly defined. It's at the margins, where the sectors collide, that things get fuzzy.

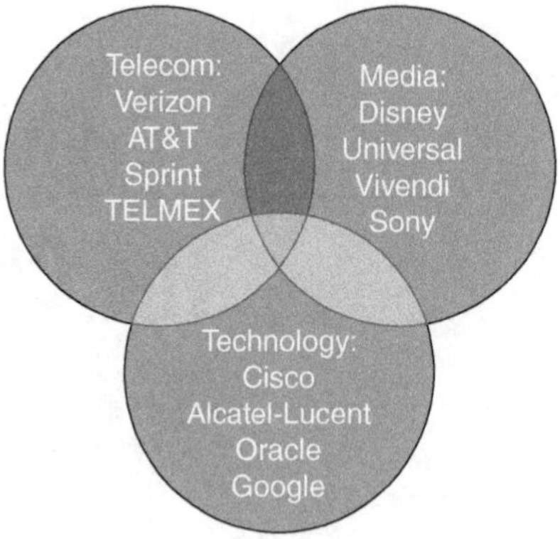

Figure 1.1 Convergence of the telecom, content, and IT domains.

The Telecom Sector

In spite of the remarkable advances in communications technologies that we have seen over the last decade or so, the role of the telecom sector hasn't changed all that much since it was created in 1876. In that year, Alexander Graham Bell, working in his lab on the invention of a multichannel harmonic telegraph, accidentally spilled a beaker of hydrochloric acid into his lap, causing him to say to his loyal assistant, "Come here, Watson, I need you." With the acid eating away at his nether regions I suspect that those weren't the words he actually uttered, nor do I believe that Watson needed a phone to hear what he *did* say. Whatever the case, beginning with the invention of the telephone 138 years ago, the role of the telecom industry has always been, and will always be, access and transport. *Access* defines the ability of the customer, the user, to gain access to the network and the connected resources that make it possible for them to make a fixed or wireless telephone call, check email, download a song or movie, set up a video conference, view a digital x-ray, and so on.

Transport refers to the responsibility that the network has to connect a user to whatever far-flung resource they wish to access. This is the *tele-* part of telecommunications; the word comes from the Greek and means over a distance. So the word *telecommunications* literally means communicating over distance.

The Technology Sector

The technology sector, sometimes called IT (for information technology), is home to a vast array of companies that have relatively well-defined roles. These companies include software creators, hardware manufacturers, cloud and hosting providers, application developers, and the more difficult-to-define companies like Google, Yahoo!, and AOL.

This sector has a broader responsibility than its telecom counterpart. Its responsibilities include:

- Design and manufacture of semiconductors, computer device sub-components (hard drives, mother boards, memory arrays, FPGAs, ASICs), monitors, televisions, computers, tablets, terminal devices, machine-to-machine communications devices, mobile phones, central office and data center equipment, and outside plant infrastructure elements.
- Design and development of operating system software (LINUX, OSX, Windows), application software, network management software, and specialized applications that serve vertical industries such as healthcare, education, government, manufacturing, and transportation.
- All elements of the burgeoning world of cloud services, including data center operations, content hosting, online cloud services such as storage-as-a-service, platform-as-a-service, infrastructure-as-a-service, and software-as-a-service, customer analytics, and monetization or billing.

The Media Sector

The media sector is all about content, whether it be commercially developed and sold, user-generated or otherwise. The media sector is responsible for creating, funding, advertising, distributing, producing, and monetizing of commercial properties such as feature films, TV programs, and music; providing distribution platforms for user-generated content, such as YouTube, Facebook, Google+, Flickr, Baidu, and Orkut; and developing interactive capability to drive a richer and more nuanced understanding of the customer so that advertising efforts have more impact.

Interdependencies

Clearly the roles of these sectors are well-delineated; each is equally as important as the other two. But there is one significant observation about them that I want to make here; an observation that drives the philosophy behind this new edition of the book and that must drive the thinking of anyone who works, sells, designs, markets, or strategizes in the TMT world. Each of these three sectors, telecom, media and technology, is fundamentally important in its own right, employing large numbers of talented people, driving technology forward with a vengeance, and changing every industry on earth. The real value, however, the real *power* in these industries, lies in the places where they overlap, the places where one empowers another. This is the technology world's manifestation of the phrase: "the whole is greater than the sum of its parts."

Consider the following: The media industry, with all of its feature films, independent productions, music videos, TV shows, songs and albums, and video games, is remarkable, but has zero value without a dependable broadband network over which to transport it and provide access to it and an IT infrastructure that can store it, monitor its use, analyze customer interaction with the content, and monetize it.

The IT industry is a modern miracle, but without a dependable broadband network infrastructure that connects its infrastructure elements to a customer and a range of content that it can store, analyze, and monetize, it is nothing more than a high-tech warehouse full of hardware and software.

And the telecom industry? Its optical infrastructure is capable of transporting the entire contents of the United States' Library of Congress from coast to coast in a matter of seconds. But without content to transport and an IT resource that can turn that transported content into revenue, that impressive network is nothing more than a dry pipe.

My point with this diatribe is that it isn't enough to simply know one of the three sectors. To truly be effective in this industry, it is critical to be a tripod—that is, to have a foot in all three. Hence, the changed focus of the *Telecom Crash Course.*

The Changing Competitive Paradigm

Remember how simple it used to be to understand the competitive landscape? Phone companies battled phone companies; Disney went at it with Warner Brothers; Google and Yahoo! fought for mindshare; IBM, DEC and Amdahl fought like sworn enemies; Dell and Gateway debased each other; and Apple and Microsoft crafted hysterically funny and hard-hitting commercials designed to cast each other in a bad light. Ah, the good old days of a decade ago: today not so simple. These players still compete against each other as they always have, but as they do battle they uneasily cast a look behind them now and again to determine what new and unanticipated player has entered the fray and that will tip the scales of battle. Disney and Warner are still at each other's throats, but they worry as much about YouTube as they do about each other. IBM, DEC, and Amdahl are no longer an issue for each other since DEC was absorbed by HP as part of its Compaq acquisition and Amdahl barely exists, but IBM has a different concern about cloud players and virtualization companies. Dell doesn't worry so much about Gateway anymore, but they *do* worry about Apple and Samsung with their very successful tablets and about Google with its growing netbook business. Apple and Microsoft will always have a love-hate relationship because of their shared history (read Walter Isaccson's *Steve Jobs* for a better understanding of the relationship between these two siblings), but they are more concerned today with the rise of Android and Google's eminence in their space.

Other forces are shaping things as well that are causing concern and consternation. For example,

The market's growing power base. A shift of power is taking place from the provider of technology services to the consumer of technology services, largely the result of choice. Customers now have the ability to pit one provider against another in a constant search for the better product, experience, or price. Furthermore, the market is becoming increasingly demanding and increasingly mobile, a fact that complicates the equation of service delivery and customer experience management. Key to this is the fact that work styles are changing, largely at the insistence of enterprises that employ large numbers of people. Because communications technologies make it possible to work remotely—at home, in a moving vehicle, at a local coffee shop, at a satellite office—employers now insist that their employees move their work location from a centralized office location downtown to their home. It's good for everybody, they claim; employees are happier, more productive, and safer, because they don't have to commute every day. Employers benefit because they can reduce their real estate holdings (and therefore tax liabilities), shrink their parking lots, and contribute to the demand for a smaller carbon footprint.

Executive leadership and abdication. Increasingly, chief executives in the tech sector (CIOs, CTOs) allow themselves to be informally and unwittingly demoted. Many of them "grew up" professionally in the technology world with time spent in data centers and central offices. They're comfortable with the technology and enjoy working with it. As a result they often allow themselves to slip into a form of complacency in which they begin to worry more about the day-to-day tactical considerations of their organization (network or data center operational excellence, e.g., something they forget they have people for) and abdicate the strategic thinking and vision development that they are actually responsible for. As a consequence, nobody does it. In fairness, technology has become a support function in many companies, a fact that we'll discuss in greater detail in Chap. 11.

The need for balance between control *and* influence. When the seat of power lay almost exclusively with the provider of services, the provider *controlled* the relationship with the customer, dictating the rules of engagement for the relationship between them. Today, however, with more and more power shifting from the provider to the consumer, the relationship becomes one based more on *influence* rather than control. This is one of the most important lessons to be learned from this book: *Companies looking for long-term relevance in the TMT sector must understand that if they want to gain influence in a market, they must give up control of that market.*

An example is in order. Apple (and Google Android) consciously and deliberately gave up control of the Apps market so that they could gain influence over the platform market. Out of the 800,000 (or so) Apps available at the Apple Apps Store, only a handful were developed by Apple—among them are Pages, Numbers, Keynote, and Garage Band. All the others were developed by third parties who pay a percentage of the take to Apple for agreeing to host their application. Everybody wins: The developers get a highly visible platform from which to display and sell their wares, and Apple makes a little jingle (ok, a *lot* of jingle) in exchange for its hosting services. Give up control; gain influence. The problem, of course, is that for too many companies today, control is deeply embedded in their behavioral DNA and they have a very difficult time making the transition. They have no choice, however. If they don't change their business model, somebody else will, and the business will migrate to a better customer experience.

QoS versus QoE: A fundamental change. For years, IT and technology companies have measured their goodness, their effectiveness, their customer-centricity, on the basis of a measure called quality of service (QoS). QoS includes such vaunted measures as mean time to repair, mean time between failures, percentage of packets lost during transmission, average uptime, churn, delay, and my personal favorite, jitter.[1] These are all important indicators, measuring as they do the performance and therefore the health and welfare of the network or infrastructure in the data center. The problem is that 100% of these measures are inward-looking: They focus on the network or data center, not on the customer. But that's only half of the equation, and by the way, the far less important half.

To bring the customer into the equation we must add a new measure called quality-of-experience (QoE). QoE measures "was it good for you." In other words, while interacting with my network or with my data center services, did you have an acceptable experience?

This is a far more difficult thing to measure. Companies often conduct customer service surveys, which do an adequate but rarely insightful job of understanding the experience delivered to the customer. A far better measure is one used by Canadian carrier TELUS, which they call "likelihood to recommend (L2R)." According to TELUS senior executive Josh Blair, "L2R gets to the heart of the single most important question we can ask a customer: After engaging with us and our delivered services would they be willing to recommend us to a friend, family member, or colleague?"

Willingness to take on risk. It's an interesting fact. Large, legacy companies in the TMT space will often spend hundreds of millions if not billions of dollars to acquire a company that resembles them (another telephone company, media provider, or data

---

[1]Jitter is a measure of the variability of delay during packet transmission. When a steady flow of packets occurs while watching a YouTube video, there is no jitter. But when unpredictable delay begins to make its way into the transmission stream because of network or server problems upstream, the screen freezes, stopping and starting as packets trickle in, resulting in a poor viewing experience.

center provider) as a way to increase their footprint or eliminate a competitor. Yet they won't spend $100,000 on an entrepreneurial opportunity to try something new that could result in greater relevance in the mind of the customer and a position in a nascent but growing industry. Equally troubling is the fact that their decision-making processes are often so stilted and fossilized that even if they wanted to invest, they wouldn't be able to. The amounts involved are not enough to draw the attention of senior managers, but too much for lower-level managers who have little M&A experience to take this on. As a result the opportunity languishes and ultimately goes to someone else.

The message here is that the old rules no longer apply. The lines of competition, operation, and cooperation are blurring rapidly, leading to a new model of industry analysis and competitive engagement. The model is not new; in fact, it was alive and well in Japan in the 1950s and 1960s in the form of the *Keiretsu*, a conglomeration of interdependent and highly complementary businesses that had many of the characteristics of a mutual fund, in the sense that the *whole* (the Keiretsu) protected the individual companies that comprised it through mutual support and recognized interdependency.

We're moving toward a new Keiretsu today, one that recognizes the interdependencies between the three TMT sectors and the fact that the most successful companies in the industry today aren't companies at all, but are rather ecosystems of companies and functions that work together toward a common competitive goal. Competition, then, is not between individual companies but between ecosystems. In some cases, ecosystem members may be owned by another member, as is the case with Google's acquisition of Motorola and Microsoft's purchase of Nokia. In other cases, they are simply members of the same group whose products and services complement each other, as is the case with Apple and its army of developers, not to mention the vast array of third-party manufacturers that sell products that complement Apple products. I always find it fascinating that Apple never attends the Consumer Electronics Show (CES) in Las Vegas, choosing instead to put on their own events at which they can carefully and closely control the customer experience. Yet Apple is one of the most glaringly present companies at CES, thanks to the scads of companies that make things that plug into, surround, augment, disguise, or enhance Apple devices.

Some of the Money...

Another outcome of this changing business model is a shift in how revenue is acquired and distributed. Standalone companies beware: The days of doing it all, of being so large and multifunctional that a single company can garner all of the revenue, are over. Today the revenue accrues to the ecosystem, resulting in a new truth. Under this new competitive paradigm, companies have two choices when it comes to revenue. They can have *some of the money* or *none of the money*. Those are the choices. That means new revenue models and a willingness to engage with and trust partners. This is a nontrivial item. It has been a common practice among legacy technology companies to enter into partnerships with the purpose of being able to control the partner. That's not a partnership; that's hegemony, and it fails every time it has been attempted. This goes back to the earlier observation about control versus influence: Companies entering into an ecosystem must be willing to give up control in exchange for a degree of organizational influence. And they are going to have to be ok with that.

Closing Thoughts

> Recognize that every out-front maneuver you make is going to be lonely and a little bit frightening. If you feel entirely comfortable, then you're not far enough ahead to do any good. That warm sense of everything going well is usually the body temperature at the center of the herd.
>
> (Anonymous)

There is change in the wind, and the wind is blowing fiercely. The TMT industry, and its component Telecom, Media, and Technology sectors, are found at the functional heart and deeply embedded in the nervous system of every successful company and industry on the planet. Without them there is no entertainment, no electronic communication or knowledge sharing, and no drive to create new applications that serve every corner of the world.

But there is also competition afoot that is different than anything these companies have ever experienced before. These competitors are nontraditional, small and fleet, customer-centric, driven to provide a better experience rather than focusing exclusively on better infrastructure, willing to take on risk, younger yet wiser, and ready to shake up the world.

That said, the services provided by legacy companies remain crucial, central, and necessary. The issue is not so much about the products and services they provide, but rather about *why* and *how* they provide them. We'll talk more about this later in the book, but for now, let's start paving our technology foundation with the next chapter: "The Standards that Guide Us."

Chapter One Questions

1. What are the relative benefits and liabilities of the convergence of the three TMT sectors?
2. What will the competitive landscape look like in five years?
3. What is the difference between QoS and QoE in your own organization?

CHAPTER 2

The Standards That Guide Us

My favorite technology joke: The nice thing about standards is that there are so many to choose from.

My second favorite technology joke: There are only two things you never want to watch being made. The first is sausage; the second is standards.

The world of technology is filled with apocryphal stories, technical myths, and fascinating legends. Everyone in the world of telecom knows someone who knows the outside plant repair person who found the poisonous snake in the equipment box in the manhole;[1] the person who was on the cable-laying ship when they pulled up the cable that had been bitten through by some species of deep water shark; some collection of seriously evil hackers, or the backhoe driver, who cut the cable that put Los Angeles off the air for 12 hours; or they know the person working in the data center who was there the day the Halon went off; or was privy to the inside conversations between Steve Jobs and Disney during the early days of Pixar.

There is also a collection of techno jargon that pervades the technology industry and often gets in the way of the relatively straightforward task of learning how all this stuff actually works. To ensure that such things don't get in the way of absorbing what's in this book, I'd like to begin with a discussion of some of them. I am a writer, after all; words are important to me.

This is a book about three technology areas: media, IT, and telecom. Telecom is the science of communicating over distance. It is, however, fundamentally dependent upon *data communications,* the science of moving traffic between communicating devices, so that the traffic can be manipulated in some way to make it useful. Data, in and of itself, is not particularly useful, consisting as it does of a stream of ones and zeroes that is only meaningful to the computing device that transmits it or the device on the other end that will receive and

[1]I realize that this term has fallen out of favor today, but I use it here for historical accuracy.

manipulate it. The data does not really become useful until it is converted by some application into *information,* because a human can generally understand information. The human then acts upon the information using a series of intuitive processes that further convert the information into *knowledge,* at which point it becomes truly useful. Ultimately, experience gives the person who is interpreting the data the ability to convert it into wisdom. There's a wonderful expression that says, "Good decisions come from experience, and experience comes from a lot of bad decisions." True enough: Wisdom is all about falling back on what we have learned over a long period of time to inform our interpretation of events.

Here's an example: a computer generates a steady stream of data (ones and zeroes) in response to a series of computer-related business activities. Those ones and zeroes are fed into another computer, where an application converts them into a spreadsheet of sales figures (information) for the store from which they originated. A financial analyst studies the spreadsheet, calculates a few ratios, examines some historical data (including not only sales numbers but demographics, weather patterns, and political trends), and makes an informed prediction about future stocking requirements and advertising focal points for the store based on the knowledge that the analyst was able to create from the distilled information. Once decisions have been made and the results of those decisions have been analyzed, *wisdom* allows experienced decision-makers to observe that a transient weather event in the countries of western South America caused the numbers to be lower than they should be, but they will recover. Data, information, knowledge, and wisdom—a critical continuum.

Terminology

Before we get too deep in the technological muck, I'd like to use a few paragraphs to define some fundamental terms that we will engage with over and over throughout this book, and which are therefore rather important to our conversation. Those terms are (in addition to telecom, which we've already defined as the practice of communicating over distance) *data communications, protocol, procedures,* and *standards.* These concepts apply equally well to all three of the TMT sectors. For the sake of simplicity, I'm going to use telecom examples to set the stage and will insert media and IT examples as required.

Data Communications

Data communications is generally (and boringly) defined as electronic information that has been transformed into a digital signal that can then be transported between digital devices such as computers, mobile phones, tablets, and the like. For this process to work, a stringent set of rules must be agreed and adhered to. Think of them as the digital equivalent of rules of engagement, or social mores.

What's important to note at this early stage of the game is that data communications as a practice is dependent on *standards* that govern universal and common implementation guidelines; *protocols*, in turn, are the technical specifications that ensure everything works properly. Failure to adhere to the standards, and failure to implement the proper protocols, can result in a malfunctioning network, data center, or media-deployment strategy.

Protocols

In the world of data communications, the rules that govern things are called *protocols*. A protocol is defined as a standard procedure for regulating the transmission of data between computers, which is itself a code of correct conduct. These protocols, which will be discussed in detail in Chap. 3, provide a widely accepted and standardized methodology for everything from the pin assignments on physical connectors to the sublime encoding techniques used in secure transmission systems. Simply put, they represent the rules that govern the game. Many countries play some kind of football, for example, but the rules are all slightly different. In the United States, players are required to weigh more than a car, yet be able to run faster than one. In Australian football, the game is declared forfeit if it fails to produce at least one body part amputation on the field or if at least one player doesn't eat another. In most parts of the world, *football* is what Americans call soccer. They are all variations on a similar theme, however. In data communications, the problem is the same: there are many protocols out there that accomplish the same thing. Data, for example, can be transmitted from one side of the world to the other using a variety of media options, including T1, E1, microwave, optical fiber, satellite, gigabit ethernet, coaxial cable, and even through water. The end result is identical: the data arrives at its intended destination, correctly, without errors, ready to interpret. Different protocols, however, govern the process in each case.

Standards

A discussion of protocols would be incomplete without a simultaneous discussion of *standards*. If protocols are the sets of rules by which the game is played, standards govern which set of rules will be applied for a particular game. For example, let's assume that we need to move traffic between a PC and a printer. We agree that in order for the PC to be able to transmit a printable file to the printer, both sides must agree on a common representation for the zeroes and ones that make up the transmitted data. They agree, for example, (and this is *only* an example) that they will both rely on a protocol that represents a zero as the *absence* of 3-V pulse and a one as the *presence* of a 3-V

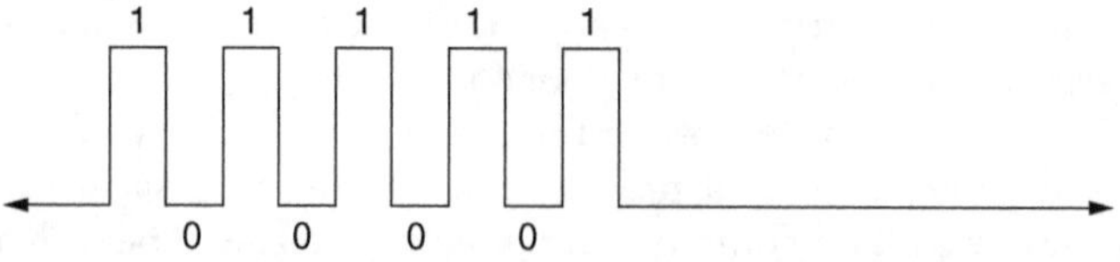

Figure 2.1 Voltage pulses on a line, indicating the presence or absence of data. In this case zero volts represents a zero, while a positive pulse represents a one.

pulse on the line, as shown in Fig. 2.1. Because they agree on the representation of data, the printer knows when the PC is sending a one and when the PC is sending a zero. Imagine what would happen if they failed to agree on such a simple thing beforehand? If the transmitting PC decides to represent a one as a 300-V pulse, and the printer expects a 3-V pulse, the two devices will have a brief (albeit exciting!) conversation, the ultimate result of which will be the release of a small puff of silicon smoke from the printer. And a "failed to print" message appears on the screen of the PC.

Next, the two devices have to decide on a standard that they will both agree to accept for originating and terminating the data that they will exchange. In this example, they are connected by a cable that has nine pins on one end and nine jacks on the other. Logically, the internal wiring of the cable looks like Fig. 2.2. Incidentally, this could just as easily be a USB cable or a Wi-Fi or Bluetooth connection.

Data communications standards have the responsibility to define the strict rules by which two or more devices will communicate with one another, down to the most picayune details. Let's assume that in this example the two devices agree that the standard calls for them to always transmit data on pin 2 of the connector. That way, they always know which pin to monitor for incoming data. However, if we stop to think about it, this one-to-one correspondence of pin-to-socket won't work. Why not? Well, if the PC transmits on pin 2, which in our

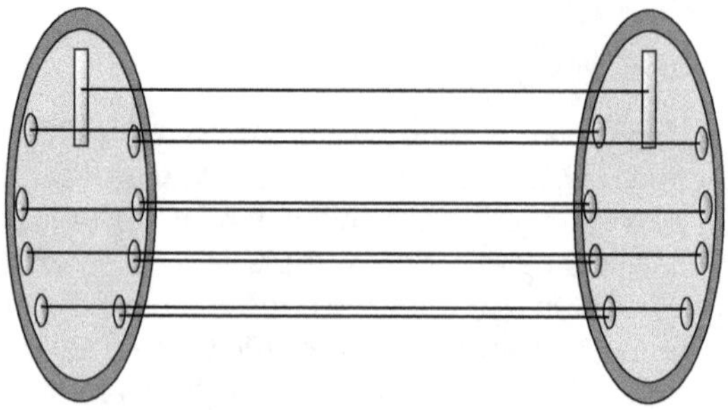

Figure 2.2 A schematic diagram of a nine-pin cable, showing the connections between each end of the cable.

example is identified as the *send data* lead, it will arrive at the printer on pin two, which is, of course, the *send data* lead for the other end. This would be analogous to holding two telephone handsets together so that two people can talk. It won't work without a great deal of hollering, since the handsets are oriented microphone-to-microphone and speaker-to-speaker! Instead, an agreement must be forged to ensure that the traffic placed on the *send data* lead somehow arrives on the *receive data* lead, and vice versa. Similarly, the other leads must be able to convey information to the far end of the connection so that normal transmission can be started and stopped.

Here's a realistic example of a typical data communications scenario in which a PC wants to send a document to a printer.

When the PC is ready to send the document, it first checks to make sure that the printer has put voltage on the *data terminal ready* (DTR) lead. The presence of voltage (known as being high) indicates that the remote device (the printer, in this case) is ready to receive the print job.

The printer responds by setting its own DTR lead high as a form of acknowledgement, immediately after which it begins to transmit the file.

Keep one thing in mind: printers are mechanical devices, PCs are not. Printers are made up of a collection of physical devices that have to move, which means that the printer takes much, much longer to print the file than it takes the PC to send it. The implication of this? It's likely that the PC will overwhelm the printer because the printer can't keep up with the incoming flood of data, and its buffer gets dangerously close to overflowing.

When this occurs the printer drops the voltage on its DTR lead indicating that it is no longer in a position to continue receiving. Recognizing this, the PC immediately stops transmitting.

As soon as the printer is ready to receive again, that is, when it has flushed its buffer, it sets the DTR lead high and printing resumes.

This agreed-upon procedure that governs the relationship between the PC and the printer is a protocol. As long as both transmitter and receiver abide by this standardized set of rules, data communications works.

Incidentally, this process of interconnecting the send and receive data leads is done by the modem, or by a *null modem cable,* that makes the communicating devices think they are communicating via a modem. The null modem cable is wired so that the send data lead on one end is connected to the receive data lead on the other end, and vice-versa; similarly, a number of control leads such as the carrier detect (CD) lead, the DTR lead, and the data set ready (DSR) leads are wired so that they give false indications to each other to indicate that they are ready to proceed with the transmission, when in fact no response from the far-end modem has actually been received.

Of course, standards don't exist solely in the world of telecom. For example, standards in the IT realm might include

- Electronic data–transport requirements.
- Standards that govern the fair use of copyrighted technical works.
- Standards for organizational Web pages (other than the corporate site).
- Fair use policies for the use of social media in the workplace.
- E-mail management and retention guidelines.
- Workstation security guidelines.
- Physical site security guidelines.
- LAN and intranet security guidelines.
- Laptop and portable device security guidelines.
- Bring your own device (BYOD) standards and policies.
- Password change mandates.
- Standard procedures for handling confidential and sensitive information.
- Technical standards for data backup.

Standards in the media world are somewhat harder to come by because of the creative nature of content creation. In the spirit of the book's overall message, however, it is clear that IT standards that govern cloud services, transmission requirements, and the like hold sway in the media domain as well. For example, standards that relate to the availability of online content are fundamentally important to the folks at Netflix, iTunes, and other online services, even though they are not technically standards that govern media.

Data Communications Standards: Where Do They Come From?

Physicists, electrical engineers, and computer scientists are generally responsible for the design of data-communications protocols. For example, the transmission control protocol (TCP) and the Internet protocol (IP), both discussed in detail later in the book, when we get into the genesis of the Internet, were written during the heady days of the Internet back in the 1960s by such early pioneers as Vinton Cerf and the late John Postel (I want to say "back in the last century" to make them seem like real pioneers!). Let me say that again: the *protocols* were written by Cerf and Postel. Standards, on the other hand, are created as the result of an ongoing consensus-building process that can take years to complete. By design, standards must meet the requirements of the entire data and telecommunications industry, which is of course global. It makes sense, therefore, that some international body be responsible for overseeing the creation of international standards. One such body is (and by all rights, should be) the United Nations (UN). Its 192 member nations work together in an attempt to harmonize whatever differences they have at various levels of interaction, one of which is international telecommunications. The International Telecommunications Union (ITU), a sub-organization of the UN, is responsible for not only coordinating the creation of worldwide standards, but also publishing them under the auspices of its *own* sub-organizations. These include the Telecommunications Standardization Sector (TSS, sometimes called the ITU-T, and formerly the Consultative Committee on International Telegraphy and Telephony, the CCITT), the Telecommunications Development Sector (TDS), and the Radio Communication Sector (RCS, formerly the Consultative Committee on International Radio, the CCIR). The organizational structure is shown in Fig. 2.3.

Of course, the UN and its sub-organizations cannot perform this task alone, nor should they. Instead, they rely upon the input of hundreds of industry-specific organizations as well as local, regional, national, and international standards bodies that feed information, perspectives, observations, and technical direction to the ITU, which serves as the coordination entity for the overall international standards creation process. These include the American National Standards Institute (ANSI); the European Telecommunications Standards Institute (ETSI, formerly the Conference on European Post and Telegraph, CEPT); Telcordia (formerly Bellcore, now part of SAIC); the International Electrotechnical Commission (IEC); the European Computer Manufacturers Association (ECMA); and a host of others.

As an example of the ITU's role in the establishment of international standards, consider the conference that occurred in Dubai in 2012. In an attempt to further standardize the extent to which telecom networks around the world can easily exchange information with one another, the ITU member nations set out to harmonize certain aspects of the manner in which they interconnect with one another. To wit:

> While the sovereign right of each state to regulate its telecommunications is fully recognized, the provisions of the present International Telecommunication Regulations (hereafter referred to as *regulations*) complement the constitution and the convention of the ITU, with a view to attaining the purposes of the ITU in promoting the development of telecommunication services and their most efficient operation while harmonizing the development of facilities for worldwide telecommunications.
>
> Member states affirm their commitment to implement these regulations in a manner that respects and upholds their human rights obligations.

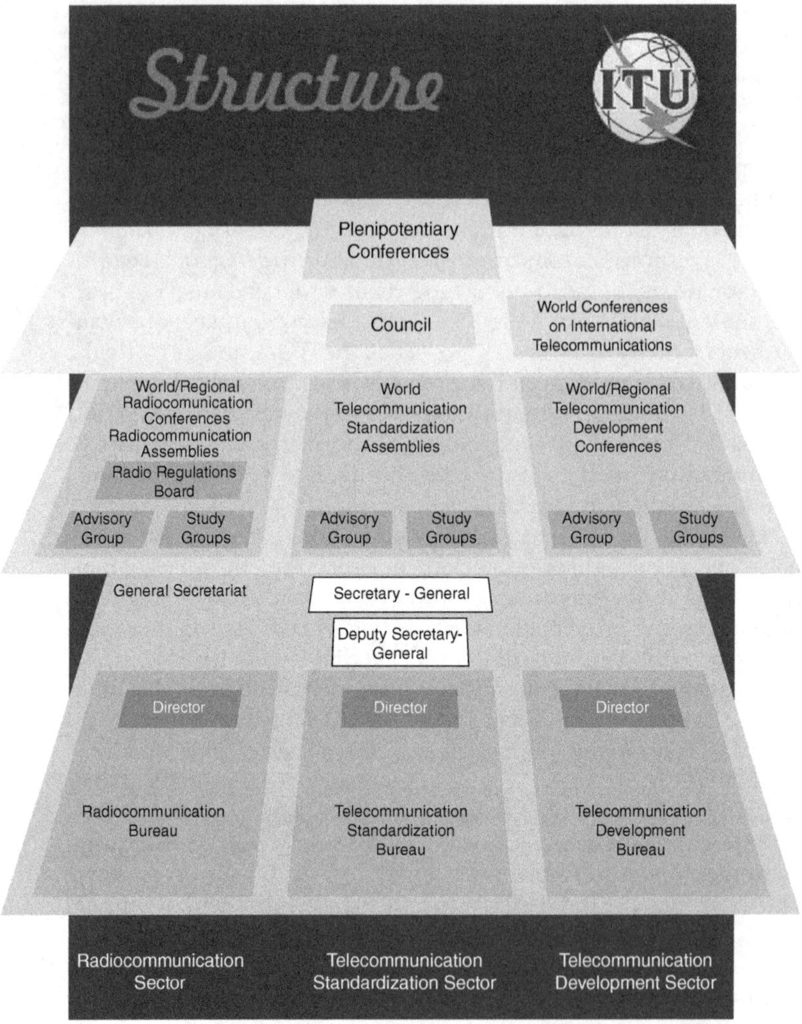

FIGURE 2.3 The overall structure of the ITU.

These regulations recognize the right of access of member states to international telecommunication services.

Furthermore, the delegates went on to define their intent in a more granular fashion:

- Member states shall promote the development of international telecommunication services and shall foster their availability to the public.

- Safety-of-life telecommunications, such as distress telecommunications, shall be entitled to transmission as of right and, where technically practicable, have absolute priority over all other telecommunications.
- Government telecommunications, including telecommunications relative to the application of certain provisions of the UN Charter, shall, where technically practicable, enjoy priority over telecommunications other than those referred to in the prior article.

Furthermore:

- Member states shall individually and collectively endeavor to ensure the security and robustness of international telecommunication networks.
- Member states should endeavor to take necessary measures to prevent the propagation of unsolicited bulk electronic communications and minimize its impact on international telecommunication services.
- The terms and conditions for international telecommunication service arrangements may be established through commercial agreements or through accounting-rate principles established pursuant to national regulation.

My intent here is not to bury you in "administrivia," but rather to point out the enormous complexity of the global telecommunications world and the extraordinarily difficult environment within which standards organizations are forced to operate.

The International Telecommunications Union

It is worthwhile to mention a bit about the ITU as a representative standards body. Founded in 1947, as part of the UN, it descended from a much older body called the Union Telegraphique, founded in 1865 and chartered to develop standards for the emerging telegraph industry. Over the years since its creation, the ITU and its three principal member bodies have developed three principal goals:

- To maintain and extend international cooperation for the improvement and interconnectivity of equipment and systems through the establishment of technical standards.
- To promote the development of the technical and natural facilities (read spectrum) for most efficient applications.
- To harmonize the actions of national standards bodies to attain these common aims, and most especially to encourage the development of communications facilities in developing countries.

The Telecommunications Standardization Sector

The goals of the TSS, according to the ITU, are as follows:

- To fulfill the purposes of the union relating to telecommunication standardization by studying technical, operating, and tariff questions and adopting formal recommendations on them with a view to standardizing telecommunications on a world-wide basis.
- To maintain and strengthen its pre-eminence in international standardization by developing recommendations rapidly.
- To develop recommendations that acknowledge market and trade-related consideration.
- To play a leading role in the promotion of cooperation among international and regional standardization organizations and forums and consortia concerned with telecommunications.
- To address critical issues that relate to changes due to competition, tariff principles, and accounting practices.
- To develop recommendations for new technologies and applications such as appropriate aspects of the global information infrastructure (GII) and global multimedia and mobility.

The Telecommunications Standardization Bureau

The Telecommunication Standardization Bureau (TSB) provides secretarial support for the work of the ITU-T Sector and services for the participants in ITU-T, diffuses information on international telecommunications worldwide, and establishes agreements with many international standards development organizations. These functions include:

- *Study group management*: The management team of the study groups is composed of the chairman, vice-chairmen of the study group, chairmen of the working parties, and the TSB counselor/engineer.
- *Secretarial support and meeting organization*: TSB provides secretariat services for ITU-T assemblies and study group meetings. TSB counselors and engineers coordinate the work of their study group meeting, and their assistants ensure the flow of meeting document production.
- *Logistics services*: The TSB provides services such as meeting-room allocation, registration of participants, document distribution, and facilities for meeting participants.
- *Approval of recommendations and other texts*: The TSB organizes and coordinates the approval process of recommendations.

- *Access to ITU-T documents for ITU-T members*: The TSB organizes and controls the dispatch of documents in paper form to participants in ITU-T work and provides electronic document–handling services (EDH) that allow easy and rapid exchange of documents, information, and ideas among ITU-T participants in order to facilitate the work of standards development. The ITU-T participants can have electronic access, via Telecommunication Information Exchange Service (TIES), to study group documents such as reports, contributions, delayed contributions, temporary and liaison documents, and the like.

The TSB also provides the following services:

- Maintenance of the ITU-T Web site and distribution of information about the activities of the sector, including the schedule of meetings, TSB circulars, collective letters, and all working documents.
- Update services for the list of ITU-T recommendations, the ITU-T work program database, the ITU-T patent statements database, and the ITU-T terms and definitions database (SANCHO), as well as update services for other databases as required.
- Country code number assignment for telephone, data, and other services.
- Registrar services for universal international freephone numbers (UIFN).
- Technical information on international telecommunications and collaborates closely with the ITU Radiocommunication Sector and with the ITU Telecommunication Development Sector for matters of interest to developing countries.
- Administrative and operational information through the ITU *operational bulletin*.
- Coordination of the editing, publication, and posting of the recommendations.

The Radio Bureau

The functions of the Radio Bureau include:

- Administrative and technical support to radiocommunication conferences, radiocommunication assemblies and study groups, including working parties and task groups.
- Application of the provisions of the radio regulations and various regional agreements.
- Recording and registration of frequency assignments and also orbital characteristics of space services, and maintenance of the master international frequency register.

- Consulting services to member states on the equitable, effective, and economical use of the radio-frequency spectrum and satellite orbits, and investigates and assists in resolving cases of harmful interference.
- Preparation, editing, and dispatch of circulars, documents, and publications developed within the sector.
- Delivery of technical information and seminars on national frequency management and radiocommunications, and works closely with the Telecommunication Development Bureau to assist developing countries.

The Development Sector

The role of the ITU's development branch (ITU-D) is to raise awareness of decision-makers about the role that telecommunications plays in national economic and social development; to offer telecommunications policy advice; to promote and catalyze the development, expansion and operation of telecom networks and services, particularly in developing countries; and to work closely with regional and global financing institutions, telecom organizations, and others to promote telecom development.

Another goal of the ITU-D is to encourage industry participation in telecom development through technology transfer incentives and technical assistance.

The ITU-D Sector has two study groups which include member states and sector members that work together on Internet infrastructure rollout, broadband deployment, network migration and interconnection, new technologies for rural applications, digital broadcasting technologies, and a wide array of other topics.

Every four years, the ITU-D hosts a World Telecommunication Development Conference (WTDC). The WTDC establishes the work programs and guidelines for defining telecom development study questions and sector priorities. In some cases it creates or changes study groups. The WTDC is designated to serve as a forum for the study of policy, organizational, operational, regulatory, technical, and financial questions related to telecommunications development.

The Standards Themselves

A word about the publications of the ITU. First, they are referred to as *recommendations,* because the ITU has no enforcement authority over the member nations that use them. Its strongest influence is exactly that—the ability to *influence* its member telecommunications authorities to use the standards because it makes sense to do so on a global basis.

The standards are published every four years, following enormous effort on the part of the representatives who sit on the organization's task forces. These representatives hail from all corners of the industry; most countries designate their national telecommunications company (where they still exist) as the representative to the ITU-T, while others designate an alternate, known as a recognized private operating agency (RPOA). The United States, for example, has designated the Department of State as its duly elected representative body. Other representatives may include manufacturers (Alcatel-Lucent, Cisco, Ericsson, Fujitsu), research and development organizations (Bell Laboratories, Xerox PARC), and other international standards bodies.

The efforts of these organizations, companies, governments, and individuals result in the creation of a collection of new and revised standards recommendations published on a four-year cycle. Historically, the standards were color-coded, published in a series of large format soft cover books, differently colored on a rotating basis. For example, the 1984 books were red; the 1988 books, blue; the 1992 books, white. It is common to hear network professionals talking about "going to the blue book." They are referring (typically) to the generic standards published by the ITU for that particular year. It is also common to hear people talk about the CCITT. Old habits die hard: the organization ceased to exist in the early 1990s, replaced by the ITU-T. The name is still used, however.

The ITU-T continues to be an organization with significant influence in the world of telecom, and by extension, in the media and IT domains as well. The activities of the ITU-T are parceled out according to a cleverly constructed division of labor. Three efforts result: study groups, which create the actual recommendations for telecomm equipment, systems, networks, and services (currently 15 study groups); plan committees, which develop plans for the intelligent deployment and evolution of networks and network services; and specialized autonomous groups (currently three) that produce resources which support the efforts of developing nations. The study groups are listed below.

- *SG 2:* Operational aspects of service provision, networks, and performance
- *SG 3:* Tariff and accounting principles including related telecommunications economic and policy issues
- *SG 4:* Telecommunication management, including Telecommunication Management Network (TMN)
- *SG 5*: Protection against electromagnetic environment effects
- *SG 6*: Outside plant
- *SG 7*: Data networks and open system communications
- *SG 9*: Integrated broadband cable networks and television and sound transmission

- *SG 10*: Languages and general software aspects for telecommunication systems
- *SG 11*: Signaling requirements and protocols
- *SG 12*: End-to-end transmission performance of networks and terminals
- *SG 13*: Multiprotocol and IP-based networks and their internetworking
- *SG 15*: Optical and other transport networks
- *SG 16*: Multimedia services, systems, and terminals
- *SG 17*: Security, languages, and telecommunication software
- *SG 18*: Mobile telecommunication networks
- *SSG*: Special study group "IMT-2000 and beyond"

The Standards

The standards are published in a series of alphabetically arranged documents, available as books, online resources, and CDs. They are functionally arranged according to the alphabetical designator of the standard as follows:

A-Organization of the work of ITU-T

B-Means of expression: definitions, symbols, and classification

C-General telecommunication statistics

D-General tariff principles

E-Overall network operation, telephone service, service operation, and human factors

F-Non-telephone telecommunication services

G-Transmission systems and media, digital systems, and networks

H-Audiovisual and multimedia systems

I-Integrated services digital network

J-Transmission of television, sound program, and other multimedia signals

K-Protection against interference

L-Construction, installation, and protection of cables and other elements of outside plant

M-TMN and network maintenance: international transmission systems, telephone circuits, telegraphy, facsimile, and leased circuits

N-Maintenance: international sound program and television transmission circuits

O-Specifications of measuring equipment

P-Telephone transmission quality, telephone installations, local line networks

Q-Switching and signaling
R-Telegraph transmission
S-Telegraph services terminal equipment
T-Terminals for telematic services
U-Telegraph switching
V-Data communication over the telephone network
X-Data networks and open-system communication
Y-Global information infrastructure and Internet protocol aspects
Z-Languages and general software aspects for telecommunication systems

Within each letter designator can be found specific, numbered recommendations. For example, recommendation number 25 in the X book contains the specifications for transporting packet-based data across a public network operating in packet mode. This, of course, is the legacy X.25 packet switching standard which served as the basis for many packet networking standards that followed. Similarly, Q.931 provides the standard for signaling in ISDN networks, and so on. The documents are remarkably easy to read and contain vast amounts of information. I am always surprised to discover how many people who work in telecommunications have never read the ITU standards. Take this, then, as *my* recommendation: Find them, find a quiet corner, and flip through them. They can be very useful.

I remember very well an experience I had years ago with the ITU standards. My family was visiting friends in California, but because of work commitments I stayed behind in Vermont preparing to travel. It was a dark snowy January night and I was sitting in front of a roaring fire, drinking a warm brandy and reading *Q.931*. The worst part was that I was really into it, enjoying the fact that I was understanding the techno jargon. And then it hit me: I had become *one of them*. I was a propellerhead, a bit-weenie, a techno-dweeb.

I spent some time writing about the ITU and its standards activities simply to explain the vagaries of the process and the role of these bodies. The ITU is representative of the manner in which all standards are developed, although the frequency of update, the cycle time, the relative levels of involvement of the various players, and the breadth of coverage of the documents vary dramatically.

Other Important Organizations

In addition to the formal standards bodies there are industry groups that are worth paying attention to. Some of them, like the United States Telecom Association (USTA), the Telecommunications Industry Association (TIA), and the National Telecommunications and Information Administration (NTIA), are trade organizations that promote the efficient and effective operation of the industry. Others, like the Cellular Telecommunications and Internet Association (CTIA) focus on specific sectors of the marketplace. Still others, like the Consumer Electronics Show (CES), or the National Association of Broadcasters (NAB), are industry groups that sponsor annual trade shows where vendors gather to display their wares. These shows are worth attending, particularly if you are running low on key chains, t-shirts, and pens (only kidding).

There are also technology-specific groups that focus specifically on optical, wireless, component, switching, ATM, and so on. All are reachable online; depending on the nature of your interest in the industry, they are worth getting in touch with.

Closing Thoughts

As we noted earlier, one of the few things you never want to watch being made is standards. That may be the case, but the importance of these governing decisions cannot be ignored, nor can their importance to the greater TMT industry be overestimated. They may not be sexy, or even particularly interesting, but they are central to the ability of the industry to function. And that makes all the difference.

Chapter Two Questions

1. What is the difference between telecommunications and data communications?

2. What is the difference between a protocol and a standard?

3. Why do you suppose the UN has global responsibility for the development of telecommunications standards?

4. Why are there so many different standards bodies? How do they work together to ensure global acceptance of new standards?

CHAPTER 3

Data Communications Protocols

Putt's law: Technology is dominated by two types of people: those who understand what they do not manage, and those who manage what they do not understand.

One simple action that kicks off a complex series of events that results in the transmission of an e-mail message, the creation and sharing of a digital medical image, the establishment of a videoconference between a child and a grandmother. The process through which this happens is a remarkable symphony of technological complexity, and it is all governed by a collection of rules called protocols. This chapter is dedicated to them. For the etymologists in the room, a protocol is defined as "A set of rules that facilitate communications."

Data Communications Systems and Functions

If I were to walk up to you on the street in pretty much any western country and extend my hand in greeting, you would quite naturally reach your hand out, grab mine, and shake it—in most parts of the world. There is a commonly accepted set of social rules that we agree to abide by, one of which is shaking hands as a form of greeting. It doesn't work everywhere: In Tibet, it is customary to extend one's tongue as far as it can be extended as a form of greeting (clearly a sign of a great culture!). In China, unless you are already friends with the person you are greeting, it is not customary to touch in any fashion.

You, of course, had a choice when I extended my hand. You could have hit it, licked it, spit in it. But because of the accepted social rules that govern western society, you took my hand in yours and shook it. These rules that govern communication—*any form of communication*—are called protocols. And the process of using protocols to convey information is called data communications. It's no accident, incidentally, that the obnoxious racket that analog modems make when they are attempting to connect to each other is called a handshake. The noise they make is their attempt to negotiate a common set of rules that works for both of them for that particular session.

The Science of Communications

Data communications is the science behind the procedures required to collect, package, and transmit data from one computing device to another, typically (but not always) over a wide area network. It is a complex process with many layers of functionality. To understand data communications, we must break it into its component parts and examine each part individually, relying on the old adage that "the only way to eat an elephant is one bite at a time." Like a Russian Matruschka doll, data communications comprises layer upon layer of operational functionality that work together to accomplish the task at hand—namely, the communication of data. These component parts are known as *protocols*, and they have one responsibility: to ensure the integrity of the data that they transport from the source device to the receiver. This integrity is measured in a variety of ways including *bit-level integrity*, which ensures that the bits themselves are not changed in value as they transit the network; *data integrity*, which guarantees that the bits are recognizable as packaged entities called frames or cells; *network integrity*, which provides for the assured delivery of those entities, now in the form of packets, from a source to a destination; *message integrity*, which not only guarantees the delivery of the packets, but in fact their *sequenced* delivery to ensure the proper arrival of the entire message; and finally, *application integrity*, which provides for the proper execution of the responsibilities of each application. This is shown graphically in Fig. 3.1.

Protocols exist in a variety of forms and are not limited to data communications applications. Military protocols define the rules of engagement that modern armies agree to abide by; diplomatic protocols define the manner in which nations interact and settle their political and geographic differences; and medical protocols document the manner in which medications are used to treat illness. As we said earlier, the word *protocol* is defined as a set of rules that facilitates communication. Data communications, then, is the science built around the protocols that govern the exchange of digital data between computing systems.

Data Communications Networks

Data communications networks are often described in terms of their architectures, as are protocols. Protocol architectures are often said to be *layered* because they are carefully divided into highly related but

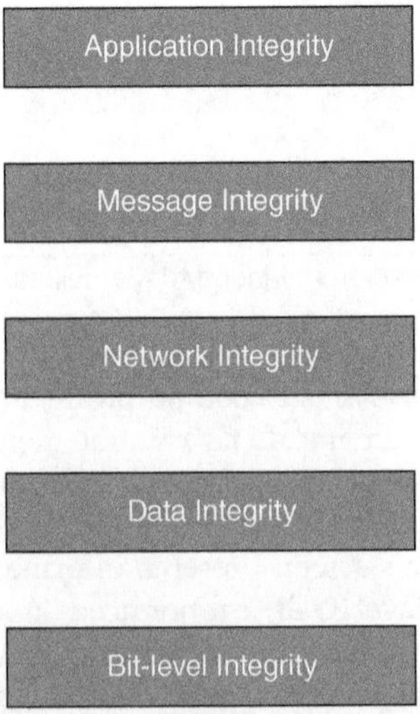

FIGURE 3.1 The various types of integrity guaranteed by a layered protocol model.

non-overlapping functional entities. This *division of labor* not only makes it easier to understand how data communications works, but also makes the deployment, maintenance, and troubleshooting of complex networks far easier.

Here's an example. The amount of code (lines of programming instructions) required to successfully carry out the complex task of data transmission is quite large. If the program that carries out all of the functions in that process were written as a single, large, monolithic chunk of code, it would be difficult to make a change to the program when updates are required because of the monolithic nature of the program. Furthermore, there's always the risk that an innocent and minor change to a line of code at the beginning of the program will cause an unintended and catastrophic failure in a line of code deep in the program.

Now imagine the following: Instead of a single set of code, we break the program into functional pieces, each of which handles a particular specific function required to carry out the transmission task properly. With this model, changes to a particular module of the overall program can be accomplished in a way that only affects that particular module, making the process far more efficient. This modularity is one of the great advantages of layered protocols.

Consider the following simple scenario, shown in Fig. 3.2: A PC-based e-mail user in Madrid with an account at Telkom wants to send a large, confidential message to another user in Marseilles. The Marseilles user is attached to a mainframe-based corporate e-mail system.

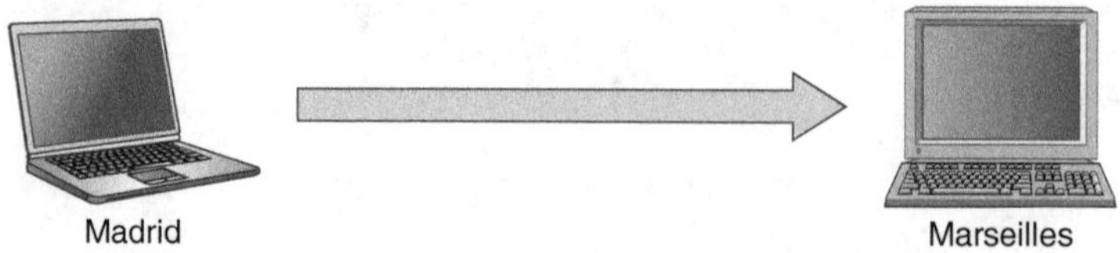

FIGURE 3.2 A transmission from Madrid to Marseilles.

In order for the two systems to communicate, a complex set of challenges must first be overcome. Let's examine them a bit more closely.

The first and most obvious challenge that must be overcome is the difference between the actual user interfaces on the two systems. The PC-based system's screen presents information to the user in a graphical user interface (GUI, pronounced 'gooey') format that is carefully designed to make it intuitively easy to use.

The mainframe system was created with intuitive ease of use in mind, but because a different company designed the interface for a mainframe host, under a different design team, it bears minimal resemblance to the PC system's interface. Both are equally capable, but look completely different.

As a result of these differences, if we were to transmit a screen of information from the PC directly to the mainframe system, it would be unreadable for no reason other than the fact that the two interfaces do not share common field names or locations.

The next problem that must be addressed is security, illustrated in Fig. 3.3. We mentioned earlier that the message that is to be sent from the user in Madrid is confidential, which means that it should probably be encrypted to protect its integrity. And because the message is large because it includes a number of photographs, the sender will probably compress it to reduce the time it takes to transmit it. Compression, which will be discussed in more detail later, is simply the process of eliminating redundant information from a file before it is transmitted or stored to make it easier to manage.

Another challenge has to do with the manner in which the information being transmitted is represented. The PC-based Outlook Express message encodes its characters using a 7-bit character set called

FIGURE 3.3 Managing security in a data transmission.

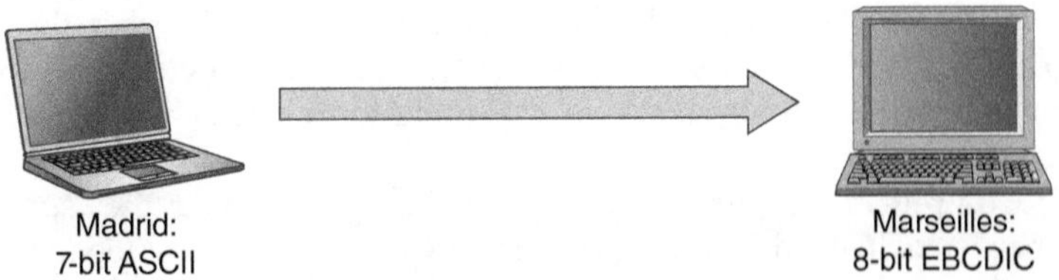

FIGURE 3.4 ASCII to EBCDIC conversion.

ASCII, the American Standard Code for Information Interchange. Mainframes, however, use a different codeset called the Extended Binary Coded Decimal Interchange Code (EBCDIC); the ASCII traffic must be converted to EBCDIC if the mainframe is to understand it, and vice-versa, as shown in Fig. 3.4.

Binary Arithmetic Review

It's probably not a bad idea to review binary arithmetic for just a moment since it seems to be one of the least understood details of data communications. I promise this will not be painful; I just want to offer a quick explanation of the numbering scheme and the various code sets that result.

Modern computers are *digital* because the values they use to perform their many functions have discrete values. (The word *digital* means discrete.) Those values are binary: in other words, they can have one of two possible values. They can either be one or zero, on or off, positive or negative, presence of voltage or absence of voltage, presence of light or absence of light. There are two possible values for any given situation. (Binary implies a system that comprises two distinct components or values.) Computers operate using base-two arithmetic, while humans use base 10. So for just a moment, let me take you back to second grade.

When we count, we arrange our numbers in columns that have values based on multiples of the number 10, as shown in Fig. 3.5.

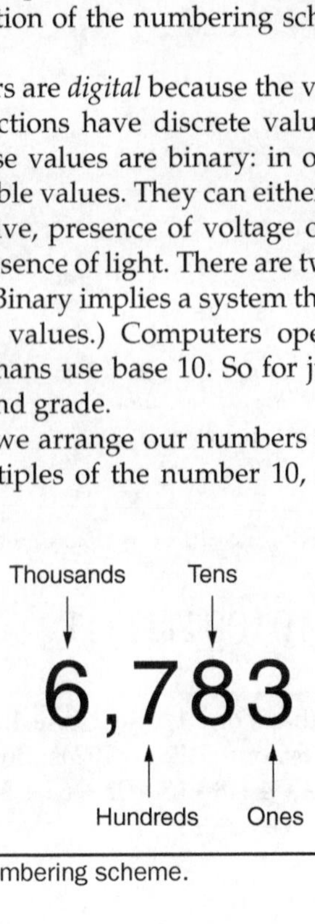

FIGURE 3.5 Base ten numbering scheme.

Here we see the number six thousand, seven hundred eighty-three, written using the decimal numbering scheme. We easily understand the value of number because we are taught to count in base ten from an early age.

Computers, however, don't speak in base ten. Instead, they speak in base two. Instead of having columns that are multiples of ten (ones, tens, hundreds, thousands, etc.), they use columns that are multiples of two, as shown in Fig. 3.6. In base ten, the columns are (reading from the right):

Ones
Tens
Hundreds
Thousands
Ten thousands
Hundred thousands
Millions
Etc.

In base 2, the columns are:

Ones
Twos
Fours
Eights
Sixteens
Thirty-twos
Sixty-fours
One hundred twenty-eights
Two hundred fifty-sixes
Five hundred twelves
One thousand twenty-fours
Etc.

So our number, 6,783, would be written as follows in base 2:

1101001111111

From right to left that's one 1, one 2, one 4, one 8, one 16, one 32, one 64, no 128s, no 246s, one 512, no 1024s, one 2048, and one 4096. Add them all up (1 + 2 + 4 + 8 +16 + 32 + 64 + 512 + 2048 + 4096) and you *should* get 6,783.

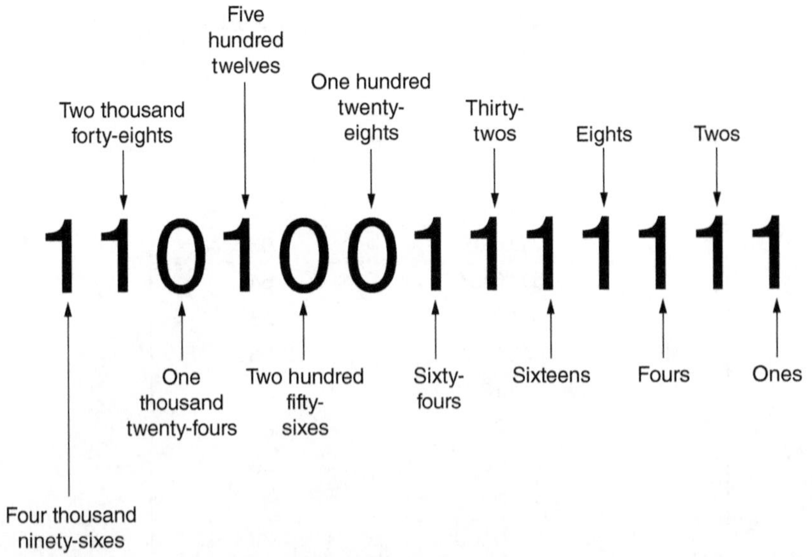

FIGURE 3.6 Base two numbering scheme.

That's binary arithmetic. Most PCs today use the seven-bit ASCII character set shown in Table 3.1. The mainframe, however (remember the mainframe?), uses the eight-bit code EBCDIC. What happens when a seven-bit ASCII PC sends information to an EBCDIC mainframe system that only understands eight-bit characters? Clearly, interpretation problems will result. Something, therefore, has to take on the responsibility of translating between the two coding schemes so that they can intelligibly transfer data between each other.

Another challenge that arises has to do with the logical relationship between the applications running in the two systems. While the PC most likely supports the e-mail account of a single user, the mainframe undoubtedly hosts hundreds, perhaps thousands of accounts, and must therefore ensure that each user receives their mail and *only* their mail. Some kind of user-by-user and process-by-process differentiation is required to maintain the integrity of the system and its applications. This is illustrated graphically in Fig. 3.7.

The next major issue has to do with the network over which the information is to be transmitted from Madrid to Marseilles. In the past, information was either transmitted via a dedicated and very expensive point-to-point circuit, or over the relatively slow public switched telephone network (PSTN). Today, however, most modern networks are packet-based, meaning that messages are broken into small, easily routable pieces, called packets, prior to transmission. Of course, this adds an additional layer of complexity to the process.

Character	ASCII Value	Decimal Value
0	0110000	48
1	0110001	49
2	0110010	50
3	0110011	51
4	0110100	52
5	0110101	53
6	0110110	54
7	0110111	55
8	0111000	56
9	0111001	57
A	1000001	65
B	1000010	66
C	1000011	67
D	1000100	68
E	1000101	69
F	1000110	70
G	1000111	71
H	1001000	72
I	1001001	73
J	1001010	74
K	1001011	75
L	1001100	76
M	1001101	77
N	1001110	78
O	1001111	79
P	1010000	80
Q	1010001	81
R	1010010	82
S	1010011	83
T	1010100	84
U	1010101	85
V	1010110	86
W	1010111	87
X	1011000	88
Y	1011001	89
Z	1011010	90

TABLE 3.1 The ASCII Character Set

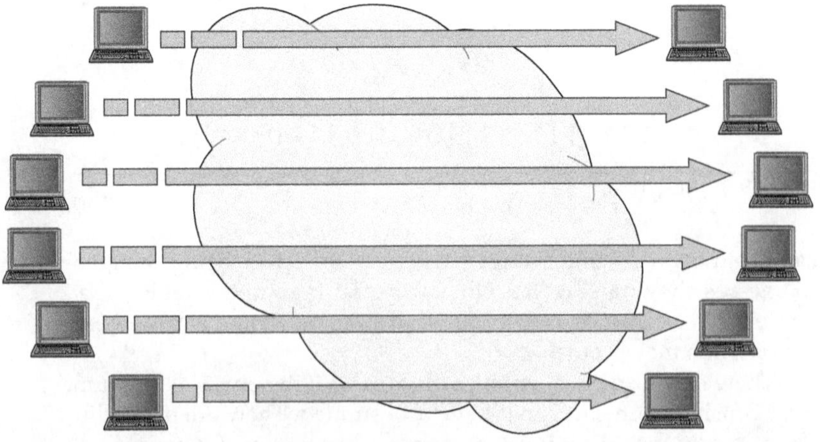

FIGURE 3.7 Logical differentiation in networks.

What happens if one of the packets fails to arrive at its destination? Or, what if the packets arrive at the destination out of order? Some process must be in place to manage these challenges and overcome the potentially disastrous results that could occur.

Computer networks have a lot in common with modern freeway systems, including the tendency to become congested. Congestion results in delay, which some applications do not tolerate well. What happens if some or all of the packets are badly delayed, as shown in Fig. 3.8? What is the impact on the end-to-end quality of the service?

Another vexing problem that often occurs is errors in the bitstream. Any number of factors, including sunspot activity, the presence of electric motors, and the electrical noise from fluorescent lights

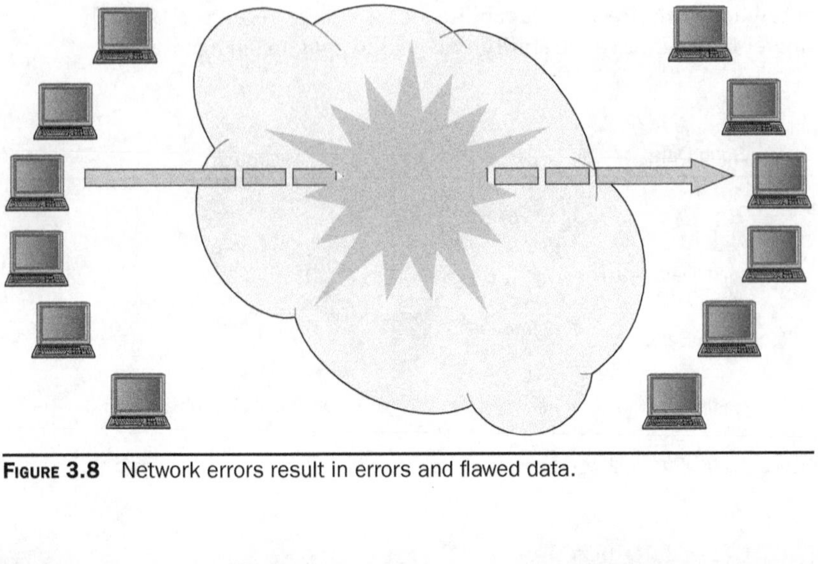

FIGURE 3.8 Network errors result in errors and flawed data.

Figure 3.9 Bit errors.

can result in ones being changed to zeroes and zeroes being changed to ones, as shown in Fig. 3.9. Obviously, this is an undesirable problem, and there must be a technique in place to detect and correct these errors when they occur.

There may also be inherent challenges with the physical medium over which the information is being transmitted. There are many different media including twisted copper wires, optical fibers, coaxial cables, and wireless systems, to name a few. None of these are perfect transmission media; they all suffer from the vagaries of backhoes, lightning strikes, sunlight, earth movement, squirrels with sharp teeth, kids with BB guns, and other impairments far too numerous to name. When these problems occur, how are they detected? Equally important, how are the transmission impairments that they inevitably cause reported and corrected?

There must also be an agreed upon set of rules that define exactly how the information is to be physically transmitted over the selected medium. For example, if the protocol to be used dictates that information will *always* be transmitted on pin two of a data cable, such as that shown in Fig. 3.10, then the other end will have a problem since its received signal will arrive on the same pin that it wants to *transmit* on! Furthermore, there must be agreement on how information is to be physically represented, how and when it is to be transmitted, and how it is to be acknowledged. What happens if a very fast transmitter overwhelms the receive capabilities of a slower receiver? Does the slower receiver have the ability, or even the *right,* to tell it slow down?

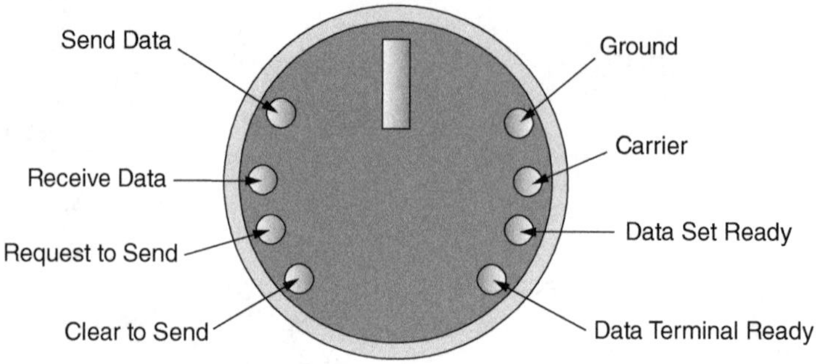

Figure 3.10 Physical agreements.

Believe it or not, there are systems in place that use a protocol that dictates that if the sender overwhelms the receiver, the receiver has the right to simply throw the excess traffic away—without informing the sender! We'll see examples of that later in the book, but believe it or not, it's actually an efficient and effective way to handle excess traffic in high-volume networks.

Collectively, all of these problems pose what seem to be insurmountable challenges to the transmission of data from a source to a receiver. And while the process is obviously complex, steps have been taken to simplify it by breaking it into logical pieces. Those pieces, as we described earlier, are called protocols. Collections of protocols, carefully selected to form functional groupings, are what make data communications work properly. And data communications, in turn, is what makes telecommunications, IT, and media transport work seamlessly.

The Network

For years now, communications networks have been functionally depicted as shown in Fig. 3.11: a big, fluffy, opaque cloud, into which lines disappear that represent circuits which magically reappear on the other side of the cloud. I'm not sure why we use clouds to represent networks; knowing what I know about their complex innards and how they work, a hairball would be a far more accurate representation.

In truth, clouds are pretty good representations of networks from the point of view of the customers that use them. Internally, networks are remarkably complex assemblages of hardware and software, as you will see in the telephony chapter. Functionally, however, they are straightforward and, regardless of the technology on which they are built, fundamentally they all work the same way: customer traffic

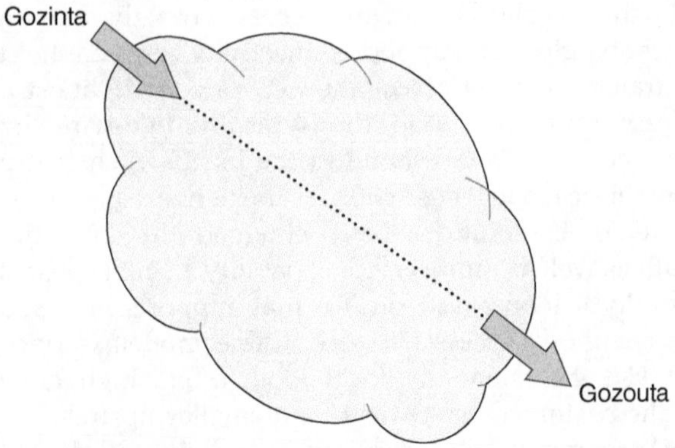

FIGURE 3.11 The network cloud.

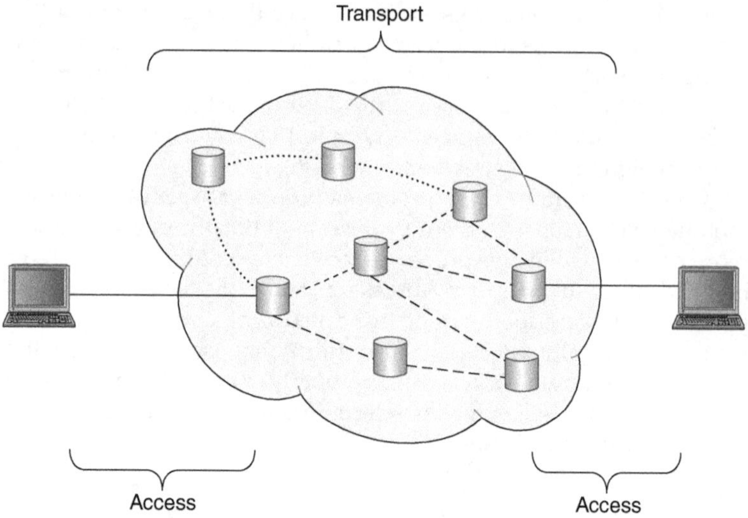

FIGURE 3.12 Access and transport regions of the network.

goes into the network on the *Gozinta;* the traffic then emerges, unchanged, on the *Gozouta.* How it happens is unimportant to the customer; all they care about—or *want* to care about—is that the network receives, interprets, transports, and delivers their voice, video, images, data, and music to the destination in a correct, timely, and cost-effective fashion. Later in the book we will discuss the various technologies that live within the network, but for now suffice it to say that its responsibilities fall into two categories: access and transport, as illustrated in Fig. 3.12.

Network Access

As the illustration shows, network access is exactly that: the collection of technologies that support connectivity between the customer and the transport resources of the network. At its most common level, access is the local loop, the (typically) two-wire circuit that connects a customer's telephone to the local switch that provides telephony service to that customer. As the network has become more data-aware, other solutions have emerged that provide greater bandwidth as well as multiservice capability. Digital subscriber line, (DSL) is a local loop-based service that supports high-speed data over the telephony access network. Cable modem service, which does *not* use the telephony local loop, offers high downstream (toward the customer) bandwidth and smaller upstream (from the customer) capacity. Wireless services, including Wi-Fi, WiMAX, LTE, satellite, microwave, and others, represent alternative options for access connectivity. All of these will be discussed in greater detail later in the book.

Miscellaneous Additional Terms

A number of other terms need to be introduced here as well, the first of which are *data terminal equipment* (DTE) and *data circuit terminating equipment* (DCE). DTE is exactly that—a device that a user employs to gain access to the network. It might be a PC, laptop, mobile handset, tablet, or telemetry device used in a machine-to-machine communications application.

A DCE is the device that actually terminates the circuit at the customer's premises, typically a modem such as DSL modem or a cable modem. The relationship between these network elements is shown in Fig. 3.13. One important point is that because much of the usage of the public switched telephone network (PSTN) is for data, and the PSTN is optimized (meaning designed) for voice, the primary role of the DCE is to make the customer's DTE look and smell and taste and feel like a telephone to the network. For example, if the DTE is a PC, then the modem's job is to collect the high-frequency digital signals being produced by the PC and modulate them into a range of frequencies that are acceptable to the bandwidth-limited voiceband of the telephone network. That's where the name comes from, incidentally—modulate/demodulate (MO-DEM). We'll explain this in greater detail later in Chap. 8.

Another pair of terms that must be introduced here is *parallel* and *serial*. You have undoubtedly seen the ribbon cables that are used to transport data inside a PC (Fig. 3.14), or the parallel wires etched into the motherboard inside the PC (Fig. 3.15). These parallel conductors are called a *bus*, and are used for the high-speed transport of multiple simultaneous bits in parallel fashion from one device inside the computer to another. Serial transmission, on the other hand, is used for the *single-file* transport of multiple bits, one after the other, usually deployed *outside* a computer.

Finally, we offer *simplex, half-duplex,* and *full-duplex transmission.* Simplex transmission means one-way only, like a radio broadcast. Half-duplex transmission means two-way, but only one way at a time, like push-to-talk radio. Finally, full-duplex means two-way

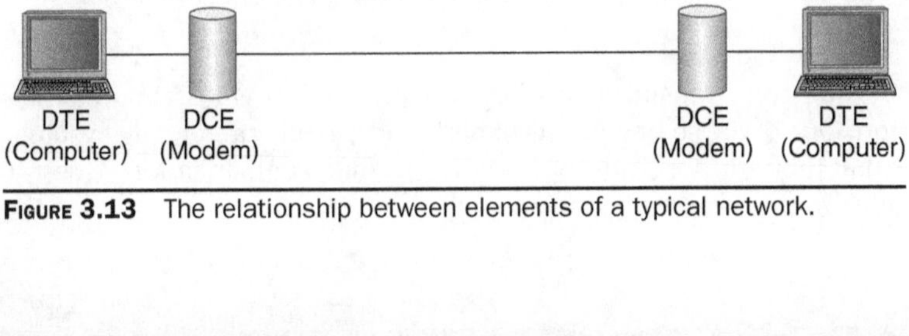

FIGURE 3.13 The relationship between elements of a typical network.

FIGURE 3.14 A parallel bus inside a computer. These can be ribbon cables, shown, or etched into the surface of the motherboard.

FIGURE 3.15 Parallel bus etched into the motherboard of the computer.

simultaneous transmission, like telephony or two-way data transmission. One of the selling points of high-end videoconferencing systems is that they support full-duplex voice, so that a person talking doesn't get stomped on if someone interrupts.

Network Transport

The fabric of the network cloud is a rich and unbelievably complex collection of hardware and software that moves customer traffic from an ingress point to an egress point, anywhere in the world. It's a function that we take entirely for granted because it is so ingrained in day-to-day life, but stop for a moment to think about what the network actually does. Not only does it deliver voice and data traffic between end points, but it does so easily and seamlessly, with various levels of service quality as required to any point on the globe (and in fact beyond!) in a matter of seconds—and with *zero* human involvement. It is the largest fully automated machine on the planet, and represents one of the greatest technological accomplishments of all time. Think about that: I can pick up a handset here in Vermont, dial a handful of numbers, and seconds later a telephone rings in Ouagadougou, Burkina Faso, in North Central Africa. How that happens borders on the miraculous. We will explore it in considerably greater detail later in the book.

Transport technologies within the network cloud fall into two categories: fixed transport and switched transport. Fixed transport, sometimes called private line or dedicated facilities, includes such technologies as T-1, E-1, DS-3, SONET, SDH, dedicated optical channels, some forms of gigabit ethernet, and microwave. Switched technologies include modem-based telephone-network transport, IP packet switching, frame relay, switched ethernet, and asynchronous transfer mode (ATM). Together with the access technologies described previously and customer premises technologies such as ethernet, Wi-Fi, and bluetooth, transport technologies offer the infrastructure components required to craft an end-to-end solution for the transport of customer information.

The Many Flavors of Transport

Over the last few years *the network* has been functionally segmented into a collection of loosely defined *regions* that define unique service types. These include the local area (sometimes referred to as the premises region), the metropolitan area, and the wide area, sometimes known as the core. Local area networking has historically defined a network that provides services to an office, a building, or even a campus. Metro networks generally provide connectivity within a city, particularly to multiple physical locations of the same company. They are usually deployed across an optical ring architecture that has the ability to detect failures in the physical medium of the ring and *self-heal* to avoid outages. Wide area networks, often called core transport, provide long distance transmission and are typically deployed using a mesh-networking model.

Transport Channels

The physical circuit over which customer traffic is transported in a network is often referred to as a *facility*. Facilities are characterized by a number of qualities such as distance, quality (signal level, noise, and distortion coefficients), and bandwidth. Distance is an important criterion because it places certain design limitations on the network, making it more expensive as the circuit length increases. This has become less of an issue in modern enterprise networks because of the ubiquity of the Internet as a universal transport mechanism; the Internet is completely distance-independent.

Over distance, signals tend to weaken and become noisy, and specialized devices (amplifiers, repeaters, or regenerators) are required to periodically clean up the signal quality and maintain the proper level of *loudness* to ensure intelligibility and recognizability at the receiving end.

Quality is related to distance in the sense that many of the same factors affect them. Signal level is clearly important, as is noise. Distortion is a slightly different beast and must be dealt with equally carefully. Noise is a random event in networks caused by lightning, fluorescent lights, electric motors, sunspot activity, and squirrels chewing on wires, and is unpredictable and largely random. Noise, therefore, cannot be anticipated with any degree of accuracy; its effects can only be recovered from.

Far more important to media transport, specifically voice, video, and music, are delay and jitter. Delay is exactly what the word implies: A problem that occurs when data packets are inadvertently delayed between the transmitter and receiver, resulting in a video that freezes, voice that sounds distorted or music that stops playing altogether.

Jitter is a measure of *the variability of the delay.* This can be even more vexing for customers because they can't anticipate when the problem will occur or how long it will last. Last night, for example, I was watching a movie with my wife on Netflix. Every 45 seconds or so, like clockwork, there was a momentary pause in the show, probably caused by a full video server buffer somewhere in the connection. We quickly became accustomed to it because it was predictable. Had it occurred randomly, however, we would have become increasingly annoyed and would have most likely turned the movie off because of a poor service quality experience.

Distortion, on the other hand, is a measurable, predictable characteristic of a transmission channel and is usually frequency-dependent. For example, certain frequencies transmitted over a particular channel will be weakened, or attenuated, more than other frequencies. Think about the last time you attended a parade. How did you know the band was approaching? If you think about it, you'll realize it was because you heard the low-frequency tones of the bass drums and tubas long before you saw the band—or heard the higher frequencies of the trumpets and trombones. Higher-frequency signals weaken faster than lower frequencies, and therefore don't travel as far without amplification and regeneration.

If we can measure this, then we can condition the channel to equalize the treatment that all frequencies receive as they are transmitted down that particular channel. This process is known as *conditioning* and is part of the higher cost involved in buying a dedicated circuit for data transmission. This is one of the reasons that the Internet is in such high demand as a transport mechanism. Even though it offers no quality of service (QoS) guarantees, it's pretty good, and both the network and the devices connected to it have become good at identifying and fixing transmission problems at the edge of the network.

Bandwidth is the last characteristic that we will discuss here, and the quest for more of it is one of the great challenges of telecommunications. Bandwidth is a measure of the number of bits that can be transmitted down a facility in any one-second period. In most cases, it is a fixed characteristic of the facility and is the characteristic that most customers are willing to pay more for. The measure of bandwidth is bits-per-second, although today the measure is more typically thousands (kilobits), millions (megabits), or billions (gigabits) per second.

The bandwidth challenge is particularly problematic and challenging in the world of wireless services, where spectrum is the desirable asset for higher and higher levels of broadband. I often hear network professionals observe "the problem we have is that we don't have enough spectrum." This is patently false. We have all the bandwidth there is—all of it! The *real* problem is that much of the available spectrum operates over a very short transmission distance, measured in a few feet; it's not very helpful for roaming users. Another significant chunk of available spectrum has an annoying tendency to cook people when they walk in front of a device that's transmitting in that range. It smells good, but it's terrible for quality of experience. So the real problem isn't spectrum *availability*; it's spectrum *management*.

Facilities are often called *channels* because physical facilities are often used to carry multiple streams of user data through a process called *multiplexing*. Multiplexing is the process of allowing multiple users to share access to a transport facility, either by taking turns or using separate frequencies within the channel. If the users take turns, as shown in Fig. 3.16, the multiplexing process is known as *time division multiplexing*, because time is the variable that determines when each user gets to transmit through the channel. If the users share the channel by occupying different frequencies, as shown in Fig. 3.17, the process is called *frequency division multiplexing*, because frequency is the variable that determines who can use the channel. It is often said that in time division multiplexing, users of the facility are given *all* of

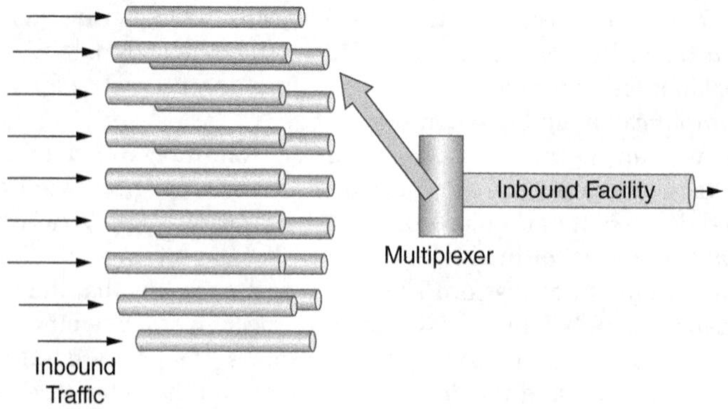

FIGURE 3.16 Time-division multiplexing. Users are given "all of the frequency some of the time."

FIGURE 3.17 Frequency division multiplexing. Users are given "some of the frequency all of the time."

the frequency *some* of the time, because they are the only one using the channel during their timeslot. In frequency division multiplexing, users are given *some* of the frequency *all* of the time, because they are the only one using their particular frequency *band* at any point.

Analog versus Digital Signaling: Dispensing with Myths

Frequency division multiplexing is normally considered to be an *analog technology,* while time division multiplexing is a *digital technology.* The word *analog* means something that bears a similarity to something else, while the word *digital* means discrete. Analog data, for example, typically illustrated as some form of sine wave such as that shown in Fig. 3.18, is an exact representation of the values of the data

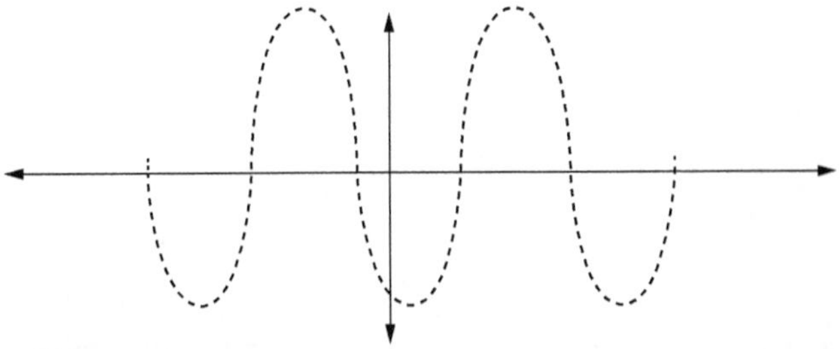

FIGURE 3.18 A typical sine wave.

being transmitted. The process of using manipulable characteristics of a signal to represent data is called *signaling*.

We should also introduce a few terms here just to keep things marginally confusing. When speaking of signaling, the proper term for digital signaling is *baseband,* while the term for analog signaling is *broadband.* This is where confusion can creep in: When talking about data (not signaling), the term broadband means *big channel,* not analog signaling.

The sine wave, undulating along in real time in response to changes in one or more parameters that control its shape, represents the exact value of each of those parameters at any point in time. The parameters are amplitude, frequency, and phase. We will discuss each in turn. Before we do, though, let's relate analog waves to the geometry of a circle. Trust me, this helps.

Consider the diagram shown in Fig. 3.19. This shape is called a sine wave. If we examine this waveform carefully, we notice some interesting things about it. First of all, every time the circle completes a full revolution (360 degrees), it draws the shape shown in Fig. 3.20. Thus halfway through its path, indicated by the zero point on the graph, the circle has passed through 180 degrees of travel. This makes sense, since a circle circumscribes 360 degrees.

The reason this is important is because we can manipulate the characteristics of the wave created in this fashion to cause it to carry

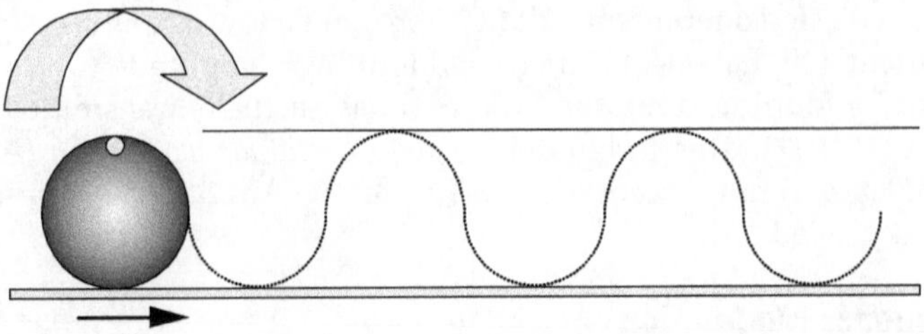

FIGURE 3.19 The details of a sine wave.

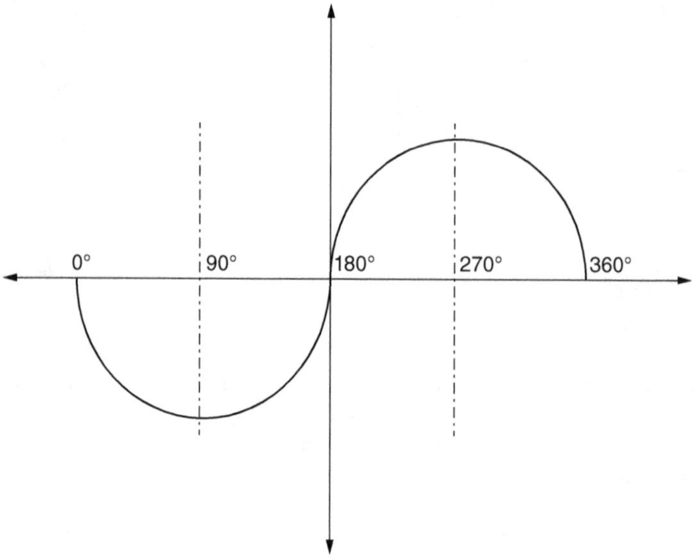

FIGURE 3.20 The path of a typical sine wave.

varying amounts of information. Those characteristics, amplitude, frequency, and phase, can be manipulated as shown in the following sections.

Amplitude Modulation

Amplitude is a measure of the loudness of a signal. A loud signal, such as that currently thumping through the window of my office from the sub-woofer in the car across the street, has high amplitude components (the sound level is extremely loud), but very low-frequency components (as evidenced by the fact that I can hear him coming from blocks away and things on my desk start to jump as he gets close). Lower-volume signals are lower in amplitude. Examples are shown in Fig. 3.21. The dashed line represents a high-amplitude signal, while the solid line represents a lower-amplitude signal. And how could this be used in the data communications realm? Simple, let's let a high-amplitude signal represent a digital zero, and a low-amplitude signal represent a digital one. If I then send four high-amplitude waves followed by four low-amplitude waves, I have actually transmitted the series 00001111. This technique is called *amplitude modulation* (AM); modulation simply means to vary. This is how AM radio works—and why it's called "AM."

Frequency Modulation

Frequency modulation (FM) is similar to amplitude modulation except that instead of changing the loudness of the signal, we change the

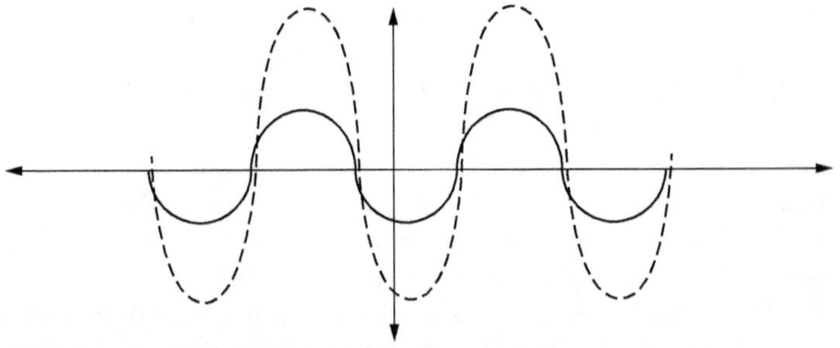

FIGURE 3.21 Examples of amplitude modulation (AM).

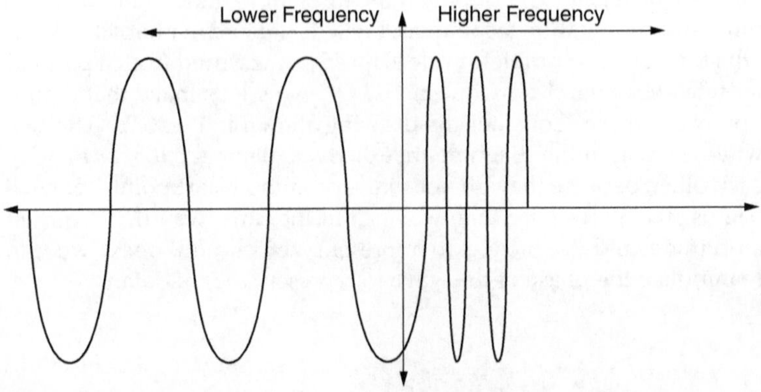

FIGURE 3.22 Examples of frequency modulation (FM).

number of signals that pass a point in a given second, illustrated in Fig. 3.22. The left side of the graph contains a lower-frequency signal component, while a higher-frequency component appears to its right. We can use this technique in the same way we used AM: If we let a high-frequency component represent a zero, and a low-frequency component represent a one, then I can broadcast our 00001111 series by transmitting four high-frequency signals followed by four low-frequency signals.

An interesting historical point about FM: the technique was invented by radio pioneer Edwin Armstrong in 1933. Armstrong created FM as a way to overcome the problem of noisy radio transmission. Prior to FM's arrival AM was the only technique available and it relied on modulation of the loudness of the signal *and* the inherent noise to make it stronger. FM did not rely on amplitude, but rather on frequency modulation, and was therefore much cleaner and offered significantly higher fidelity than AM radio. Keep in mind that signals pick up noise over distance; when an amplifier amplifies a signal, it also amplifies the noise.

Many historians of World War II technology believe that Armstrong's invention of FM transmission played a pivotal role in the outcome of the war. When WWII was in full swing, FM technology was only available to Allied forces. AM radio, the basis for most military communications at the time, could be jammed by simply transmitting a more powerful signal that overloaded the transmissions of military radios. FM, however, was not available to the Axis powers and therefore could not be jammed as easily.

We'll see another example of a wireless technology that changed the course of the war when we get to the wireless chapter.

Phase Modulation

Phase modulation (PM) is a little more difficult to understand than the other two modulation techniques. Phase is defined mathematically as "the fraction of a complete cycle elapsed as measured from a particular reference point." Any questions? Ok. Now let's make that definition make sense. Consider the drawing shown in Fig. 3.23. The two waves shown in the diagram are exactly 90 degrees *out of phase* with each other, because they do not share a common start point—wave B begins 90 degrees later than wave A. In the same way that we used amplitude and frequency to represent zeroes and ones, we can manipulate the phase of the wave to represent digital data.

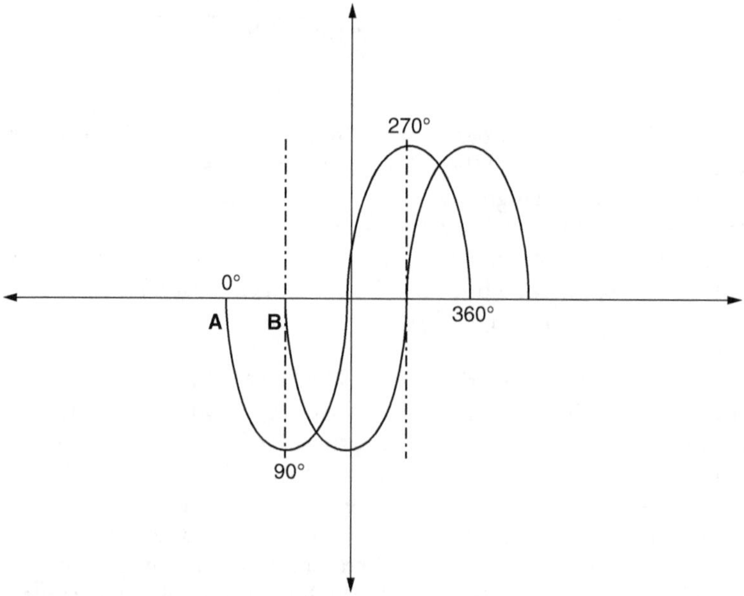

FIGURE 3.23 Examples of phase modulation (PM).

A few years ago, I saw a very graphic (and way too cool) example of how phase modulation can be used in a practical way. I was in Arizona with a film crew, shooting a video for one of my corporate clients. The theme of the video was based on the statement that if "we don't intelligently deploy technology, the competition will leave us in the dust." Building on that phrase about dust, we rented a ghost town in Arizona and created a video metaphor around it. In the last scene of the show, a horse-drawn wagon loaded with boxes of technology products disappears into a bloody Arizona sunset, and just before the wagon disappears over the hill, it skyrockets into the sky on a digital effect that looks like a rocket trail. We loved it. The only problem was that when we got back into the post-production studio and began to assemble the final show, we discovered to our horror that the sound of an airplane could be heard behind our narrator, and since airplanes didn't exist in the old west, we had a big problem. Not to be bested, our audio engineer asked us to pipe the sound into the audio booth. Listening to the recording, he went to his wall of audio CDs and selected a collection of airplane sounds. He listened to several of them until he found one that was correct. Setting the levels correctly so that they matched those of the video soundtrack, he inverted the CD signal (180 degrees out of phase with the soundtrack signal) and electronically added it to the soundtrack.

The airplane noise disappeared from the narration.

Digital Signaling

Data can be transmitted in a digital fashion as well. Instead of a smoothly undulating wave crashing on the computer beach, we can use an approximation of the wave to represent the data. This technique is called *digital signaling*. In digital signaling, an interesting mathematical phenomenon called the *Fourier series* is called into play to create what most people call a square wave, shown in Fig. 3.24. In the case of digital signaling, the Fourier series is used to approximate the *square* nature of the waveform. The details of how the series actually works are beyond the scope of this book, but suffice it to say that by mathematically combining the infinite series of odd harmonics of

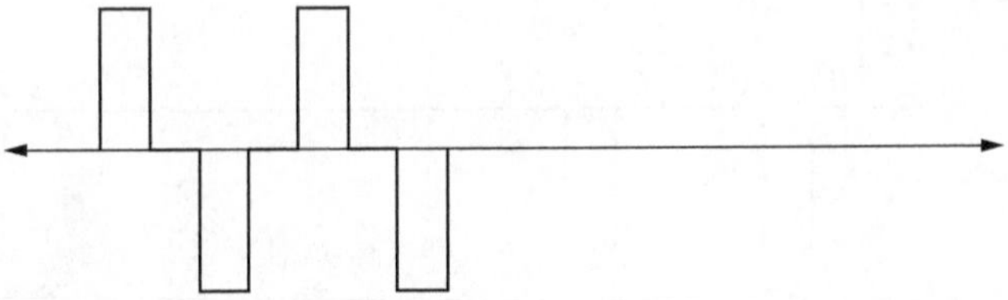

FIGURE 3.24 A square wave, generated by manipulating the Fourier series.

a fundamental wave, the ultimate result is a squared off shape that approximates the square wave that commonly depicts digital data transmission. This technique is called digital signaling, as opposed to the amplitude, frequency, and phase-dependent signaling techniques used in analog systems.

In digital signaling, zeroes and ones are represented as either the absence or presence of voltage on the line, and in some cases by either positive or negative voltage, or both. Figure 3.25, for example, shows a technique in which a zero is represented by the presence of positive voltage, while a one is represented as zero voltage. This is called a *unipolar signaling scheme.* Figure 3.26 shows a different technique, in which a zero is represented as positive voltage, while a one is represented as negative voltage. This is called a *non-return to zero signaling scheme,* because zero voltage has no meaning in this technique. Finally, Fig. 3.27 demonstrates a *bipolar signaling system.* In this technique, the presence of voltage represents a one, but notice that every

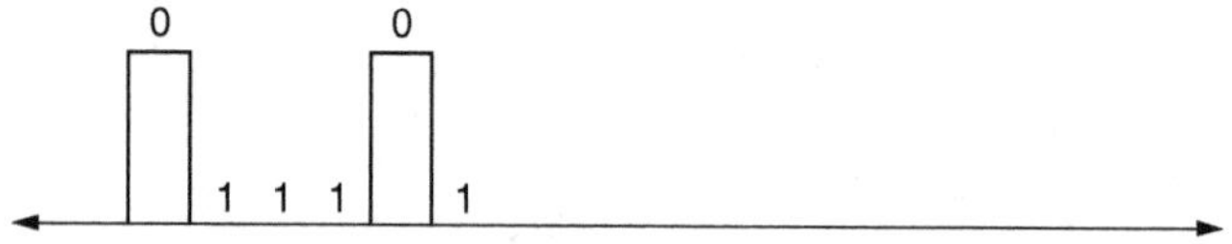

FIGURE 3.25 Unipolar signaling scheme.

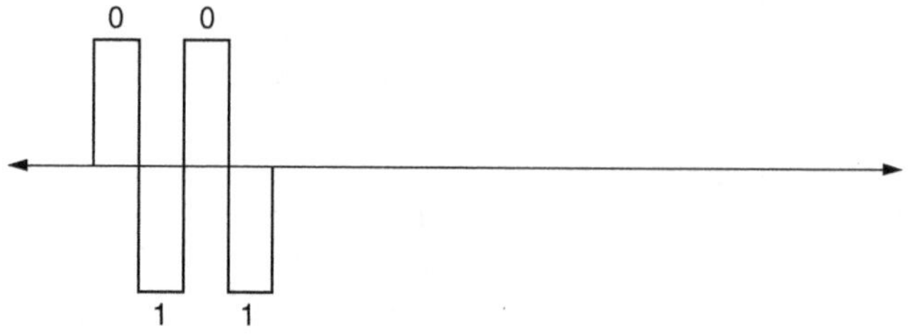

FIGURE 3.26 Non-return to zero signaling scheme.

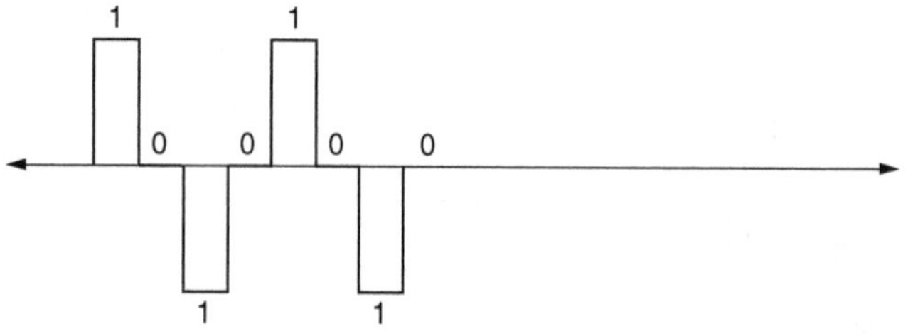

FIGURE 3.27 Bipolar signaling scheme.

other one is opposite in polarity from the one that precedes it *and* the one that follows it. Zeroes, meanwhile, are represented as zero voltage. This technique, called *alternate mark inversion* (AMI) is commonly used in T- and E-Carrier systems for reasons that will be discussed in Chap. 4. There are other techniques in use, but these are the most common.

Clearly, both analog and digital signaling schemes can be used to represent digital data depending upon the nature of the underlying transmission system. It is important to keep the difference between *data* and *signaling* clearly separate. Data is the information that is being transported, and it can be either analog or digital in nature. For example, music is a purely analog signal because its values constantly vary over time. It can be represented, however, through both analog and digital signaling techniques. The zeroes and ones that spew forth from a computer are clearly digital information, but they, too can be represented either analogically or digitally. For example, the broadband access technology known as *digital subscriber line (DSL)* is not digital at all: there are analog modems at each end of the line, which means that *analog signaling techniques* are used to represent the *digital data* that is being transmitted over the local loop.

Combining Signaling Techniques for Higher Bit Rates

Let's assume that we are operating in an analog network. Under the standard rules of the analog road, one signaling event represents one bit. For example, a high-amplitude signal represents a one, and a low-amplitude signal represents a zero. But what happens if we want to increase our bit rate? One way is to simply signal faster. Unfortunately, the rules of physics limit the degree to which we can do that. In the 1920s, a senior researcher at Bell Laboratories who has now become something of a legend in the field of communications came to the realization that the bandwidth of the channel over which information is being transmitted has a direct bearing on the speed at which signaling can be done across that channel. According to Harry Nyquist, the broader the channel, the faster the signaling rate can be. In fact, Nyquist determined that the signaling rate can never be faster than two times the highest frequency that a given channel can accommodate. And this is important … why?

Unfortunately, the telephone local loop was historically engineered to support the limited bandwidth requirements of the human voice. Even though the human voice comprises a rich mélange of frequencies (just listen to James Earl Jones), it can be understood and recognized as long as a certain range of frequencies are included—specifically the range of frequencies between 300 and 3,300 Hz that has come to be known as the voice band. The traditional voice network was engineered, therefore, to deliver 4 KHz of bandwidth to

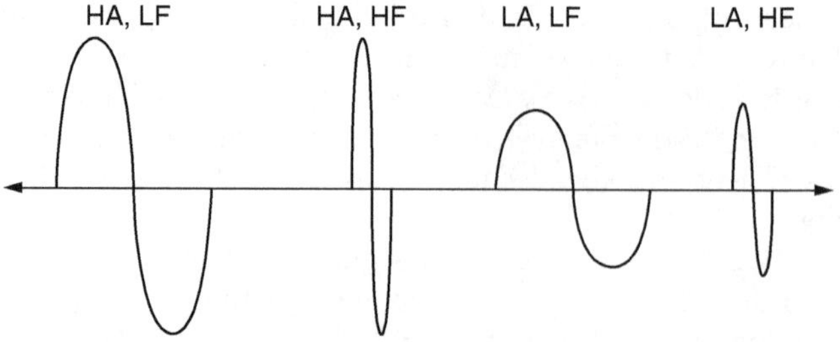

FIGURE 3.28 Multibit signaling scheme.

each local loop,[1] which means that the fastest signaling rate achievable over a telephony local loop is 8,000 signals per second. The measure of signals-per-second is called baud (Bd), so a 2,400 bit-per-second modem was also called a 2,400-Bd modem.

Yet during the late 1980s and the early-to-mid 1990s, it was common to see advertisements for 9,600-Bd modems. This is where the confusion of terms becomes obvious: as it turns out, these were *9,600 bit-per-second modems*—a big difference. Modems of 9,600 Bd used on the PSTN were patently impossible, right? It's impossible to signal over the bandwidth-limited voice network at speeds higher than 8,000 Bd. This, however, introduces a whole new problem: How do we create higher bit rates over signaling rate-limited (and therefore bandwidth limited) channels?

To achieve higher signaling rates, one of two things must be done: either broaden the channel, which is not always feasible, or figure out a way to have a single signaling event convey more than a single bit.

Consider the following example. We know from our earlier discussion that we can represent two bits by sending a high-amplitude signal followed by a low-amplitude signal (high-amplitude signal represents a zero, low-amplitude signal represents a one). What would happen though, if we were to combine amplitude modulation with frequency modulation? Consider the four waveforms shown in Fig. 3.28. By combining the two possible values of each characteristic (high and low frequency or amplitude), we create four possible states, each of which can actually represent two bits as shown in Fig. 3.29. We can have a high-bandwidth, high-frequency signal; a high-bandwidth, low-frequency signal; a low-bandwidth, low-frequency signal; and a low-bandwidth, high-frequency signal. Consider what we have just done: We have created a system in which each signaling

[1]One way in which this was done was through the use of load coils. Load coils are electrical traps that tune the local loop to a particular frequency range, only allowing certain frequencies to be carried. This created a problem later for digital technologies, as we will discuss.

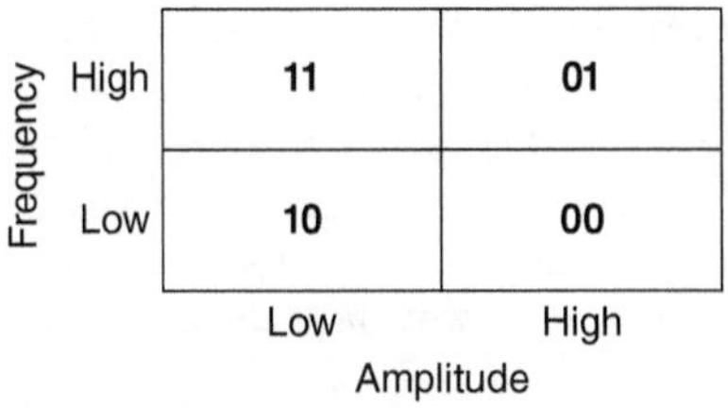

Figure 3.29 Representing multiple bits with a single event.

event represents two bits, which means that our bit rate is twice our signaling rate. And as we said earlier, the signaling rate is the baud. It may or may not be the same as the bit rate, depending on the scheme being used.

Let's take our concept one step farther. Figure 3.30 shows a system in which we are encoding four bits for each signal, a technique known as quad-bit encoding. This scheme, sometimes called *quadrature amplitude modulation* (QAM, pronounced "Kwăm") permits a single signal to represent four bits, which means that there is a 4:1 ratio between the bit rate and the signaling rate (baud). Thus, it is possible

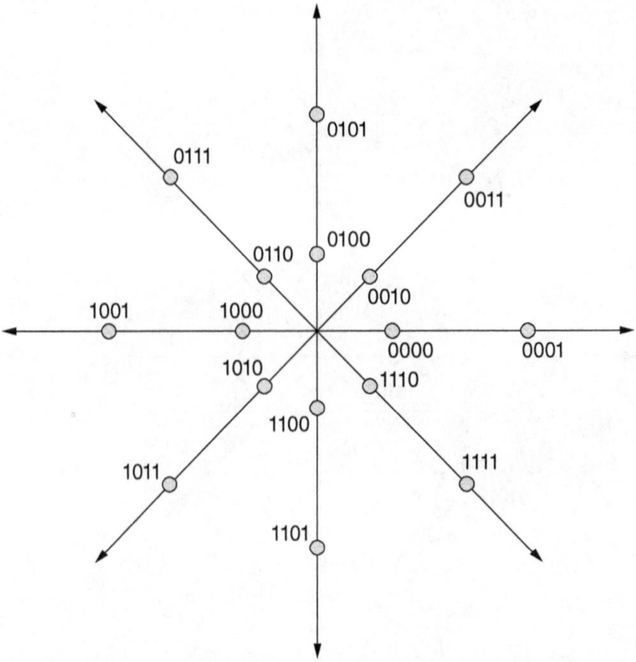

Figure 3.30 Quadrature amplitude modulation (QAM), sometimes called trellis coding.

to achieve higher bit rates in the bandwidth-limited telephony local loop by using multibit encoding techniques such as QAM. The first high bit rate modems (9,600 bits-per-second) used this technique or a variation of it to overcome the design limitations of the network. In fact, these multibit schemes are also used by the cable industry to achieve the high bit rates they need to operate their multimedia broadband networks.

There is one other limitation that must be mentioned: noise. Look at Fig. 3.31. Here we have a typical QAM graph, but now we have added noise, in the form of additional points on the graph that have no implied value. When a receiver sees them, however, how does it know which points are noise, and which are data? Similarly, the oscilloscope trace shown in Fig. 3.32 of a high-speed transmission would be difficult to interpret if there were noise spikes intermingled with the data. There is, therefore, a well-known relationship between the noise level in a circuit and the maximum bit rate that is achievable over that circuit, a relationship that was first described by Bell Labs researcher Claude Shannon who is widely known as the father of information theory. In 1948, Shannon published *A Mathematical Theory of Communication* that is now universally accepted as the framework for modern communications theory. We won't delve into the complex

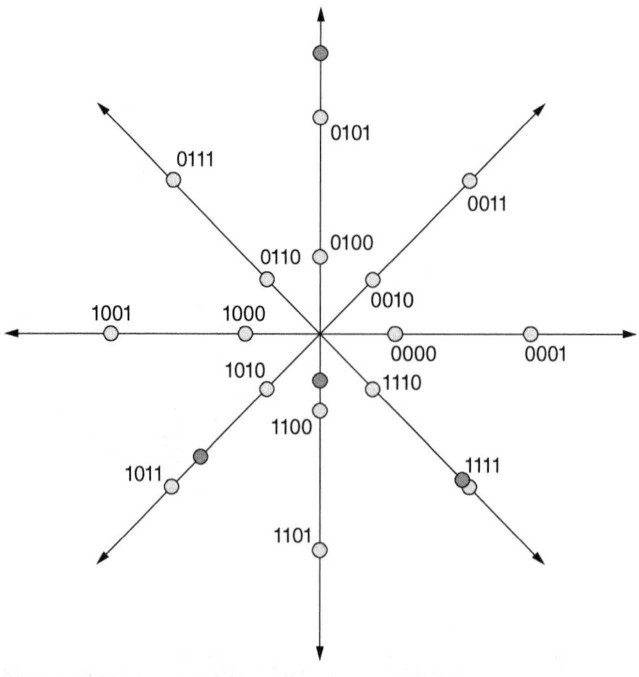

FIGURE 3.31 The impact of noise in a QAM system.

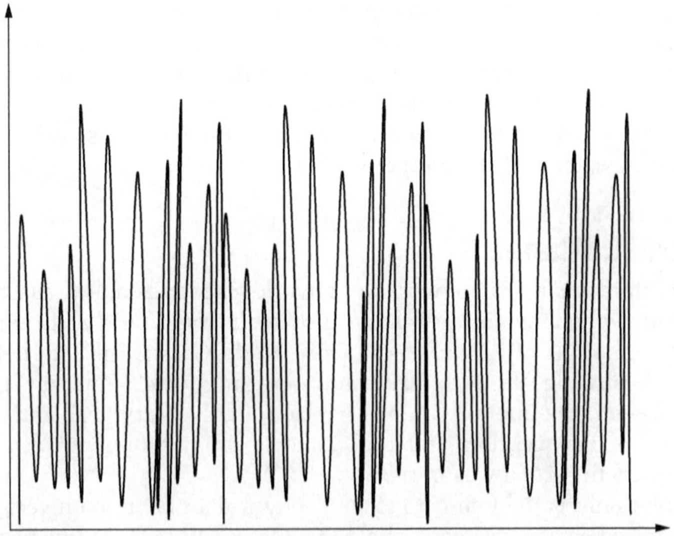

Figure 3.32 A typical oscilloscope trace showing high frequency signal elements. How do you separate noise from signal?

(but fascinating) mathematics that underlie Shannon's theorem, but suffice it to say that his conclusions are seminal: the higher the noise level in a circuit, the lower the achievable bandwidth.

The bottom line? *Noise matters.* It matters so much, in fact, that network designers and engineers make its elimination the first order of business in their overall strategies for creating high-bandwidth networks. This is one of the reasons that optical fiber-based networks have become so critically important in modern transport systems—they are far less subject (but not immune!) to noise, and absolutely immune to the electromagnetic interference that plagues metallic networks. Cable companies that now offer data services have the same issues and concerns. Every time a customer decides to play installer by adding a cable spur for a new television set in their home and crimping the connector on the end of the cable with a pair of pliers instead of a tool specifically designed for the purpose, they create a point where noise can leak into the system, causing problems for everyone. And it gets even more melodramatic than that: Unauthorized connection to the cable network can cause problems that go way beyond unauthorized access to service. "Cable networks are high-frequency systems," one technician observes. "Some of the harmonics created in cable networks just happen to fall within the range of frequencies used in avionics, and therefore have the potential to affect aviation communications and navigation. So when you see the cable truck that looks like a commercial fishing boat cruising the neighborhood with all the antennas on top, they're looking for signal leakage from unauthorized taps. They *will* find them and they *will* come in and fix them, and you *will* get a bill for it. So if you want to add a connection in the house, call us. It's cheaper."

That completes our introduction of common terms, with one notable exception: The Internet.

The Internet: What Is It?

The Internet is a vast network of networks, recognized as the fastest growing phenomenon in human history. In the words of Douglas Adams, author of *A Hitchhiker's Guide to the Galaxy,* the Internet is "Big. Really big. Vastly, hugely, mind-bogglingly big." And, it is getting bigger: the Internet doubles in size roughly every year, and has been since its inception. As Table 3.2 shows, today's Internet is almost 200 times the size it was in 1995.

Not only is the Internet global in physical scope, it is universally recognized. *Everybody* knows about the Internet. In 1993, it came booming into the public consciousness, put down roots, spread like a biological virus, and flourished. Like other famous public figures, it has been on the cover of every major magazine in the world, has been the star of books, articles, TV shows, and movies, has been praised as the most significant social force in centuries, and debased as the source of a plethora of worldwide ills. And yet, for all this fame and notoriety, little is actually known about the Internet itself—at least, its private side. It is known to be a vast network of interconnected networks, with new appendages connecting *approximately* every 10 minutes. According to World Wide Web Size (*http://www.worldwidewebsize.com*), it hosts 1.4 billion Web pages and connects to approximately 9 billion devices worldwide.

The World Wide Web (WWW)

The World Wide Web was first conceived by Tim Berners-Lee, considered to be the "Father of the World Wide Web." A physicist by training, Berners-Lee began his career in the computer and telecommunications industries following graduation from Oxford, before accepting a consulting position as a software engineer with the European Organization for Nuclear Research (CERN) during the late 1970s.

Date	Users	% of Global Population	Source
December, 1995	16 million	0.4%	IDC
March, 2013	2.8 billion	39%	ITU

Table 3.2 The Extraordinary Growth of the Global Internet

During his stint in Geneva, Berners-Lee observed that CERN suffered from the problems that plague most major organizations: information location, management, and retrieval. CERN is a research organization with large numbers of simultaneous ongoing projects, a plethora of internally published documentation, and significant turnover of people. Much of the work conducted at CERN involves collaboration about large-scale, high-energy physics research (the Hadron Super Collider that tentatively identified the Higgs Boson is one of their projects) that demands instantaneous information sharing between physicists all over the world. Berners-Lee found that his ability to quickly locate and retrieve specific information was seriously impaired by the lack of a single common search capability and the necessarily dispersed nature of the organization. To satisfy this need, he collaborated with Robert Cailliau to write the first World Wide Web client, a search and archive program that they called *Enquire*. Enquire was never published as a product although Berners-Lee, Cailliau, and the CERN staff used it extensively. It did, however, prove to be the foundation for the World Wide Web.

In May of 1990, Berners-Lee published *Information Management: A Proposal*, in which he described his experiences with hypertext systems and the rationale for Enquire. He described the system's layout, feel, and function as being similar to Apple's Hypercard, or the old "Adventure" game in which players moved from "page to page" as they navigated through the game.[2] Remember this? Some of you will:

>YOU FIND YOURSELF IN A SMALL ROOM. THERE IS A DOOR TO THE LEFT.

>>OPEN DOOR

Enquire had no graphics and was therefore rudimentary compared to modern Web browsers. To its credit, the system ran on a multiuser platform and could therefore be accessed simultaneously by multiple users. To satisfy the rigorous demands of the CERN staff, Berners-Lee and Cailliau designed the system around the following parameters:

- It had to offer remote access from across a diversity of networks.
- It had to be system and protocol independent, since CERN was home to a wide variety of system types—VM/CMS, Mac, VAX/VMS, and Unix.
- It had to run in a distributed processing environment.

[2]You'll love this: I recently discovered that some kind soul has ported Adventure to the PC platform, and the game, called WinFrotz, is downloadable from *http://www.pcworld.com/downloads/file_description/0,fid,22456,00.asp*.

- It had to offer access to all existing data types as well as to new types that would follow.
- It had to support the creation of personal, private links to new data sources as each user saw fit to create them.
- It had to support, in the future, diverse graphics types.
- It (ideally) had to support a certain amount of content and data analysis.

In November 1990 Berners-Lee wrote and published, with Robert Cailliau, *WorldWide Web: A Proposal for a HyperText Project.* In it the authors described an information-retrieval system in which large and diverse compendia of information could be searched, accessed, and reviewed freely, using a standard user interface based on an open, platform-independent design. This paper relied heavily on Berners-Lee's earlier paper.

In *WorldWide Web: A Proposal for a HyperText Project*, Berners-Lee and Cailliau proposed the creation of a *World Wide Web* of information that would allow the various CERN entities to access the information they needed based on a common and universal set of protocols, file exchange formats, and keyword indices. The system would also serve as a central (although architecturally distributed) repository of information and would be totally platform-independent. Furthermore, the software would be available to all and distributed free of charge.

Once the paper had been circulated for a time, the development of what we know today as the Web occurred with remarkable speed. The first system was developed on a NeXT platform (a tip of the hat to Steve Jobs). The first general release of the WWW inside CERN occurred in May of 1991, and in December, the world was notified of the existence of the World Wide Web (known then as W3) thanks to an article in the CERN computer newsletter.

Over the course of the next few months, browsers began to emerge. Erwise, a GUI client, was announced in Finland, and Viola was released in 1992 by Pei Wei of O'Reilly & Associates. NCSA joined the W3 consortium, but didn't announce the Mosaic browser until February of 1993.

Throughout all of this development activity, W3 servers, based on the newly released hypertext transfer protocol (HTTP) that allowed diverse sites to exchange information, continued to proliferate. By January of 1993, there were 50 known HTTP servers; By October, there were over 200, and WWW traffic comprised 1 percent of aggregate NSFNet backbone traffic. Very quietly, the juggernaut had begun.

In May 1994, the first international WWW conference was held at CERN in Geneva, and from that point on they were organized routinely, always to packed houses and always with a disappointed cadre of oversubscribed would-be attendees left out in the cold.

From that point on, the lines that clearly define "what happened when" begin to blur. NCSA's Mosaic product, developed largely by Marc Andreessen at the University of Illinois in Chicago, hit the mainstream and brought the World Wide Web to the masses. Andreessen, together with Jim Clark, would go on to found Netscape Corporation shortly thereafter.

The following timeline, included courtesy of PBS, shows the highlights of the Internet's colorful history (as well as a few other great unrelated moments).

Internet Timeline (1960–2013)

1960 There is no Internet . . .

1961 Still no Internet . . .

1962 The RAND Corporation begins research into robust, distributed communication networks for military command and control.

1962–1969 The Internet is first conceived in the early '60s. Under the leadership of the Department of Defense's Advanced Research Project Agency (ARPA), it grows from a paper architecture into a small network (ARPANET) intended to promote the sharing of super-computers amongst researchers in the United States.

1963 Beatles play for the Queen of England.

1964 *Dr. Strangelove* portrays nuclear holocaust which new network must survive.

1965 The DOD's Advanced Research Project Association begins work on "ARPANET." ARPA sponsors research into a *cooperative network of time-sharing computers.*

1966 *U.S. Surveyor* probe lands safely on moon.

1967 First ARPANET papers presented at *Association for Computing Machinery* symposium. Delegates at a symposium for the Association for Computing Machinery in Gatlingberg, Tenn. discuss the first plans for the ARPANET.

1968 First generation of networking hardware and software designed.

1969 ARPANET connects first four universities in the United States. Researchers at four U.S. campuses create the first hosts of the ARPANET, connecting Stanford Research Institute, UCLA, UC Santa Barbara, and the University of Utah.

1970 ALOHANET developed at the University of Hawaii.

1970–1973 The ARPANET is a success from the very beginning. Although originally designed to allow scientists to share data and access remote computers, email quickly becomes the most popular application. The ARPANET becomes a high-speed digital post office as people use it to collaborate on research projects and discuss topics of various interests.

1971 The ARPANET grows to 23 hosts connecting universities and government research centers around the country.

1972 The InterNetworking Working Group (INWG) becomes the first of several standards-setting entities to govern the growing network. Vinton Cerf is elected the first chairman of the INWG, and later becomes known as a *Father of the Internet.*

1973 The ARPANET goes international with connections to University College in London, England and the Royal Radar Establishment in Norway.

1974–1981 Bolt, Beranek, and Newman open Telenet, the first commercial version of the ARPANET. The general public gets its first vague hint of how networked computers can be used in daily life as the commercial version of the ARPANET goes online. The ARPANET starts to move away from its military-research roots.

1975 Internet operations transferred to the Defense Communications Agency.

1976 Queen Elizabeth goes online with the first royal email message.

1977 UUCP provides email on THEORYNET.

1978 TCP checksum design finalized.

1979 Tom Truscott and Jim Ellis, two grad students at Duke University, and Steve Bellovin at the University of North Carolina establish the first USENET newsgroups. Users from all over the world join these discussion groups to talk about the net, politics, religion, and thousands of other subjects.

1980 Mark Andreesen turns 8. In 14 more years, he will revolutionize the Web with the creation of Mosaic.

1981 ARPANET has 213 hosts. A new host is added approximately once every 20 days.

1982–1987 The term *Internet* is used for the first time. Bob Kahn and Vinton Cerf are key members of a team which creates TCP/IP, the common language of all Internet computers. For the first time, the loose collection of networks which made up the ARPANET is seen as an internet, and the Internet as we know it today is born. The mid-80s marks a boom in the personal computer and super-minicomputer industries. The combination of inexpensive desktop machines and powerful, network-ready servers allows many companies to join the Internet for the first time. Corporations begin to use the Internet to communicate with each other and with their customers.

1983 TCP/IP becomes the universal language of the Internet.

1984 William Gibson coins the term "cyberspace" in his novel *Neuromancer*. The number of Internet hosts exceeds 1,000.

1985 Internet e-mail and newsgroups now part of life at many universities

1986 Case Western Reserve University in Cleveland, Ohio creates the first *Freenet* for the Society for Public Access Computing.

1987 The number of Internet hosts exceeds 10,000.

1988–1990 Internet worm unleashed. The Computer Emergency Response Team (CERT) is formed to address security concerns raised by the Worm. By 1988, the Internet is an essential tool for communications, however, it also begins to create concerns about privacy and security in the digital world. New words, such as "hacker," "cracker" and" electronic break-in", are created. These new worries are dramatically demonstrated on Nov. 1, 1988 when a malicious program called the "Internet Worm" temporarily disables approximately 6,000 of the 60,000 Internet hosts. System administrator turned author, Clifford Stoll, catches a group of Cyberspies, and writes the best-seller *The Cuckoo's Egg*. The number of Internet hosts exceeds 100,000. A happy victim of its own unplanned, unexpected success, the ARPANET is decommissioned, leaving only the vast network-of-networks called the Internet. The number of hosts exceeds 300,000.

1991 The World Wide Web is born!

1991–1993 Corporations wishing to use the Internet face a serious problem: Commercial network traffic is banned from the National Science Foundation's NSFNET, the backbone of the Internet. In 1991, the NSF lifts the restriction on commercial use, clearing the way for the age of electronic commerce. At the University of Minnesota, a team led by computer programmer Mark MaCahill releases "gopher," the first point-and-click way of navigating the files of the Internet in 1991. Originally designed to ease campus communications, gopher is freely distributed on the Internet. MaCahill calls it "the first Internet application my mom can use." 1991 is also the year in which Tim Berners-Lee, working at CERN in Switzerland, posts the first computer code of the World Wide Web in a relatively innocuous newsgroup, "alt.hypertext." The ability to combine words, pictures, and sounds on Web pages excites many computer programmers who see the potential for publishing information on the Internet in a way that can be as easy as using a word processor. Marc Andreesen and a group of student programmers at NCSA (the National Center for Supercomputing Applications located on the campus of University of Illinois at Urbana Champaign) will eventually develop a graphical browser for the World Wide Web called Mosaic. Traffic on the NSF backbone network exceeds one trillion bytes per month. One million hosts have multimedia access to the Internet over the MBONE. The first audio and video broadcasts take place over a portion of the Internet known as the "MBONE." More than 1,000,000 hosts are part of the Internet. Mosaic, the first graphics-based Web browser, becomes available. Traffic on the Internet expands at a 341,634% annual growth rate.

1994 The Rolling Stones broadcast the Voodoo Lounge tour over the M-Bone. Marc Andreesen and Jim Clark form Netscape Communications Corp. Pizza Hut accepts orders for a mushroom, pepperoni with extra cheese over the net, and Japan's Prime Minister goes online at *www.kantei.go.jp*. Backbone traffic exceeds 10 trillion bytes per month.

1995 NSFNET reverts back to a research project, leaving the Internet in commercial hands. The Web now comprises the bulk of Internet traffic. The Vatican launches *www.vatican.va.* James Gosling and a team of programmers at Sun Microsystems release an Internet programming language called Java, which radically alters the way applications and information can be retrieved, displayed, and used over the Internet.

1996 Nearly 10 million hosts online. The Internet covers the globe. As the Internet celebrates its 25th anniversary, the military strategies that influenced its birth become historical footnotes. Approximately 40 million people are connected to the Internet. More than $1 billion per year changes hands at Internet shopping malls, and Internet related companies like Netscape are the darlings of high-tech investors. Users in almost 150 countries around the world are now connected to the Internet. The number of computer hosts approaches 10 million (interesting, considering that today there are more than 300 million!).

Within 30 years, the Internet has grown from a Cold War concept for controlling the tattered remains of a post-nuclear society to the *information superhighway.* Just as the railroads of the nineteenth century enabled the *Machine Age*, and revolutionized the society of the time, the Internet takes us into the *Information Age*, and profoundly affects the world in which we live.

The Age of the Internet arrives.

1997 Some people telecommute over the Internet, allowing them to choose where to live based on quality of life, not proximity to work. Many cities view the Internet as a solution to their clogged highways and fouled air. Schools use the Internet as a vast electronic library, with untold possibilities. Doctors use the Internet to consult with colleagues half a world away. And even as the Internet offers a single global village, it threatens to create a second class citizenship among those without access. As a new generation grows up as accustomed to communicating through a keyboard as in person, life on the Internet will become an increasingly important part of life on Earth.

1998–2004 The Internet bubble climbs to insane levels and money flows into the telecom and IT industries like water over Niagara. The bubble rises, and falls, and $7 trillion in market value evaporates. Recovery begins in early 2004, just in time for the next big bubble to begin.

2004–2013 Online spending reaches a record high of $117 billion, a 26% increase over 2003. In 2005, YouTube.com is launched. By 2006 there are 100 million reachable Web sites online, including Twitter and Facebook. By 2007, the number of legal music downloads climbs to 6.7 million per week. That same year, World of Warcraft surpasses 9 million subscribers worldwide.

Things continue to happen. In 2008, Microsoft offers to buy Yahoo for $44.6 billion in an ill-fated attempt to go up against Google. Instagram launches in 2010; Google Plus arrives in 2011.

By 2012, the Internet is being used as a platform for global political activism. Facebook issues its IPO (and probably wishes it hadn't).

Needless to say, the history of the Internet will continue to develop, and we will continue to be surprised by the events that unfold. For now, however, let's get back into the mud a bit with our study of protocols by introducing the OSI model.

The Open Systems Interconnection Reference Model

Perhaps the best-known "family" of protocols is the International Organization for Standardization's *Open Systems Interconnection Reference Model*, usually called the *OSI Model* for the sake of simplicity. Shown in Fig. 3.33 and comprising seven layers, it provides a logical way to study and understand data communications and is based on the following simple rules. First, each of the seven layers

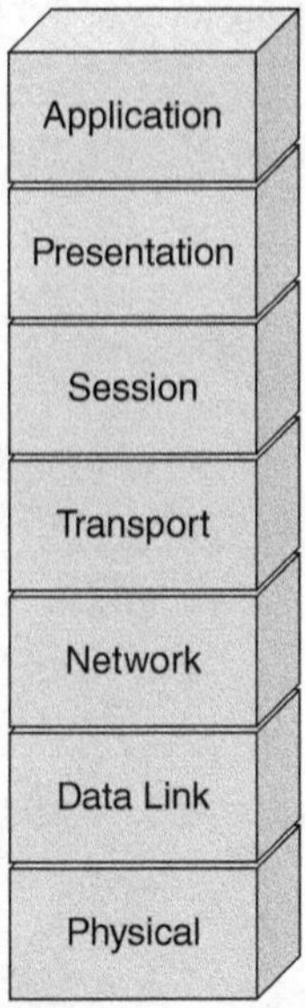

Figure 3.33 The open systems interconnection (OSI) reference model.

must perform a clearly defined set of responsibilities which are unique to that layer, and guarantee the requirement of functional modularity. Second, each layer depends upon the services of the layers above and below to do its own job, as we would expect, given the modular nature of the model. Third, the layers have no idea how the layers around them do what they do; they simply know that they do it. This is called *transparency*. Finally, there is nothing magic about the number seven. If the industry should decide that the model needs an eighth layer, or that layer five is redundant (and there are those who do), then the model will be changed. The key is functionality. There is an ongoing battle within the ranks of OSI model pundits, for example, over whether there is actually a requirement for *both* layers six and seven, because many believe them to be so functionally similar that one or the other is redundant. Others question whether there is *really* a need for layer five, the functions of which are considered by many to be superfluous and redundant. To these people I recommend the purchase of a dog. Whether the addition or elimination of a layer ever actually happens is not important. The fact that it *can* is what matters.

It is important to understand that the OSI model is nothing more than a conceptual way of thinking about data communications. It isn't hardware; it isn't software. It merely simplifies and groups the processes of data transmission so that they can be easily understood and manipulated.

Let's look at the OSI model in a little more detail. (Please refer to Fig. 3.34). I tend to think of it as a seven-drawer chest. In each drawer, there is a collection of standards, and when network implementers

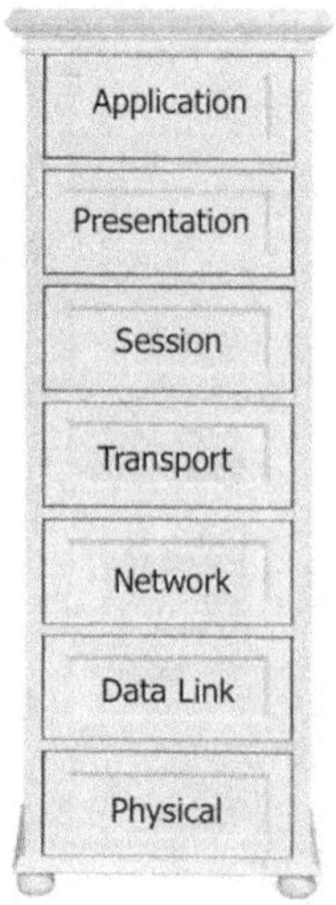

FIGURE 3.34 OSI as a "seven-drawer chest."

set up a network they rummage through each drawer, select the most appropriate standard for their requirements, and set up the network.

As we mentioned earlier, the model is a seven-layer construct within which each layer is tightly dependent upon the layers surrounding it. The application layer, at the top of the model, *speaks* to the actual application process that creates the information to be transported by the network; it is closest to the customer and the customer's processes, and is therefore the most customizable and manipulable of all the layers. It is highly open to interpretation. On the other end of the spectrum, the *physical layer* dwells within the confines of the actual network, and is totally standards-dependent. There is minimal room here for interpretation; a pulse is either a one or a zero—there's nothing in between. Physical layer standards therefore tend to be highly commoditized, while application layer standards tend to be highly specialized. This becomes extremely important as the service delivery model shifts from delivering commodity bandwidth to providing customized services—even if they're mass customized—to the customer base. Service providers are clawing their way up the OSI food chain to get as close to the application layer as they can, because of the Willie Sutton rule.

The Willie Sutton Story

Willie Sutton became famous in the 1930s for a series of outrageous bank robberies during which he managed to outwit the police at every turn. During his career he had two nicknames, "The Actor" and "Slick Willie," because of his ingenious tendency to use a wide array of disguises during his robberies. A sucker for expensive clothes, Sutton was known to be an immaculate dresser. Although he was a bank robber, he had the reputation of being a gentleman. One teller remembers him entering the bank dressed to the nines carrying flowers, which he presented to her in exchange for her money. Another victim said Sutton's robberies were like attending the movies, except that the usher had a gun.

Sutton was finally caught (the first time, anyway) and sentenced to 30 years in prison. He escaped on December 11, 1932 by climbing a prison wall. Two years later, he was recaptured and sentenced to serve 25 to 50 years in Eastern State Penitentiary, Philadelphia, for the robbery of the Corn Exchange Bank.

Sutton's career was not over yet, however. On April 3, 1945, Sutton was one of 12 convicts who escaped from Eastern State Penitentiary through a tunnel. He was recaptured the same day by Philadelphia police officers and sentenced to life imprisonment as a fourth-time offender. He was then transferred to the Philadelphia County Prison in Homesburg, Pennsylvania to live out the rest of his days. On February 10, 1947, Sutton tired of prison life. He and several other prisoners, dressed as prison guards, carried two ladders across the prison yard to the wall shortly after dark. When the searchlights froze them in its glare, Sutton yelled, "It's okay," and no one stopped him. They climbed the wall under the watchful eye of the guards and disappeared into the night.

On March 20, 1950, Willie Sutton was added to the FBI's *Ten Most Wanted* list. Because of his expensive clothing habit, his photograph was given to tailors all over the country in addition to the police. On February 18, 1952, a tailor's 24-year-old son recognized Sutton on the New York subway and followed him to a local gas station. The man reported the incident to the police, who later arrested Sutton.

He did not resist his arrest by New York City Police but denied any robberies or other crimes since his 1947 escape from the Philadelphia County Prison. When he was arrested, Sutton owed one life sentence plus 105 years to the people of Pennsylvania. Because of his new transgressions (mostly making the police look remarkably incompetent), his sentence was augmented by an additional 30 years to life in New York State Prison.

Shortly after his final incarceration, a young reporter was granted a rare interview with Sutton in his prison cell. When they met, Sutton shook his hand and asked, "What can I do for you, young man?" The reporter, nervous, stammered back, "M-M-Mr. Sutton, why do you rob banks?" Sutton sat back and replied with a smile, "Because, young man, that's where the money is."

That's not the end of the story, however. In 1969, the New York State prison authority decided that Sutton did not have to serve his entire sentence of two life sentences plus 105 years, because of failing health. So, on Christmas Eve, 1969, Sutton, now 68, was released from Attica State Prison. And in 1970, Sutton did a television commercial to promote the New Britain, Connecticut, Bank and Trust Company's new photo credit card program. You have to love the little ironies in life.

Sutton died in 1980 in Spring Hill, Florida, at the age of 79.

So what does Willie Sutton have to do with the OSI model and data communication protocols? Not much, but his career does. Today's service providers are striving to climb the services food chain because the money is up there with the customers. Yes, of course there is money to be made at the physical layer end of the model, but the sustainable, growable revenues, and the far more valuable customer relationships that are based on trust and loyalty, are up where services can be customized endlessly to meet the changing needs of customers.

Back to the Model

More now about the inner workings of the OSI model.

The functions of the model can be broken into two pieces, as illustrated by the dashed line in Fig. 3.35 between layers three and four

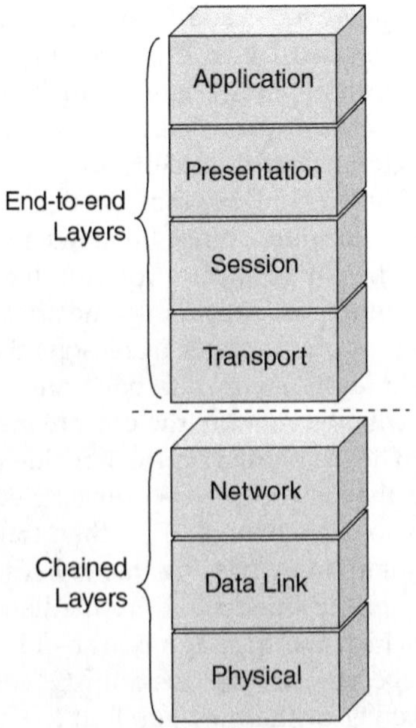

FIGURE 3.35 The OSI model showing the chained layers and the end-to-end layers.

that divides the model into the *chained layers* below and the *end-to-end layers* above.

The chained layers comprise layers one through three: the physical layer, the data link layer, and the network layer. They are responsible for providing a service called *connectivity.* This is the traditional domain of the telephone company—in fact, they actually operate in layers one and two—the layers required for transport and switching functions. The end-to-end layers, on the other hand, comprise the transport layer, the session layer, the presentation layer, and the application layer. They provide a service called *interoperability.* The differences between connectivity and interoperability are important.

Connectivity is the process of establishing a physical connection so that electrons can flow correctly from one end of a circuit to the other. There is little intelligence involved in the process; it occurs, after all, pretty far down in the primordial protocol ooze of the OSI model. Connectivity is critically important to traditional network providers—it represents their lifeblood. Customers, on the other hand, are typically only aware of the criticality of connectivity when it isn't there for some reason. No dial tone? Visible connectivity. Can't connect to the ISP? Visible connectivity. Dropped call on a cell phone? Visible connectivity.

Interoperability, however, is something that customers are much more aware of on a regular basis. Interoperability is the process of guaranteeing *logical* (not physical) *connectivity* between two communicating processes over a physical network. It's wonderful that the lower three layers give a user the ability to spit bits back and forth across a wide area network. But what do the bits mean? Without interoperability, that question cannot be answered. For example, in our e-mail scenario, the e-mail application running on the PC and the e-mail application running on the mainframe are logically incompatible with each other for any number of reasons that we will discuss shortly. They can certainly swap bits back and forth, but without some form of protocol intervention, the bits are meaningless.

Think about it: The PC shown on the left side of Fig. 3.36 creates an e-mail message that is compressed, encrypted, ASCII-encoded, and shipped across logical channel 17. It then transits that message toward its final destination by passing through a series of intermediate switches along the way. But do the intermediate switches that create the path over which the message is transmitted care about the details of the message they are transferring? Of course not. Only the transmitter and receiver of the message that house the applications that will have to interpret it care about such things. The intermediate switches care that they have electrical connectivity, that they can see the bits, that they can determine whether they are the *right* bits, and whether they are the intended recipient of those bits—or not. Therefore the end devices, the sources and sinks of the message, must implement all seven layers of the OSI model, because they must not only concern themselves with *connectivity* issues, but also with issues of *interoperability*. The intermediate devices, however, only care about the functions and responsibilities provided by the lower three layers. Interoperability, because it only has significance in the end devices, is provided by the end-to-end layers—layers four through seven. Connectivity on the other hand is provided by the chained layers, layers one through three, because those functions are required in every link of the network chain—hence the name. Let me say this one more time: *The chained layers are the historical domain of the telephone company.* They are trying, however, to escape that domain and move into more lucrative climes.

FIGURE 3.36 The transmission of an e-mail message.

Layer by Layer

The OSI model relies on a process called *enveloping*, illustrated in Fig. 3.37, to perform its tasks. If we return to our earlier e-mail example, we find that each time a layer invokes a particular protocol, it wraps the user's data in an *envelope* of overhead information that tells the receiving device about the protocol that was employed at that layer. For example, if a layer uses a particular compression technique to reduce the size of a transmitted file, and a specific encryption algorithm to disguise the contents of the file, then it is important that the receiving device be made aware of the techniques employed so that it knows how to decompress and decrypt the file when it receives it. Needless to say, quite a bit of overhead must be transmitted with each piece of user data. The overhead is needed, however, if the transmission is to work properly. So as the user's data passes down the so-called stack from layer to layer, additional information is added at each step of the way as illustrated by the series of envelopes. In summary then, the message to be transported is handed by the originating application to layer seven, which performs application layer functions and then attaches a header to the beginning of the message that explains the functions performed by that layer so that the receiver can interpret the message correctly. In our illustration, that header function is represented by information written on the envelope at each layer. When the receiving device is finally handed the message at the physical layer, each succeeding layer must open its own envelope until the kernel, the message, is exposed for the receiving application. Thus OSI protocols really do work like a nested Russian doll. After peeling back layer after layer of the network onion, the core message is exposed.

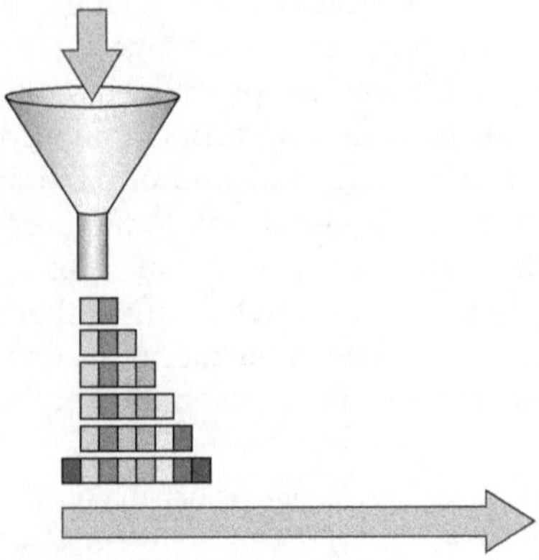

FIGURE 3.37 Enveloping.

Let's now go back to our e-mail example, but this time, we'll describe it within the detailed context of OSI's layered architecture. But first, a language lesson.

Esperanto

There is an old and somewhat comforting cliché which observes, "Wherever one goes, people speak English." In fact, less than 10 percent of the world's population speaks English, and to their credit many of them speak it as a second language.[3] Many believe there is a real need for a truly international language. In 1887, Polish physician Ludwig L. Zamenhof published a paper on the need for a universally spoken tongue. He believed that most of the world's international diplomacy disputes resulted from a communication failure between monolingual speakers and the inevitable misunderstandings of nuance that occur when one language is translated into another. Zamenhof set out to solve this "Tower of Babel" problem (origin of the word "babble," by the way), the result of which was the creation of the international language called *Esperanto.* In Esperanto, the word Esperanto means *one who hopes.*

Since its creation, Esperanto has been learned by millions, and believe it or not is widely spoken—current estimates are approximately two million speakers. And its use is far from being purely academic: Meetings are held in Esperanto, advertising campaigns use it, hotels and restaurants publish literature using it, and professional communities such as health care and scientific research now use Esperanto widely as a way to universally communicate information. Second only to English, it is the *lingua franca* of the international world. It is most commonly spoken in Central and Eastern Europe, East Asia (particularly mainland China), South America, and Southwest Asia. It is less commonly spoken in North America, Africa, and the Middle East.

Esperanto's questionable success as the language of international communication results from three advantages. It is easy to learn; it is politically neutral; and, there are practical reasons to learn it. The structure of the language is so simple and straightforward that it can typically be learned in less than a quarter of the time it takes to learn a traditional language. For example, all letters have one sound and one sound *only.* There are only 16 grammar rules to learn, compared to the hundreds that pervade English and other Romance, Indo-European, or Germanic languages. Furthermore, there are no irregular verb forms (you have to love that!). Even the vocabulary is simple to

[3]There is an old joke among seasoned international travelers that goes likes this: "What do you call someone who speaks three languages?" *Trilingual.* "OK, what do you call someone who speaks *two* languages?" *Bilingual.* OK, what do you call someone who speaks *one* language? *American.*

learn; many words are instantly recognizable (and learnable), such as these:

- Telefono (telephone)
- Biciclo (bicycle)
- Masxino (machine)
- Reto (network)
- Kosmo (outer space)
- Plano (plan)

Speakers of languages other than English will recognize the roots of these words; Reto, for example, is similar to the Spanish word *red* (network). A pretty good Esperanto-English dictionary can be found at *http://esperanto-panorama.net/vortaro/eoen.htm.*

So what does this have to do with telecommunications and the transmission of e-mail messages? Read on.

Layer 7: The Application Layer

The network user's application (Outlook, Outlook Express, Apple Mail, etc.) passes data down to the uppermost layer of the OSI model, called the application layer. The application layer provides a set of highly specific services to the application above that have to do with the *meaning* or *semantic content* of the data. These services include, among others, file transfer, remote file access, terminal emulation, network management, mail services, and data interoperability. This interoperability is what allows our PC user and our mainframe-based user to communicate: the application layer converts the application-specific information into a common, *canonical* form that can be understood by both systems. A canonical form is a form that can be understood universally. The word comes from *canon,* which refers to the body of officially established rules or laws that govern the practices of a church. The word also means an accepted set of principles of behavior that all parties in a social or functional grouping agree to abide by; hence, the applicability of Esperanto. Its creators believed that if everyone on Earth were to learn to speak Esperanto, most of the world's ills would disappear, stemming as they did, they believed, from miscommunication. Needless to say, the problem of multinational interoperability was more complicated than that, with cultural mores being the greatest impediments. Try to get the French to give up French, for example, or the Spaniards to give up Spanish. Not gonna happen!

Let's examine a real-world example of a network-oriented canonical form.

When network hardware manufacturers build components—switches, multiplexers, cross-connect systems, DSLAMs, cable modem termination systems—to sell to their customers, they do so knowing that one of the most important aspects of a successful hardware sale is the inclusion of an element management system that will allow the customer to manage the device within the network. The only problem is that today, most networks are made up of equipment purchased from a variety of vendors. Each vendor develops its own element managers on a device-by-device basis, which work exceptionally well for each device. This does not become a problem until it comes time to create a management hierarchy for a large network, shown in Fig. 3.38, at which time the network management center begins to look a lot like a Best Buy or Future Shop television department. Each device or set of devices requires its own display monitor, and when one device in the network fails, causing a waterfall effect, the network manager must reconstruct the entire chain of events to discover what the original causative factor was. This is sometimes called the "three mile island effect." Back in the 1970s, when the three mile island nuclear power plant went critical and tried to make Pennsylvania glow in the dark, it became clear to the Monday morning quarterbacks trying to reconstruct the event and create the "How could this have been prevented" document that all the information required to turn the critical failure of the reactor into a nonevent was in fact in the control room—buried somewhere in the hundreds of pages of fanfold paper that came spewing out of the high-speed printers scattered all over the control room.

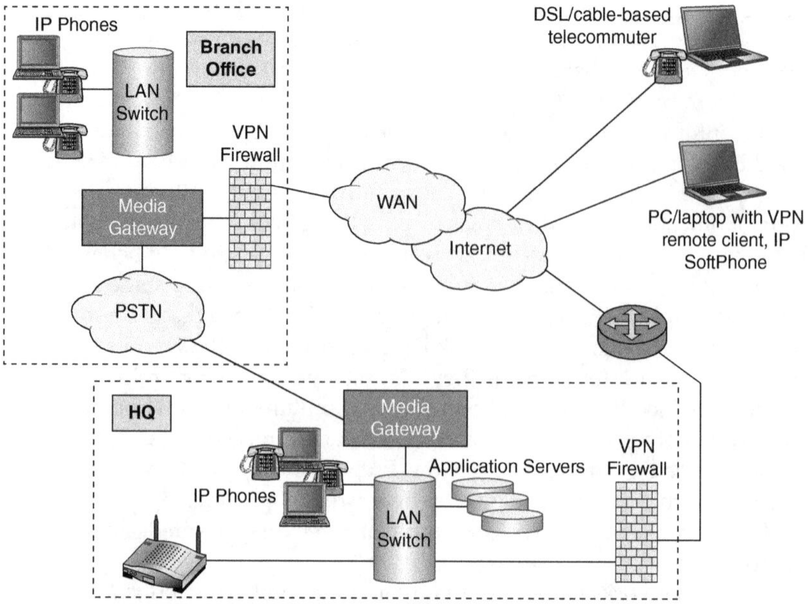

FIGURE 3.38 The complexity of network management.

There was no procedure in place to receive the output from the many managed devices and processes involved in the complex task of managing a nuclear reactor, analyze the output, and hand a simple, easy-to-respond-to decision to the power plant operators.

The same problem exists in complex networks. Most of them have hundreds of managed devices with simple associated element management systems (EMS) that generate relatively primitive data about the health and welfare of each device. The information from these element managers is delivered to the network management center, where it is displayed on one of many monitors that the network managers themselves use to track and respond to the status of the network. What they *really* need is a single map of the network that shows all of the managed devices in green if they are healthy. If a device begins to approach a pre-established threshold of performance, the icon on the map that represents that device turns yellow, and if it fails entirely it turns red, yells loudly and automatically reports and escalates the trouble. In one of his many books on America management practices, USC Professor Emeritus Warren Bennis observed that "the business of the future will be run by a person and a dog. The person will be there to feed the dog; the dog will be there to make sure the person doesn't touch anything." Clearly that model applies here.

So how can this ideal model of network management be achieved? First of all, every vendor will tell you that their element management system is the best element manager ever created. None of them are willing to change the user interface that they have created so carefully. Using a canonical form, however, there is no reason to. All that has to be done is to exact an agreement from every vendor that stipulates that while they do not have to change their user interface, they must agree to speak some form of technological Esperanto on the back side of their device. That way the users still get to use the interface they have grown accustomed to, but on the network side every management system will talk to every other management system using a common and widely accepted form. Again, it's just like a canonical language: If people from five different language groups need to communicate, they have a choice. They can each learn everybody else's language (four additional languages per person) or they can agree on a canonical language (Esperanto), which reduces the requirement to a single language each.

Here's another example to help you understand this concept, this time from the world of photography. I am an avid commercial photographer, specializing in nature, landscape, and travel images. I happen to be a Nikon shooter; I own about a dozen camera bodies, 30 or so lenses, and countless other accessories that connect to the Nikon system.

My best friend Dennis McCooey is equally passionate about photography, but he is a Canon shooter. Like me, he has an impressive pile of gear for his system. When we are together shooting, we spend endless hours arguing about the relative merits of Nikon versus Canon just to pass the time of day. It's a silly argument, because both systems are equally good.

There is one difference, however, that is important to our topic of canonical forms. We shoot different things, which means that while there is some overlap in our systems, we also each have lenses that the other doesn't have. The problem is that Nikon lenses won't mount on a Canon body, and Canon lenses won't mount on a Nikon body. So we're out of look due to the proprietary directions that each manufacturer took.

Behind the scenes, however, there is a canonical form that saves the day. The protocol used by both Canon and Nikon to read and write to the memory cards that go into the cameras is the same. So if I want to use one of Dennis' lenses, I can't mount it on my body, but I *can* put my memory card into his camera and shoot. And he can do the same with my equipment. Problem solved.

In the world of network management there are several canonical forms. The most common are ISO's common management information protocol (CMIP), the IETF's simple network management protocol (SNMP), and the object management group's common object request brokered architecture (CORBA). As long as every EMS agrees to use one of these on the network side of the device, every system can talk to every other system. Problem solved. A single monitor in the network management center can display a single, holistic view of the network and thus resolve problems much faster than before.

Other canonical forms found at the application layer include ISO's X.400/X.500 message handling service (Quiz: Where would you find these if you wanted to read them on a dark snowy night?) and the IETF's simple mail transfer protocol (SMTP) for e-mail applications; ISO's file transfer, access, and management (FTAM), the IETF's file transfer protocol (FTP), and the hypertext transfer protocol (HTTP) for file transfer; and a host of others. Note that the services provided at this layer are highly specific in nature: they perform a limited subset of tasks.

Layer 6: The Presentation Layer

For our e-mail example, let's assume that the application layer converts the PC-specific information to X.400 format and adds a header that tells the receiving device to look for X.400-formatted content. This is not a difficult concept; think about the nature of the information that must be included in any e-mail encoding scheme. Every system must have a field for:

- Sender (From)
- Recipient (To)

- Date
- Subject
- Cc
- Bcc
- Attachment
- Copy
- Message body
- Signature (optional)
- Priority
- Various other miscellaneous fields

The point is that the number of defined fields is relatively small, and as long as each mail system knows what the fields are and where they exist in the coding scheme of the canonical standard, it will be able to map its own content to and from X.400 or SMTP. Problem solved.

Once the message has been encoded properly as an e-mail message, the application layer passes the now slightly larger message down to the presentation layer. It does this across a layer-to-layer interface using a simple set of commands called *service primitives*.

The presentation layer provides a more general set of services than the application layer that have to do with the structural *form* or *syntax* of the data. These services include code conversion, such as 7-bit ASCII to 8-bit EBCDIC translation; compression, using such services as ZIPfile, British Telecom Lempel-Ziv, the various releases of the Moving Picture Experts Group (MPEG), the Joint Photographic Experts Group (JPEG), and a variety of others; and encryption, including pretty good privacy (PGP), public key cryptography (PKC), defense encryption standard (DES), and the like. Note that these services can be used on any form of data: spreadsheets, word processing documents, images and rock music can all be compressed and encrypted. Compression is typically used to reduce the number of bits required to represent a file through a complex manipulative mathematical process that identifies redundant information in the image, removes it, and sends the resulting smaller file off to be transmitted or archived. To explain how compression works, Let's examine JPEG.

JPEG

JPEG was developed jointly by ISO and the ITU-T as a technique for the compression of still images while still retaining varying degrees of quality as required by the user's application. Here's how it works. Please refer to Fig. 3.39*a* and *b*, which are photographs of my daughter, Cristina.

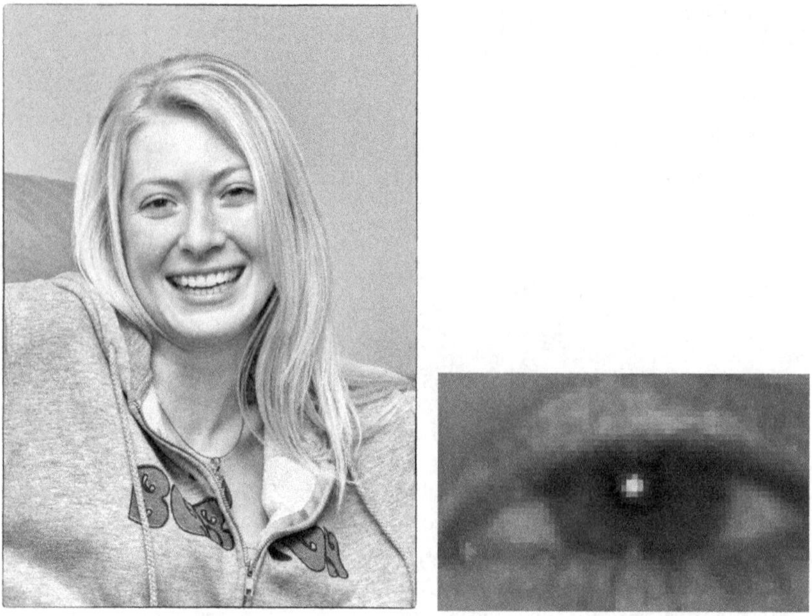

(a) (b)

FIGURE 3.39 My daughter, Cristina, showing the details of JPEG compression.

Figure 3.39*a* shows the original photograph, a reasonably good quality picture that has in fact been substantially compressed using JPEG. Figure 3.39*b* comprises a small portion of the image on the left, specifically Cristina's right eye. Notice the small boxes that make up the image. Those boxes are called *picture elements,* or pixels. Each pixel requires substantial computer memory and processing resources: 8 bits for storage of the red components of the image, 8 bits for green, and eight bits for blue, the three primary colors (and the basis for the well-known RGB color scheme). That's 24-bit color, and every pixel on a computer screen requires them. Furthermore, there are a lot of pixels on a screen: Even a low-resolution monitor that operates at 640 × 480 has 307,200 pixels, with 24 bits allocated per pixel. That equates to 921,600 bytes of information, or roughly 1 megabyte (MB). And that's just color information. Just for fun, let's see what happens when we make the image move, as we will do if we're transporting video. Since typical video generates 30 frames per second, that's 221,184,000 bits that have to be allocated per second—a 222 Mbps signal. That's faster than a 155 Mbps SONET OC-3c signal! The point? We'd better be doing *some* kind of compression!

JPEG uses an ingenious technique to reduce the bit count in still images. First, it clusters the pixels in the image (look at Fig. 3.39*b*) into 16-pixel × 16-pixel groups, which it then reduces to 8 × 8 groups by eliminating every other pixel. The JPEG software then calculates an average color, hue, and brightness value for each 8 × 8 block, which it encodes and transmits to the receiver. In some cases, the image can be further compressed, but the point is that the number of bits required to reconstruct a high-quality image is dramatically reduced by using JPEG. The downside is that every time the image is opened by a viewing or editing program it must be decompressed and then recompressed, resulting in the loss of further data in the image. As a consequence, JPEG is called a *lossy* compression scheme. As a photographer, I rarely use JPEG for my images, preferring to rely on lossless techniques such as TIFF (more later).

More about Compression

Compression schemes do a good job of compressing and faithfully reconstituting images, particularly when the image being compressed is a photograph or video clip. To understand the dynamics of this relationship, let's take a moment to consider what it takes to create a digital photograph displayed on a computer screen.

A laptop computer display is often referred to as being 640 × 480, 800 × 600, 1024 × 768, or in some cases even higher resolution. These numbers refer to the number of *picture elements,* more commonly called *pixels,* which make up the display. Look closely at the screen of your computer and you will find that it is made up of thousands of tiny spots of light (the pixels), each of which can take on whatever characteristics are required to correctly and faithfully paint the image on the screen. Figure 3.40 shows a close-up of the pixels on the screen

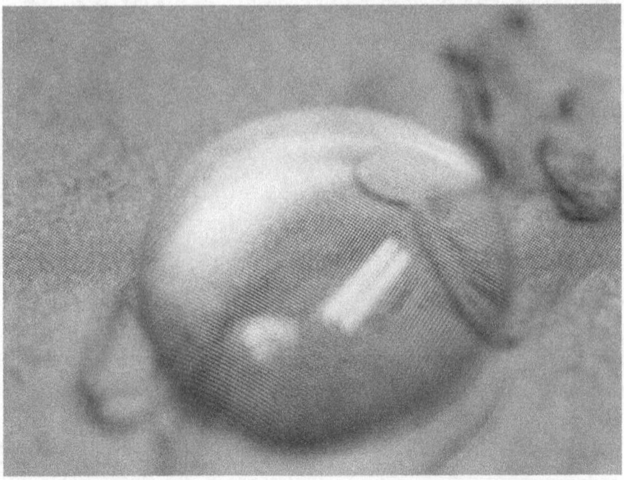

FIGURE 3.40 iPad pixels magnified through a water drop on the screen.

of my iPad, magnified through a water drop. These characteristics include color components (sometimes called chrominance), black and white components (sometimes called luminance), brightness (the intensity of the signal), and hue (the actual wavelength of the color). These characteristics are important in video and digital imaging systems because they determine the quality of the final image. The image, then, is a mosaic of light; the *tiles* that comprise the mosaic are light-emitting diodes that create the proper light at each pixel location.

Apple's Retina Display

A discussion of screen resolution wouldn't be complete without a brief mention of Apple's high-resolution retina display. This new screen technology has a 4:3 aspect ratio and offers a resolution called *Quad Extended Graphics Array* (QXGA). The display resolution offered by these devices is a very impressive 2048 × 1536 pixels at a density of 264 pixels-per-inch.

What makes the retina display unique (and impressive) is that Apple has managed to fit four times as many pixels into the display's real estate as a comparable device. Normally this *pixel cramping* would lead to color fringing and fuzziness, but Apple figured out a way to eliminate that problem. Using a technology developed by Sharp and JSR called super high aperture (SHA), they add vertical separation between the pixels and the electronics that carry the signal by inserting a 3-μm-thick acrylic resin on top of the thin film transistors that make up the display. This eliminates crosstalk and fuzziness, resulting in the well-known sharpness that characterizes the retina display.

Each pixel has a red, green, and blue *light generator*, as shown in Fig. 3.41. Red, green, and blue are called the *primary colors*, because as colors they form the basis for the creation of all other colors. It is a

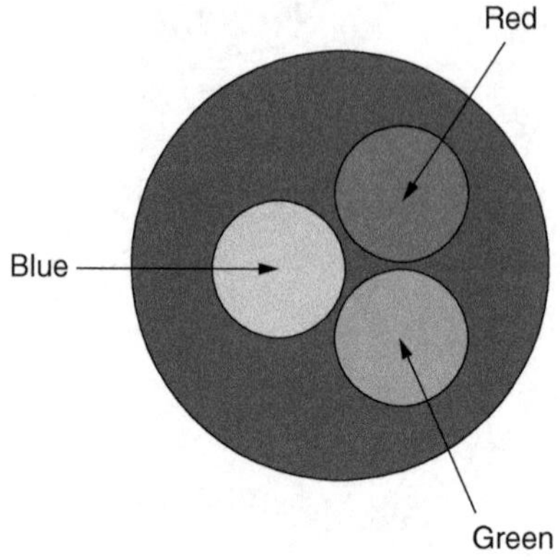

FIGURE 3.41 The pixels in a typical RGB system.

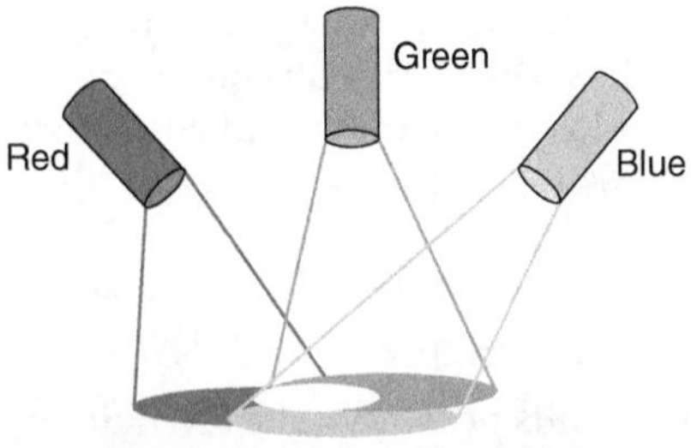

FIGURE 3.42 The red, green, and blue components yield white light when combined.

well-known fact that if three white lights are covered with red, green, and blue color gels, respectively, and the lights are shined at roughly the same spot as shown in Fig. 3.42, the result will be a light spot for each color, but where the three colors intersect the result will be white light. *The combination of the three primary colors creates white.*

Each primary color also has a *complimentary color* in the overall spectrum. As Fig. 3.43 shows, the complementary color for red is cyan, while the complimentary colors for green and blue are magenta and yellow, respectively. Table 3.3 shows the relationships that exist between the primary and complementary colors.

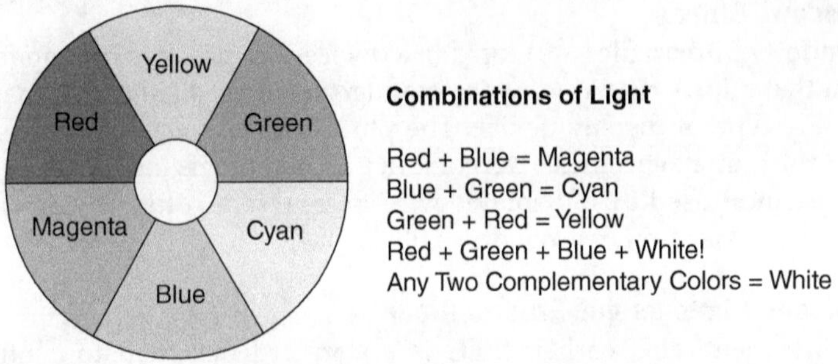

FIGURE 3.43 The color wheel, showing the complements to the primary colors.

If You Combine	The Result Is
Red + blue	Magenta
Green + red	Yellow
Blue + green	Cyan
Red + green + blue	White
Any two complementary colors	White

TABLE 3.3 Primary and Complementary Color Combinations

For full-color, uncompressed images, each of the red, green, and blue elements requires 8 bits, or 24 bits per pixel. This yields what is known as "256-bit color" (2^8). Now consider the storage requirements for an image that is 640×480 pixels in size:

$$640 \times 480 = 307{,}200 \text{ pixels}$$
$$307{,}200 \text{ pixels} \times 24 \text{ bits/pixel} = 7{,}372{,}800 \text{ bits}$$
$$7{,}372{,}800 \text{ bits}/8 \text{ bits per byte} = 921{,}600 \text{ bytes per image}$$

In other words, an uncompressed, relatively low-quality image requires one megabyte of storage. A larger 1024×768-pixel image requires 6.3 MB of storage capacity in its uncompressed form. Given today's relatively low bandwidth access solutions (DSL, cable modems, and 3G wireless), we should be thankful that compression technologies exist to reduce the bandwidth required to move them through a network!

Image Coding Schemes

Images are encoded using a variety of techniques. These include Windows bitmap (BMP), the joint photographic experts group (JPEG), the graphical interchange format (GIF), and the tag image file format (TIFF). They are described in detail in the following sections.

Windows Bitmap

Windows bitmap files are stored in a device-independent bitmap format that allows the Windows operating system to display the image on any type of display device. The phrase *device independent* means that the bitmap specifies pixel color in a format that is independent of the method used by a display device to represent color. The default filename extension of a Windows file is .BMP.

The Joint Photographic Experts Group

As we mentioned earlier, JPEG is a standard designed to control image compression. The acronym stands for *Joint Photographic Experts Group*, the original name of the international body that created the standard. Comprising both technologists and artists, the JPEG committee created a highly complete and flexible standard that compresses both full-color and gray-scale images. It is effective when used with photographs, artwork, and medical images, and works less effectively on text and line drawings. JPEG is designed for the compression of still images, although there is a related standard called Motion JPEG 2000 (MJP2). Motion JPEG was original developed for PC applications that required rudimentary support for early video applications; today it is used in digital cameras, webcams, and by nonlinear video editing systems. It continues to enjoy native support by the QuickTime Player, the PlayStation console, and browsers such as Safari, Google Chrome, and Mozilla Firefox. MPEG, discussed later, is designed for the compression of moving images such as multimedia, movies, and real-time diagnostic images. However, according to the MJP2 committee, MJP2 will be the compression technology of choice for medical imaging, security systems, and digital cameras.

As we noted earlier, JPEG is a *lossy* compression scheme, meaning that once the original image has been compressed, the decompressed image loses a slight degree of integrity when it is viewed. There are, of course, lossless compression algorithms; JPEG, however, achieves more efficient compression than is possible with competing lossless techniques. On first blush this would appear to be a problem for many applications since the compression process actually *squeezes* information out of the original image, leaving a slightly inferior artifact. For example, a diagnostician examining a medical image might be concerned with the fact that the compressed image is not as good as the original. Luckily, this is not a problem for one very simple reason: the human eye is an imperfect viewing device. JPEG is designed to take advantage of well-understood limitations in the human eye, particularly the fact that small color changes are not perceived as discretely as small brightness changes. Clearly, JPEG is designed to compress images that will be viewed by people and therefore do not have to be absolutely faithful reproductions of the original image from which they were created. A machine analysis of a JPEG image would certainly yield inferior results, but to the human eye it is perfectly acceptable.

Of course, the degree of loss that occurs during a JPEG transformation can be controlled by adjusting a variety of compression parameters. For example, file size and image quality can be adjusted. A medical image, which requires extremely high quality on the image output side, requires a large file size while a compressed text document could easily suffer significant loss without losing readability, resulting in a very small file.

The hardware or software coders that create (compress) and expand (decompress) JPEG images are called coder-decoders (CODECs). If high image quality is not critically important, a low-cost, higher-loss CODEC can be used, thus reducing the overall cost of the deployed hardware or software solution.

Tag Image File Format

Tag Image File Format is a tag-based image format designed to promote the interoperability of digital images. The format came into being in 1986 when Aldus Corporation, working with leading scanner vendors, created a standard file format for images to be used in desktop publishing applications. The first version of the specification was published in July 1986; more current versions are released regularly by Adobe and are available at their Web sites.

The format that defines a file specifies both the structure of the file and its content. TIFF content consists of a series of definitions of individual fields. The structure, on the other hand, describes how to actually find the fields. These *pointers* are called tags.

TIFF provides a general-purpose format that is compatible with a variety of scanners and image-processing applications. As a professional photographer, I rely almost exclusively on TIFF as the format in which I export images from my digital cameras for printing or transmission to a publisher. It is device-independent and is acceptable to most operating systems, including Windows, MacOS, and UNIX/LINUX.

Adobe continues to enhance TIFF within publishing applications and maintains backward compatibility whenever possible.

Compressing Moving Images

Growth in videoconferencing, on-demand Web-based training, Web-based video applications such as Netflix and YouTube, and gaming is fueling the growth of digital video technology. But the problems mentioned above that challenge still images still loom large.

The most widely used compression standard for video is MPEG, created by the Moving Pictures Expert Group, the joint ISO/IEC/ITU-T organization that oversees standards development for video. MPEG is relatively straightforward. There are three types of MPEG frames created during the compression sequence. They are intra (I)-frames, predicted (P) frames, and bidirectional (B) frames. An intra-frame or I-frame is nothing more than a frame that is coded as a still image and used as a reference frame. Predicted frames, on the other hand, are predicted from the most recently reconstructed I or P frame. B (bidirectional) frames are predicted from the closest two I or P frames, one from the past and one in the future. Imagine the following scenario: You are converting a piece of video that you shot at the beach to MPEG. The scene, shown in Fig. 3.44, lasts six seconds, and is nothing more than footage of the fishing boat moving slowly in front of the camera, which is locked down on a tripod. Standard video captures a series of still frames, one every 1/30th of a second (30 frames per second). MPEG performs an analysis of the video based on the reference (I)-frames, the predicted frames, and the bidirectional frames. From the image shown it should be clear that very little changes from one frame to another in a 1/30th second interval. The boat may move slightly (but very slightly), and the foam that the propeller is churning up will change. Other than that, very little in the scene changes. Without going into too much technical detail, what MPEG does is reuse those elements of the I-frame that don't change—or that change infrequently—so that it does not have to recreate them, thus reducing overall compression time. In our fishing boat scene, it should be fairly obvious that the background certainly won't change much (unless a bird flies into it), the immediate foreground won't change, and the

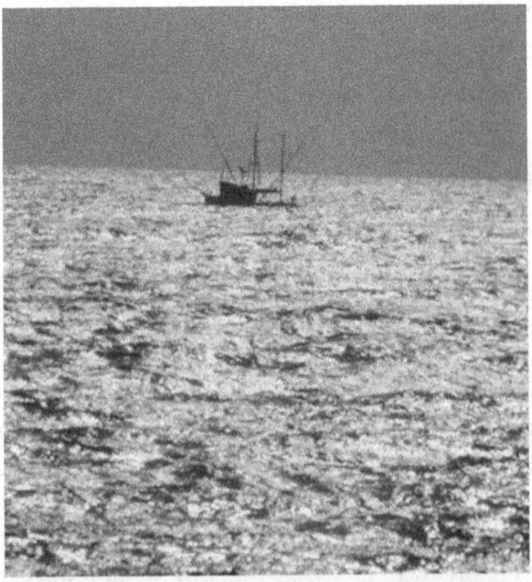

FIGURE 3.44 A scene to be converted to MPEG.

color and shape of the boat are constant. This is a fairly predictable scene. As a result, the number of I-frames that will be interspersed among the P and B frames is relatively small. If the scene were different—a fishing boat being tossed about on rough seas, for example—then the number of minimally uncompressed reference frames would be greater because of the constantly changing point of reference. MPEG looks backward to establish patterns of behavior of past frames, then looks at the reference frame, and finally predicts what future frames will probably look like based on past history. Ultimately, the sequence of frames is as follows:

IBBPBBPBBPBBIBBPBBPB...

There are 12 frames from I to I, and the ratio of P-frames to B-frames is based on experience. Sequentially the frames would look like the sequence shown in Fig. 3.45.

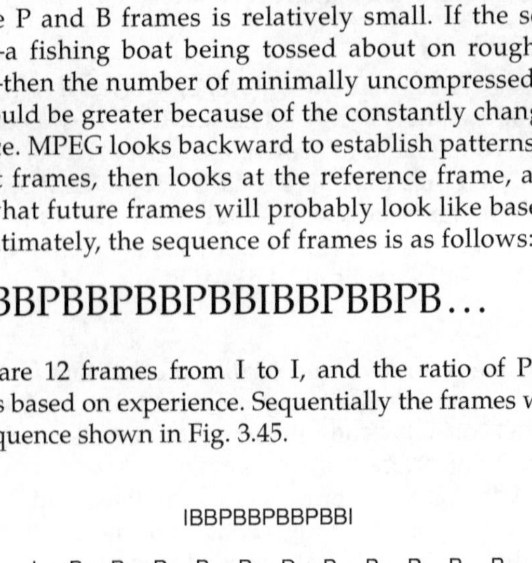

FIGURE 3.45 The relationship between I, B, and P frames in an MPEG series.

The MPEG Standards

There are several different versions of MPEG compression. They are described in the following sections:

MPEG-I

MPEG-I was created to solve the transmission and storage challenges associated with relatively low bandwidth situations, such as for PC-to-CD-ROM transmission or low bit rate data circuits.

MPEG-II

MPEG-II is designed to address much more sophisticated transmission schemes in the 6 to 40 Mb/s range. This positions it to handle applications such as broadcast television and low-end HDTV, as well as variable bit rate video delivered over packet, and ethernet networks.

MPEG-II initially gained the attention of both the telephone and the cable television industries as the initial facilitator of video-on-demand services. Today, it is widely used in DVD production because it has the ability to achieve high compression levels without loss of quality, as well as in cable television, digital broadcasting system (DBS), and high-definition TV (HDTV). MPEG-II has also found a home in corporate videos and training products because of its ability to achieve CD-compatible compression levels.

MPEG-III

MPEG-III is not to be confused with MP3, which is a completely different standard that is described below. MPEG-III was designed to handle the requirements of emerging 1080p HDTV signals that operated in the range of 20 to 40 Mb/s. However, it soon became redundant; standards developers and their engineering counterparts realized that MPEG-II could easily handle these requirements, so in 1992 MPEG-III was absorbed by MPEG-II.

MPEG-IV

MPEG-IV is the result of an international effort that involves hundreds of researchers from all over the world. Its development was finalized in October 1998 and became an International Standard in the first months of 1999.

MPEG-IV builds on the proven success of three fields: digital television, interactive graphics applications, and interactive multimedia such as the distribution of HTML-encoded Web pages.

M4a

M4a is an extension of MPEG-IV that is used to encode audio content. Technically, it is a subset of MPEG-IV called MPEG-IV Part 14. It is usually compressed using a lossy compression scheme called Advance Audio Coding (AAC), but it can also be formatted using the

Apple lossless format. Its close cousin is M4p, which incorporates DRM to prevent piracy and unauthorized copying.

MPEG-VII

MPEG-VII, officially known as the *multimedia content description interface,* is a markup language similar to HTML (in fact, it uses XML) used to describe multimedia content data that allows some degree of interpretation of the information's actual meaning, which can be accessed by a computer. MPEG-VII does not target any particular application; instead it attempts to standardize the interpretation of image elements so that it can support a broad range of applications.

Consider the following scenario. A video editor, attempting to assemble a final cut of a video, finds that she needs a specific piece of B-roll (fill-in footage) to serve as a segue and to cover a voice-over during one particular scene in the video. She knows that the footage she needs exists in the company's film vault; she just can't remember what project it was part of. Using MPEG-VII, she requests the following: "I need a seven-second clip that has the following characteristics."

- Bright blue sky, mid-day
- White beach in foreground, light surf
- Red and green beach umbrella in lower-left corner of shot
- Sea gull enters from upper left, exits upper right
- Ship on horizon moving slowly from right to left

With MPEG-VII, the clip can be found using these descriptors.

The elements that MPEG-VII attempts to standardize support a broad range of applications such as digital libraries, the selection of broadcast media, postproduction and multimedia editing, and home entertainment devices. MPEG-VII also adds a new level of search capability to the Web, making it possible to search for specific multimedia content as easily as text can be searched today. This capability is particularly valuable to media managers and users of large content archives and multimedia catalogues that allow people to select content for purchase. The same information used for successful content retrieval will also be used to select and filter *push* content for highly targeted advertising applications.

Any domain that relies on the use of multimedia content will benefit from the deployment of MPEG-VII. Consider the following list of examples:

- Architecture, real estate, and interior design (e.g., searching for ideas)
- Broadcast media selection (e.g., radio channel, TV channel)
- Cultural services (history museums, art galleries, and the like)

- Digital libraries (e.g., image catalogue, musical dictionary, bio-medical imaging catalogues, film, video, and radio archives)
- E-Commerce (e.g., personalised advertising, on-line catalogues, directories of e-shops)
- Education (e.g., repositories of multimedia courses, multimedia search for support material)
- Home entertainment (e.g., systems for the management of personal multimedia collections, including manipulation of content, like home video editing, searching a game, karaoke)
- Investigation services (e.g., human characteristics recognition, forensics)
- Journalism (e.g., searching speeches of a certain politician using his name, his voice, or his face)
- Multimedia directory services (e.g., yellow pages, tourist information, geographical information systems)
- Multimedia editing (e.g., personalised electronic news service, media authoring)
- Remote sensing (e.g., cartography, ecology, natural resources management)
- Shopping (e.g., searching for clothes that you like)
- Social (e.g., dating services)
- Surveillance (e.g., traffic control, surface transportation, non-destructive testing in hostile environments)

The standard also lists the following examples of how the capabilities of MPEG-VII might be used:

- A musician plays a few notes on a keyboard and retrieve a list of musical pieces similar to the required tune, or images that match the notes in a certain way, as described by the artist. Ever use the application, Shazam?
- An artist draws a few lines on a screen and finds a set of images containing similar graphics, logos, and ideograms.
- A clothing designer selects objects, including colour patches or textures and retrieve examples among which you select the interesting objects to compose your design.
- Using an excerpt of Pavarotti's voice, an opera fan retrieves a list of Pavarotti's records and video clips in which Pavarotti sings a particular piece, as well as photographs of Pavarotti.

This is a remarkable (and remarkably complex!) standard that continues to evolve today and that will no doubt see greater integration as a result of the stronger focus on customer experience management.

MP3

At the risk of repeating, myself, let's be clear: MP3 and MPEG-III are not the same thing. The formal name for MP3 is actually MPEG-I or MPEG-II audio layer 3, which is a signal encoding scheme that compresses audio using a lossy compression technique. It has become the de facto audio encoding standard for MP3 players such as the iPod.

MP3 was actually developed by the MPEG organization's MPEG-I team and was later expanded to also encompass the audio compression requirements of MPEG-II. It is a perfect compromise with the human ear, which as we've already seen is an imperfect listening device. A file encoded using the MP3 technique set at a 128 Kbps signal rate is approximately $1/11^{th}$ the size of the original uncompressed file that is created for a CD.

Encryption

The preceding section focused on the layer six function known as compression. Encryption, on the other hand, is used when the information contained in a file is deemed sensitive enough to require that it be hidden from all but those eyes with permission to view it.

Encryption is one aspect of an ancient science called cryptography. Cryptography is the science of writing in code; its first known use dates to 1900 B.C. when an Egyptian scribe used nonstandard hieroglyphs to capture the private thoughts of a customer. Some historians feel that cryptography first appeared as the natural result of the invention of the written word; its use in diplomatic messages, business strategy, and battle plans certainly supports the theory.

In data communications and telecommunications, encryption is required any time the information to be conveyed is sensitive and the possibility exists that the transmission medium is insecure or that the transmitted information might be intercepted. This can occur over any network, although the Internet is most commonly cited as being the most insecure of all networks.

All secure networks require a set of specific characteristics if they are to be truly secure. The most important of them are listed as follows:

- *Privacy or confidentiality:* The ability to guarantee that no one can read the message except the intended recipient.
- *Authentication:* The guarantee that the identity of the recipient can be verified with full confidence.
- *Message integrity:* Assurance that the receiver can confirm that the message has not been changed in any way during its transit across the network.
- *Nonrepudiation:* A mechanism to prove that the sender really sent this message and was not sent by someone pretending to be the sender.

Cryptographic techniques, including encryption, have two responsibilities: to ensure that the transmitted information is free from theft or any form of alteration, and to provide authentication for both senders and receivers. Today, three forms of encryption are most commonly employed: secret key (or symmetric) cryptography, public-key (or asymmetric) cryptography, and hash functions. How they work is beyond the scope of this book, but there are numerous resources available on the topic (see the Bibliography in the Appendix). One of the best resources is *An Overview of Cryptography* by my good friend Gary Kessler. The paper, which Gary updates routinely, is available at *http://www.garykessler.net/library/crypto.html.*

Another interesting encryption technology that has taken on an air of importance in the last few years is a technique called steganography. *Steganography* is a technique used to surreptitiously hide one message inside another. The name derives from Johannes Trithemius's *Steganographia,* published in 1621, a work on cryptography and steganography that was in fact disguised as a book on black magic.

Steganographic messages are typically encrypted before being embedded in another message, after which a *covertext* is created to contain the encrypted message. This is called ***stegotext***. For example, consider a JPEG image, which comprises groups of 8-bit bytes that ultimately encode the image that will be displayed on-screen. Because JPEG is a compression technique, some quality of the original image is lost in the process, although with modern compression algorithms the loss is kept to an absolute minimum. Like all bytes, those found in a JPEG image have both a least significant bit and a most significant bit. In fact, changing the least significant bit from a one to a zero or vice-versa results in no appreciable change to the output image, because of the limitations of the human eye. It makes sense, therefore, that someone could write a relatively simple routine to selectively encode the least significant bit of each byte, collect all of the encoded bits, and group them into an embedded message. Make sense? Naturally, the larger the image the more data can be hidden within it.

Consider, for example: a 24-bit bitmap which has 8 bits representing each of the three primary colors (red, green, and blue, RGB) for each pixel. The red color alone, represented by 8 bits, has 256 possible values (2^8). The difference between a red value of 01111111 and a red value of 01111110 will be undetectable by the human eye. Therefore, the least significant bit can be used for something other than color information. If we then encode the green and the blue as well, we will have 3 encodable bits, which means that 3 pixels will yield 9 bits, one more than what is required to encode an 8-bit character.

So why do we care about this? In October 2001, the *New York Times* published an article claiming that terrorists were using steganographic techniques to encode messages into images (by some accounts, pornographic messages at adult Web sites). While these claims were largely dismissed by security experts, it is certainly a possibility. Needless to say, security analysts have created steganalysis tools which can be used to detect and read messages embedded steganographically in images and other file types.

From a more practical perspective, steganographic techniques can be used to embed important information such as digital watermarks, copyright information, and the like.

Let's turn our attention now back to our e-mail message.

Layer 5: The Session Layer

We have now left the presentation layer behind. Our e-mail message is encrypted, compressed, and may have gone through an ASCII-to-EBCDIC code conversion before descending into the complexity of the session layer. As before, the presentation layer added a header containing information about the services it employed.

For being such an innocuous layer, the session layer certainly engenders a lot of attention. Some believe that the session layer could be eliminated by incorporating its functions into the layer above or the layer below, thus simplifying the OSI model. Whatever, the bottom line is that it *does* perform a set of critical functions that cannot be ignored.

First, the session layer ensures that a logical relationship is created between the transmitting and receiving applications. It guarantees, for example, that our PC user in Madrid receives his or her mail and *only* his or her mail from the mainframe, which is undoubtedly hosting large numbers of other e-mail users. This requires the creation and assignment of a logical session identifier.

Many years ago, I recall an instance when I logged into my e-mail account during my years at the telephone company and found to my horror that I was actually logged into my vice-president's account. Needless to say I back-peddled out of there as fast as I could. Today I know that this occurred because of an execution glitch in the session layer.

Layer five also shares responsibility for security with the presentation layer. You may have noticed that when you log in to your e-mail application the first thing the system does is ask for a login ID, which you dutifully enter. The ID appears in the appropriate field on the screen. When the system asks for your password, however, the password does not appear on the screen—The field remains blank or is filled with stars, shown graphically in Fig. 3.46. This is because the session layer knows that the information should not be displayed. When it receives the correct login ID, it sends a command to the terminal (your PC) asking you to enter your password. It then immediately sends a second message to the PC telling it to turn off *local echo* so that your keystrokes are not echoed back on to the screen. As soon as the password has been transmitted, the session layer issues a command to turn local echo back on again, allowing you to once again see what you type.

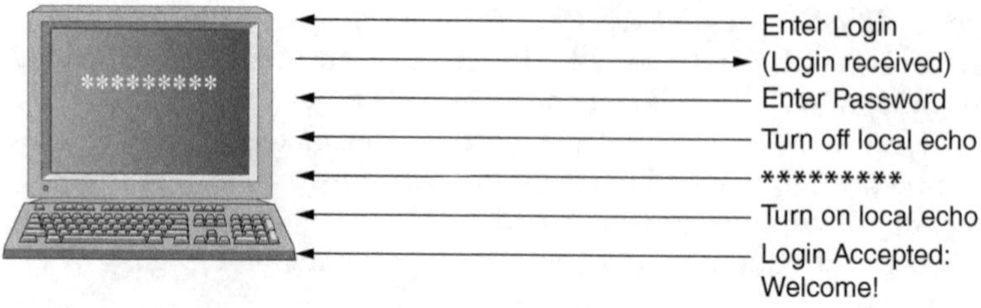

FIGURE 3.46 How the session layer handles passwords.

Another responsibility of the session layer that is particularly important in mainframe environments is a process called *checkpoint restart*. This is a process that is analogous to the autosave function available on many PCs today. I call it the Hansel and Gretel function: as the mainframe performs its many tasks during the online day, the session layer keeps track of everything that has been done, scattering a trail of digital bread crumbs along the way as processing is performed. Should the mainframe fail for some reason (the dreaded abnormal end, or ABEND), the session layer will provided a series of recovery checkpoints. As soon as the machine has been rebooted, the session layer performs the digital equivalent of walking back along its trail of bread crumbs. It finds the most recent checkpoint, and the machine uses that point as its most recent recovery data, thus eliminating the possibility of losing huge amounts of recently processed information. Modern backup programs such as Apple's Time Machine or commercial programs like Chronos offer this as part of their service.

So, the session layer may not be the most glamorous of the seven layers, but its functions are clearly important. As far as standards go, the list is fairly sparse; see the ITU-T's X.225 standard for the most comprehensive document on the subject.

After adding a header, layer five hands the steadily growing protocol data unit (PDU) down to the transport layer. This is the point where we first enter the network. Until now, all functions have been software-based, and in many cases a function of the operating system.

Layer 4: The Transport Layer

The transport layer's job is simple: to guarantee end-to-end, error-free delivery of the entire transmitted message: not bits, not frames or cells, not packets, but the entire message. It does this by taking into account the nature and robustness of the underlying physical network over which the message is being transmitted, including the following characteristics:

- Class of service required
- Data-transfer requirements

- User interface characteristics
- Connection management requirements
- Specific security concerns
- Network management and reporting status data

Based on the robustness of the underlying network and the level of service integrity demanded by the application, the transport layer will determine how much of its considerable power needs to be brought to bear on the transmission. Before we do that, however, let's take a side trip to discuss the types of networks we typically encounter when we study transmission protocols.

Network Topologies

There are two basic network types: dedicated and switched. We will examine each in turn before discussing transport layer protocols.

Dedicated networks are exactly what the name implies: an always-on network resource, often dedicated to a specific customer, which provides very high-quality transport service. That's the good news. The bad news is that dedicated facilities tend to be expensive, particularly because the customer pays for them whether they use the facility or not. Unless they are literally using it 100% of the time, the network is costing them money. The other downside of a dedicated facility is susceptibility to failure: should an unwitting backhoe driver decide to take the cable out, there is no alternative route for the traffic to take. It requires some sort of intervention on the part of the service provider that is largely manual. Furthermore, dedicated circuits tend to be inflexible because again, they are dedicated.

Switched resources, on the other hand, work in a different fashion and have their own set of advantages and disadvantages to consider. First and foremost, they require an understanding of the word *virtual.*

When a customer purchases a dedicated facility, they literally *own* the transmission resources between the two communicating entities as shown in Fig. 3.47. Either the circuit itself is physically dedicated to them (common in the 1980s), or one or more timeslots on a shared physical resource such as T-Carrier or SONET/SDH is dedicated to them. Data is placed on the timeslot or the circuit and it travels to the other end of the established facility—very simple and

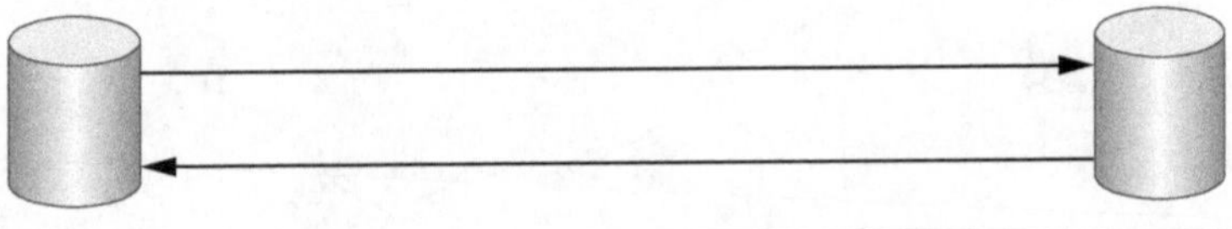

FIGURE 3.47 A dedicated facility provisioned for a customer.

straightforward. There is no possibility of a misdelivered message because the message only has a single possible destination. Imagine turning on the spigot in your front yard to water the plants and having water pour out of your neighbor's hose—it would be about that ridiculous.

In a switched environment, things work differently. In switched networks, the only thing that is actually dedicated is a timeslot, because everything in the network that is physical is shared among many different users. Imagine what a wonderful boon to the service providers this technology is: it gives them the ability to properly support the transport requirements of large numbers of customers while selling the same physical resources to them, over and over and over again. Imagine!

To understand how this technology works, please refer to Fig. 3.48. In this example, device D on the left needs to transmit data to device K on the right. Notice that access to the network resources (the switches in the cloud) is shared with three other machines. In order for this to work, each device must have a unique identifier so that the switches in the network can distinguish among all the traffic streams that are flowing through them. This identifier, often called a *Virtual Circuit Identifier,* is assigned by the Transport Layer as one of its many responsibilities. As examples, legacy X.25, frame relay, and ATM are all switched network technologies that rely on this technique. In X.25 the identifier was called a virtual circuit identifier; in frame relay, a data link connection identifier (DLCI, pronounced 'Delsey'); and in ATM it is called a virtual circuit identifier as well. These will be described later in Chap. 9.

When device D generates its message to device K for transport across the network, the transport layer *packages* the data for transmission. Among other things, it assigns a logical channel that the ingress switch uses to uniquely identify the incoming data. It does this by creating a unique combination of the unique logical address

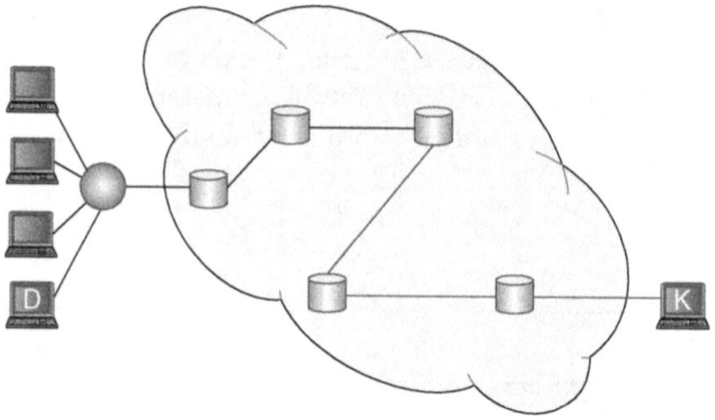

FIGURE 3.48 Virtual circuit identifiers in transmission systems.

with the shared physical port to create an entirely unique virtual circuit identifier.

When the message arrives at the first switch, the switch enters the logical channel information in a routing table that it uses to manage incoming and outgoing data. There can be other information in the routing table as well, such as quality of service indicators. These details will be covered later when we discuss the network layer (layer three).

The technology that a customer uses in a switched network is clearly not dedicated, but it gives the appearance that it is. This is called *virtual circuit service,* because it gives the appearance of being there when in fact it isn't. Virtual private networks (VPNs), for example, give a customer the appearance that they are buying private network service. In a sense they are: They do have a dedicated logical facility; the difference is that they share the highly secure physical facilities with many other users, which allows the service provider to offer the transport service for a lower cost. Furthermore, secure protocols protect each customer's traffic from interception. VPNs are illustrated in Fig. 3.49.

As you may have intuited by now, the degree of involvement that the transport layer has varies with the type of network. For example, if the network consists of a single, dedicated, point-to-point circuit, then there is very little that could happen to the data during the transmission, since the data would consist of an uninterrupted, *single-hop* stream—there are no switches along the way that could cause pieces of the message to go awry. The transport layer, therefore, would have little to do to guarantee the delivery of the message. The only real problem that might occur is a failure of the physical facility.

However, what if the architecture of the network is not as robust as a private line circuit? What if this is a packet network, in which

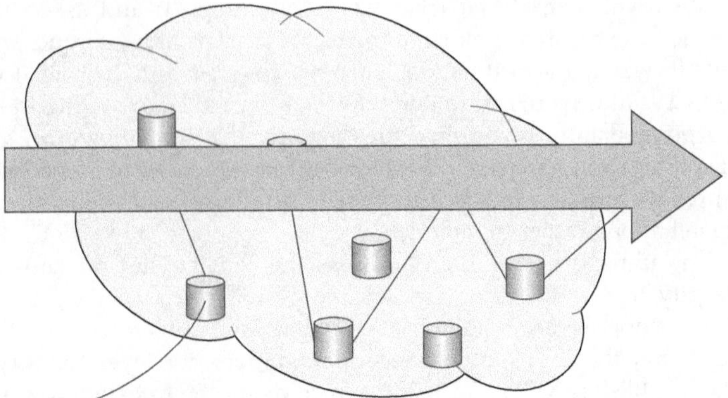

FIGURE 3.49 A typical virtual private network.

case the message is broken into segments by the transport layer, which are independently routed through the fabric of the network? Furthermore, what if there is no guarantee that all of the packets will take the same route through the network wilderness? In that case, the *route* actually consists of a *series* of routes between the switches, like a string of sausage links. In this situation, there is no guarantee that the components of the message will arrive in sequence—in fact, there is no guarantee that they will arrive at all! The transport layer therefore has a major responsibility to ensure that all of the message components arrive and that they carry enough additional information in the form of yet another header—this time on each packet—to allow them to be properly resequenced at the destination. The header, for example, contains sequence numbers that the receiving transport layer can use to reassemble the original message from the stream of random packets.

Consider the following scenario: A transmitter fires a message into the network, where it passes through each of the upper layers until it reaches the originating transport layer, which segments the message into a series of five packets, labeled one of five, two of five, three of five, and so on. The packets enter the network and proceed to make their way across the wilderness of the network fabric. Packets one, two, three, and five arrive without incident, although they do arrive out of order. Packet four, unfortunately, gets caught in a routing loop in New Mexico. The receiving transport layer, tasked with delivering a complete, correct message to the layers above, puts everything on hold while it awaits the arrival of the errant packet. The layer, however, will only wait so long. It has no idea where the packet is; it does, however, know where it is *not*. After some predetermined period of time the receive transport layer assumes that the packet isn't going to make it and initiates recovery procedures that result in the retransmission of the missing packet.

Meanwhile, the lost packet has finally stopped and asked for directions, extricated itself from the traffic jams of Albuquerque, and made its way to the destination. It arrives, covered with dust, an "I've Seen Crystal Caverns" bumper sticker on its trailer, expecting to be incorporated into the original message. By this time, however, the packet has been replaced with the resent packet. Clearly, some kind of process must be in place to handle duplicate packet situations, which happen rather frequently.

The transport layer then becomes the center point of message integrity.

Transport layer standards are diverse and numerous. ISO, the ITU-T and the IETF publish recommendations for layer four. The ITU-T publishes X.224 and X.234, which detail the functions of both connection-oriented and connectionless networks. ISO publishes ISO

8073, which defines a transport protocol with five layers of functionality ranging from TP0 through TP4, as given in the following list:

- Class 0 (TP0): Simple class
- Class 1 (TP1): Basic error recovery class
- Class 2 (TP2): Multiplexing class
- Class 3 (TP3): Error recovery and multiplexing class
- Class 4 (TP4): Error detection and recovery class

TP0 has the least capability; it is roughly equivalent to the IETF's user datagram protocol (UDP) which will be discussed in Chap. 10. TP4 is the most common ISO transport layer protocol and is equivalent in capability to the IETF's transmission control protocol (TCP). It provides an ironclad transport function and operates under the assumption that the network is wholly unreliable and must therefore take extraordinary steps to safeguard the user's data.

Before we descend into the wilds of the network layer, let's introduce the concepts of switching and routing.

Switching and Routing

Modern networks are often represented as a cloud filled with boxes representing switches or routers. Depending upon such factors as congestion, cost, number of hops between routers, and other considerations, the network selects the optimal end-to-end path for the stream of packets created by the transport layer. Depending upon the nature of the network layer protocol that is in use, the network will take one of two actions. It will either establish a single path over which all the packets will travel, in sequence; or, the network will simply be handed the packets and told to deliver them as it sees fit. The first technique, which establishes a seemingly dedicated path, is called *connection-oriented service*; the other technique, which does *not* dedicate a path, is called *connectionless service*. We will discuss each of these in turn. Before we do, however, let's discuss the evolution of switched networks.

Evolution of Switched Networks

Modern switched networks typically fall into one of two major categories: *circuit switched*, in which the network preestablishes a path for transport of traffic from a source to a destination, as is done in the traditional telephone network; and *store-and-forward networks*, where the traffic is handed from one switch to the next as it makes its way across the network fabric. When traffic arrives at a switch in a store-and-forward network, it is stored, examined for errors and destination information, and forwarded to the next hop along the path—hence the name, store and forward. Packet switching is one form of store-and-forward technology.

Store-and-Forward Switching

Thus, from some far-away and beleaguered island, where all day long the men have fought a desperate battle from their city walls, the smoke goes up to heaven; but no sooner has the sun gone down than the light from the line of beacons blazes up and shoots into the sky to warn the neighboring islanders and bring them to the rescue in their ships.

The Iliad by Homer, circa 700 BC

The first store-and-forward networks were invented and used by the early Greeks and Romans. Indeed, Mycenae learned of the fall of Troy because of a line of signal towers between the two cities that used fire in each tower to transmit information from tower to tower. Openings on the side of each tower could be alternately opened and blocked, and using a rudimentary signaling code, short messages could be sent between towers in a short period of time. A message could be conveyed across a large country such as France in a matter of hours, as Napoleon discovered and used to his great advantage.

The earliest *modern* store-and-forward networks were the telegraph networks. When a customer handed over a message in the form of a yellow paper flimsy that was to be transmitted, the operator would transmit the message in code over the open wire telegraph lines to the next office, where the message printed out on a streaming paper tape. On the tape would appear a sequence of alternating pencil marks and gaps, or spaces, combinations of which represented characters—a mark represented a one, while a space represented a zero. A point of historical interest is that the terminology *mark* and *space* is common parlance in modern networks: in T-Carrier, the encoding scheme (remember our discussion about signaling in the last chapter?) is called alternate mark inversion (AMI), because every other one alternates in polarity from the ones that surround it. Similarly, alternate space inversion is used in other signaling schemes such as on the legacy ISDN D-Channel.

At any rate, the entire message would be delivered in this fashion, from office to office to office, ultimately arriving at its final destination, a technique called *message switching*. Over time, of course, the process became fully mechanized and the telegraph operators disappeared.

There was one major problem with this technique. What happened if the message, upon arrival, was found to be corrupt, or if it simply did not arrive for some odd reason? In that case, the entire message would have to be re-sent at the request of the receiver. This added overall delay in the system and was awfully inefficient since in most cases only a few characters were corrupted. Nevertheless, the entire message was retransmitted. Once the system was fully mechanized it meant that the switches had to have hard drives on which to store the incoming messages, which added more delay since hard drives are mechanical devices and by their very nature relatively slow.

Improvements didn't come along until the advent of *packet switching.*

Packet Switching

With the arrival of low-cost, high-speed solid-state microelectronics in the 1970s, it became possible to take significant steps forward in switching technology. One of the first innovative steps was *packet switching*. In packet switching, the message that was transmitted in its entirety over the earlier message-switched store-and-forward networks is broken into smaller, more manageable pieces that are numbered by the transport layer before being passed into the network for routing and delivery. This innovation offered several advantages. First, it eliminates the need for the mechanical, switch-based hard drive because the small packets can now be handled blindingly fast by solid-state memory. Second, should a packet arrive with errors, it and it alone can be discarded and replaced. In message-switched environments, an unrecoverable bit error resulted in the inevitable retransmission of the entire message—not a particularly elegant solution. Packet switching, then, offers a number of distinct advantages.

As before, of course, there are also disadvantages. There is no longer (necessarily) a physically or logically dedicated path from the source to the destination, which means that the ability to guarantee quality of service on an end-to-end basis is severely restricted. There are ways around this as you will see in the section that follows, but they are often costly and *always* complex.

Packet switching can be implemented in two very different ways. We'll discuss them now.

Connection-Oriented Networks

> *Capt. Lewis is brave, prudent, habituated to the woods, & familiar with Indian manners & character. He is not regularly educated, but he possesses a great mass of accurate observation on all the subjects of nature which present themselves here, & will therefore readily select those only in his new route which shall be new. He has qualified himself for those observations of longitude & latitude necessary to fix the line he will go over.*
>
> Thomas Jefferson to Dr. Benjamin Rush of Philadelphia on why he picked Meriwether Lewis for the Corps of Discovery.

> *Six papers of ink powder; sets of pencils; "Creyons," two hundred pounds of "best rifle powder;" four hundred pounds of lead; 4 Groce fishing Hooks assorted; twenty-five axes; woolen overalls and other clothing items, including 30 yds. Common flannel; one hundred flints; 30 Steels for striking or making fire; six large needles and six dozen large awls; three bushels of salt.*
>
> Partial list of items purchased by Lewis for the trip.

When Meriwether Lewis and William Clark left St. Louis with the Corps of Discovery in 1803 to travel up the Missouri and Columbia Rivers to the Pacific Ocean, they had no idea how to get where they were going. They traveled with and relied on a massive collection of maps, instruments, transcripts of interviews with trappers and Native American guides, an awful lot of courage, and the knowledge of Sacajawea, the wife of independent French-Canadian trader Toussaint Charbonneau, who accompanied them on their journey. As they made their way across the wilderness of the northwest, they marked trees every few hundred feet by cutting away a large and highly visible swath of bark, a process known as *blazing*. By blazing their trail, others could easily follow them without the need for maps, trapper lore, or guides. They did not need to bring compasses, sextants, chronometers, or to hire local guides; they simply followed the well-marked trail. Today, trail markers use brightly colored paint instead of an axe to mark a route through the forest.

If you understand this concept then you also understand the concept of connection-oriented switching, sometimes called *virtual circuit switching*, one of the two principal forms of switching technologies. When a device sends packets into a connection-oriented network, the first packet, often called a *call setup packet* or *discovery packet*, carries embedded in it the final destination address that it is searching for. It might be referred to as the *Meriwether Lewis packet*. Upon arrival at the first switch in the network, the switch examines the packet, looks at the destination address, and selects an outgoing port that will get the packet closer to its destination. It has the ability to do this because presumably, somewhere in the recent past, it has recorded the *port of arrival* of a packet from the destination machine, and concludes that if a packet arrived on that port from the destination host, then a good way to get closer to the destination is to go out the same port that the arriving packet came in on. The switch then records in its routing tables an entry that dictates that all packets originating from the same source (the source being a virtual circuit address/physical port address combination that identifies the logical source of the packets) should be transmitted out the same switch port. This process is then followed by every switch in the path, from the source to the destination. Each switch makes table entries, similar to the blazes left by the Corps of Discovery.[4]

With this technique, the only packet that requires a complete address is the initial one that blazes the trail through the network wilderness. All subsequent packets carry nothing more than a short identifier—a virtual circuit address —that instructs each switch they pass through how to handle them. The address is sort of an "I'm with Meriwether" identifier.

---

[4]The alternative is to have a network administrator manually preconfigure the routes from source to destination. This guarantees a great deal of control, but also obviates the need for an intelligent network.

Thus all the packets with the same origin will follow the same path through the network. Consequently, they will arrive in order, and will all be delayed the same amount of time as they traverse the network. The service provided by connection-oriented networks is called *virtual circuit service*, because it simulates the service provided by a dedicated network. The technique is called connection-oriented because the switches perceive that there is a relationship, or connection, between all of the packets that derive from the same source.

As with most technologies, there is a downside to connection-oriented transmission: in the event of a network failure or heavy congestion somewhere along the predetermined path, the circuit is interrupted and will require some form of intervention to correct the problem because the network is not self-healing from a protocol point-of-view. There are certainly network management schemes and stopgap measures in place to reduce the possibility that a network failure might cause a service interruption, but in a connection-oriented network these measures are external. Nevertheless, because of its ability to emulate the service provided by a dedicated network, connection-oriented services are widely deployed and very successful. Examples include frame relay and ATM. They will be discussed in detail in Chap. 9.

Connectionless Networks

The alternative to connection-oriented switching is *connectionless switching*, sometimes called *datagram service*. In connectionless networks, there is no predetermined path from the source to the destination. There is no call setup packet; all data packets are treated independently, and the switches perceive no relationship between them as they arrive—hence the name, *connectionless*. Every packet carries a complete destination address, since it cannot rely on the existence of a pre-established path created by a call setup packet.

When a packet arrives at the ingress switch of a connectionless network, the switch examines the packet's destination address. Based on what it knows about the topology of the network, congestion, cost of individual routes, distance (sometimes called *hop count*), and other factors that affect routing decisions, the switch will select an outbound route that optimizes whatever parameters the switch has been instructed to concern itself with. Each switch along the path does the same thing.

For example, let's assume that the first packet of a message, upon arrival at the ingress switch, would normally be directed out physical port number seven, because based upon current known network conditions that port provides the shortest path (lowest hop count) to the destination. However, upon closer examination, the switch realizes that while port seven provides the shortest hop count, the route beyond the port is severely congested. As a result, the packet is routed out port thirteen, which results in a longer path but avoids the congestion.

And because there is no preordained route through the network, the packet will simply have to get directions when it arrives at the next switch.

Now, the second packet of the message arrives. Because this is a connectionless environment, however, the switch does not realize that the packet is related to the packet that preceded it. The switch examines the destination address on the second packet, then proceeds to route the packet as it did with the preceding one. This time, however, upon examination of the network, the switch finds that port seven, the shortest path from the source to the destination, is no longer congested. It therefore transmits the packet on port seven, thus ensuring that packet two will in all likelihood arrive before packet one! Clearly, this poses a problem for message integrity, and illustrates the criticality of the transport layer, which, you will recall, provides end-to-end message integrity by reassembling the message from a collection of out-of-order packets that arrive with varying degrees of delay because of the vagaries of connectionless networks.

Connectionless service is often called *unreliable* because it fails to guarantee delay minimums, sequential delivery, or for that matter, any kind of delivery. This causes many people to question why network designers would rely on a technology that guarantees so little. The answer lies within the layered protocol model. While connectionless networks do not guarantee sequential delivery or limits on delay, they *will* ultimately deliver the packets. Because they are not required to transmit along a fixed path, the switches in a connectionless network have the freedom to *route around* trouble spots by dynamically selecting alternate pathways, thus ensuring delivery, albeit somewhat unpredictable. If this process results in out-of-order delivery, no problem: that's what the transport layer is for. Data communications is a team effort and requires the capabilities of many different layers to ensure the integrity and delivery of a message from the transmitter to the receiver. Thus, even an *unreliable* protocol has distinct advantages.

An example of a well-known connectionless layer three protocol is the Internet Protocol (IP). It relies on TCP a transport layer protocol, to guarantee end-to-end correctness of the delivered message. There are times, however, when the foolproof capabilities of TCP and TCP-like protocols are considered overkill. For example, network management systems tend to generate large volumes of small messages on a regularly scheduled basis. These messages carry information about the health and welfare of the network and about topological changes that routing protocols need to know about if they are to maintain accurate routing tables in the switches. The problem with these messages is that they (1) are numerous, and (2) often carry information that hasn't changed since the *last* time the message was generated, thirty seconds ago.

TCP and TP4 protocols are extremely overhead-heavy compared to their lighter-weight cousins UDP and TP1. Otherwise, they would not be able to absolutely, positively guarantee the delivery of the message. In some cases, however, there may not be a need to absolutely, positively guarantee delivery. After all, if I lose one of those status messages, no problem: it will be generated again in 30 seconds anyway. The result of this is that some networks choose not to employ the robust and capable protocols available to them, simply because the marginal advantage they provide doesn't merit the transport and processing overhead they create in the network. Thus connectionless networks are extremely widely deployed. After all, those billions (or so) of Internet users must be *reasonably* happy with the technology.

Let's now examine the network layer. Please note that we are now entering the realm of the chained layers, which you will recall are used by all devices in the path—end-used devices as well as the switches or routers themselves.

Layer 3: The Network Layer

The network layer, which is the uppermost of the three chained layers, has two key responsibilities: routing and congestion control. We will also briefly discuss switching at this layer, even though many books consider it to be a layer-two process. So for the purists in the audience, please bear with me—there's a method to my madness.

When the telephone network first started its remarkable growth during the sunrise of the twentieth century, there was no concept of switching. If a customer wanted to speak with another customer, they had to have a phone in their house with a dedicated path to that person's home. Another person required another phone and phone line—and you quickly begin to see where this is leading. Figure 3.50

FIGURE 3.50 If the telephone network were any more successful, its access lines would block the sun, bringing on the next ice age! (*Courtesy Lucent Technologies.*)

illustrates the problem: the telephone network's success would bring on the next ice age, blocking the sun with all the aerial wire the telephone network would be required to deploy. Consider this simple mathematical model: In order to fully interconnect, or mesh, five customers as shown in Fig. 3.51 so that any one of them can call any other, the telephone company would have to install 10 circuits, according to the equation $n(n-1)/2$, where n is the number of devices that wish to communicate. Extrapolate that out now to the population of even a small city—say, 2,000 people. That boils down to 3,997,999 circuits that would have to be installed—all to allow 2,000 people to call each other. Obviously, some alternative solution was greatly needed. That solution was the switch.

The first *switches* did not arrive until 1878 with the near-disastrous hiring of young boys to work the cord boards in the first central offices. John Brooks, author of *Telephone: The First 100 Years,* offers the following.

The year of 1878 was the year of male operators, who seem to have been an instant and memorable disaster. The lads, most of them in their late teens, were simply too impatient and high-spirited for the job, which, in view of the imperfections of the equipment and the inexperience of the subscribers, was one demanding above all patience and calm. According to the late reminiscences of some of them, they were given to lightening the tedium of their work by roughhousing, shouting constantly at each other, and swearing frequently at the customers. An early visitor to the Chicago exchange said of it, "The racket is almost deafening. Boys are rushing madly hither and thither, while others are putting in or taking out pegs from a central framework as if they were lunatics engaged in a game of fox and cheese. It was a perfect bedlam."

Later in 1878, the boys were grabbed by their ears and removed from their operator positions, replaced quickly—and, according to a

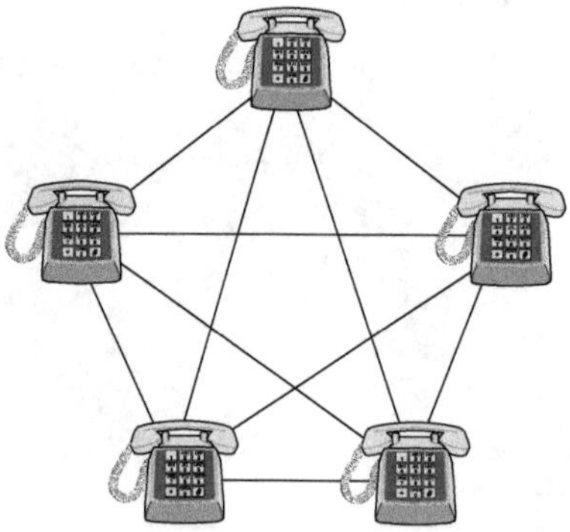

FIGURE 3.51 A full-mesh network of five customers.

FIGURE 3.52 The first female operators.

multitude of accounts, to the enormous satisfaction of the customers —by women, shown in Fig. 3.52.

These operators were in fact the first switches. Instead of needing a dedicated circuit running from every customer to every other customer, each subscriber needed a single circuit that ran into the central exchange, where it appeared on the jack field in front of an operator (each operator managed approximately 100 positions, the maximum number according to Bell System studies). When a customer wanted to make a call they would crank the handle on their phone, generating current which would cause a flag to drop, a light to light or a bell to ring in front of the operator. Seeing the signal the operator would plug a headset into the customer's jack appearance and announce that they were ready to receive the number to be dialed. The operator would then simply *cross-connect* the caller to the called party then wait for the receiver to be picked up on the other end. The operator would periodically monitor the call and pull the patch cord down when the call was complete. Incidentally, note the roller skates on the feet of the supervisors. The cord boards were so long that the process of collecting tickets took far too long, so they added mobility!

This model meant that instead of needing 3,997,999 circuits to provide universal connectivity for a town of 2,000 people, 2,000 were needed—a rather *significant* reduction in capital outlay for the telephone company, wouldn't you say? The network now looked like Fig. 3.53.

Over time, manual switching slowly disappeared, replaced by mechanical switches initially followed by more modern all-electronic switches. The first true mechanical switches didn't arrive until 1892 when Almon Strowger's Step-by-Step switch was first installed by his company, Automatic Electric. We'll discuss Strowger's remarkable invention in Chap. 4.

FIGURE 3.53 Simplification of the network, thanks to a central operator.

The bottom line to all this is that switches create temporary end-to-end paths between two or more devices that wish to communicate with each other. They do it in a variety of ways; circuit-switched networks create a *virtually dedicated path* between the two end points and offer constant end-to-end delay, making them acceptable for delay-sensitive applications or connections with long hold times. This is important: circuit-switched networks don't eliminate delay; they simply make the delay consistent.

Store-and-forward networks, particularly packet networks, work well for short, bursty messages with minimal delay sensitivity.

To manage all this, however, is more complicated than it would seem at first blush. First of all, the switches must have the ability to select not only a path, but the *best* path, based on QoS parameters. This constitutes intelligent routing. Second, they should have some way of monitoring the network so that they always know its current operational conditions. Finally, should they encounter unavoidable congestion, the switches should have one or more ways to deal with it.

Routing Decisions

So, how are routing decisions made in a typical network? Whether connectionless or connection-oriented, the routers and switches in the network must take into account a variety of factors to determine the best path for the traffic they manage. These factors fall into a broad category of rule sets called *routing protocols*. For reference purposes, please refer to the *tree* shown in Fig. 3.54.

Once the transport layer has taken whatever steps are necessary to prepare the packets for their transmission across the network, they are passed to the network layer.

The network layer has two primary responsibilities in the name of network integrity: *routing* and *congestion control*. Routing is the process of intelligently selecting the most appropriate route through the

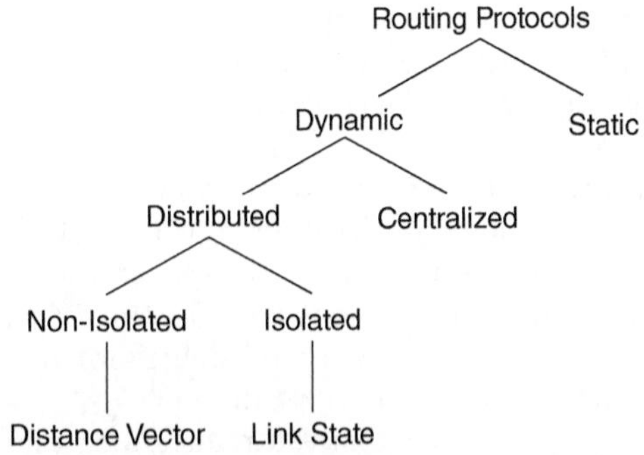

FIGURE 3.54 The routing protocol hierarchy.

network for the packets; congestion control is the process that ensures that the packets are minimally delayed (or at least equally delayed) as they make their way from the source to the destination. We will begin with a discussion of routing.

Routing Protocols

Routing protocols are divided into two main categories—*static routing protocols* and *dynamic routing protocols*. Static routing protocols are those that require a network administrator to establish and maintain them. If a routing table change is required, the network administrator must manually make the change. This ensures absolute security, but is labor intensive and therefore less frequently used other than in highly secure environments (military, health care) or network architectures that are designed around static routing because the routes are relatively stable anyway (IBM's Systems Network Architecture, SNA, for example).

More common are dynamic routing protocols, where network devices make their own decisions about optimum route selection. They do this in the following general way: they pay attention to the network around them; collect information from their neighbors about best routes to particular destinations based on such parameters as least number of hops, least delay, lowest cost, or highest bandwidth; archive those bits of information in tables; and then selectively flush the tables periodically to ensure that the information contained in them is always as current as possible. Because dynamic routing protocols assume intelligence in the switch and can therefore reduce the amount of human intervention required, they are commonly used and are in fact far and away the most widely deployed routing protocols.

Dynamic routing protocols are further divided into two subcategories, *centralized* and *distributed*. Centralized routing protocols concentrate the route decision-making processes in a single node, thus ensuring that all nodes in the network receive the same and most current information possible. When a switch or router needs routing information that is not contained in its own table, it sends a request to the root node asking for direction. There are significant downsides to this technique; by concentrating the decision-making capability in a single node, the likelihood of a catastrophic failure is dramatically increased. If that node fails, the entire network's ability to seek optimal routing decisions fails. Second, because all nodes in the network must go to that central device for routing instructions, a significant choke point can result.

There are several options that can reduce the vulnerability of a single point of failure. The first, of course, is to distribute the routing function. This conflicts with the concept of centralized routing, but only somewhat. Consider the Internet, for example. It uses a sort of hybrid of centralized and distributed routing protocols in its domain name server (DNS) function. A limited number of devices are tasked with the responsibility of maintaining knowledge of the topology of the network and the location of domains, providing something of an AAA or CAA trip planning service for data packets.

Another option is to designate a backup machine tasked with the responsibility to take over in the event of a failure of the primary routing machine. This technique, used in the early Tymnet packet networks, relied on the ability of the primary machine to send a *sleeping pill packet* to the backup machine, with these instructions: "Pay attention to what I do, memorize everything I learn, but just sit in the corner and be a potted plant. Take no action, make no routing decisions—just learn. Let the sleeping pill keep you in a semi-comatose state. If for some reason, I fail, I will stop sending the sleeping pills, at which time you will wake up and take over until I recover": an ingenious technique, but overly complex and far too failure-prone for modern network administrators. Distributed routing protocols are far more common today.

Distributed Routing Protocols

In distributed routing protocol environments, each device collects information about the topology of the network and makes independent routing decisions based upon the information they accumulate. For example, if a router sees a packet from source *X* arrive on port 12, it knows that somewhere out port 12 it will find destination *X*. It doesn't know how far out there necessarily, just that the destination is somewhere out there over the digital horizon. Thus if a packet arrives on another port looking to be transmitted to *X*, the router knows that by sending the packet out port 12 it will at least get closer to its destination. It therefore makes an entry in its routing tables to that effect, so that the next time a packet arrives with the same destination, the switch can consult its table and route the packet quickly.

These routing protocols are analogous to the process of stopping and asking for directions on a road trip (or not), reputedly one of the great male-female differentiators—right after who controls the TV remote. Anthropologists must have a field day with this kind of stuff. According to apocryphal lore, women have no problem whatsoever stopping and asking for directions, while men are loathe to do it—one of those silly threats to the manhood things. Anyway, back to telecom. If you were planning a road trip across the country you could do so using one of two philosophies. You could go to AAA or a travel agent and have them plan out the entire route for you, or you could do the Jack Kerouac thing and simply get in the car and drive. Going to AAA seems to be the simplest option because once the route is planned, all you have to do is follow the directions—Lewis' blazed trail, as it were. The downside is that if you make it as far as Scratch Ankle, Alabama (yes, it's a real place) and the road over which you are supposed to travel is closed, you are stuck—you have to stop and ask for directions anyway.

The alternative is to simply get in the care, drive to the nearest gas station, and tell them that you are trying to get to Dime Box, Texas. The attendant will no doubt tell you the following: "I don't know where Dime Box is, but I know that Texas is down in the southwest. So if you take this highway here to Kansas City, it'll get you closer. But you'll have to ask for better directions when you get there." The next gas station attendant may tell you, "Well, the quickest way to central Texas is along this highway, but it's rush hour and you'll be stuck for days if you go that way. I'd take the surface streets. It's a little less comfortable, but there's no congestion." By stopping at a series of gas stations as you traverse the country and asking for help you will eventually reach your destination, but the route may not be the most efficient. That's okay, though, because you will never have to worry about getting stuck with bad directions. Clearly the first example of these is connection-oriented; the second is connectionless. Connection-oriented travel is a far more comfortable, secure way to go; connectionless is riskier, less sure, more flexible, and much more fun. Obviously, distributed routing protocols are centrally important to the traveler as well as to the gas station attendant who must give them reliable directions, and equally important to routers and switches in connectionless data networks.

Distributed routing protocols fall into two categories: *distance vector* and *link state*. Distance vector protocols rely on a technique called *table swapping* to exchange information about network topology with each other. This information includes destination-cost pairs that allow each device to select the least-cost route from one place to another. On a scheduled basis, routers transmit their entire routing tables on all ports to all adjacent devices. Each device then adds any newly arrived information to its own tables, thus ensuring currency.

The only problem with this technique is that it results in a tremendous amount of management traffic (the tables) being sent between network devices, and if the network is relatively static—that is, changes in topology don't happen all that often—then much of the information is unnecessary and can cause serious congestion. In fact, it is not uncommon to encounter networks that have more management traffic traversing their circuits than actual user traffic. What's wrong with this picture? Distance vector works well in small networks where the number of multihop traverses is relatively low, thus minimizing the impact of its bandwidth-intensive nature.

Distance vector protocols have that name because of the way they work. Recovering physicists will remember that a vector is a measure of something that has both direction and magnitude associated with it. The name is appropriate in this case, because the routing protocol optimizes on a direction (port number) and a magnitude (hop count).

Because networks are growing larger, traffic routinely encounters route solutions with large hop counts. This reduces the effectiveness of distance vector solutions. A better solution is the link state protocol. Instead of transmitting entire routing tables on a scheduled basis, link state protocols use a technique called *flooding* to only transmit changes that occur to adjacent devices *as they occur*. This results in less congestion and more efficient use of network resources, and reduces the impact of multiple hops in large-scale networks.

Both distance vector and link state protocols are in widespread use today. The most common distance vector protocols are the routing information protocol (RIP) Cisco's interior gateway routing protocol (IGRP), and border gateway protocol (BGP). Link state protocols include open shortest path first (OSPF) commonly used on the Internet.

Clearly, both connection-oriented and connectionless transport techniques, as well as their related routing protocols, have a place in the modern telecommunications arena. As quality of service becomes a critical component of the service offered by network providers, the importance of both routing and congestion control becomes apparent. We now turn our attention to the second area of responsibility at the network layer, *congestion control.*

Congestion Control

At its most fundamental level, congestion control is a mechanism for reducing the volume of traffic on a particular route through some form of load balancing. No matter how large, diverse, or capable a network is, some degree of congestion is inevitable. It can result from sudden unexpectedly high utilization levels in one area of the network, from failures of network components, or from poor engineering. In the telephone network, for example, the busiest calling day of the year in the United States is Mother's Day. To reduce the probability that a caller will not be able to complete a call to Mom, network traffic engineers take extraordinary steps to load balance the network. For example, when subscribers on the east coast are making long distance calls at 9:00 in the morning, west coast subscribers haven't even turned on their latte machines yet. Network resources in the west are underutilized during that period, so engineers route east coast traffic westward and then hairpin it back to its destination, to spread the load across the entire network. As the day gets later, they reduce the volume of westward-bound traffic to ensure that California has adequate network resources for its own calls.

There are two terms that are important in this discussion. One is congestion; the other is delay. The terms are often used interchangeably, but they are not the same thing, nor are they always related to one another.

Years ago, I lived in the San Francisco Bay Area where traffic congestion is a way of life. I often had to drive across the many bridges that crisscross San Francisco Bay, the Suisun Straits, or the Sacramento River. Many of those bridges require drivers to stop and pay a toll, resulting in localized delay. The time it takes to stop and pay the toll is mere seconds, yet traffic often backs up for miles as a result of this local phenomenon, causing a global effect.

This is the relationship between the two: local delay often results in widespread congestion. And congestion is usually caused by inadequate buffer or memory space. Increase the number of buffers—the lanes of traffic on the bridge, if you will—and congestion drops off. Open another line or two at Home Depot ("no waiting on line seven!"), and congestion drops off.

The various players in the fast food industry manage congestion in different ways, and with dramatically different results. Without naming them, some use a single queue with a single server to take orders, a technique that works well until the lunch rush begins. Then things back up dramatically. Others use multiple queues with multiple servers, a technique that is better except that one queue can experience serious delays should someone place an order for a nonstandard item or try to pay with a credit card. That line then experiences serious delay. The most effective restaurants stole an idea from the airlines, and use a single queue with multiple servers. This keeps things moving because the instant a server is available, the next person in line is served.

Remember when Jeff Goldblum, the chaos theorist in *Jurassic Park,* talked about the *butterfly effect*? How a butterfly flapping its wings in the Amazon Basin can kick off a chain of events that affect weather patterns in New York City? That aspect of *chaos theory* contributes greatly to the manner in which networks behave, and the degree to which their behavior is immensely difficult to predict under load.

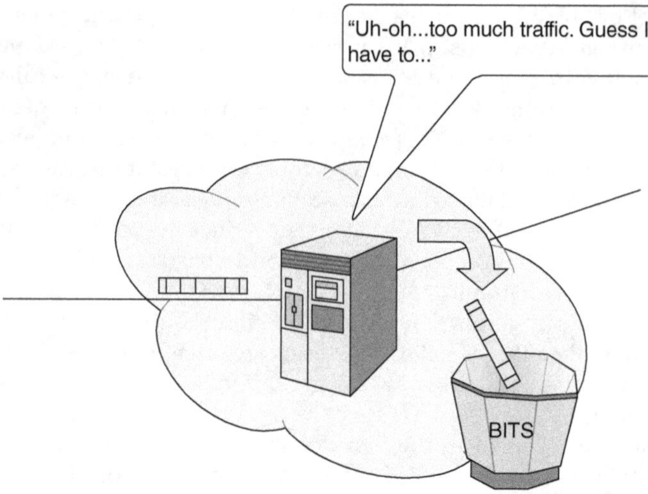

FIGURE 3.55 Packet discard—the simplest congestion control mechanism.

So how is congestion handled in data networks? The simplest technique, used by both frame relay and ATM, is called packet discard, shown in Fig. 3.55. In the event that traffic gets too heavy based on pre-established management parameters, the switches simply discard excess packets. They can do this because of two facts: First, networks are highly capable and the switches rarely have to resort to these measures; and second, the devices on the ends of the network are intelligent and will detect the loss of information and take whatever steps are required to have the discarded data resent. As drastic as this technique seems, it is not all that catastrophic. Modern networks are heavily dependent on optical fiber and highly capable digital switches; as a result, packet discard, while serious, does not pose a major problem to the network. And even when it is required, recovery techniques are fast and accurate.

Because congestion occurs primarily as the result of inadequate buffer capacity, one solution is to preallocate buffers to high-priority traffic. Another is to create multiple queues with varying priority levels. If a voice or video packet arrives that requires high-priority, low-delay treatment, it will be deposited into the highest priority queue for instantaneous transmission. This technique holds great promise for the evolving all-services Internet, because it's greatest failing is its inability to deliver multiple, dependable, sustainable grades of service quality.

Other techniques are available that are somewhat more complex than packet discard, but do not result in loss of data. For example, some devices, when informed of congestion within the network, will delay transmission to give the network switches time to process their overload before receiving more. Others will divert traffic to alternate,

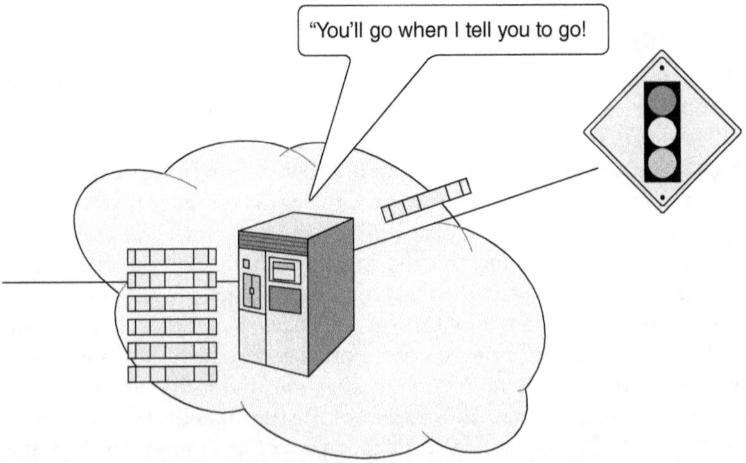

FIGURE 3.56 Using choke packets to control network congestion.

less-congested routes, or *trickle* packets into the network, a process known as *choking*. This is shown in Fig. 3.56. Frame relay, for example, has the ability to send what is called a *choke packet* into the network, implicitly telling the receiving device that it should throttle back to reduce congestion in the network. Whether it does or not is another story, but the point is that the network has the intelligence to invoke such a command when required. This technique is now used on freeways in large cities: traffic lights are installed on major onramps that meter the traffic onto the roadway. This results in a dramatic reduction in congestion. Other networks have the intelligence to diversely route traffic, thus reducing the congestion that occurs in certain areas.

Clearly, the network layer provides a set of centrally important capabilities to the network itself. Through a combination of network protocols, routing protocols, and congestion control protocols, routers and switches provide granular control over the integrity of the network.

Back to the E-Mail Message

Let us return now to our e-mail example. The message has been divided into packets by the transport layer and delivered in pieces to the network layer, which now takes whatever steps are necessary to ensure that the packets are properly addressed for efficient delivery to the destination. Each packet now has a header attached to it that contains routing information and a destination address. The next step is to get the packet correctly to the next link in the network chain—in this case, the next switch or router along the way. This is the responsibility of the data link layer.

The Data Link Layer

The data link layer is responsible for ensuring bit-level integrity of the data being transmitted. In short, its job is to make the layers above believe that the world is an error-free and perfect place. When a packet is handed down to the data link layer from the network layer, it wraps the packet in a *frame*. In fact, the data link layer is sometimes called the *frame layer.* The frame built by the data link layer comprises several fields, shown graphically in Fig. 3.57, that give the network devices the ability to ensure bit-level integrity and proper delivery of the packet, now encased in a frame, *from switch to switch*. Please note that this is different from the network layer, which concerns itself with routing packets *to the final destination*. Even the addressing is unique: packets contain the address of the ultimate destination, used by the network to route the packet properly; frames contain the address of the next link in the network chain (the next switch), used by the network to move the packet along, switch by switch.

As the diagram illustrates, the beginning and end fields of the frame are called *flags*. These fields, made up of a unique series of bits (0111110), can only occur at the beginning and end of the frame—they are never allowed to occur within the bitstream inside the frame through a process that we will describe momentarily. These flags are used to signal to a receiving device that a new frame is arriving or that it has reached the end of the current frame,[5] which is why their unique bit pattern can never be allowed to occur naturally within the data itself because it could indicate to the receiver (falsely) that this is the end of the current frame. If the flag pattern *does* occur within the bit-stream, it is disrupted by the transmitting device through a process called *bit stuffing* or *zero-bit insertion*, in which an extra zero is inserted in the middle of the flag pattern, based on the following rule set. When a frame of data is created at an originating device, the very last device to touch the frame—indeed, the device that actually adds the flags—is called a *universal synchronous/asynchronous receiver-transmitter (USART)*.

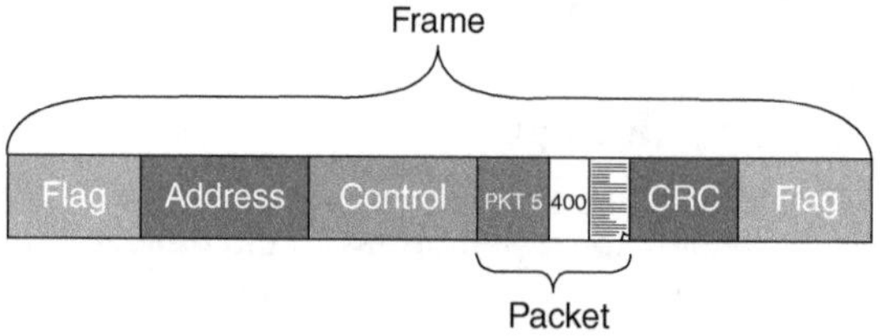

FIGURE 3.57 A typical data link layer frame showing the fields that it comprises.

[5]Indeed, the final flag of one frame is often the beginning flag of the *next* frame.

The USART, sometimes called an integrated data link controller (IDLC), is a chipset that has a degree of embedded intelligence. This intelligence is used to detect (among other things) the presence of a false flag pattern in the bitstream around which it builds the frame. Since a flag comprises a zero followed by six ones and a final zero, the IDLC knows that it can never allow that particular pattern to exist between any two real flags. So, as it processes the incoming bitstream, it looks for that pattern and makes the following decision: *If I see a zero followed by five ones, I will automatically and without question insert a zero into the bitstream at that point.* This is illustrated in Fig. 3.58.

This of course destroys the integrity of the message, but it doesn't matter. At the receive device, the IDLC monitors the incoming bits. As the frame arrives it sees a *real* flag at the beginning of the frame, an indication that a frame is beginning. As it monitors the bits flowing by it *will* find the zero followed by five bits, at which point it knows, beyond a shadow of a doubt, that the very next bit is a zero—which it will promptly remove, thus restoring the integrity of the original message.

The receiving device has the ability to detect the extra zero and remove it before the data moves up the protocol stack for interpretation. This bit stuffing process guarantees that a *false flag* will never be interpreted as a final flag and acted upon in error.

The next field in the frame is the *address field*. This field identifies the address of the next switch in the chain to which the frame is directed, and changes at every node. The only address that remains constant is the destination address, safely embedded in the packet itself.

The third field found in many frames is called the *control field*. It contains supervisory information that the network uses to control the integrity of the data link. For example, if a remote device is not responding to a query from a transmitter, the control field can send a *mandatory response required* message that will allow it to determine the nature of the problem at the far end. It is also used in hierarchical multipoint networks to manage communications functions. For example, a multiplexer may have multiple terminal devices attached to it, all of which routinely transmit and receive data. In some systems, only a single device is allowed to *talk* at a time. The control field

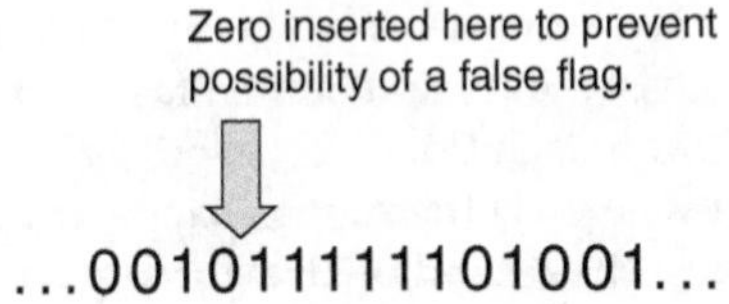

FIGURE 3.58 Zero-bit insertion, sometimes called bit stuffing.

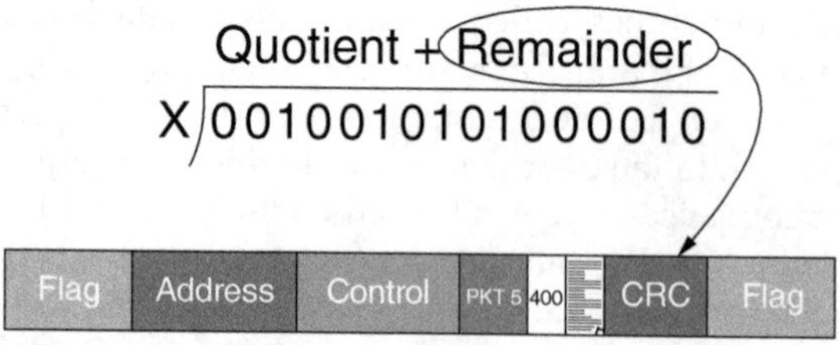

FIGURE 3.59 Calculating the cyclic redundancy check (CRC) for error detection.

can be used to force these devices to take turns. This field is optional; some protocols do not use it.

The final field we will cover is the *cyclic redundancy check* (CRC) field. The CRC is a mathematical procedure used to test the integrity of the bits within each frame. It does this by treating the zeroes and ones of data as a binary number (which, of course, it is) instead of as a series of characters. It divides the *number*, shown in Fig. 3.59, by a carefully crafted polynomial value that is designed to *always* yield a remainder following the division process. The value of this remainder is then placed in the CRC field and transmitted as part of the frame to the next switch. The receiving switch performs the same calculation, then compares the two remainders. As long as they are the same, the switch knows that the bits arrived unaltered. If they are different, the received frame is discarded and the transmitting switch is ordered to resend the frame, a process that is repeated until the frame is received correctly. This process can result in transmission delay, because the data link layer will not allow a bad frame to be carried through the network. Thus, the data link layer converts errors into delay.

Error Recovery Options

There are a number of techniques in common use that allow receiving devices to recover from bit errors. The simplest of these is frame discard, the technique used by frame relay and ATM networks. In frame discard environments, an errored frame is simply discarded—period. No other form of recovery takes place within the network. Instead, the end devices (the originator and receiver) have the end-to-end responsibility to detect that a frame is missing and take whatever steps are necessary to generate a second copy. There are reasons for this strategy that will be discussed in the section on fast packet services.

A second common technique is called *forward error correction* (FEC). FEC is used when: (1) there is no backward channel available over which to request the resend of an errored packet, or (2) the transit delay is so great that a resend would take longer than the application would allow, such as in a satellite hop over which an application is transmitting delay-sensitive traffic. Instead, FEC systems transmit the application data with additional information that allows a receive device to not only determine that an error has occurred, but to fix it. No resend is required.

The third and perhaps most common form of error detection and correction is called *detect and retransmit.* Detect and retransmit systems use the CRC field to detect errors when they occur. The errored frames are then discarded, and the previous switch is ordered to resend the errored frame. This implies a number of things: the frames must be numbered, there must be some form of positive and negative acknowledgment system in place, the transmitter must keep a copy of the frame until its receipt has been acknowledged, and there must be some facility in place to allow the receiver to communicate upstream to the transmitter.

Two recovery techniques are commonly utilized in synchronous systems. To understand them, we must first introduce a couple of transmission protocols used to meter the transmission of frames between switches.

In early communications systems (1970s), the network was known to be relatively hostile to data transmission. After all, if noise occurred during a voice conversation, no problem—it was a simple matter to ignore it provided it wasn't too bad. In data, however, a small amount of noise could be catastrophic, easily capable of destroying a long series of frames in a few milliseconds. As a result, early data systems such as IBM's *binary synchronous communications* (BISYNC) used a protocol called *stop-and-wait* that would only permit a single frame at a time to be outstanding without acknowledgment from the receiver. This was obviously terribly inefficient, but in those early days was as good as it got. Thus if a major problem occurred, the maximum number of frames that would ever have to be resent was one.

As time passed and network quality improved, designers got brave and began to allow multiple unacknowledged frames to be outstanding, a process called *pipelining*. In pipelined systems, a maximum *window size* is agreed upon that defines the maximum number of frames that can ever be allowed to be outstanding at any point in time. If the number is reached, the window closes, closing down the pipeline and prohibiting other frames from being transmitted. As soon as one or more frames clear the receiver, that many new frames are allowed into the pipeline by the sliding window. Obviously this protocol is reliant on the ability to number the frames so that the system knows when the maximum window size has been reached.

Ok, back to error recovery. We mentioned earlier that there are two common techniques. The first of these is called *selective retransmit.* In selective retransmit environments, if an error occurs, the transmitter is directed to resend *only the errored frame.* This is a complex technique, but is quite efficient.

The second technique is called *go back N*. In go back *N* environments, if an error occurs, the transmitter is directed to go back to the errored frame *and retransmit everything from that frame forward.* This technique is less efficient but is far simpler from an implementation point-of-view.

Let's look at an example. Let's assume that a transmitter generates five frames, numbered one of five, two of five, three of five, and so on. Now let's assume that frame three arrives and is found to be errored. In a selective retransmit environment, the transmitter is directed to resend frame three, and *only* frame three. In a go back *N* environment, however, the transmitter will be directed to resend everything from frame three going forward, which means that it will send frames three, four, and five. The receiver will simply discard frames four and five.

So let's review the task of the data link layer. It frames the packet so that it can be checked for bit errors; provides various line control functions so that the network itself can be managed properly; provides addressing information so that the frame can be delivered appropriately; and performs error detection and (sometimes) correction.

Practical Implementations

A number of widely known network technologies are found at the data link layer of the OSI model. These include the access protocols used in modern wireless networks such as CDMA and GSM, as well as those used in modern local area networks, such carrier sense, multiple access with collision detection (CSMA/CD, used in ethernet), and token ring. CSMA/CD is a protocol that relies on contention for access to the shared backbone over which all stations transmit, while token ring is more civilized—stations take turns sharing access to the backbone. This is also the domain of frame relay and ATM technologies, which provide high-speed switching.

Frame relay is a high-speed switching technology that emerged as a good replacement for private line circuits. Today it is considered a legacy technology but at the time it offered a wide range of bandwidth and while switched, delivers service quality that is equivalent to that provided by a dedicated facility.

Asynchronous transfer mode (ATM) has become one of the most important technologies in the service provider pantheon today because it provides the ability to deliver true, dependable, and granular quality of service over a switched network architecture, thus giving service providers the ability to aggregate multiple traffic types on a single network fabric. This means that IP, which is a network layer protocol, can be transported across an ATM backbone, allowing for the smooth and service-driven migration to an all-IP network. Eventually, ATM's overhead-heavy quality of service capabilities will be replaced by more elegant solutions, but until that time comes it still plays a central role in the delivery of service.

We will discuss all of these services in later chapters.

The Physical Layer

Once the CRC is calculated and the frame is fully constructed, the data link layer passes the frame down to the *physical layer*, the lowest layer in the networking food chain. This is the layer responsible for the physical transmission of bits, which it accomplishes in a wide variety of ways. The physical layer's job is to transmit the bits, which includes the proper representation of zeroes and ones, transmission speeds, and physical connector rules. For example, if the network is electrical, then what is the proper range of transmitted voltages required to identify whether the received entity is a zero or a one? Is a one in an optical network represented as the presence of light or the absence of light? Is a one represented in a copper-based system as a positive or as a negative voltage, or both? Also, where is information transmitted and received? For example, if pin two is identified as the transmit lead in a cable, what lead is data received over? All of these physical parameters are designed to ensure that the individual bits are able to maintain their integrity and be recognized by the receiving equipment.

Many transmission standards are found at the physical layer including T-1, E-1, SONET, SDH, DWDM, and the many flavors of digital subscriber line (DSL). *T-1* and *E-1* are longtime standards that provide 1.544 and 2.048 Mbps of bandwidth respectively; they have been in existence since the early 1960s and occupy a central position in the typical network. The synchronous optical network (SONET) and the synchronous digital hierarchy (SDH) provide standards-based optical transmission at rates above those provided by the traditional carrier hierarchy. *Dense wavelength division multiplexing* (DWDM) is a frequency division multiplexing technique that allows multiple wavelengths of light to be transmitted across a single fiber, providing massive bandwidth multiplication across the strand. It will be discussed in detail later in the chapter on optical networking. And *DSL* extends the useful life of the standard copper wire pair by expanding the bandwidth it is capable of delivering as well as the distance over which that bandwidth can be delivered.

OSI Summary

We have now discussed the functions carried out at each layer of the OSI model. Layers six and seven ensure application integrity; layer five ensures security and logical integrity; and layer four guarantees the integrity of the transmitted message. Layer three ensures network integrity, layer two, data integrity, and layer one, the integrity of the bits themselves. Thus transmission is guaranteed on an end-to-end basis through a series of protocols that are interdependent upon each other and which work closely to ensure integrity at every possible level of the transmission hierarchy.

So let's now go back to our e-mail example and walk through the entire process.

The e-mail application running on the PC creates a message at the behest of the human user[6] and passes the message to the application layer. The application layer converts the message into a format that can be universally understood as an e-mail message, in this case X.400. It then adds a header that identifies the X.400 format of the message.

The X.400 message with its new header is then passed down to the presentation layer which encodes it as ASCII, encrypts it using PGP, and compresses it using a British Telecom Lempel-Ziv compression algorithm. After adding a header that details all this, it passes the message to layer five.

The session layer assigns a logical session number to the message, glues on a packet header identifying the session ID, and passes the steadily growing message down to the transport layer. Based on network limitations and rule sets, the transport layer breaks the message into 11 packets and numbers them appropriately. Each packet is given a header with address and quality-of-service information.

The packets now enter the chained layers where they will first encounter the network. The network layer examines each packet in turn, and, based on the nature of the underlying network (connection-oriented? Connectionless?) and congestion status, queues the packets for transmission. After creating the header on each packet they are handed individually down to the data link layer.

The data link layer proceeds to build a frame around each packet. It calculates a CRC, inserts a data link layer address, inserts appropriate control information, and finally adds flags on each end of the frame. Note that all other layers add a header *only;* the data link layer is the only layer that also adds a trailer.

Once the data link frame has been constructed it is passed down to the physical layer, which encodes the incoming bitstream according to the transmission requirements of the underlying network. For example, if the data is to be transmitted across a T- or E-Carrier network, the data will be encoded using alternate mark inversion and will be transmitted across the facility to the next switch at either 1.544 Mbps or 2.048 Mbps, depending on whether the network is T-1 or E-1.

---

[6]I specifically note *human user* here because some protocols do not recognize the existence of the human in the network loop. In IBM SNA environments, for example, users are devices or processes that *use* network resources. There are no humans.

When the bitstream arrives at the next switch (not the destination), the bits flow into the physical layer, which determines that it can read the bits. The physical layer hands the bits up to the data link layer, which proceeds to find the flags so that it can frame the incoming stream of data and check it for errors. If we assume that it finds none, it strips off the data link frame surrounding the packet and passes the packet up to the network layer. The network layer examines the destination address in the packet, at which point it realizes that it is not the intended recipient. So, it passes it back to the data link layer, which builds a new frame around it, calculating a new CRC and adding a new data link layer address as it does so. It then passes the frame back to the physical layer for transmission. The physical layer spits the bits out the facility to the next switch, which for our purposes we will assume is the intended destination. The physical layer receives the bits and passes them to the Data Link Layer, which checks them for errors. If it finds an errored frame it requests a resend, but ultimately receives the frame correctly. It then strips off the header and trailer, leaving the original packet. The packet is then passed up to the network layer, which after examining the packet address determines that it is in fact the intended recipient of the packet. As a result it passes the packet up to the transport layer after stripping off the network layer header.

The transport layer examines the packet and notices that it has received packet 3 of 11 packets. Because its job is to assemble and pass entire messages up to the session layer, the transport layer simply places the packet into a buffer while it waits for the other 10 packets to arrive. It will wait as long as it has to; it knows that it cannot deliver a partial message because the higher layers are not smart enough to figure out the missing pieces.

Once it has received all 11 packets, the transport layer reassembles the original message and passes it up to the session layer, which examines the session header created by the transmitter and notes that this is to be handed to whatever process cares about logical channel number seven. It then strips off the header and passes the message up to the presentation layer.

The presentation layer reads the presentation layer header created at the transmit end of the circuit and notes that this is an ASCII message that has been encrypted using PGP and compressed using BTLZ. It decompresses the message using the same protocol; decrypts the message using the appropriate public key; and, because it is resident in a mainframe, converts the ASCII message to EBCDIC. Stripping off the presentation layer header, it hands the message up to the application layer. The application layer notes that the message is X.400 encoded, and is therefore an e-mail message. As a result it passes the message to the e-mail application that is resident in the mainframe system.

The process just described happens every time you hit the *send* button.

Click.

Other Protocol Stacks

Of course, OSI is not the only protocol model. In fact, for all its detail, intricacy and definition, it is rarely used in practice. Instead, it serves as a true model for comparing disparate protocol stacks. In that regard it is unequalled in its value to data communications.

The most commonly deployed protocol stack is that used in the Internet, the so-called *TCP/IP stack*. Instead of seven layers, TCP/IP comprises four. The bottom layer, called the *network interface layer*, includes the functions performed by OSI's physical and data link layers. It includes a wide variety of protocols as shown in Fig. 3.60. We describe them here briefly, because we will describe the functions and relationships of these protocols in a later chapter in greater detail.

The *internet protocol layer* is roughly functionally equivalent to the OSI model's network layer. It performs routing and congestion control functions, and as the diagram illustrates includes some of the protocols we mentioned earlier: RIP, OSPF, and a variety of address conversion protocols.

The *transmission control protocol layer* is responsible for message integrity, similar to the service provided by OSI's transport layer. It is extremely capable and has the ability to recover from virtually any network failure imaginable to ensure the integrity of the messages it is designed to protect. For situations where the high degree of protection provided by TCP (and its attendant overhead) is considered to be overkill, a corollary protocol, called *user datagram protocol* (UDP) is also available at this layer. It provides a connectionless network service and

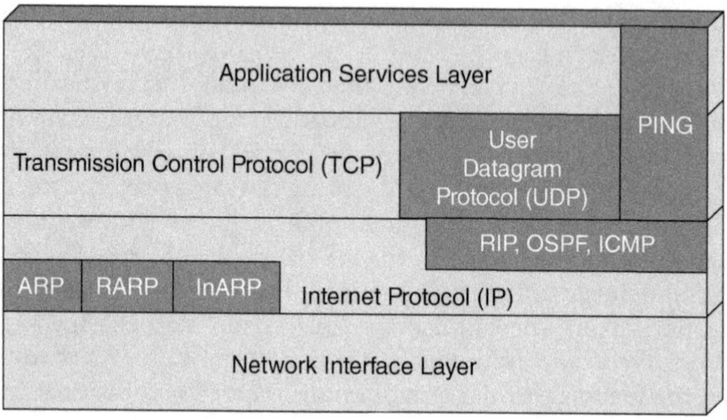

FIGURE 3.60 Dueling protocol stacks: OSI vs. TCP/IP.

is used in situations where the transported traffic is less critical and where the overhead inherent in TCP poses a potential problem due to congestion.

The uppermost layer in the TCP/IP stack is called the *application services layer*. This is where the utility of the stack becomes obvious because this is where the actual applications are found such as HTTP, FTP, Telnet, and the other utilities that make the Internet useful to the user. We will discuss the TCP/IP protocols stack in more detail later in the book, but for now the information provided will suffice.

The point of all these protocols is to give a designer the ability to create a network that will transport the customer's voice, data, video, images, or MP3 files with whatever level of service quality the traffic demands. We now know that data communications protocols make it possible to transport all types of traffic with guaranteed service. Let's turn our attention now to the network itself. In the chapters that follow we will discuss the history of the most remarkable technological achievement on earth—the telephone network. Later we will discuss the anatomy of a typical data network and the technologies found there.

Chapter Summary

This chapter is designed to acquaint the reader with the fundamental terms and concepts that characterize the data and telecommunications worlds today, which in turn help us understand the complex magic that gives life to the IT and media worlds. Now we can move deeper into the magic. In the next chapter, we introduce the history, design, philosophy, structure, and use of the most complicated machine ever built—the global telephone network.

Chapter Three Questions

1. Why are ASCII and EBCDIC important? Are there other character codes that are still in use? What are they?

2. Convert the number $2{,}387_{10}$ to its base two equivalent.

3. Convert the binary number 1101001101010_2 to its base ten equivalent.

4. Give three examples of noise in a telephone network.

5. Explain the difference between noise and distortion.

6. The OSI model is referred to as a layered protocol model. Why are layered models advantageous?

7. Explain the difference between layers six and seven. Are they both necessary? Why?

8. What is the difference between connectivity and interoperability? Is one more important than the other? Why or why not?

9. What layer of OSI cares most about each of the following: bits, messages, cells, frames, packets?

10. Explain the difference between a connection-oriented and a connectionless network, including appropriate applications for each.

11. What is the simplest form of error correction?

CHAPTER 4

Telephony

Any sufficiently advanced technology is indistinguishable from Magic.

—A. C. Clarke

It's an easy thing to kick a large, slow-moving creature, as legacy telephone companies are often perceived to be. An enterprise as large as a telephone company is easy to criticize, not only because they are large, but also because they don't move quickly—nor can they. They have many moving parts and they are so deeply embedded in society, commerce, global affairs, politics, and crucial vertical industries like healthcare, education, national defense, and energy, that rapid change could be catastrophic to any or all of them.

That said, when push comes to shove, they have the ability to move the Earth if need be, as we'll see in a few pages.

In this chapter, we will explore telephony and the telephone network. Described as "the largest and most complex machine ever built," the telephone network is an impressive thing. We begin with some history, then explore the network itself, and finally explore telephony service—how it works, how it makes its way across the network, how it is managed. We will then get our hands dirty as we examine the process of voice digitization followed by the multiplexing and high-speed transport of voice through such technologies as PCM (the basis for T1 and E1), SONET, and synchronous digital hierarchy (SDH).

Some people feel that voice is something of an anachronism and has no place in a book about the new, slick world of telecommunications. I disagree: without an understanding of how the voice network operates it is impossible to understand or appreciate how data networks operate. Furthermore, as voice moves inexorably toward an all-IP model, the voice stream becomes nothing more than one element of the overall corporate data stream. And since the IT organization manages data in most corporations, voice is rapidly becoming just another corporate data application. If for no other reason, this convergence of voice and data, this ongoing merger of the voice and data worlds, is reason enough to spend some time on telephony. This chapter bridges the gap between the two, and we begin our tale in New York City.

Miracle on Second Avenue

February 26, 1975 was a business-as-usual day at New York Telephone's lower Manhattan switching center. Located at 2nd Avenue and 13th Street, the massive 11-story building was the telephony nerve center for 12 Manhattan exchanges that provided service to 300 New York City blocks that included 104,000 subscribers and 170,000 telephones. Within the service area were 6 hospitals, 11 firehouses, 3 post offices, 1 police station, 9 schools, and 3 universities. The building was massive, but like most central offices was completely invisible to the public. It was just a big, windowless structure that belonged to the telephone company. No one really knew what went on in there; nobody cared.

When night fell, most of the building's employees went home, leaving a small crew to handle maintenance tasks and minor service problems. The night was quiet; work in the building was carried out routinely. It was going to be another boring evening—or so they thought.

At 12:30, just after midnight, a relatively inconsequential piece of power equipment in the building's sub-basement cable vault shorted and spit a few errant sparks into the air around it. It caused no alarms because it didn't actually fail. One of the sparks, however, fell on a piece of plastic insulation and began to smolder. The insulation melted and began to burn, changing from a smoldering spot on the surface of a cable to a full-blown fire. Soon the entire cable vault was ablaze.

The cables in a central office exit the cable vault on their way to upper floor switching equipment through 6-in-diameter holes cored in the concrete floors of the office. The fire burned its way up the cables, exited the cored holes, and spread from floor to floor. Soon all of the lower floors of the building were engulfed, and New York Telephone was on its way to hosting the single worst disaster that any telephone company at the time had ever experienced.

Emergency service vehicles converged on the building. They evacuated everyone and began flooding the lower floors with hundreds of thousands of gallons of water. Smoke and steam billowed up and out of the structure (see Fig. 4.1), destroying equipment on the upper floors that had not as yet been affected by the fire.

Two days later the fire was finally extinguished and telco engineers were able to re-enter the building to assess the damage. On the first floor, the 240-ft-long main distribution frame was reduced to a melted puddle of iron. Water had ruined the power equipment in the basement. Four panel-switching offices on the lower floors were completely destroyed. Cable distribution ducts between floors were deformed by the heat and deemed useless. Carrier equipment on the second floor was destroyed. Switching hardware on the fourth, fifth and sixth floors were smoke- and water-damaged and would require massive restoration and cleaning before they could be used. *And 170,000 telephones were out of service.*

FIGURE 4.1 Smoke and steam, billowing from the 2nd Avenue office.

Within an hour of discovering the fire, the Bell System mobilized its forces in a massive effort to restore service to the affected area. New York Telephone, AT&T Long Lines (the company's long-distance arm), Western Electric (which became Lucent and is now part of Alcatel-Lucent), and Bell Laboratories converged on the building and commandeered a vacant storefront across the street to serve as their base of operations. Lee Oberst, New York City Area Vice President, coordinated the nascent effort that would cost more than $90 million.

The Bell System—and it really was a system—immediately kicked off a host of parallel efforts to restore the central office. Calling upon the organization's widespread resources, Chairman of the Board John deButts put out an urgent demand for personnel, and within hours 4,000 employees from across the globe descended upon New York City and began to work 12-hour shifts, with 2,000 employees in the building on each shift. Meanwhile, central office engineers reviewed the original drawings of the building's complex infrastructure to determine what they would need in the way of new equipment to restore service. All other Bell Operating Companies (BOCs) were placed on indefinite equipment holds pending the restoration of the New York office. Nassau Recycle Corporation, a Western Electric subsidiary, began the removal and recycling process of the 6,000 tons of debris that came out of the building, much of it toxic. Mobile telephone subsidiaries from across the country sent mobile radio units to the city to provide emergency telephony services (cellular telephony was still a dream at this point). The units were installed all over the affected area and announcements were posted throughout the city to inform residents where they were located.

Simultaneously, Bell Laboratories scientists studied the impact of the fire and crafted the best technique for cleaning the equipment that could be salvaged. One thousand three hundred fifty quarts (qt) of specially formulated cleaning fluid were mixed and shipped to the building along with thousands of toothbrushes and hundreds of thousands of Q-Tips. And a high-capacity transmission facility between New York and New Jersey, still in the planning stages, was accelerated, installed, and put into service to carry the traffic that would normally have passed through the 2nd Avenue office.

Within 24 hours, miracles were underway. Service had been restored to all affected medical, police, and fire facilities. The day after the fire, a new main distribution frame had been located at Western Electric's Hawthorne manufacturing facility and was shipped by cargo plane to New York. Luckily the third floor of the building had been vacant and was therefore available as a staging area to assemble and install the 240-ft-long iron frame. Under normal circumstances, from the time a frame is ordered, shipped, installed, and tested in an office, six months elapse. *This frame was ordered, shipped, installed, wired, and tested in four days.*

It is almost impossible to understand the magnitude of the restoration effort, but the following numbers may help. 6,000 tons of debris was removed from the building and 3,000 tons was installed including 1.2 billion feet of underground wire, 8.6 million feet of frame wire, 525,000 linear feet of exchange cable, and 380 million conductor feet of switchboard cable. Five million underground splices were completed to connect it all together. And 30 trucking companies, 11 airlines, and 4,000 people were pressed into service.

Just after midnight on March 21, 22 days following the fire, service from the building was restored to 104,000 subscribers. Normally, a job of this magnitude would have taken more than a year to complete. But because of the Bell System's remarkable ability to marshal resources during times of crisis, the building was restored in less than a month. AT&T Chairman John DeButts:

> "In the last couple of weeks I have had the opportunity to observe at first hand the strength of the organization structure that the Justice Department's antitrust suit seeks to destroy. This service restoration has been called a dramatic accomplishment—rightly. But only in its urgency

does the teamwork demonstrated on this occasion differ from the teamwork that characterizes the Bell System's everyday job."

Of course, the antitrust suit, now known universally as divestiture, went forward as planned. In 1984, the Bell System became a memory, with just as many people hailing its death as mourning it. And while there is much to be said for the innovation, competitive behavior, and reduced prices for telecommunications services that came from the breakup of the Bell System, it is hard to read an account such as the one above without a small lump in the throat. In fact, most technology historians conclude that we could not build the Bell System network if we were tasked to do so today.

To understand the creation of a system that could accomplish something as remarkable as the fire recovery described above, let's take a few pages to describe the history of this marvelous industry. I think you'll find it rather interesting.

The History of Telephony

As early as 1677, inventor Robert Hooke, who among other things formulated the theory of planetary movement, demonstrated that sound vibrations could be transmitted over a piece of drawn iron wire, received at the other end, and interpreted correctly. His work proved to be something of a bellwether and in 1820, 143 years later, Charles Wheatstone transmitted interpretable sound vibrations through a variety of media including thin glass and metal rods. In 1831, Michael Faraday demonstrated that vibrations such as those made by the human voice could be converted to electrical pulses. Needless to say, this was fundamental to the eventual development of the telephone.

Bernard Finn is the former curator of the Smithsonian's National Museum of American History. His knowledge of the world of science and its vanguard during the heady times of telephony's arrival was unparalleled. During an interview, he observed to me that America was a bit of a backwater as far as science was concerned. "The inventive climate of that early part of the century was largely [focused on] mechanical inventions," he noted. "Even the early electrical stuff was basically mechanical and the machine shop was the center of inventive activity."

With the invention of the telegraph, which occurred simultaneously in the United States, England, and Germany in the 1830s, it became possible to communicate immediately across significant distances. By the second half of the nineteenth century, three individuals were working on practical applications of Faraday's observations. Alexander Graham Bell, a speech teacher for the hearing impaired, and a part-time inventor; Elijah Gray, the inventor of the telegraph, and an employee of Western Electric; and Thomas Edison, an inventor and former employee of Western Union. All three were working on the same problem: How to send multiple simultaneous messages across a single pair of wires. And as it turned out, all three invented devices that they called harmonic telegraphs; they were nothing more than frequency division multiplexers, but for the time were remarkably technologically advanced.

The original devices that they created performed the multiplexing task mechanically by vibrating metallic reeds at different frequencies. The frequencies would pass down the line and be received correctly at the other end. The only problem with this technique was channel separation, because if the frequency bands were relatively close to each other, they would eventually interfere, causing problems at the receiving end of the circuit. For telegraph, therefore, harmonic multiplexing was deemed impractical.

All three inventors—Gray, Edison, and Bell—were in a dead heat to create the first device that would transport voice across a long-distance network. None of them had really tested their inventions extensively in the field; they had been demonstrated in the laboratory only. Nevertheless, on February 14, 1876, Alexander Graham Bell filed a notice of invention on his own untested device. Simultaneously, Elijah Gray filed a notice of invention on *his* device. There is some argument as to who filed first; many believe that Gray was actually the first to get to the patent office and file. Whatever the case, three days later Bell became the first person to transmit voice electronically across a network. And in 1877, following numerous experiments and modifications to his original device, Bell founded the Bell Telephone Company.

Meanwhile, Thomas Edison continued his work, and in 1876 was hired by Western Union along with the device he patented the year before. The arrival of the telephone was seen as a major problem by Western Union but they had no intention of sitting idly by while Gray and Bell destroyed their business with their new-fangled invention. In 1877, Edison crafted a very good transmitter based on compressed carbon powder (still in widespread use today, by the way; see Fig. 4.2) that would eventually replace the far less capable transmitter invented by Bell. For the experimenters among my readers out there, it's a fun exercise to build a compressed carbon microphone. In a mortar and pestle (OK, a plastic bag and a hammer) grind two charcoal briquettes to a fine powder. Pour the powder into a jar lid until the powder is even with the lip of the lid. Stretch a piece of plastic wrap tightly over the lid and secure it with a rubber band. Next, drill two holes in the lid and insert in each an insulated wire that has been stripped back about a 1/2 in to expose the copper wire. Next connect the wires to a speaker and battery as shown in the illustration. When you speak into the *microphone*, the sound waves impinging upon the stretched plastic wrap cause the carbon powder to be compressed and expanded, which in turn changes the resistance between the two

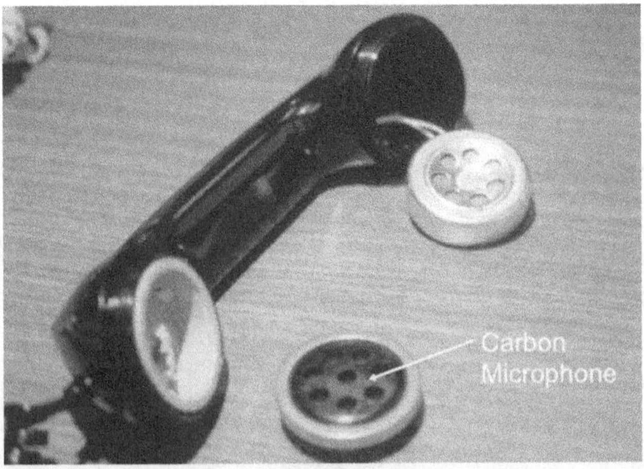

FIGURE 4.2 The current version of Bell's compressed carbon microphone, still used today.

conductors inserted into the powder. Your voice will magically emanate from the speaker.

Naturally, tensions ran high between the Bell Telephone Company and Western Union during this period. In 1877, Bell offered to sell his patent to Western Union for $100,000. Western Union turned down Bell's offer, deciding instead to buy Gray and Edison's patents and form the American Speaking Telephone Company.

Bell sued Western Union for patent infringement in 1878 and the case was settled in 1879 to everyone's satisfaction. Bell won the rights to Western Union's telephone patents, provided he pay royalties for the duration of the 17-year patent life of each device. Meanwhile, Bell Telephone Company agreed to stay out of the telegraph business and Western Union agreed to stay away from telephony.

Once the legal wrangling had ended, telephone service growth moved along rather quickly. By the spring of 1880, the United States had 138 exchanges and 30,000 subscribers. By 1887, 146,000 miles of wire had been strung to connect 150,000 subscribers to nearly 750 main offices and 44 branch offices.

There were no switches initially (they hadn't been invented yet) so telephones were sold in pairs, one for each end of the connection. Wires were strung in a haphazard fashion attached to the outside of buildings and strung across neighbors' rooftops; they threatened to fill the sky, as shown in Fig. 4.3.

The problem was the obvious lack of a central switch. The role of the switch in a network is to create and maintain temporary connections between two people who wish to speak with one another. The alternative was to create multiple one-to-one connections. Obviously,

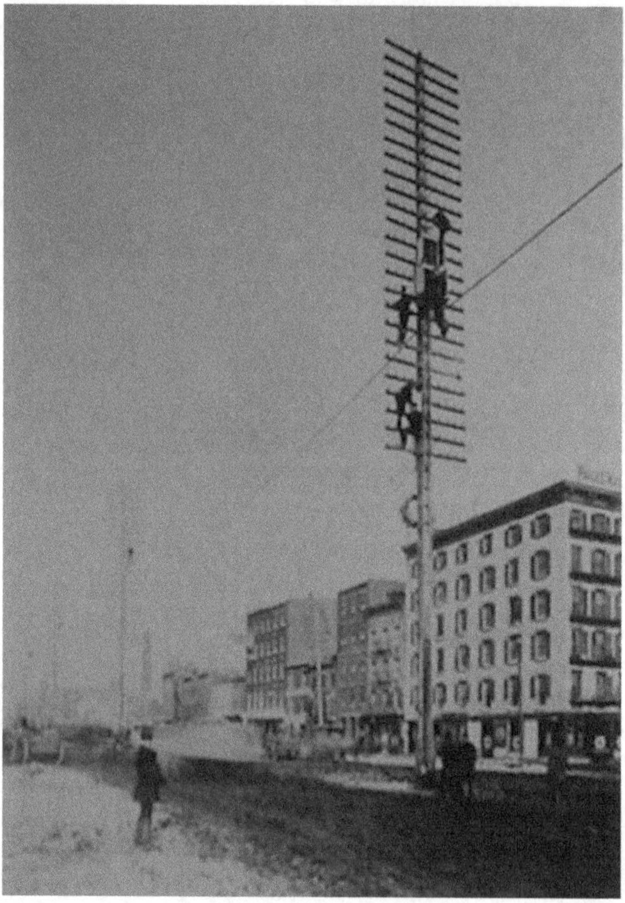

Figure 4.3 Access lines, filling the sky with copper.

this one-to-one relationship was limiting and over-the-top expensive. A technique was needed that would allow one phone to connect to many in an on-demand fashion.

The answer came in the form of the central office. Connections were installed from all the phones in a local area to the central exchange where operators could monitor the lines for service requests. The customer would then tell the operator whom they wanted to speak with, and the operator would set up the call using patch cords. When the parties had finished speaking, the operator would pull down the patch cord, freeing up the equipment for another call. This entire function was the first form of switching. The design soon became problematic, however, because an operator could typically monitor about 100 positions effectively.

The answer, of course, came with the arrival of the first switch, developed by Almon Strowger. Strowger was not an inventor, nor was he a telephone person. He was, in fact, an undertaker in a small town in Missouri. One day, he came to the realization that his business was (ok, I won't say dying) declining, and upon closer investigation determined that the town's operator was married to his competitor! As a result, any calls that came in for the undertaker naturally went to her husband—and *not* to Strowger.

To equalize the playing field, Strowger called upon his considerable talents as a tinkerer and designed a mechanical switch and dial telephone, shown in Fig. 4.4, which is still in use today in a number of developing countries.

The first semi-automatic switch was installed in Newark, New Jersey in 1914, and the first truly automatic switch went live in Omaha, Nebraska in 1921. For all this innovation, however, direct long-distance dialing did not become a reality until 1951.

Soon the network had evolved to a point where it could handle thousands of simultaneous calls. As the technology evolved, so too did the business. In 1892 Alexander Graham Bell inaugurated a 950-mile circuit between New York and Chicago, ushering in the ability to carry long-distance telephony. Meanwhile the company continued to evolve; after acquiring Western Electric it changed its name to American Telephone and Telegraph, and in an ongoing attempt to

FIGURE 4.4 The Strowger phone.

protect its fiefdom, began buying up small competitors as quickly as they came into being. In 1909, the company purchased a controlling interest in Western Union, thus garnering the attention of the Interstate Commerce Commission, the federal agency responsible (as of 1909) for the oversight of American wire and radio-based communications. That responsibility would later be passed on to the newly created Federal Communications Commission.

The Kingsbury Commitment

In 1913, in the first of a series of anti-trust decisions, the *Kingsbury Commitment* forced AT&T to divest its holdings in Western Union, stop buying its competition, and provide free interconnection with other network providers. AT&T did not realize it then, but the Kingsbury Commitment was the harbinger of drastic things to come—namely the divestiture of the company in 1984 that would break AT&T into pieces.

In the years that followed, a number of significant technological events took place in telephony. Load coils were created and installed on telephone lines, extending signal strength dramatically and reducing bandwidth requirements for each customer. In 1915, the first repeaters were installed on long-distance telephone circuits, making it possible to deploy circuits of essentially unlimited length. In fact, later that same year a circuit was successfully installed between New York and San Francisco.

The FCC Arrives

In 1934, President Roosevelt created the Federal Communications Commission with the signing into law of the Communications Act of 1934. This act moved industry oversight from the Interstate Commerce Commission to the FCC and recognized the importance of telephone service, mandating that it would be both affordable and universal.

By now, AT&T provided local and long-distance service and offered central office equipment and telephones to more than 90 percent of the market. Not surprisingly, that smacked of monopoly to the federal government, and from that point on AT&T was squarely in the FCC's gun sights.

The Hush-a-Phone Decision

In 1948, AT&T sued the Hush-a-Phone Corporation for manufacturing a device that physically connected to the telephone network. The device, shown in Fig. 4.5, was nothing more than a butterfly-shaped metal box that fit over the mouthpiece of the telephone, blocking out extraneous noise and providing a degree of privacy to the person using the phone. It had no electrical component to it; it was merely a box. AT&T sued on the grounds that anything connected to the AT&T network could mean the end civilization as we know it. When it

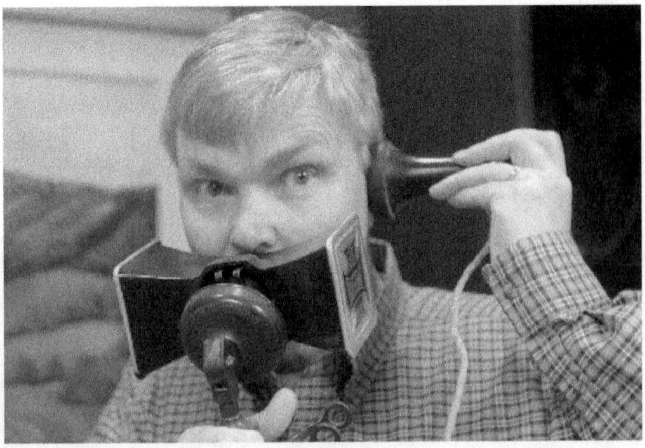

FIGURE 4.5 The Hush-a-Phone.

became clear to the judge that the device was not electrical, Hush-a-Phone won the case, and the District of Columbia Court of Appeals ruled that AT&T could not prohibit people from making money from devices that did not electrically connect to AT&T's network.

The Antitrust Filing

In 1949, the justice department filed an anti-trust suit against AT&T and Western Electric. The result was the consent decree of 1956, which allowed AT&T to keep Western Electric but restricted them to the delivery of common carrier services.

The Carterphone Decision

In the early 1960s, inventor Tom Carter created a device called the Carterphone, shown in Fig. 4.6. The Carterphone allowed mobile car radios to connect to the telephone network, and *did* require electrical connectivity to AT&T. Initially AT&T prohibited Carter from connecting his device under any circumstances, but when he appealed to the FCC, he was given permission, opening the customer-provided equipment (CPE) market for the first time. AT&T's concerns about possible damage to the network were reasonable, but by the time AT&T vs. Carter came to trial the device had been in use for several years and clearly hadn't done any damage whatsoever.

The MCI Decision

In 1969, the very next year, the FCC gave Microwave Communications Incorporated (MCI) permission to offer private line services between St. Louis and Chicago over its privately owned microwave system. And in 1971, the court extended its 1956 Consent Decree

FIGURE 4.6 The Carterfone.

mandate, ordering AT&T to allow non-Bell companies such as MCI to connect directly to their network, ending AT&T's stranglehold on the private line market. The decision was based on a far-reaching FCC policy designed to create nationwide, full-blown competition in the private line marketplace.

The Execunet Decision

It didn't end there, however. In 1972, the FCC mandated that satellites could be used to transport voice and compete with AT&T, and that value-added carriers could resell AT&T services. In 1977, in the now famous Execunet decision, MCI won a legal battle that allowed them to compete directly with AT&T in the switched long-distance services market, AT&T's bastion of revenue. In 1974, they extended their attack, filing an antitrust suit against AT&T and charging them with unfairly restricting competition and dominating the marketplace. That same year the justice department filed an antitrust suit of their own, signaling the beginning of the end for the Bell System.

Computer Enquiry II

In 1980, another crack appeared in AT&T's armor when the FCC consciously recognized the difference between basic and enhanced services. *Computer Enquiry II* stipulated that basic services would be regulated and therefore tightly controlled. Enhanced services, including CPE, would be deregulated and made completely competitive. AT&T was told that while it could sell enhanced services but only through fully separate subsidiaries to ensure no mixing of revenues between the two sides of the business.

Enter Harold Greene and the Modified Final Judgment

In 1981, the United States versus AT&T came to trial in Judge Harold Greene's court. In the months that followed, it became painfully clear to AT&T that it would not win this game. On January 8, 1982 the two sides announced that they had reached a mutually acceptable agreement. The agreement, known as the Modified Final Judgment, stipulated that AT&T would divest itself of its 22 local telephone companies, the Bell Operating Companies (BOCs). The company would retain ownership of Western Electric, Bell Laboratories and AT&T Long Lines and would be allowed to enter non-carrier lines of business such as computers and the like. Meanwhile, the BOCs were gathered into seven regional BOCs called RBOCs, tasked with providing local service to the regions they served. The RBOCs were further subdivided into local access and transport areas (LATAs), which defined the areas within which they were allowed to provide transport.

The Equal Access Decision

While the best-known impact of divestiture was the breakup of AT&T—one result of which was the liberalization of the telecommunications marketplace in the United States—a second decision that was tightly intertwined with the Bell System's breakup was largely invisible to the public, yet was *at least* as important to AT&T competitors MCI and Sprint as the breakup itself. This decision, known as *equal access*, had one seminal goal: to make it possible for customers to take advantage of one of the products of divestiture, the ability to select one's long-distance provider from a pool of available service providers, in this case AT&T, MCI, or Sprint. This, of course, was the realization of a truly competitive marketplace in the long-distance market segment.

Since then, things have changed. Long-distance as a separate sub-industry no longer exists, and certainly is no longer the cash cow it once was for the service provider industry. Voice remains important, but the ability to generate large-scale revenue from it is elusive as voice over Internet protocol (VoIP) and various free, Internet-based services have arrived. The sale of private line service, while still important, is becoming increasingly elusive as a major revenue component because of the rise of IP-based VPNs over the Internet, the increasing popularity of gigabit ethernet, and the growing use of broadband wireless as a solution.

To understand this evolution, it is helpful to have a high-level understanding of the overall architecture of the network. In the pre-divestiture (prior to 1984) world, AT&T was *the* provider for local and long-distance service and communications equipment. An AT&T central office, therefore, was awash in AT&T hardware—switches, cross-connect devices, multiplexers, amplifiers, repeaters, and myriad other devices.

Figure 4.7 shows a typical network layout in the pre-divestiture world. A customer's telephone is connected to the service provider's

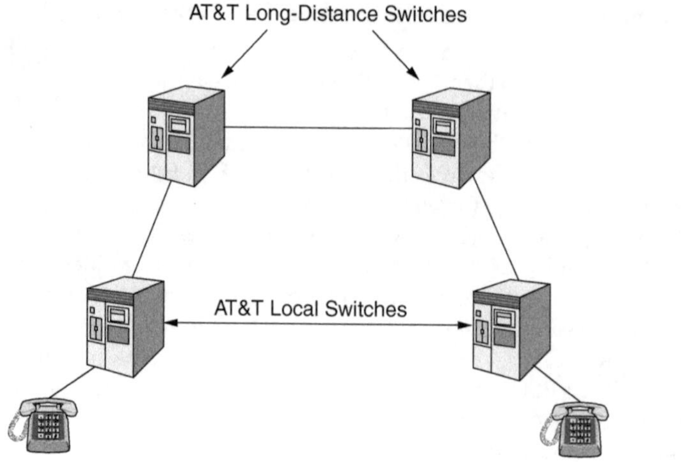

Figure 4.7 The layout of the pre-divestiture Bell System network.

network by a local loop connection (so-called twisted pair wire). The local loop, in turn, ultimately connects to the local switch in the central office. This switch is the point at which customers first touch the telephone network and it has the responsibility to perform the initial call setup, maintain the call while it is in progress, and tear it down when the call is complete. It also has the responsibility to create a call record that is converted into a billing record, as required. This switch is called a local switch because its primary responsibility is to set up local calls that originate and terminate within the same switch. It has one other responsibility, though, and that is to provide the necessary interface between the local switch and the long-distance switch, so that calls between adjacent local switches (or between far-flung local switches) can be established. The process, then, goes something like this. When a customer lifts the handset and goes off-hook, a switch in the telephone closes, completing a circuit that allows current flow that in turn brings dial tone to the customer's ear. Upon hearing the dial tone the customer enters the destination address of the call (otherwise known as a telephone number). The switch receives the telephone number and analyzes it, determining from the area code and prefix information whether the call can be completed within the local switch or must leave the local switch for another one. If the call is indeed local, it merely burrows through the crust of the switch and then reemerges at the receiving local loop. If the call is a toll or long-distance call, it must burrow through the hard crunchy coating of the switch, pass through the soft chewy center, and emerge again on the other crunchy side on its way to a long-distance switch. Keep in mind that the local switch has no awareness of the existence of customers or telephony capability beyond its own cabinets. Thus, when it receives a telephone number that it is incapable of processing, it hands it off to a higher-order switch with the implied message, "Here, I have no idea what to do with this, but I assume that you do."

The long-distance switch receives the number from the local switch, processes the call, establishes the necessary connection, and passes the call on to the remote long-distance switch over a long-distance facility. The remote long-distance switch passes the call to the remote local switch, which rings the destination telephone, and ultimately the call is established.

Please note that in this pre-divestiture example, the originating local loop, local switch, long-distance switch, remote local loop, and all of the interconnect hardware and wiring belonged to AT&T. They were mostly manufactured by Western Electric based on a set of internal manufacturing standards that, were there other manufacturers in the industry, would be considered proprietary. Because AT&T was the only game in town (in the United States at least) prior to divestiture, AT&T created the standard for transmission interfaces.

Fast forward now to January 1, 1984, and put yourself into the mind of Bill McGowan, the Chairman of MCI, whose company's very survival depended upon the successful implementation of equal access. Unfortunately, equal access had one very serious flaw. Keep in mind that because the post-1984 network was emerging from the darkness of monopoly control, all of the equipment that comprised the network infrastructure was bought at the proverbial company store—and was, by the way, proprietary.

Consider the newly recreated post-divestiture network model shown in Fig. 4.8. At the local switch level, little has changed—at this point in time there is still only a single local service provider. At the

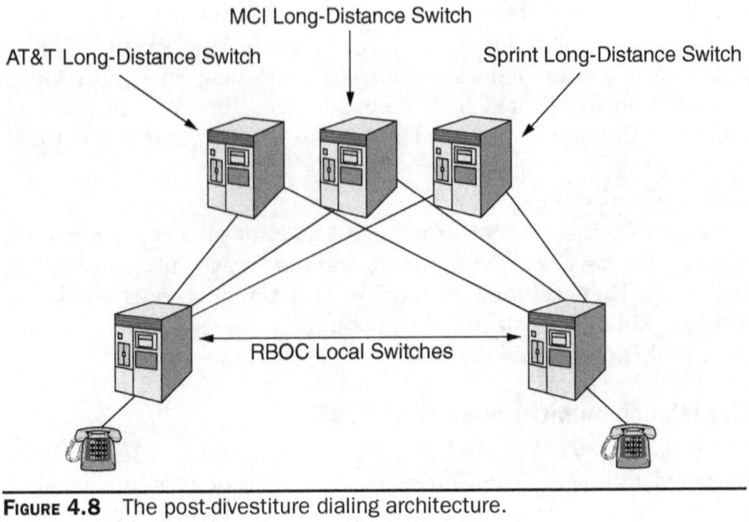

FIGURE 4.8 The post-divestiture dialing architecture.

long-distance level, however, there is a significant change. Instead of a single long-distance service provider called AT&T, there are now three—AT&T, MCI, and Sprint. The competitive mandate of equal access was designed to guarantee that a customer could freely select their long-distance provider of choice. If they wanted to use MCI's service instead of AT&T's, a simple call to the local telephone company's service representative would result in the generation of a service order that would cause the customer's local service to be logically disconnected from AT&T and reconnected to MCI so that long-distance calls placed by the subscriber would automatically be handed off to MCI. The problem of "equal access" to customers for the three long-distance providers was solved.

Since 1984, the telecommunications marketplace has continued to evolve, sometimes in strange and unpredictable ways. Harold Greene oversaw the remarkable transformation of the industry until his death in January of 2000. Over time, the seven RBOCs slowly accreted to a smaller number as SBC, Pacific Bell, and Ameritech joined forces, sucking SNET into their midst in the process; NYNEX and Bell Atlantic danced around each other until they became Verizon, pulling GTE into the fray; and Qwest acquired US West, which in turn was acquired by Century Link in 2011. Bellsouth acquired Cingular Wireless, and was then acquired by AT&T in 2006. Today the major players are Verizon, AT&T, Sprint (which was acquired by Japan's SoftBank in mid-2013), and CenturyLink.

Not to Be Forgotten: Cable

Let's not forget that the cable television providers are also players in this space. Led by aggressive players like TCI in the mid-1990s, cable went through a wrenching reinvention that included full digitization of their networks and the installation of large, redundant optical rings that gave them the ability to not only offer more diverse content but to also enter the voice and data realms as well. Today the largest cable providers in the United States are Comcast, Time-Warner, Verizon FiOS, Cox Communications, AT&T U-verse, Charter, and Cablevision.

...And the Others

We also must mention the growing presence of smaller independent players that have entered the voice and data marketplace in the last few years. They include a plethora of VoIP providers such as Skype, Vonage, Mitel, MagicJack, and Phone.com. There are literally *hundreds* of them. We'll discuss VoIP service later in the book.

The Telecommunications Act of 1996

In 1996, the FCC released the Telecommunications Act of 1996, designed to revamp the Communications Act of 1934 and make it friendlier to the services carried by network providers today.

When the Telecom Act of 1996 began to take root, a fundamental shift in the telecom industry began to take place. Initially, competing with the seven Regional Bell Operating Companies in major metropolitan markets were companies like Teleport Communications Group (TCG) and Metropolitan Fiber Systems (MFS), the first pejoratively named *bypassers*, soon relabeled *alternative* or *competitive access providers*. These companies built optical ring networks in the central business districts of large cities and competed effectively with the incumbent providers by offering redundant dual entry networks and an all-optical transport infrastructure. They flourished because they had the luxury to cherry-pick their markets. They did not labor to develop a presence in Dime Box, Texas, or Scratch Ankle, Alabama; they did, however, establish beachheads in Dallas and New York City, Minneapolis-St. Paul and San Francisco, Seattle and Miami—because, as bank robber Willie Sutton was wont to observe, "That's where the money is."

Over time, these local loop competitors became known as competitive local exchange carriers (CLECs), while the RBOCs became known as incumbent local exchange carriers (ILECs). The line, however, blurred somewhat: some ILECs announced their intentions to enter the territories of other ILECs, effectively becoming de facto CLECs in their own right. Over time, the ranks of the CLECs swelled to several hundred in the United States alone. Meanwhile, other segments emerged from the service shadows. Beginning with the arrival of the Internet and Web in 1993, internet service providers (ISPs) grew like mushrooms on a summer lawn. Independent wireless providers like Winstar and Teligent sprouted in cities. A new market segment, the data local exchange carriers (DLECs), emerged with names like Northpoint, Rhythms NetConnections, and Covad. Taking advantage of a regulatory decision intended to further increase competition at the local loop level, they offered DSL service over existing local loops. In other words, a customer could purchase his or her phone service from the local ILEC, but had the option to buy broadband access from someone else—broadband access service that was carried over the ILEC's local loop. Pure bandwidth providers (sometimes called bandwidth barons) like Qwest, Global Crossing, and 360Networks, riding on massive installed bases of long-haul optical fiber, offered ridiculously low prices on transport. Cable MSOs, having largely completed the digital and optical conversion of their local distribution networks, began to offer two-way interactive services including Internet access.

The act (which soon came to be known as TA96) was also designed to address the requests by ILECs to become long-distance providers through a 14-point checklist that they were required to complete before being considered for entry into the long-distance market. None of them ever complied totally with the demands.

Let's turn our attention now to the network itself.

The Telephone Network

"Sure, I know how it works. You pick up the phone, dial the numbers, and wait. A little man grabs the words, runs down the line, and delivers them to whomever I'm calling. It seems just about that simple. I mean, come on—how complicated can it be? It's just a telephone call."

Thus was described to me the process of placing a telephone call by a fellow on the street whom I once interviewed for a video I was creating about telephony. His perception of the telephone network, how it works, and what it requires to work is similar to most peoples'. Yet the telephone network is without question the single greatest and most complex agglomeration of technology on the planet. It extends its web seamlessly to every country on earth, providing instantaneous communication for not only voice, but for video, data, television, medical images, sports scores, music, high-quality graphics, secure military intelligence, banking information, and teleconferences. Yet it does so with almost complete transparency and with virtually 100 percent availability. In fact, the only time its presence is noticed is when it isn't there—as happened on 2nd Avenue in New York, in Hinsdale, Illinois following a major central office fire, and in Chicago following a flood that isolated the Mercantile Exchange and placed hundreds of customers out of service.

How the network works is something of a mystery to most people, so we're going to dissect the typical telephone network and examine its anatomy, complete with pictures. This section is not for the squeamish.

The Modern Telephone System

The best way to gain an understanding of how the telephone network operates is by studying a modern railroad system. Consider Fig. 4.9, which is a route map for the mythical Midland, Amarillo, and Roswell Railroad. Rail yards in each of the three cities are interconnected by high-volume trunk lines. The train yards, also known as switchyards, are used as aggregation, storage, and switching facilities for the cargo that arrives on trains entering each yard. A 90-car train from El Paso, for example, may arrive in Midland as planned. Half the cars are destined for Midland, while the other half are destined for Amarillo. At the Midland yard, the train is stored, disassembled, reassembled for transport, and switched as required to move the Amarillo traffic on to Amarillo. Switches in the yards create temporary paths from one side of the yard to the other. Meanwhile, route bosses in the yard towers analyze traffic patterns and route trains across a variety of alternative paths to ensure the most efficient delivery. For example, traffic from Amarillo to Roswell could be routed through Midland, but it would obviously be more efficient to send it on the direct line that runs

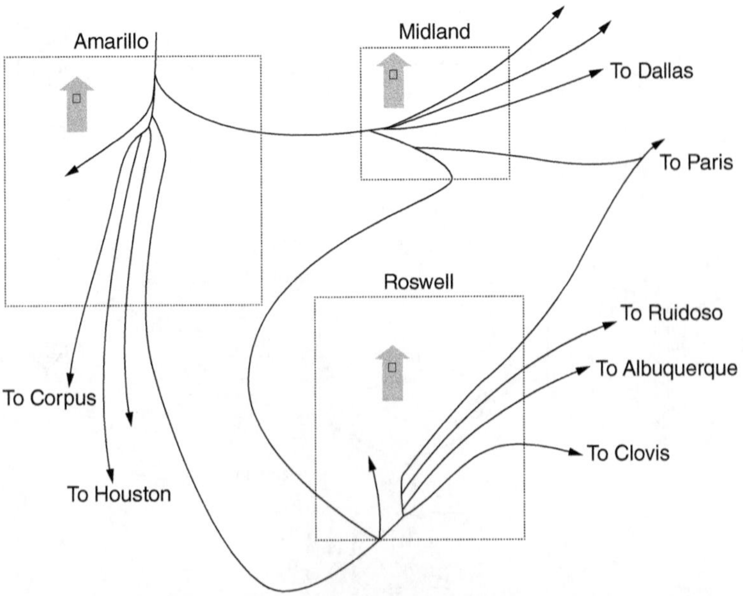

Figure 4.9 The mythical Midland, Amarillo, and Roswell railroad.

between Amarillo and Roswell. Assuming that the direct route is available and is not congested, the route boss might very well choose that alternative.

Notice also that there are short local lines, called feeder lines, which pump local traffic into the switchyards. These lines typically run shorter trains destined most likely for local delivery. Some of them, however, carry cargo destined for distant cities. In those cases the cargo would be combined with that of other trains to create a large, efficient train with all cargo bound for the same destination.

Trunks, lines, feeders, local access spurs, switches, routers—all terms used commonly in the telecommunications industry. Trunks are high-bandwidth, long-distance transport facilities. Lines, feeders, and local access spurs are local loops. Switches and routers are—well, switches and routers. And the overall function of the two is exactly the same, as is their goal of delivering their transported payload in the most efficient fashion possible.

When Bell's contraption first arrived on the scene it wasn't a lot more complicated than a railroad, frankly. As we noted earlier, there was no initial concept of switching. Instead, the plan was to gradually evolve the network to a full mesh architecture as customers demanded more and more connections. Eventually, operators were added to provide a manual switching function of sorts, and over time they were replaced with mechanical, then electromechanical, then all-electronic switches. Ultimately intelligence was overlaid in the form

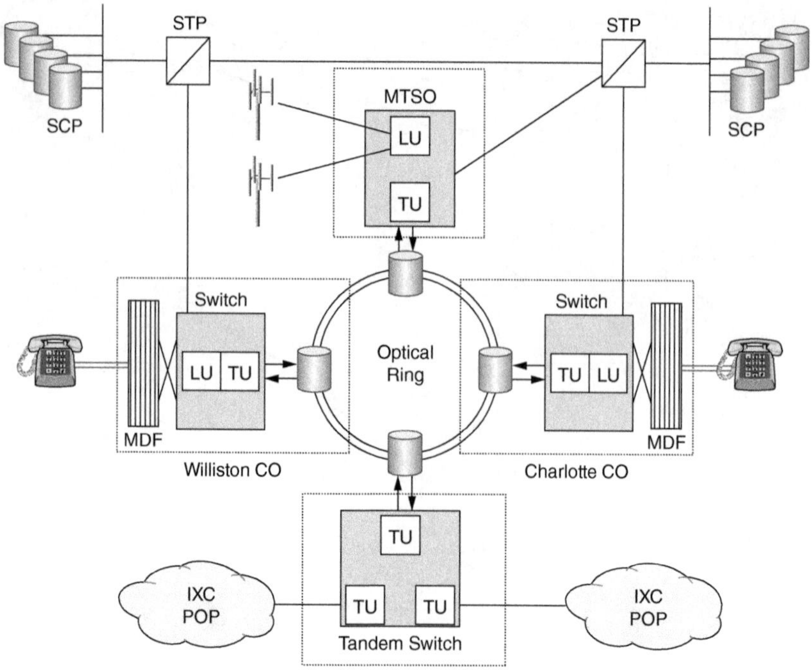

Figure 4.10 Dick Pecor's famous drawing of the telephone network.

of signaling, giving the network the ability to make increasingly informed decisions about traffic handling as well as to provide value-added services.

Now that you understand the overall development strategy that Bell and his cohorts had in mind, as well as the basics of switching, we turn our attention to the inner workings of the typical network, shown in Fig. 4.10. I must take a moment here to thank my good friend Dick Pecor, who created this drawing— from his head, I might add. Thanks, Dick.

Scalpel, please.

In the Belly of the Beast

When a customer makes a telephone call, a complex sequence of events takes place that ultimately leads to the creation of a temporary end-to-end, service-rich path. Let's follow a typical phone call through the network. Cristina in Huntington is going to call Jess in Michigan. This is a high-level explanation; we'll add more detail later.

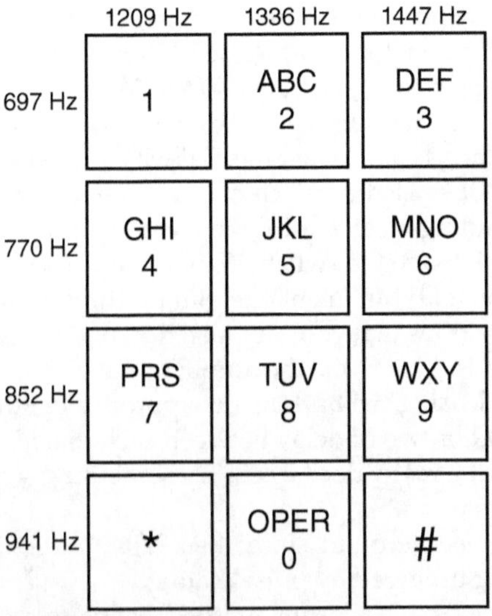

FIGURE 4.11 DTMF tone pairs.

When Cristina picks up the telephone in her home, the act of lifting the handset[1] closes a circuit that allows current to flow from the switch in the local central office that serves her neighborhood to her telephone. The switch electronically attaches an oscillator to the circuit called a *dial tone generator*, which creates the customary sound that we all listen for when we wish to place a call.[2] The dial tone serves to notify the caller that the switch is ready to receive the dialed digits.

Cristina now dials Jess's telephone number by pressing the appropriate buttons on the phone. Each button generates a *pair of tones* (listen carefully—you can hear them) that are slightly dissonant. This is done to prevent the possibility of a human voice randomly generating a frequency that could cause a misdial. The tone pairs, shown in Fig. 4.11, are carefully selected so that they cannot be naturally generated by the human voice. This technique is called dual tone, multi-frequency (DTMF). It is not, however, the only way.

---

[1]It doesn't matter whether the phone is corded or cordless. If cordless, pushing the TALK button creates a radio link between the handset and the base station, which in turn closes the circuit.

[2]It's interesting to note that all-digital networks, such as those that use ISDN as the access technology, don't generate dial tone because packets don't make any noise! Instead, the ISDN signaling system (Q.931 for those of you collecting standards) transmits a "ring the phone" packet to the distant telephone. The phone itself generates the dial tone—and the ringing bell!

You may have noticed that there is a switch on many phones with two positions labeled Tone and Pulse. When the switch is set to the Tone position, the phone generates DTMF. When it is set on Pulse, it generates the series of clicks that old dial telephones made when they were used to place a call.[3] When the dial was rotated—let's say to the number three—it caused a switch contact to open and close rapidly three times, sending a series of three electrical pulses (or more correctly, interruptions) to the switch. Want to impress your friends? Try pulse dialing on a DTMF telephone. Simply find a phone that has a real button or buttons that you can push down to hang up the phone. Pick up the handset, then dial the number you wish to call by rapidly pushing and releasing the buttons the appropriate number of times, leaving a second or two of delay between each dialed digit. It takes a little practice but it really does work. Try something simple like Information (411) first.

DTMF has been around since the 1970s, but switches are still capable of being triggered by pulse dialing.

Back to our example. Cristina finishes dialing Jess's number and a digits collector in the switch receives the digits. The switch then performs a rudimentary analysis of the number to determine whether it is served out of the same switch. It does this by looking at the area code (numbering plan area, or NPA) and Prefix (NXX) of the dialed number.

A telephone number comprises three sections, shown below.

802-555-7837

The first three digits are the NPA, which identifies the called region. For example, I live in Vermont where the entire state is served by a single NPA—802. Other states, such as California, have a dozen or more NPAs to serve its much denser population base.

The second three digits are the NXX, which identifies the exchange, or office, that the number is served from, and by extension the switch to which the number is connected. Each NXX can serve 10,000 numbers (555-0000 through 555-9999). Remember the New York fire? Twelve exchanges were lost, under which 104,000 subscribers were operating (out of a possible 120,000). A modern central office switch can typically handle as many as 15 to 20 exchanges of 10,000 lines each.

In our example, Jess is served out of a different switch than Cristina so the call must be routed to a different central office. Before that routing can take place, however, Cristina's local switch routes a query to the signaling network, known as signaling system 7 (SS7).

---

[3]About 15 years ago I took a trip to Seattle to teach a class, and my family accompanied me. One evening while strolling along Pike Street we came upon a bank of modern pay phones, all of which had dials instead of DTMF buttons. My kids, still little back then, begged me for quarters so that they could call their friends on those cool new phones.

SS7 provides the network with intelligence. It is responsible for setting up, maintaining, and tearing down a call, while at the same time providing access to enhanced services such as caller ID, 800 number portability, local number portability (LNP), line information database (LIDB) lookups for credit card verification, and other enhanced features. In a sense it makes the local switch's job easier by centralizing many of the functions that formerly had to be performed locally.

The original concept behind SS7 was to separate the actual calls on the public telephone network from the process of setting up and tearing down those calls as a way to make the network more efficient. This had the effect of moving the intelligence out of the PSTN and into a separate network where it could be somewhat centralized and therefore made available to a much broader population. The SS7 network, shown in Fig. 4.12, consists of packet switches (signal transfer points, or STPs) and intelligent database engines (service control points, or SCPs), interconnected to each other and to the actual telephone company switches (service switching points, or SSPs) via digital links, typically operating at anywhere from 56 Kbps to 2.048 Mbps (E1).

When a customer places a call, the local switching infrastructure issues a software interrupt via the SSP so that the called and calling party information can be handed off to the SS7 network, specifically an STP. The STP in turn routes the information to an associated SCP, which performs a database lookup to determine whether any special call-handling instructions apply. For example, if the calling party has chosen to block the delivery of caller ID information, the SCP query will return that fact.

Once the SCP has performed its task, the call information is returned to the STP packet switch, which consults routing tables and then selects a route for the call. Upon receipt of the call, the destination switch will cooperate with SS7 to determine whether the called party's phone is available, and if it is, it will ring the phone. If the

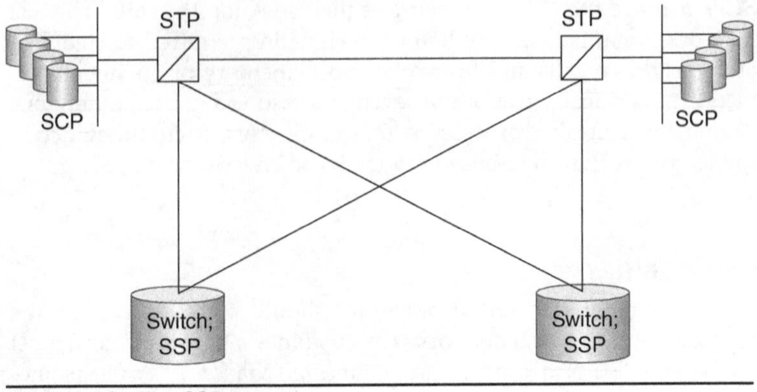

FIGURE 4.12 The SS7 network.

customer's number is not available due to a busy condition or some other event, a packet will be returned to the source indicating the fact and SS7 will instruct the originating SSP to put busy tone or reorder in the caller's ear.

At this point, the calling party has several options, one of which is to invoke one of the many Class services such as *automatic ringback*. With automatic ringback, the network will monitor the called number for a period of time, waiting for the line to become available. As soon as it is, the call will be cut through, and the calling party will be notified of the incoming call via some kind of distinctive ringing.

Thus, when a call is placed to a distant switch, the calling information is passed to SS7, which uses the caller's number, the called number, and SCP database information to choose a route for the call. It then determines whether there are any special call-handling requirements to be invoked such as Class services, and instructs the various switches along the way to process the call as appropriate.

These features comprise a set of services known as the *advanced intelligent network (AIN)* a term coined by Telcordia. Telcordia began its life as Bellcore, originally created as a common services organization to provide support for the seven regional BOCs that resulted from the divestiture of AT&T. Telcordia was sold to Science Applications International Corporation (SAIC) in 1996. In June 2011, Ericsson announced an agreement to acquire Telcordia, a transaction that completed in January 2012. Because Telcordia dealt with a considerable amount of proprietary and often sensitive information, the acquisition had to be approved by the Committee on Foreign Investment in the United States (CFIUS). The result was the creation of Applied Communication Sciences. In February 2013, Ericsson announced that it was renaming its interconnection business to iconectiv, which offers a wide array of services related to both wireline and wireless technologies including number portability, device theft prevention, numbering and addressing, mobile messaging, and spectrum management.

The SSPs (switches) are responsible for basic calling, while the SCPs manage the enhanced services that ride atop the calls. The SS7 network, then, is responsible for the signaling required to establish and tear down calls and to invoke supplementary or enhanced services. It is critically important, even more so today as the network begins the complex process of migrating from a circuit-switched model to a VoIP packet-based model. More on that in Chap. 10.

Network Topology

Before we enter the central office, we should pause to explain the topology of the typical network. The customer's telephone, and most likely, their PC and modem, are connected via *house wiring* (sometimes called inside wire) to a device on the outside of the house or

Figure 4.13 Protector block.

office building called a *protector block* (Fig. 4.13). It is nothing more than a high-voltage protection device (a fuse, if you will) designed to open the circuit in the event of a lightning strike or the presence of some other high-voltage source. It protects both the subscriber and the switching equipment.

The protector connects to the network via *twisted pair wire.* Twisted pair is literally that—a pair of wires that have been twisted around each other to reduce the amount of crosstalk and interference that occur between wire pairs packaged within the same cable. The number of twists per foot is carefully chosen.

The twisted pair(s) that provides service to the customer arrive at the house on what is called the drop wire. It is either aerial, as shown in Fig. 4.14, or buried underground. We note that there may be multiple pairs in each drop wire because today the average household typically orders a second line for a home office, computer, fax machine, or teenager.

Once it reaches the edge of the subscriber's property the drop wire typically terminates on a terminal box, such as that shown in Fig. 4.15. There, all of the pairs from the neighborhood are cross-connected to the main cable that runs up the center of the street. This architecture is used primarily to simplify network management. When a new neighborhood is built today, network planning engineers estimate the number of pairs of wire that will be required by the homes in that neighborhood. They then build the houses; install the

Figure 4.14 Aerial drop wire.

Figure 4.15 Terminal box, sometimes called a B-box.

network and cross-connect boxes along the street; and cross connect each house's drop wire to a single wire pair. Every pair in the cable has an appearance at every terminal box along the street, as shown in Fig. 4.16. This allows a cable pair to be reassigned to another house

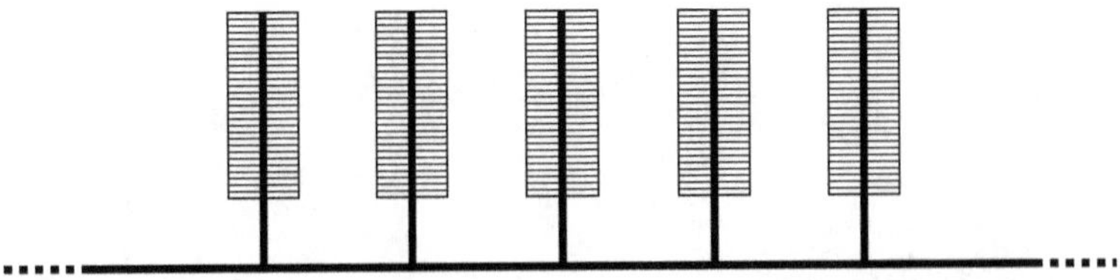

Figure 4.16 Cable pair appearances.

elsewhere on the street, should the customer move. It also allows cable pairs to be easily replaced should a pair go bad. This design dramatically simplifies the installation and maintenance process, particularly given the high demand for additional cable pairs today. This design also results in a challenge for network designers. These multiple appearances of the same cable pair in each junction box are called *bridged taps.* They create problems for digital services because electrical signal echoes can occur at the point where the wire terminates at a pair of *unterminated* terminal lugs. Suppose, for example, that cable pair number 117 is assigned to the house at #2 Blanket Flower Circle. It is no longer necessary, therefore, for that cable pair to have an appearance at other terminal boxes because it is now assigned. Once the pair has been cross-connected to the customer's local loop drop wire, the technician *should* remove the appearances at other locations along the street by terminating the open wire appearances at each box. This eliminates the possibility of a signal echo occurring and creating errors on the line, particularly if the line is digital.

Subscriber Loop Carrier

When outside plant engineers first started designing their networks, they set them up so that each customer was given a cable pair from their house all the way to the central office. The problem with this model was cost. It was very expensive to provision a cable pair for each customer. With the arrival of time division multiplexing, however, the "one dog, one bone" solution was no longer the only option. Instead, engineers were able to design a system under which customers could share access to the network as shown in Fig. 4.17. This technique, known as a *subscriber loop carrier,* uses a collection of T-1 carriers to combine the traffic from a cluster of subscribers and thus reduce the amount of wire required to interconnect them to the central office. The best-known carrier system is called the SLC-96 (pronounced *slick),* originally manufactured by Western Electric/Lucent, which transports traffic from 96 subscribers over four 4-wire T-Carriers (plus a spare). A remote SLC terminal is shown in Fig. 4.18. Thus 96 subscribers require only 20 wires between their neighborhood and the central office instead of the 192 wires that would otherwise be

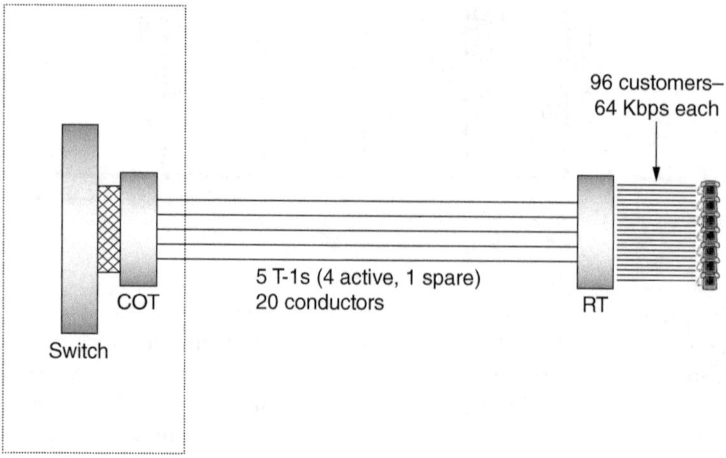

Figure 4.17 Subscriber loop carrier.

Figure 4.18 A remote SLC terminal.

required. The only problem with this model is that customers are by and large restricted to the 64 Kbps of bandwidth that loop carrier systems assign to each subscriber. That means that subscribers wishing to buy *more* than 64 Kbps—such as those that want DSL—are out of luck. And since it is estimated that as many as 70% of all subscribers in the United States are served from loop carriers, this poses a problem that service providers are scrambling to overcome. New versions of loop carrier standards and technologies such as GR-303 and optical remotes that use fiber instead of copper for the trunk line between the remote terminal and the central office terminal go a long way toward solving this problem by making bandwidth allocation far more flexible.

Typically, as long as a customer is within 12,000 ft from a central office they will be given a dedicated connection to the network, as shown in Fig. 4.19. If they are farther than 12,000 ft from the CO, however, they will normally be connected to a subscriber loop carrier system of one type or another.

Either way, the customer's local loop, whether as a standalone pair of wires or as a timeslot on a carrier system, makes its way over the physical infrastructure on its way to the network. It may be aerial, as shown in Fig. 4.20, or underground, as shown in Fig. 4.21. If it has to travel a significant distance, it may encounter a load coil along the way, which is a device that *tunes* the local loop to the range of frequencies required for voice transport and extends the distance over which the signals can travel. A *load pot*, shown in Fig. 4.22, comprises

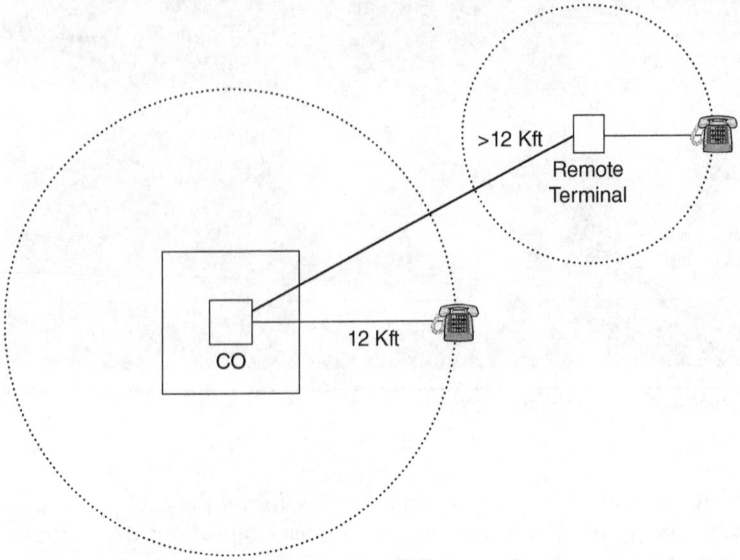

FIGURE 4.19 Dedicated network connection.

FIGURE 4.20 Aerial plant.

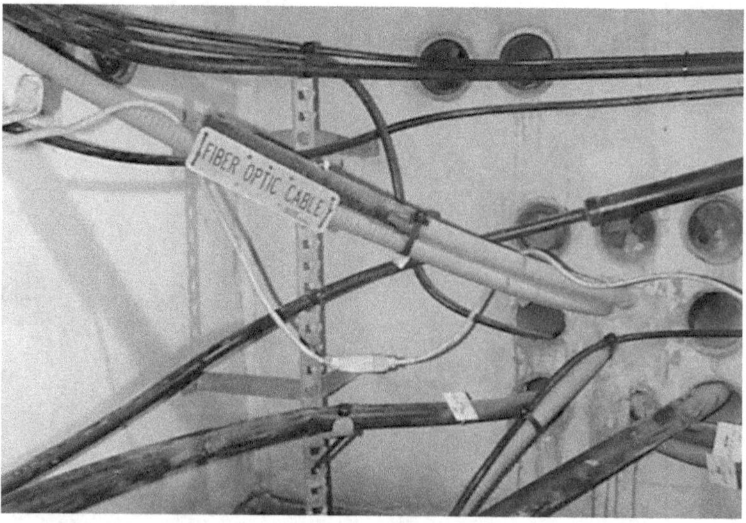

FIGURE 4.21 Underground plant.

multiple load coils and performs loading for all the cable pairs in a cable that require it. It may also encounter a repeater if it is a digital loop carrier; the repeater receives the incoming signal, now weakened and noisy from the distance it has traveled, and reconstructs it, amplifying it before sending it on its way.

FIGURE 4.22 A load pot.

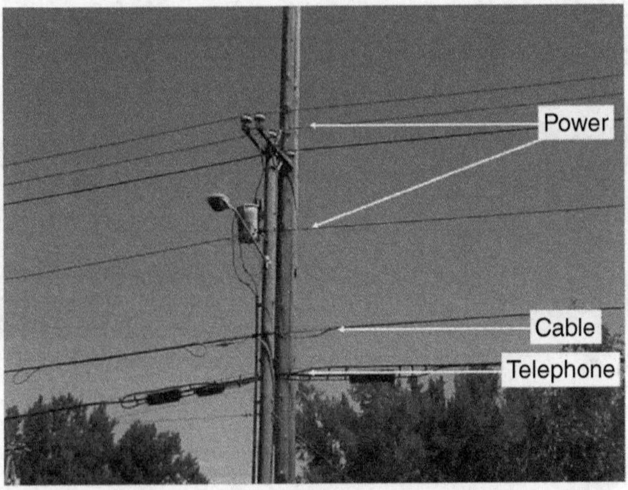

FIGURE 4.23 Sharing of a utility pole by various utilities.

One last item of interest: in spite of the fact that we call them telephone poles, the wooden poles that support telephone cables are in fact shared among telephone, power, and cable providers, as shown in Fig. 4.23. Telephony plant is the lowest, followed by cable; power is the highest, high enough that the tallest technician working on a phone or cable problem could not accidentally touch the open power conductors. Normally, the telephone company owns the pole and leases space on it to other utilities.

As a cable approaches the central office, its pairs are often combined with those of other approaching cables in a splice case (Fig. 4.24) to create a larger cable. For example, four 250-pair cables

Figure 4.24 Splice case.

Figure 4.25 The cable vault in the basement of a typical central office.

may be combined to create a 1,000 pair cable, which in turn may be combined with others to create a 5,000 pair cable that enters the central office. Once inside the office the cables are broken out for distribution. This is done in the cable vault; an example is shown in Fig. 4.25. The large cable on the left side of the splice case is broken down into the collection of smaller cables exiting on the right.

Into the Central Office

We could say a great deal more about the access segment of the network, and will later. For now, though, let's begin our tour of the central office.

By now our cable has traveled over aerial and underground traverses and has arrived at the central office. It enters the building via a buried conduit that feeds into the *cable vault*, the lowest sub-basement in the office. (Remember, this is where the New York City fire began.) Because the cable vault is exposed to the underground conduit system, there is always a danger that noxious and potentially explosive gases (methane, mostly) could flow into the cable vault and be set afire by a piece of sparking electrical equipment. These rooms are therefore closely monitored by the electronic equivalent of the canary in a coal mine, and will cause alarms to sound if gas is detected at dangerous levels.

The cables that enter the cable vault are large and encompass hundreds if not thousands of wire pairs, as shown in Fig. 4.26. Their security is obviously critical, because the loss of a single cable could mean loss of service for hundreds or thousands of customers. To help maintain the integrity of the outside plant, large cables are pressurized with very dry air, fed from a pressurization pump and manifold system in the cable vault (Fig. 4.27). The pressure serves two purposes: it keeps moisture from leaking into a minor breach in the cable, and serves as an alarm system in the event of a major cable failure. Cable plant technicians can analyze the data being fed to them from pressure transducers along the cable route and very accurately determine the location of the break. The pressure in the cables can be monitored, as shown.

As soon as the large cables have been broken down into smaller pair bundles, they leave the cable vault on their way up to the main distribution frame (MDF).

Before we leave the basement of the CO we should discuss power, a major consideration in an office that provides telecommunications services to hundreds of subscribers.

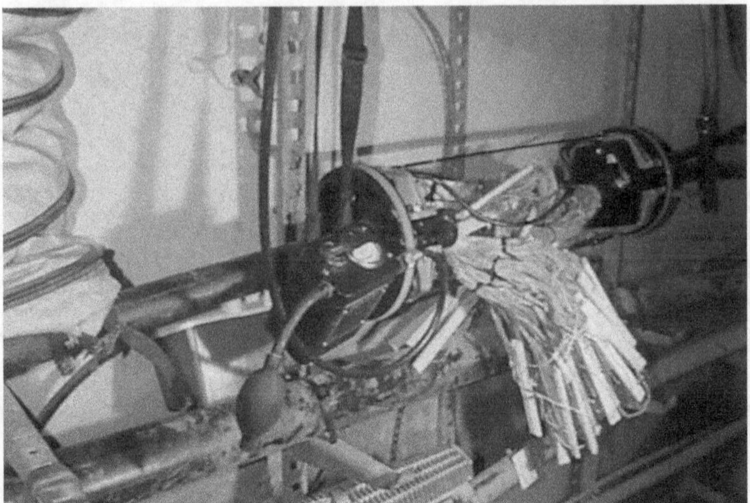

FIGURE 4.26 Cables in a cable vault.

FIGURE 4.27 Dry air pressurization system for cables.

When you plug in a laptop computer, the power supply does not actually power the laptop. The power supply's job is to charge the battery (assuming the battery is installed in the laptop); the battery powers the computer. That way, if commercial power goes away, the computer is unaffected because it was running on the battery to begin with.

Central offices work the same way. They are powered by massive wet cell battery arrays, shown in Fig. 4.28, that are stored in the

FIGURE 4.28 Battery arrays in a central office.

FIGURE 4.29 Commercial power in a CO.

basement. The batteries are quite large—about 2-ft tall—and are connected together by massive copper buss bars. Meanwhile, commercial power, shown in Fig. 4.29 (OK, not really—the real commercial power is shown in Fig. 4.30), is fed to the building from several sources on the power grid to avoid dependency on a single source of power. The AC power is fed into an inverter and filter that converts it to 48 V DC, which is then trickled to the batteries. The voltage maintains a constant charge on them. In the event that a failure of commercial power should occur, the building's *uninterruptible power supply* (UPS) equipment kicks in and begins a complex series of events designed to protect service in the office. First, it recognizes that the building is now isolated and is on battery power exclusively. The batteries have enough current to power a typical CO for several hours before things start to fail. Second, technicians begin overseeing the process of restoration and of shutting down non-essential systems to conserve battery power. Third, the UPS system initiates startup of the building's turbine, a massive jet engine that can be spun up in about a 1/2 hour. The turbine is either in a soundproof room in the basement or in an enclosure on the roof of the building (Fig. 4.31). Once the turbine has spun up to speed, the building's power load can slowly be shed onto it. It will power the building until it runs out of kerosene, but most buildings have several days' supply of fuel in their underground tanks. As one data center design engineer told me

FIGURE 4.30 The real power in a CO.

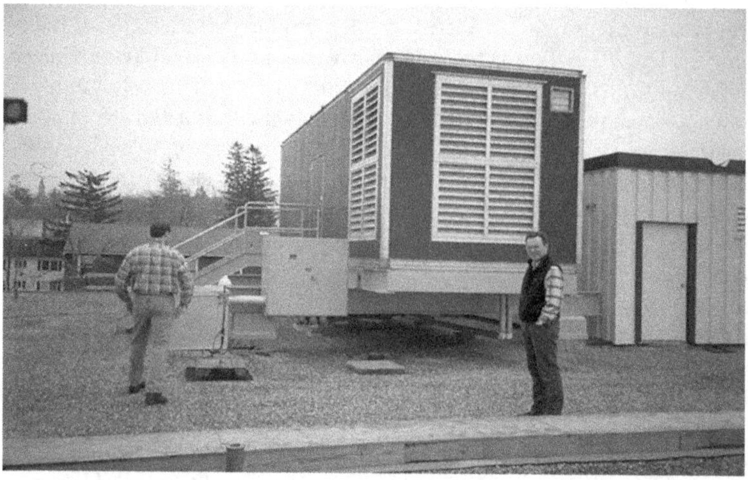

FIGURE 4.31 The turbine enclosure on the roof of a central office.

several years ago, "If we're down so long that we run out of fuel, our problems are far worse than an office that's out of service."

The cables leave the cable vault via fireproof foam-insulated ducts and travel upstairs to whatever floor houses the *MDF*. The MDF, shown in Fig. 4.32, is a large iron jungle gym sort of frame that provides physical support and technician access for the thousands of pairs of wire that interconnect customer telephones to the switch. The cables from the cable vault ascend and are hardwired to the *vertical side* of the MDF, shown in Fig. 4.33. The ends of the cables are called

FIGURE 4.32 The main distribution frame (MDF).

FIGURE 4.33 Vertical side of the MDF.

FIGURE 4.34 Carbon protectors.

the *cable heads.* The vertical side of the MDF has additional overvoltage protection. The arrays of plug-ins are called carbons; they are simply carbon fuses that open in the event of an overvoltage situation and protect the equipment and people in the office. A close-up of carbons is shown in Fig. 4.34.

The vertical side of the MDF is wired to the *horizontal side.* The horizontal side, shown in Fig. 4.35, is ultimately connected to the central office switch. Notice the mass of wire lying on each shelf of the frame: these are the cable pairs from the customers served by the office. Imagine the complexity of troubleshooting a problem in that hairball. I have spent many hours up to my shoulders in wire, tugging on a wire pair while a technician 50 ft down the frame, also up to his elbows in wire, feels for the wire that is moving so it can be traced. The horizontal side also provides craft/technician access for repair purposes. In Fig. 4.36, Dick Pecor has connected a test set to the appearance of a cable pair and is listening for noise.

New technicians are usually given a trial by fire when they arrive for work at a switching office. One of the most common pranks is to ask them to help troubleshoot a cable pair problem. They are asked to reach as far as they possibly can into the mass of wire on the horizontal side of the frame and feel for the wire that is moving, so that they can then wiggle it to help a technician farther down the MDF locate it and trace it to locate a problem. The new tech, now bent over and reaching deep into the frame holding a wire pair, doesn't realize that another technician is now on the vertical side of the frame, reaching into the wire mass with a wire lasso which is slipped over the new

FIGURE 4.35 Horizontal side of the MDF.

FIGURE 4.36 Dick Pecor, troubleshooting a cable pair on the MDF.

tech's arm and tied tightly to the far side of the frame. The technicians then go to lunch, leaving the new tech wired in place. Naturally, other technicians ignore his pleas for help; they remember their own experiences all too well.

From the MDF, the cable pairs are connected to the local switch in the office. Fig. 4.37 shows a Lucent #5ESS local switch. Take note of

FIGURE 4.37 Lucent #5ESS switch, one of the best-known examples of a large, local switch.

the size of this beast: it occupies a very large room and demonstrates why the cost of building and maintaining a traditional switched network is so high. All of this, by the way, can be replaced by a much smaller and less costly IP router, one of the reasons for the ongoing evolution to all-IP network.

Remember that the job of the local switch is to establish temporary connections between two or more customers that wish to talk, or between two computers that wish to spit bits at each other. The *line units* (LUs) on the switch provide a connection point for every cable pair served by that particular switch in the office. Conceptually, then, it is a relatively simple task for the switch to connect one subscriber in the switch with another subscriber in the same switch. After all, the switch maintains tables—a directory, if you will—so it knows the subscribers to which it is directly connected.

Far more complicated is the task of connecting a subscriber in one switch to a subscriber in another. This is where network intelligence becomes critically important. As we mentioned earlier, when the switch receives the dialed digits it performs a rudimentary analysis on them to determine whether the called party is locally hosted. If the number is in the same switch, the call is established. If it resides in another switch, the task is a bit more complex. First, the local switch must pass the call on to the local tandem switch, which provides access to the *points-of-presence* (POPs) of the various long-distance carriers' equipment that serves the area. The tandem switch typically does not connect directly to subscribers; it connects to other switches only. It also performs a database query through SS7 to determine who the subscriber's long-distance carrier-of-choice is so that it can route the call to the appropriate carrier. The tandem switch then hands the call off to the long-distance carrier, which transports it over the long-distance network to the carrier's switch in the remote (receiving) office. The long-distance switch passes the call through the remote tandem, which in turn hands the call off to the local switch that the called party is attached to.

SS7's influence once again becomes obvious here. One of the problems that occurred in earlier telephone system designs was the following. When a subscriber placed a call, the local switch handed the call off to the tandem, which in turn handed the call off to the long-distance provider. The long-distance provider seized a trunk over which it transported the dialed digits to the receiving central office. The signaling information, therefore, traveled over the path designed to produce revenue for the telephone company, a process known as *in-band signaling*—because the signaling information was carried in the voice band. As long as the called party was home, and wasn't on the phone, the call would go through as planned and revenue would flow. If they weren't home, however, or if they were on the phone, then the call would not complete and no revenue was generated. Furthermore, the network resources that another caller might have used to place a call were not available to them because they were tied up transporting call setup data, which produces no revenue.

SS7 changes all that. With SS7, the signaling data travels across a dedicated packet network from the calling party to the called party. SS7 verifies the availability of the called party's line, reserves it, and then—*and only then*— seizes a talk path. Once the talk path has been created it rings the called party's phone and places ringing tone in the caller's ear. As soon as the called party answers, SS7 silently bails out until one end or the other hangs up. At that point it returns the path to the pool of available network resources.

Interoffice Trunking

We have not discussed the manner in which offices are connected to one another. As Fig. 4.38 illustrates, an optical fiber ring with add-drop multiplexers interconnects the central offices so that interoffice traffic can be safely and efficiently transported. The switches have *trunk units (TUs)* that connect the back side, called the trunk side, of the switch to the wide area network, in this case the optical ring.

Trunks that interconnect offices have historically been four-wire copper facilities (a pair in each direction). Today they are almost exclusively optical but are still referred to as "four-wire" because of their two-way nature.

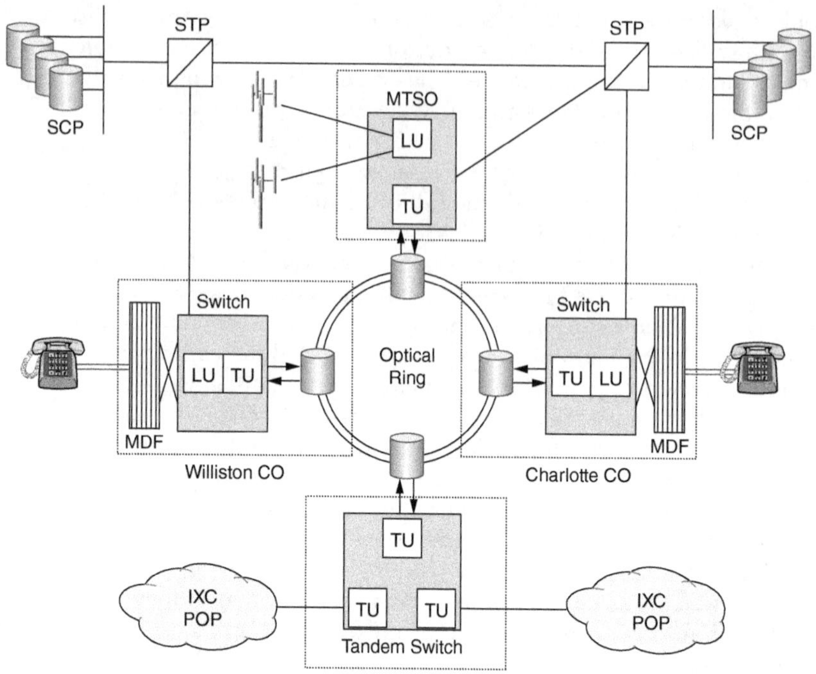

Figure 4.38 An optical ring architecture that interconnects multiple offices.

An interesting point about trunks. In the 1960s and early 1970s, most interoffice trunks were analog rather than digital. To signal to each other they used single frequency tones. Because these trunks did not *talk* directly to customers, there was no reason to worry about a human voice inadvertently emitting a sound that could be misconstrued as a dialing tone. There was therefore no reason to use DTMF dialing—instead, trunk signaling was performed using single-frequency tones. Specifically, if a switch wished to seize a long-distance trunk it issued a single-frequency 2600 Hz tone, which would signal the seizure to take place. Once the trunk seizure had occurred, the dialed digits could be out-pulsed and the call would proceed as planned.

In 1972, John Draper, better known by his hacker name Captain Crunch, determined that the plastic boson's whistle that came packed in boxes of Cap'n Crunch cereal emitted—you guessed it—2,600 Hz. Anyone who knew this could steal long-distance service from AT&T by blowing the whistle at the appropriate time during call setup. Before long, word of this capability became common knowledge and Cap'n Crunch cereal became the breakfast food of choice for hackers all over the country.

Soon hackers everywhere were constructing "blue boxes," small oscillators built from cheap parts that would emit 2,600 Hz and make possible the kind of access that Draper and his cohorts were engaged in. In fact, Steve Wozniak and Steve Jobs, both founders of Apple, were early blue box users who used the money they made building blue boxes to fund the company that became Apple. There's a fascinating account of this in Walter Isaacson's biography of Steve Jobs.

Draper, a highly skilled hacker, was eventually caught and prosecuted for his activities, but was given work furlough to continue his software design studies. He is now a successful intrusion detection software designer.

Returning to our example again, then, let's retrace our steps. Cristina dials Jess's number. The call travels over the local loop, across aerial and/or underground facilities, and enters the central office via the cable vault. From the cable vault the call travels up to the main distribution frame and then on to the switch. The number is received by the switch in Cristina's serving central office, which performs an SS7 database lookup to determine the proper disposition of the call and any special service information about Cristina that it should invoke. The local switch routes the call over intra-office facilities to a tandem switch, which in turn connects the call to the POP of whichever long-distance provider Cristina is subscribed. The long-distance carrier invokes the capabilities of SS7 to determine whether Jess's phone is available, and if it is, hands the call off to the remote local service provider. When Jess answers, the call is cut through and it progresses normally.

There are, of course, other ways that calls can be routed. The local loop could be wireless, and if it is, the call from the cell phone is received by a cell tower (Fig. 4.39) and transported to a special dedicated cellular switch in a central office called a *mobile telephone switching office* (MTSO). The MTSO processes the call and hands it off to the wireline network via interoffice facilities. From that point on the call is handled like any other. It is either terminated locally on the local switch or handed to a tandem switch for long-distance processing. The fact of the matter is that the only part of a cellular network that is truly wireless is the local loop—everything else is fixed.

Before we wrap this chapter, we should take a few minutes to discuss carrier systems, voice digitization, and multiplexed transport, all of which take place (for the most part) in the central office.

In this final section, we'll explain T- and E-Carrier, SONET, SDH, and voice digitization techniques.

Conserving Bandwidth: Voice Transport

The original voice network, including access, transmission facilities, and switching components, was exclusively analog until 1962 when T-carrier emerged as an inter-office trunking scheme. The technology was originally introduced as a short-haul, four-wire facility to serve

Figure 4.39 A typical cell tower.

metropolitan areas and was a technology that customers would *never* have a reason to know about—after all, what customer could ever need a meg-and-a-half of bandwidth? Over the years, however, it evolved to include coaxial cable facilities, digital microwave systems, fiber, and satellite, and of course became a premier access technology that customers knew a great deal about. In fact, T-carrier and E-carrier (the standard used in most of the world outside of North America) are now considered by many to be legacy technologies, replaced by IP trunking and gigabit ethernet facilities.

Consider the following scenario: A corporation builds a new headquarters facility in a new area just outside the city limits. The building will provide office space for approximately 2,000 employees, a considerable number. Those people will need telephones, computer data facilities, fax lines, videoconferencing circuits, and a variety of other forms of connectivity.

The telephone company has two options that it can exercise to provide access to the building. It can make an assessment of the number of pairs of wire that the building will require and install them; or, it can do the same assessment but provision the necessary bandwidth through carrier systems that transport multiple circuits over a shared facility. Obviously, this second option is the most cost-effective, and is in fact the option that is used for these kinds of installations. This model should sound familiar: earlier we discussed loop-carrier systems and the fact that they reduce the cost of provisioning network access to far-flung neighborhoods. This is the same concept; instead of a residential neighborhood, we're provisioning a *corporate neighborhood*.

The most common form of multiplexed access and transport is T-Carrier, or E-Carrier outside the United States. Let's take a few minutes to describe them.

Framing and Formatting in T-1

The standard T-Carrier multiplexer, shown in Fig. 4.40, accepts inputs from 24 sources, converts the inputs to PCM bytes, then time division multiplexes the samples over a shared four-wire facility, as shown in Fig. 4.41. Each of the 24 input channels yields an 8-bit sample, in round-robin fashion, once every 125 ms (microseconds) (8,000 times per second). This yields an overall bit rate of 64 Kbps for each channel (8 bits per sample $\times$ 8,000 samples per second). The multiplexer gathers one 8-bit sample from each of the 24 channels, and aggregates them into a 192-bit frame. To the frame it adds a frame bit, which expands the frame to a 193-bit entity. The frame bit is used for a variety of purposes that will be discussed in a moment.

The 193-bit frames of data are transmitted across the four-wire facility at the standard rate of 8,000 frames per second, for an overall T-1 bit rate of 1.544 Mbps. Keep in mind that 8 Kbps of the bandwidth consists of frame bits (one frame bit per frame, 8,000 frames per second); only 1.536 Mbps belongs to the user.

Beginnings: D1 Framing

The earliest T-Carrier equipment was referred to as D1, and was considerably more rudimentary in function than modern systems (see Fig. 4.42). In D1, every 8-bit sample carried 7 bits of user information (bits one through seven) and 1 bit for signaling (bit 8). The signaling bits were used for exactly that: indications of the status of the line (on-hook, off-hook, busy, high and dry, and the like), while the 7 user bits carried encoded voice information. Since only 7 of the 8 bits were available to the user, the result was considered to be less than toll quality (128 possible values, rather than 256). The frame bits, which in modern systems indicate the beginning of the next 192-bit frame of data, toggled back and forth between zero and one.

FIGURE 4.40 Standard T-carrier multiplexer.

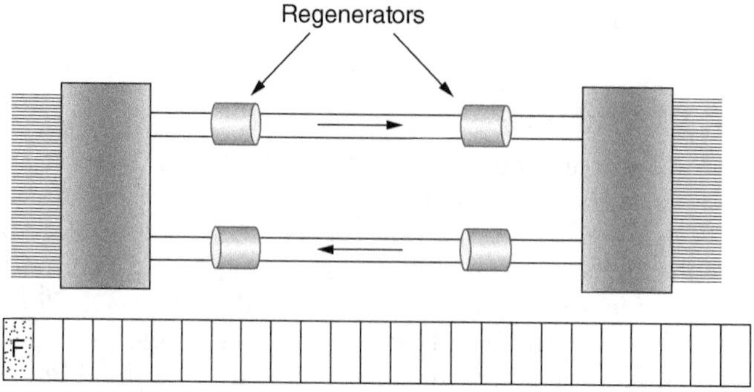

8-bits per sample, 24 samples per frame + frame bit = 193 bits.
8,000 frames are generated per second, yielding 1.544 Mbps.

FIGURE 4.41 T-carrier architecture showing four-wire facility.

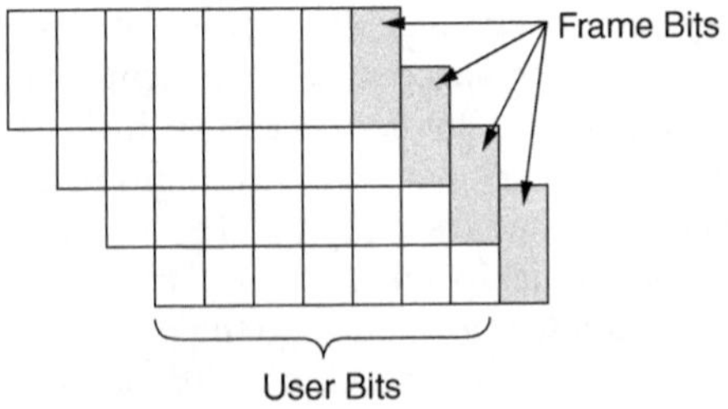

FIGURE 4.42 D1 framing.

Evolution: D4

As time went on and the stability of network components improved, an improvement on D1 was sought and found. Several options were developed, but the winner emerged in the form of the D4 or superframe format. Rather than treat a single 193-bit frame as the transmission entity, superframe *gangs together* twelve 193-bit frames into a 2,316-bit entity, shown in Fig. 4.43, that obviously includes 12 frame bits. Please note that the bit rate has not changed: we have simply changed our view of what constitutes a frame.

Since we now have a single (albeit large) frame, we clearly don't need 12 frame bits to frame it; consequently, some of them can be redeployed for other functions. In superframe, the 6 odd-numbered frame bits are referred to as terminal framing bits, and are used to synchronize the channel bank equipment. The even framing bits, on the other hand, are called signal framing bits and are used indicate to the receiving device *where* robbed-bit signaling occurs.

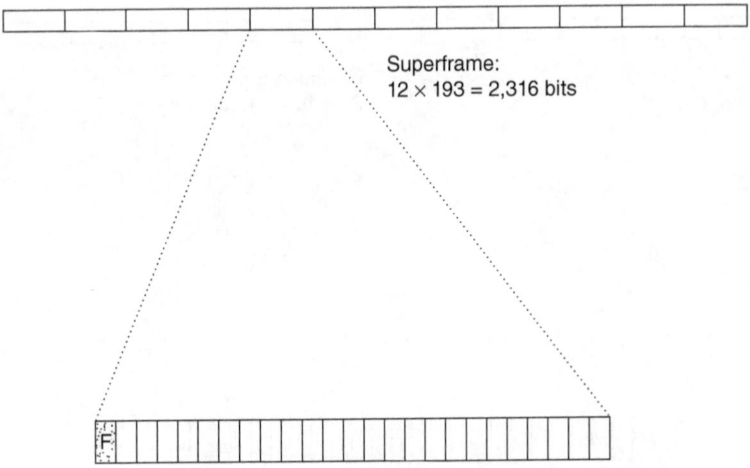

FIGURE 4.43 Superframe framing. Notice that it is more complex than D1.

In D1, the system reserved 1 bit from every sample for its own signaling purposes, which succeeded in reducing the user's overall throughput. In D4, that is no longer necessary: instead, we signal less frequently, and only occasionally rob a bit from the user. In fact, because the system operates at a high transmission speed, network designers determined that signaling can occur relatively infrequently and still convey adequate information to the network. Consequently, bits are robbed from the sixth and eighth iteration of each channel's samples, and then only the least significant bit from each sample. The resulting change in voice quality is negligible.

Back to the signal framing bits: within a transmitted superframe, the second and fourth signal framing bits would be the same, but the sixth would toggle to the opposite value, indicating to the receiving equipment that the samples in that subframe of the superframe should be checked for signaling state changes. The eighth and tenth signal framing bits would stay the same as the sixth, but would toggle back to the opposite value once again in the twelfth, indicating once again that the samples in that subframe should be checked for signaling state changes.

Today: Extended Superframe

While superframe continues to be used in some networks, an improvement came about in the 1980s in the form of *extended superframe* (ESF), shown in Fig. 4.44. ESF groups 24 frames into an entity instead of the 12 used in superframe, and like superframe, reuses some of the frame bits for other purposes. Bits 4, 8, 12, 16, 20, and 24 are used for framing, and form a constantly repeating pattern (001011...). Bits 2, 6, 10, 14, 18, and 22 are used as a 6-bit cyclic

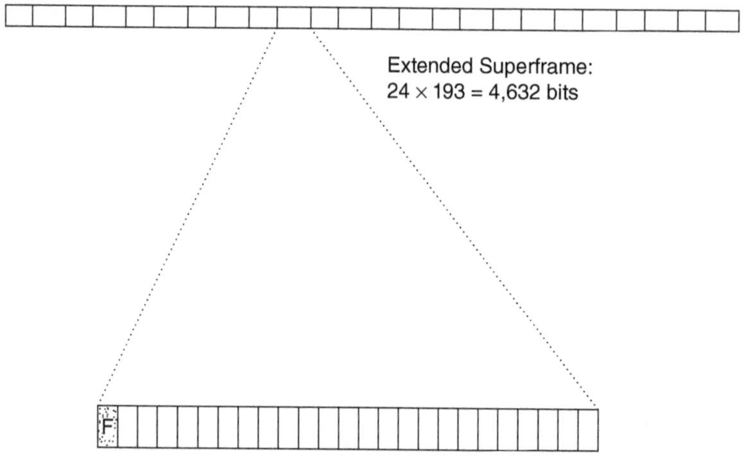

FIGURE 4.44 Extended superframe framing.

redundancy check (CRC) to check for bit errors on the facility. Finally, the remaining bits—all of the odd frame bits in the frame—are used as a 4 Kbps facility data link for end-to-end diagnostics and network management tasks. Technicians looking to troubleshoot a carrier facility will sometimes transmit the famous "quick brown fox" message down this facility to monitor for errors.

ESF provides one major benefit over its predecessors: the ability to do non-intrusive testing of the facility. In earlier systems, if the user reported trouble on the span, the span would have to be taken out of service for testing. With ESF, that is no longer necessary because of the added functionality provided by the CRC and the facility data link.

The Rest of the World: E-1

E-1, used for the most part outside of the United States and Canada, differs from T-1 on several key points. First, it boasts a 2.048 Mbps facility rather than the 1.544 Mbps facility found in T-1. Second, it utilizes a 32-channel frame rather than 24. Channel one contains framing information and a 4-bit cyclic redundancy check (CRC-4); channel 16 contains all signaling information for the frame; and channels one through 15 and 17 through 31 transport user traffic. The frame structure is shown in Fig. 4.45.

There are a number of similarities between T-1 and E-1 as well: channels are all 64 Kbps, and frames are transmitted 8,000 times per second. And whereas T-1 gangs together 24 frames to create an extended superframe, E-1 gangs together 16 frames to create what is known as an ETSI multiframe. ETSI is an acronym for the European Telecommunications Standards Institute, one of the principal regulatory bodies in the EU. The multiframe is subdivided into two sub-multiframes; the CRC-4 in each one is used to check the integrity of the sub-multiframe that preceded it.

A final word about T-1 and E-1: because T-1 is a departure from the international E-1 standard, it is incumbent upon the T-1 provider to perform all interconnection conversions between T-1 and E-1 systems. For example, if a call arrives in the United States from a European country, the receiving American carrier must convert the incoming E-1 signal to T-1. If a call originates from Canada and is

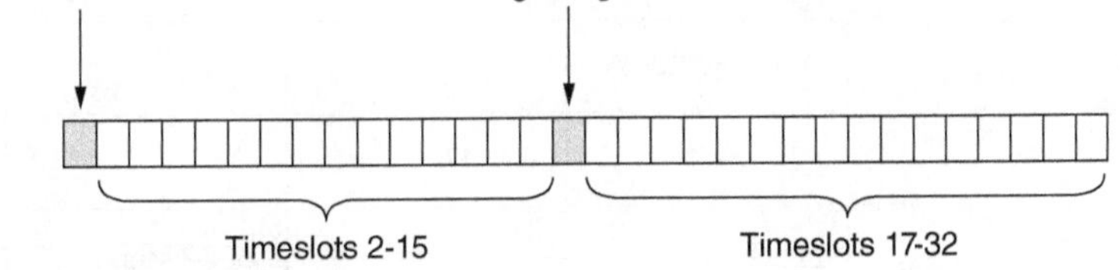

FIGURE 4.45 The E1 frame structure.

terminated in Australia, the Canadian originating carrier must convert the call to E-1 before transmitting it to Australia.

Up the Food Chain: From T-1 to DS3...and Beyond

When T-1 and E-1 first emerged on the telecommunications scene they represented a dramatic step forward in terms of the bandwidth that service providers now had access to. In fact, they were *so* bandwidth rich that there was no concept that a customer would ever need access to them. What customer, after all, could ever have a use for a-million-and-a-half bits per second of bandwidth?

Of course, that question was rendered moot in short order as increasing requirements for bandwidth drove demand that went well beyond the limited capabilities of low-speed transmission systems. As T-1 became mainstream its usage went up, and soon requirements emerged for digital transmission systems with capacities greater than 1.544 Mbps. The result was the creation of what came to be known as the North American Digital Hierarchy, shown in Fig. 4.46. The table also shows the European and Japanese hierarchy levels.

Hierarchy Level	Europe	United States	Japan
DS-0	64 Kbps	64 Kbps	64 Kbps
DS-1		1.544 Mbps	1.544 Mbps
E-1	2.048 Mbps		
DS-1c		3.152 Mbps	3.152 Mbps
DS-2		6.312 Mbps	6.312 Mbps
E-2	8.448 Mbps		32.064 Mbps
DS-3	34.368 Mbps	44.736 Mbps	
DS-3c		91.053 Mbps	
E-3	139.264 Mbps		
DS-4		274.176 Mbps	
			397.2 Mbps

FIGURE 4.46 The North American, Japanese, and European digital hierarchies.

From DS-1 to DS-3

We have already seen the process employed to create the DS-1 signal from 24 incoming DS-0 channels and an added frame bit. Now we turn our attention to higher bit rate services. As we wander our way through this explanation, pay particular attention to the complexity involved in creating higher rate payloads. This is one of the great advantages of the optical standards known as the synchronous optical network (SONET) and the synchronous digital hierarchy (SDH).

The next level in the North American digital hierarchy is called DS-2. And while it is rarely seen outside of the safety of the multiplexer in which it resides, it plays an important role in the creation of higher bit rate services. It is created when a multiplexer *bit interleaves* four DS-1 signals, inserting as it does so a control bit, known as a C-bit, every 48 bits in the payload stream. Bit interleaving is an important construct here because it contributes to the complexity of the overall payload. In a bit interleaved system, multiple bit streams are combined on a bit-by-bit basis as shown in Fig. 4.47. When payload components are bit-interleaved to create a higher rate multiplexed signal, the system first selects bit one from channel one, bit one from channel two, bit one from channel three, and so on. Once it has selected and transmitted all of the first bits, it goes on to the second bits from each channel, then the third, until it has created the super-rate frame. Along the way it intersperses C-bits, which are used to perform certain control and management functions within the frame.

Once the 6.312 Mbps DS-2 signal has been created, the system shifts into high gear to create the next level in the transmission hierarchy. Seven DS-2 signals are then bit-interleaved along with C-bits after every 84 payload bits to create a composite 44.736 Mbps DS-3 signal. The first part of this process, the creation of the DS-2 payload, is called *M12 multiplexing;* the second step, which combines DS-2s to form a DS-3, is called *M23 multiplexing.* The overall process is called *M13,* and is illustrated in Fig. 4.48.

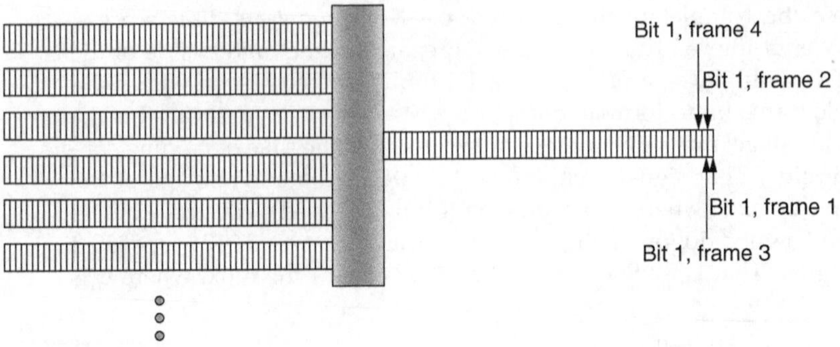

FIGURE 4.47 Bit interleaving.

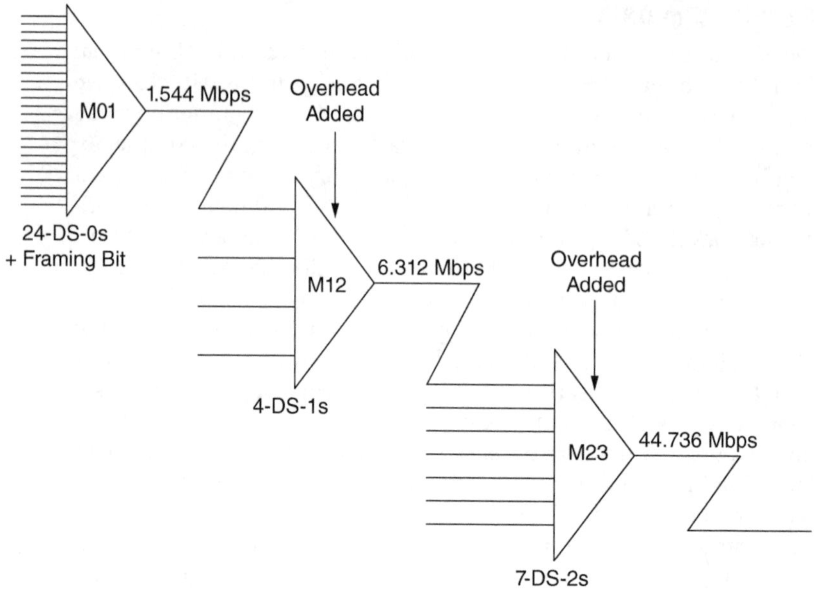

FIGURE 4.48 The M13 multiplexing process.

The problem with this process is the bit-interleaved nature of the multiplexing scheme. Because the DS-1 signal components arrive from different sources, they may be (and usually are) slightly off from one another in terms of the overall phase of the signal—in effect, their *speeds* differ slightly. This is unacceptable to a multiplexer which must rate-align them if it is to properly multiplex them, beginning with the head of each signal. In order to do this, the multiplexer inserts additional bits, known as *stuff bits*, into the signal pattern at strategic places which serve to rate align the components. The structure of a bit-stuffed DS-2 frame is shown in Fig. 4.49; a DS-3 frame is shown in Fig. 4.50.

The complexity of this process should now be fairly obvious to the reader. If we follow the left-to-right path shown in Fig. 4.51, we see the complexity that suffuses the M13 signal-building process. Twenty-four 64 Kbps DS0s are aggregated at the ingress side of the T-1[4] multiplexer, grouped into a T-1 frame, and combined with a single frame bit to form an outbound 1.544 Mbps signal. (I call this the M01 stage; that's my nomenclature, used for the sake of naming continuity.) That signal then enters the intermediate M12 stage of the multiplexer, where it is combined (bit-interleaved) with three others and a good dollop of alignment overhead to form a 6.312 Mbps DS-2 signal. That DS-2 then enters the M23 stage of the mux, where it is

[4]The process is similar for the E-1 hierarchy.

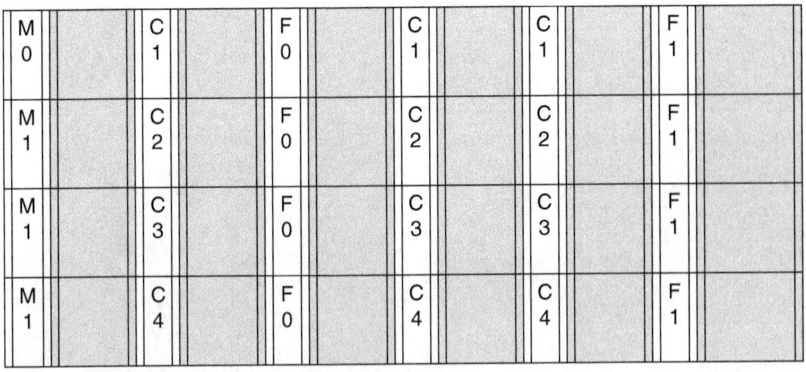

FIGURE 4.49 A bit-stuffed DS2 frame.

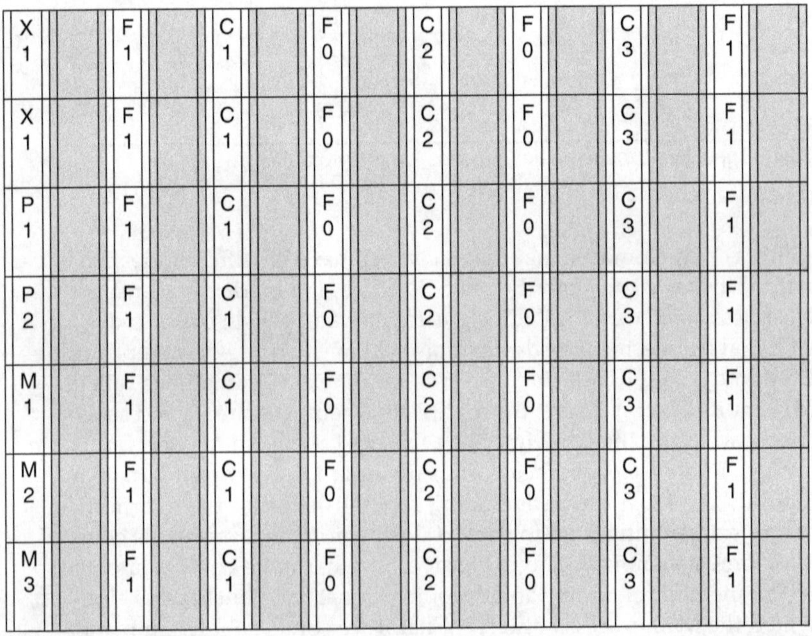

FIGURE 4.50 A DS3 frame.

bit-interleaved with six others and another scoop of overhead to create a DS-3 signal. At this point, we have a relatively high-bandwidth payload that is ready to be moved across the wide area network.

Of course, as our friends in the UK are wont to say, there is always the inevitable spanner that gets tossed into the works (those of us on the left side of the Atlantic call it a wrench). Keep in mind that the 28 (do the math) bit-interleaved DS-1s may well come from 28 different

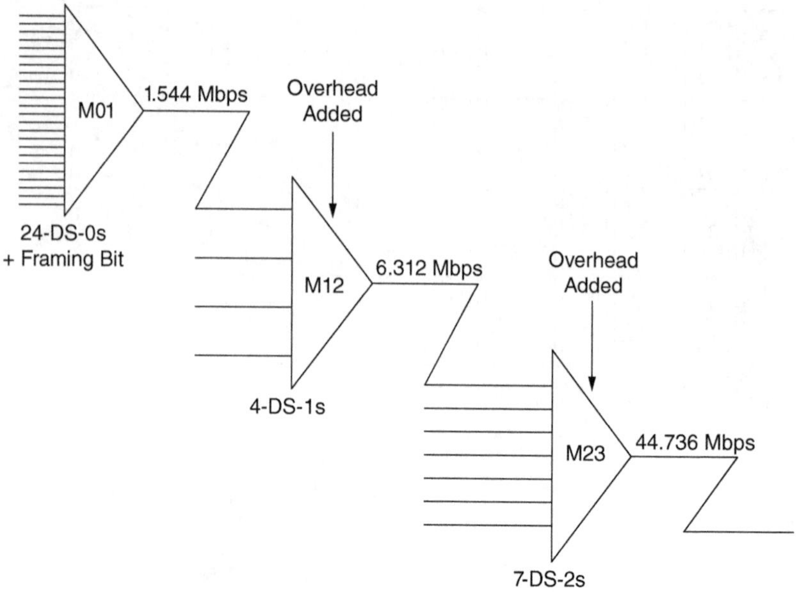

Figure 4.51 The overall "construction process" from DS0 to DS3.

sources—which means that they may well have 28 different destinations. This translates into the pre-SONET digital hierarchy's greatest weakness, and one of SONET's greatest advantages. In order to drop a DS-1 at its intermediate destination, we have to bring the composite DS-3 into a set of back-to-back DS-3 multiplexers (sometimes called M13 multiplexers). There, the ingress mux removes the second set of overhead, finds the DS-2 in which the DS-1 we have to drop out is carried, removes its overhead, finds the right DS-1, drops it out, then rebuilds the DS-3 frame, including reconstruction of the overhead, before transmitting it on to its next destination. This process is complex, time-consuming, and expensive. So what if we could come up with a method for adding and dropping signal components that eliminated the M13 process entirely? What if we could do it as simply as the process shown in Fig. 4.52?

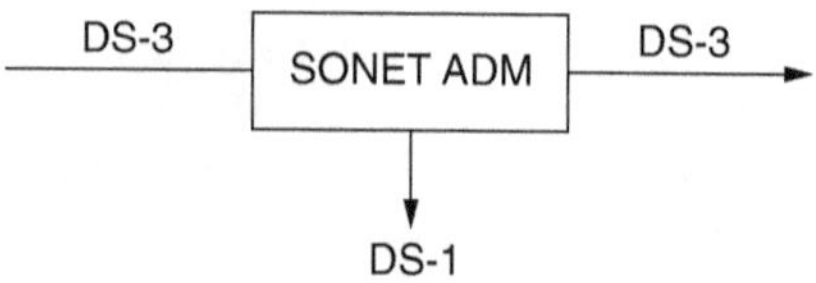

Figure 4.52 A much more simplified add-drop mechanism.

We can. It's called SONET in North America, SDH in the rest of the world, and it dramatically simplifies the world of high-speed transport.

The Synchronous Optical Network

SONET brings with it a subset of advantages that makes it stand above competitive technologies. These include mid-span meet, improved OAM&P, support for multipoint circuit configurations, non-intrusive facility monitoring, and the ability to deploy a variety of new services. We will examine each of these in turn.

SONET Advantages: Mid-Span Meet

Because of the monopoly nature of early networks, interoperability was a laughable dream. Following the divestiture of AT&T, however, and the realization of equal access, the need for interoperability standards became a matter of priority. Driven largely by MCI, the newly competitive telecommunications industry fought hard for standards that would allow different vendors' optical multiplexing equipment to interoperate. This interoperability came to be known as mid-span meet, SONET's greatest contribution to the evolving industry.

Improved OAM&P

Improved OAM&P is without question one of the most important contributions that SONET brings to the networking table. Element and network monitoring, management, and maintenance have always been something of a catch-as-catch-can effort because of the complexity and diversity of elements in a typical service provider's network. SONET overhead includes error-checking capability, bytes for network survivability, and a diverse set of clearly defined management messages.

Multipoint Circuit Support

When SONET was first deployed in the network, the bulk of the traffic it carried derived from point-to-point circuits such as T-1 and DS-3 facilities. With SONET came the ability to hub the traffic, a process that combines the best of cross-connection and multiplexing to perform a capability known as *groom and fill*. This means that aggregated traffic from multiple sources can be transported to a hub, managed as individual components, and redirected out any of several outbound paths without having to completely disassemble the aggregate payload. Prior to SONET this process required the pair of back-to-back M13 multiplexers that we discussed in the preceding section. This capability, combined with SONET's discreet and highly capable management features, results in a wonderfully manageable system.

Non-Intrusive Monitoring

SONET overhead bytes are embedded in the frame structure, meaning that they are universally transported alongside the customer's payload. Thus tight and granular control over the entire network can be realized, leading to more efficient network management and the ability to deploy services on an as-needed basis.

New Services

SONET bandwidth is imminently scalable, meaning that the ability to provision additional bandwidth for customers that require it on an as-needed basis becomes real. As applications evolve to incorporate more and more multimedia content and to therefore require greater volumes of bandwidth, SONET offers it by the bucket load. Already interfaces between SONET and gigabit ethernet are being written; interfaces to ATM and other high-speed switching architectures have been in existence for some time. New mappings emerge with regularity as higher and higher bandwidth demands are created.

SONET Evolution

SONET was initially designed to provide multiplexed point-to-point transport. However, as its capabilities became better understood and networks became *mission-critical,* its deployment became more innovative and soon it was deployed in ring architectures as shown in Fig. 4.53. These rings represent one of the most commonly deployed network topologies. For the moment, however, let's examine a point-to-point deployment. As it turns out, rings don't differ all that much—with a few staggeringly important exceptions, as we'll see.

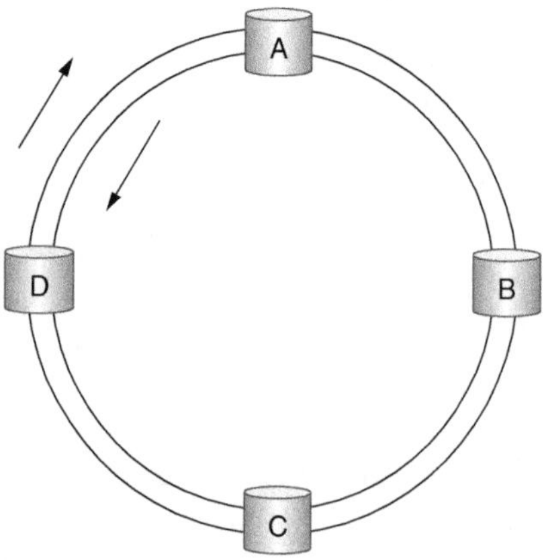

FIGURE 4.53 A typical SONET ring architecture.

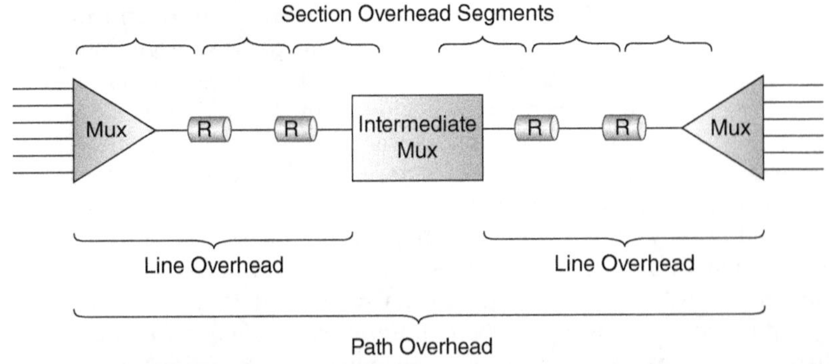

FIGURE 4.54 The functional regions of a typical point-to-point circuit.

If we consider the structure and function of the typical point-to-point circuit, we find a variety of devices and *functional regions* as shown in Fig. 4.54. The components include end devices, multiplexers in this case, which provide the *point of entry* for traffic originating in the customer's equipment and seeking transport across the network; a full-duplex circuit, which provides simultaneous two-way transmission between the network components; a series of repeaters or regenerators, responsible for periodically reframing and regenerating the digital signal; and one or more intermediate multiplexers which serve as nothing more than pass-through devices.

When non-SONET traffic is transmitted into a SONET network, it is packaged for transport through a step-by-step, quasi-hierarchical process that attempts to make reasonably good use of the available network bandwidth and ensure that receiving devices can interpret the data when it arrives. The intermediate devices, including multiplexers and repeaters, also play a role in guaranteeing traffic integrity, and to that end the SONET standards divide the network into three regions: path, line, and section. To understand the differences between the three, let's follow a typical transmission of a DS-3, probably carrying 28 T-1s, from its origination point to the destination.

When the DS-3 first enters the network, the ingress SONET multiplexer packages it by wrapping it in a collection of additional information, called *path overhead,* which is unique to the transported data. For example, it attaches information that identifies the original source of the DS-3 so that it can be traced in the event of network transmission problems; a bit-error control byte; information about how the DS-3 is actually mapped into the payload transport area (and unique to the payload type); an area for network performance and management information; and a number of other informational components that have to do with the end-to-end transmission of the unit of data.

The packaged information, now known as a *payload,* is inserted into a SONET frame and at that point another layer of control and management information is added called *line overhead*. Line overhead is responsible for managing the movement of the payload from multiplexer to multiplexer. To do this it adds a set of bytes that allow receiving devices to find the payload inside the SONET frame. As you will learn a bit later, the payload can occasionally wander around inside the frame due to the vagaries of the network. These bytes allow the system to track that movement.

In addition to these *tracking bytes*, the line overhead includes bytes that monitor the integrity of the network and have the ability to effect a switch to a backup transmission span if a failure in the primary span occurs, as well as another bit-error checking byte, a robust channel for transporting network management information, and a voice communications channel that allows technicians at either end of a the line to plug in with a handset (sometimes called a butt-in, or buttinski) and communicate while troubleshooting.

The final step in the process is to add a layer of overhead that allows the intermediate repeaters to find the beginning of a received frame of data so that it can be properly synchronized. This overhead, called the *section overhead*, contains a unique initial framing pattern at the beginning of the frame, an identifier for the payload signal being carried, another bit-error check, a voice communications channel, and another dedicated channel for network management information, similar to but smaller than the one identified in the line overhead.

The result of all this overhead, much of which seems like overkill (and in many peoples' minds *is*), is that the transmission of a SONET frame containing user data can be identified and managed with tremendous granularity from the source all the way to the destination, over vast transmission distances.

So, to summarize, the hard little kernel of DS-3 traffic is gradually surrounded by three layers of soft and chewy overhead information as shown in Fig. 4.55 that help it achieve its goal of successfully transiting

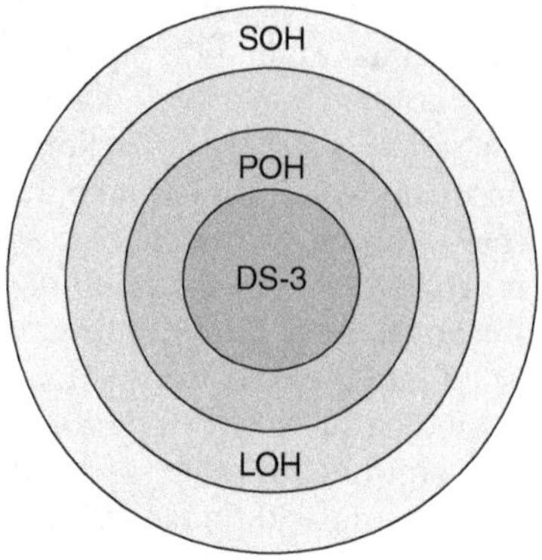

FIGURE 4.55 Overhead information in SONET.

the network. The section overhead is used at every device the signal passes through, including multiplexers and repeaters; the line overhead is only used between multiplexers; and the information contained in the path overhead is only used by the source and destination multiplexers—the intermediate multiplexers don't care about the specific nature of the payload because they don't have to terminate or interpret it.

The SONET Frame

Keep in mind once again that we are doing nothing more complicated than building a T-1 frame with an attitude. Recall that the T-1 frame comprised 24 8-bit channels (samples from each of 24 incoming data streams) plus a single bit of overhead. In SONET, we have a similar construct—a lot more channel capacity, and a lot more overhead, but the same functional concept.

The fundamental SONET frame is shown in Fig. 4.56, and is known as a *synchronous transport signal, level one,* or STS-1. It is depicted as being 9 bytes tall and 90 bytes wide, for a total of 810 bytes of transported data including both user payload and overhead. At the risk of stating the obvious, the 9-bytes-tall model is for illustrative purposes only: in reality, the SONET multiplexer transmits all 90 bytes of the first row, then wraps to transmit the second row, and so on. The SONET frame is depicted as a 9 × 90-byte frame so that it will fit on the page of a book!

The first three columns of the SONET frame are the section and line overhead, known collectively as the *transport overhead.* The bulk of the frame itself, to the left, is the *synchronous payload envelope* (SPE), which is the *container* area for the user data that is being transported.

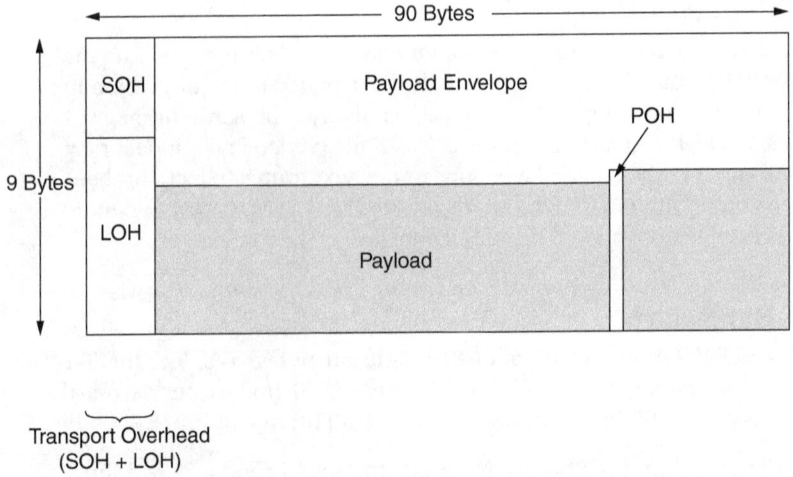

FIGURE 4.56 The fundamental SONET frame.

The data, previously identified as the payload, begins *somewhere* in the payload envelope. The actual starting point can vary, as we will see later. The path overhead begins when the payload begins: because it is unique to the payload itself, it travels closely with the payload. The first byte of the payload is in fact the first byte of the path overhead.

A word about nomenclature: Two distinct terms are used, often (incorrectly) interchangeably. The terms are *synchronous transport signal (STS)* and *optical carrier level (OC)*. They are used interchangeably because while an STS-1 and an OC-1 are both 51.84 Mbps signals, one is an electrically framed signal (STS) while the other describes an optical signal (OC). Keep in mind that the signals that SONET transports usually originate at an electrical source such as a T-1. These data must be collected and multiplexed at an electrical level before being handed over to the optical transport system. The optical networking part of the SONET system speaks in terms of OC.

The SONET frame is transmitted serially on a row-by-row basis. The SONET multiplexer transmits (and therefore receives!) the first byte of row one, all the way to the 90th byte of row one, then wraps to transmit the first byte of row two, all the way to the 90th byte of row two, and so on, until all 810 bytes have been transmitted.

Because the rows are transmitted serially, the many overhead bytes do not all appear at the beginning of the transmission of the frame—instead, they are peppered along the bitstream like highway markers. For example, the first two bytes of overhead in the section overhead are the framing bytes, followed by the single-byte signal identifier. The next 87 bytes are user payload, followed by the next byte of section overhead—in other words, there are 87 bytes of user data between the first three section overhead bytes and the next one! The designers of SONET were clearly thinking the day they came up with this, because each byte of data appears just when it is needed. Truly a remarkable thing.

Because of the unique way that the user's data is mapped into the SONET frame, the data can actually start pretty much anywhere in the payload envelope. The payload is always the same number of bytes, which means that if it starts *late* in the payload envelope it may well run into the payload envelope of the next frame. In fact, this happens more often than not, but it's ok—SONET is equipped to handle this odd behavior. We'll discuss it shortly.

SONET Bandwidth

The SONET frame consists of 810 eight-bit bytes, and like the T-1 frame is transmitted once every 125 μs (8,000 frames per second). Doing the math, this works out to an overall bit rate of

$$810 \text{ bytes/frame} \times 8 \text{ bits/byte} \times 8{,}000 \text{ frames/second} = 51.84 \text{ Mbps}$$

the fundamental transmission rate of the SONET STS-1 frame.

That's a lot of bandwidth—51.84 Mbps is slightly more than a 44.736 Mbps DS-3, a respectable carrier level by anyone's standard. But what if more bandwidth is required? What if the user wants to transmit multiple DS-3s, or perhaps a single signal that requires more than 51.84 Mbps, such as a 10 Gbps ethernet signal? Or for that matter, a payload that requires *less* than 51.84 Mbps! In those cases, we have to invoke more of SONET's magic.

The STS-N Frame

In situations where multiple STS-1s are required to transport multiple payloads, all of which fit in an STS-1's payload capacity, SONET allows for the creation of what are called STS-N frames, where *N* represents the number of STS-1 frames that are multiplexed together to create the frame. If three STS-1s are combined, the result is an STS-3. In this case, the three STS-1s are brought into the multiplexer and *byte interleaved* to create the STS-3 as shown in Fig. 4.57. In other words, the multiplexer selects the *first* byte of frame one, followed by the *first* byte of frame two, followed by the *first* byte of frame three. Then it selects the *second* byte of frame one, followed by the *second* byte of frame two, followed by the *second* byte of frame three, and so on, until it has built an interleaved frame that is now three times the size of an STS-1: 9×270 bytes instead of 9×90. Interestingly (and impressively!), the STS-3 is still generated 8,000 times per second.

The technique described above is called a *single-stage multiplexing process* because the incoming payload components are combined in a single step. There is also a two-stage technique that is commonly used. For example, an STS-12 can be created in either of two ways.

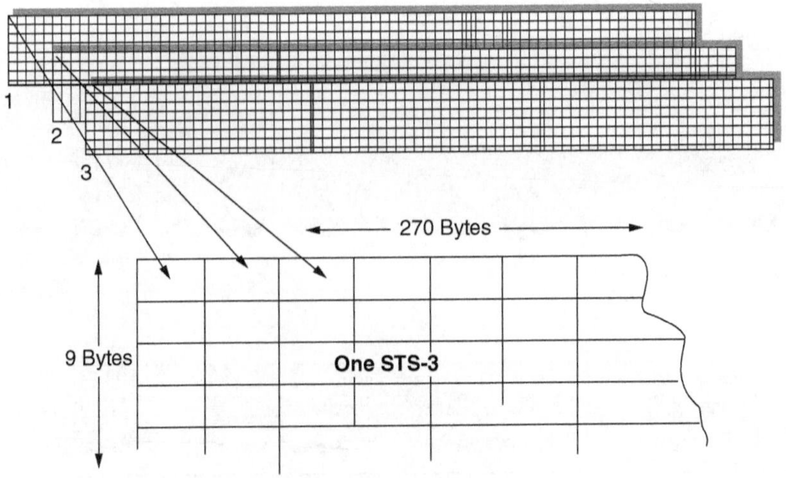

Figure 4.57 Using byte-interleaving to create an STS-3 frame.

Twelve STS-1s can be combined in a single-stage process to create the byte interleaved STS-12; alternatively, four groups of three STS-1s can be combined to form four STS-3s, which can then be further combined in a second stage to create a single STS-12. Obviously, two-stage multiplexing is more complex than its single-stage cousin, but both are used.

Special note: The overall bit rate of the STS-N system is $N \times$ STS-1. However, the *maximum* bandwidth that can be transported is STS-1—but N of them can be transported! This is analogous to a channelized *T*-1, where the overall bit rate of the carrier is 1.544 Mbps but the maximum payload that can be carried in each channel is 64 Kbps.

The STS-Nc Frame

Let's assume that we have to transport a legacy, 100 Mbps fast ethernet payload. In this case, 51.84 Mbps is inadequate for our purposes. For this we need what is known as a *concatenated signal.* One thing you can say about SONET: it doesn't hurt for polysyllabic vocabulary.

On the long, lonesome stretches of outback highway in Australia, unsuspecting car drivers often encounter a devilish vehicle known as a road train. Imagine an 18-wheel tractor-trailer (see top drawing, Fig. 4.58, for a remarkable illustration) barreling down the highway at 80 miles/ hour, but now imagine that it has six trailers—in effect, a 98-wheeler. These things give passing a whole new meaning. If a road

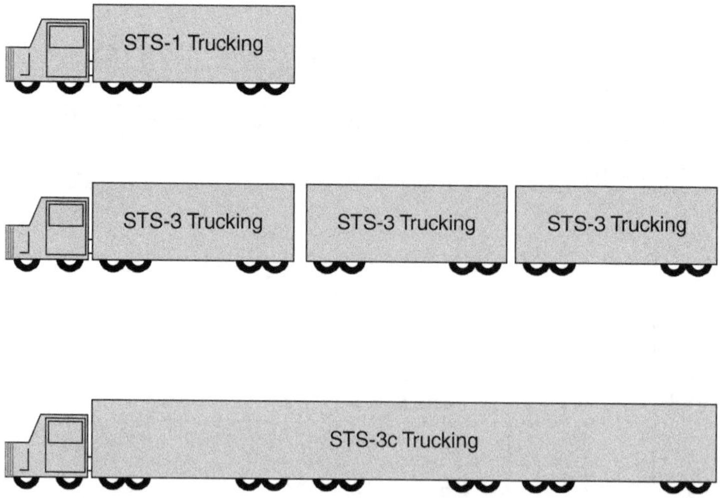

FIGURE 4.58 Building a concatenated frame.

train is rolling down the highway pulling three 50-ft trailers (middle drawing), then it has the ability to transport 150 ft of cargo—but only if the cargo is segmented into 50-ft chunks.

But what if the trucker wants to transport a 150-ft long item? In that case, a special trailer must be installed that provides room for the 150-ft payload (bottom drawing).

If you understand the difference between the second and third drawings, then you understand the difference between an STS-N and an STS-Nc. The word *concatenate* means *to string together*, which is exactly what we do when we need to create what is known as a *super-rate frame*—in other words, a frame capable of transporting a payload that requires more bandwidth than an STS-1 can provide, such as our 100 Mbps fast ethernet frame. In the same way that an STS-N is analogous to a channelized T-1, an STS-Nc is analogous to an *unchannelized* T-1. In both cases, the customer is given the full bandwidth that the pipe provides; the difference lies in how the bandwidth is parceled out to the user.

Overhead Modifications in STS-Nc Frames

When we transport multiple STS-1s in an STS-N frame, we assume that they may arrive from different sources. As a result, each frame is inserted into the STS-N frame with its own unique set of overhead. When we create a concatenated frame, though, the data that will occupy the combined bandwidth of the frame derives from the same source—if we pack a 100 Mbps fast ethernet signal into a 155.53 Mbps STS-3c frame, there's only one signal to pack. It's pretty obvious, then, that we don't need three sets of overhead to guide a single frame through the maze of the network. For example, each frame has a set of bytes that keep track of the payload within the synchronous payload envelope. Since we only have one payload, therefore, we can eliminate two of them. And the path overhead that is unique to the payload can similarly be reduced, since there is a column of it for each of the three formerly individual frames. In the case of the pointer that tracks the floating payload, the first pointer continues to perform that function; the others are changed to a fixed binary value that is known to receiving devices as a *concatenation indication*. The details of these bytes will be covered later in the overhead section.

Transporting Sub-Rate Payloads: Virtual Tributaries

Let's now go back to our Australian road train example. This time our driver is transporting individual cans of Fosters beer. From what I remember about the last time I was dragged into an Aussie pub (and it isn't much), the driver could probably transport about six cans of Fosters per 50-ft trailer. So now we have a technique for carrying payloads smaller than the fundamental 50-ft payload size. This analogy works well for understanding SONET's ability to transport payloads that require less bandwidth than 51.84 Mbps, such as T-1 or traditional 10 Mbps ethernet.

When a SONET frame is modified for the transport of sub-rate payloads, it is said to carry *virtual tributaries.* Simply put, the payload envelope is chopped into smaller pieces that can then be individually used for the transport of multiple lower-bandwidth signals.

Creating Virtual Tributaries

To create a *virtual tributary-ready* STS, the synchronous payload envelope is subdivided. An STS-1 comprises 90 columns of bytes, 4 of which are reserved for overhead functions (Section, Line, and Path). That leaves 86 for actual user payload. To create virtual tributaries, the payload capacity of the SPE is divided into seven 12-column pieces, called *virtual tributary groups.* Math majors will be quick to point out that $7 \times 12 = 84$, leaving two unassigned columns. These columns, shown in Fig. 4.59, are indeed unassigned, and are given the rather silly name of *fixed stuff.*

Now comes the fun part. Each of the VT groups can be further subdivided into one of four different VTs to carry a variety of payload types, as shown in Fig. 4.60. A VT1.5, for example, can easily transport a 1.544 Mbps signal within its 1.728 Mbps capacity, with a little room left over. A VT2, meanwhile, has enough capacity in its 2.304 Mbps structure to carry a 2.048 Mbps European E-1 signal, with a little room left over. A VT3 can transport a DS-1C signal, while a VT6 can easily accommodate a DS-2, again, each with a little room left over.

One aspect of virtual tributaries that must be mentioned is the mix-and-match nature of the payload. Within a single SPE, the seven

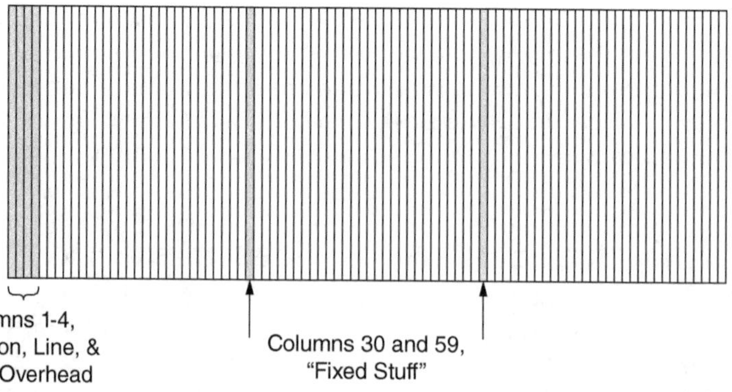

FIGURE 4.59 Fixed stuff in virtual tributaries.

VT Type	Columns/ VT	Bytes/ VT	VTs/ Group	VTs/ SPE	VT Bandwidth
VT1.5	3	27	4	28	1728
VT2	4	36	3	21	2.304
VT3	6	54	2	14	3.456
VT6	12	108	1	7	6.912

FIGURE 4.60 Virtual tributary payload types.

VT groups can carry a variety of different VTs. However, each VT group can carry only one VT type.

That "little room left over" comment (above) is, by the way, one of the key points that SONET and SDH detractors point to when criticizing them as legacy technologies, claiming that in these times of growing competition and the universal drive for efficiency they are inordinately wasteful of bandwidth, given that they were designed when the companies that delivered them were monopolies and less concerned about such things than they are now. For now, though, suffice it to say that one of the elegant aspects of SONET is its ability to accept essentially any form of data signal, map it into standardized positions within the SPE frame, and transport it efficiently and at very high speed to a receiving device on the other side of town or the other side of the world.

Creating the Virtual Tributary Superframe

You will recall that when DS-1 frames are transmitted through modern networks today they are typically formatted into extended super frames in order to eke additional capability out of the comparatively large percentage of overhead space that is available. When DS-1 or other signals are transported via an STS-1 formatted into virtual tributary groups, four consecutive STS-1s are *ganged together* to create a single VT Superframe, as shown in Fig. 4.61. To identify the fact that

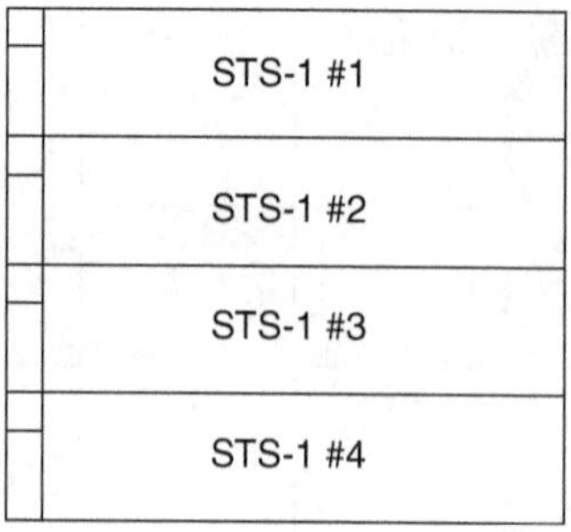

FIGURE 4.61 A VT superframe.

the frames are behaving as a VT Superframe, certain of the overhead bytes are modified to indicate the change.

Note: For those readers interested in more detail about the SONET overhead, please refer to the bibliography in the appendix. Several good books on SONET and SDH are listed there.

SONET Synchronization

SONET relies on a timing scheme called *plesiochronous timing*. As I implied earlier, the word sounds like one of the geological periods that we all learned in geology classes (jurassic, triassic, plesiochronous, and plasticene). Plesiochronous derives from the Greek and means *almost timed*. Other words that are commonly tossed about in this industry are *asynchronous* (not timed); *isochronous* (constant delay in the timing); and *synchronous* (timed). SONET is plesiochronous in spite of its name (synchronous optical network) because the communicating devices in the network rely on multiple timing sources and are therefore allowed to drift slightly relative to each other. This is fine, because SONET has the ability to handle this with its pointer adjustment capabilities.

The devices in a SONET network have the luxury of choosing from any of five timing schemes to ensure accuracy of the network. As long as the schemes have stratum 4 accuracy or better, they are perfectly acceptable timing sources. The five are discussed as follows:

Line timing: Devices in the network derive their timing signal from the arriving input signal from another SONET device. For example, an add-drop multiplexer that sits out on a customer's premises derives its synchronization pulse from the incoming bit stream, and might provide further timing to a piece of CPE that is out beyond the ADM.

Loop timing: Loop timing is somewhat similar to line timing; in loop timing, the device at the end of the loop is most likely a terminal multiplexer.

External timing: The device has the luxury of deriving its timing signal directly from a stratum 1 clock source.

Through timing: Similar to line timing, a device that is through timed receives its synchronization signal from the incoming bit stream, but then forwards that timing signal on to other devices in the network. The timing signal then passes *through* the intermediate device.

Free running: In free running timing systems, the SONET equipment in question does not have access to an external timing signal, and must derive its timing from internal sources only.

One final point about SONET should be made. When the standard is deployed over ring topologies, two timing techniques are used. Either external timing sources are depended upon to time network elements, or one device on the ring is internally timed (free running) while all the others are through-timed.

One Final Thought: Next-Generation SONET

As we have discussed, SONET and SDH were introduced in the 1980s as an optical solution for high-bandwidth vendor interoperability in optical networks. SONET transmission rates have steadily increased to OC-768 (40 Gbps), and the protocol was developed primarily to transport voice in fixed timeslots through the network, in retrospect making SONET inefficient for the transport of the increasing volumes of bursty traffic.

Beginning in the 1990s, the Internet arrived, and with it a profusion of data transferred over the existing network infrastructure. STS-1 (51.84 Mbps) is the basic rate at which SONET operates, and higher bandwidth services are delivered as multiples of STS-1. This technique, however, is less than ideal for the transport of multimedia data payloads, and when it *is* used, a variety of challenges appear, including the need to identify techniques to *rescue* the stranded bandwidth that occurs when SONET is used for data transport without having to upgrade existing SONET ADMs and cross-connect boxes; identification of the most effective way to carry data traffic over SONET; the creation and deployment of a reasonable technique for dynamically and efficiently resizing allocated bandwidth without traffic loss; methods and procedures to ensure survivability of link groups in the event of a link loss; and processes to support multiple traffic types across the same physical infrastructure. The solution to this plethora of challenges came in the form of *next-generation SONET*, and with NGS came a new set of terms: *virtual concatenation*, *link capacity adjustment scheme* (LCAS), and *generic framing procedure* (GFP).

Virtual Concatenation

Virtual concatenation is based on the concept that a group of smaller containers can be concatenated to form a larger container. The containers, typically STS-1s, are then routed across the network towards the same destination where the individual STS-1s are re-aligned and sorted to form the original payload. In OSI terms this technique sounds like a layer four, transport layer process, doesn't it?

Virtual concatenation has been defined for VT1.5, VT2, VT3, VT6, STS-1 SPE, and STS-3c SPE containers. A virtually concatenated group is referred to as a *base container-nv*, where *n* is a number that indicates the number of containers of the base container type. For example, VT1.5-4v indicates that the group consists of four VT1.5s that are virtually concatenated. This process of *payload channelization* is performed at the end points in the network.

There are two different forms of virtual concatenation. If the channel rate is STS-1 and greater for a channel in a group, then *high-order virtual concatenation* (HOVC) is used. If the channel rate is below STS-1, then *low-order virtual concatenation* (LOVC) is used.

Link Capacity Adjustment Scheme

The link capacity adjustment scheme (LCAS) is a complementary technology to virtual concatenation. It facilitates the adjustment of the size of a group of channels that have been virtually concatenated. LCAS synchronizes transmission between the sender and receiver so that the size of a virtually concatenated circuit can be increased or decreased without corrupting the data signal.

Generic Framing Procedure

Generic framing procedure (GFP) is an encapsulation technique that allows traffic from a fibre channel-based storage area network (SAN) to be carried directly onto the SONET network efficiently and cost-effectively.

Previously, storage traffic had to be converted to an intermediary protocol such as ATM, frame relay, or IP before being placed on a SONET network. This added overhead and issues related to security, resiliency, latency, and performance. GFP is an optical switch interface that lets companies use their SONET facilities more effectively. GFP can be deployed at the edge of a leased SONET network and can then be used to allocate portions of the circuit using virtual concatenation (VCAT). VCAT extends the utility of SONET transport by letting bandwidth be allocated in multiples of 50 Mbps, thus allocating only the bandwidth required by a particular application.

In combination with VCAT, the cost-effectiveness of GFP is significantly improved. Enterprises can transport data using optimal bandwidth allocations, and service providers can maximize the efficiency of their overall network. Furthermore, enterprises can deploy private storage over SONET solutions by purchasing storage-specific equipment for SONET and leasing SONET circuits from a service provider.

SONET Summary

Clearly SONET is a complex and highly capable standard designed to provide high-bandwidth transport for legacy and new protocol types alike. The overhead that it provisions has the ability to deliver a remarkable collection of network management, monitoring, and transport granularity.

The European synchronous digital hierarchy, shares many of SONET's characteristics, as we will now see. SONET, you will recall, is a limited North American standard for the most part. SDH, on the other hand, provides high-bandwidth transport for the rest of the world.

Most books on SONET and SDH cite a common list of reasons for their proliferation, including a recognition of the importance of the global marketplace and a desire on the parts of manufacturers, therefore, to provide devices that will operate in both SONET and SDH environments; the global expansion of ring architectures; a greater focus on network management and the value that it brings to the table; and massive, unstoppable demand for more bandwidth. To those, add these: increasing demand for high-speed routing capability to work hand-in-glove with transport; deployment of DS-1, DS-3, and E-1 interfaces directly to the enterprise customer as access solutions; growth in demand for broadband access technologies such as cable modems, the many flavors of DSL, and two-way satellite connectivity; the ongoing replacement of traditional circuit-switched network fabrics with packet-based transport and mesh architectures; a renewed focus on the SONET and SDH overhead with an eye toward using it more effectively; and convergence of multiple applications on a single, capable, high-speed network fabric.

SDH Nomenclature

Before launching into a functional description of SDH it would be good to first cover the differences in naming convention between the two. This will help to dispel confusion (hopefully!).

You'll recall that the fundamental SONET unit of transport uses a 9 row by 90-column frame that comprises 3 columns of section and line overhead, 1 column of path overhead, and 87 columns of payload.

The payload, which is primarily user data, is carried in a payload envelope that can be formatted in various ways to make it carry a variety of payload types. For example, multiple SONET STS-1 frames can be combined to create higher-rate systems for transporting multiple STS-1 streams, or a single higher-rate stream created from the combined bandwidth of the various multiplexed components. Conversely, SONET can transport sub-rate payloads, known as virtual tributaries, which operate at rates slower than the fundamental STS-1 SONET rate. When this is done, the payload envelope is divided into virtual tributary groups, which can in turn transport a variety of virtual tributary types.

In the SDH world similar words apply, but they are different enough that they should be discussed. As you will see, SDH uses a fundamental transport container that is three times the size of its SONET counterpart. It is a nine-row by 270-column frame that can be configured into 1 of 5 container types, typically written C-*N* (where C means container). *N* can be 11, 12, 2, 3, or 4; they are designed to transport a variety of payload types.

When an STM-1 is formatted for the transport of virtual tributaries (known as virtual containers in the SDH world), the payload pointers must be modified. In the case of a payload that is carrying virtual containers, the pointer is known as an administrative unit type 3 (AU-3). If the payload is *not* structured to carry virtual containers, but is instead intended for the transport of higher rate payloads, then the pointer is known as an administrative unit type 4. Generally speaking, an AU-3 is typically used for the transport of North American digital hierarchy payloads; AU-4 is used for European signal types.

The SDH Frame

To understand the SDH frame structure it is first helpful to understand the relationship between SDH and SONET. Functionally, they are identical: in both cases the intent of the technology is to provide a standardized transmission system for high-speed data. SONET is indeed optimized for the T-1-heavy North American market, while SDH is more applicable to Europe; beyond that, however, the overhead and design considerations of the two are virtually identical. There are, however, some key differences.

Perhaps the greatest difference between the two lies in the physical nature of the frame. A SONET STS-1 frame comprises 810 total bytes, for an overall aggregate bit rate of 51.84 Mbps—perfectly adequate for the North American 44.736 Mbps DS-3. An SDH STM-1 frame, however, designed to transport a 139.264 Mbps European E-4 or CEPT-4 signal, must be larger if it is to accommodate that much bandwidth—it clearly won't fit in the limited space available in an STS-1. An STM-1, then, operates at a fundamental rate

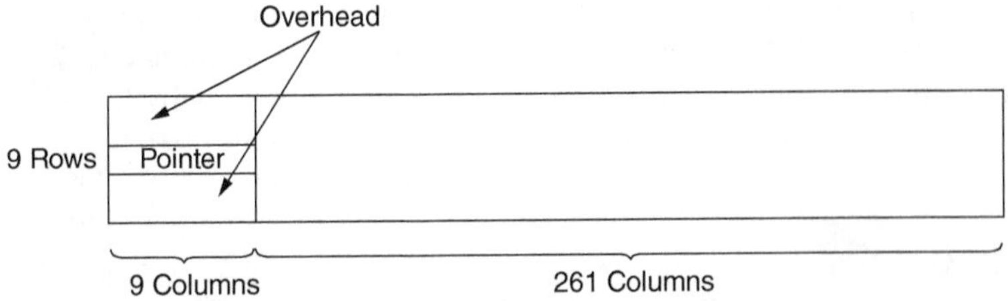

FIGURE 4.62 An STM-1 frame.

of 155.52 Mbps, enough for the bandwidth requirements of the E-4. This should be where the déjà vu starts to kick in: perceptive readers will remember that 155.52 Mbps number from our discussions of the SONET STS-3, which offers *exactly* the same bandwidth. An STM-1 frame is shown in Fig. 4.62. It is a byte-interleaved, nine-row by 270-column frame, with the first nine columns devoted to overhead and the remaining 261 devoted to payload transport.

A comparison of the bandwidth between SONET and SDH systems is also interesting. The fundamental SDH signal is exactly *three times* the bandwidth of the fundamental SONET signal, and this relationship continues all the way up the hierarchy.

STM Frame Overhead

The overhead in an STM frame is very similar to that of an STS-1 frame, although the nomenclature varies somewhat. Instead of section, line, and path overhead to designate the different regions of the network that the overhead components address, SDH uses *regenerator section, multiplex section, and path overhead,* as shown in Fig. 4.63. The regenerator section (RSOH) occupies the first 3 rows of 9 bytes,

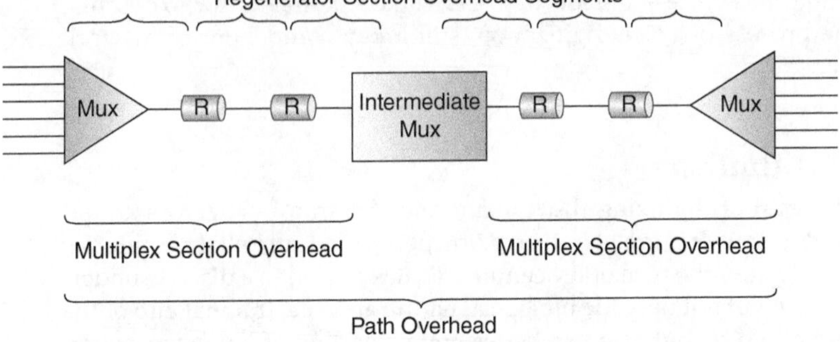

FIGURE 4.63 STM frame overhead.

the multiplex section (MSOH) the final 5. Row 4 is reserved for the pointer. As in SONET, the path overhead floats gently on the payload tides, rising and falling in response to phase shifts. Functionally, these overhead components are identical to their SONET counterparts.

Overhead Details

Because an STM-1 is three times as large as an STS-1, it has three times the overhead capacity—nine columns instead of three (plus Path Overhead). The first row of the RSOH is its SONET counterpart, with the exception of the last 2 bytes, which are labeled as being reserved for national use and are specific to the PTT administration that implements the network. In SONET, they are not yet assigned. The second row is different from SONET in that it has 3 bytes reserved for media-dependent implementation (differences in the actual transmission medium, whether copper, coaxial, or fiber) and the final 2 bytes reserved for national use. As before, they are not yet definitively assigned in the SONET realm.

The final row of the RSOH also sports two bytes reserved for media-dependent information, while they are reserved in SONET. All other regenerator section or section overhead functions are identical between the two.

The multiplex section overhead in the SDH frame is almost exactly the same as that of the SONET line overhead, with one exception: row nine of the SDH frame has 2 bytes reserved for national administration use. They are reserved in the SONET world.

The pointer in an SDH frame is conceptually identical to that of a SONET pointer, although it has some minor differences in nomenclature. In SDH, the pointer is referred to as an administrative unit pointer, referring to the standard naming convention described earlier.

SONET and SDH were originally rolled out to replace the T1 and E1 hierarchies, which were suffering from demands for bandwidth beyond what they were capable of delivering. Their principal deliverable was voice—and lots of it. Let's take a moment now to describe the process of voice digitization, still a key component of network transport.

Voice Digitization

The goal of digitizing the human voice for transport across an all-digital network grew out of work performed at Bell Laboratories shortly after the turn of the century. That work led to a discrete understanding of not only the biological nature and spectral makeup of the human voice, but also to a better understanding of language, sound patterns, and the sounded emphases that comprise spoken language.

The Nature of Voice

A typical voice signal comprises frequencies that range from approximately 30 Hz to 10 KHz. Most of the speech energy, however, lies between 300 and 3,300 Hz, the so-called voice band. Experiments have shown that the frequencies below 1 KHz provide the bulk of recognizability and intelligibility, while the higher frequencies provide richness, articulation, and natural sound to the transmitted signal.

The human voice comprises a remarkably rich mix of frequencies, and this richness comes at a considerable price. In order for telephone networks to transmit voice's entire spectrum of frequencies, significant network bandwidth must be made available to every ongoing conversation. There is a substantial price tag attached to bandwidth; it is a finite commodity within the network, and the more of it that is consumed, the more it costs.

The Network

Thankfully, work performed at Bell Laboratories at the beginning of the twentieth century helped network designers confront this challenge head-on. To understand it, let's take a tour of the telephone network.

The typical network, shown in Fig. 4.64, is divided into several regions: the access plant, the switching, multiplexing and circuit connectivity equipment (the central office), and the long-distance transport plant. The access and transport domains are often referred to as the *outside plant*; the central office is, conversely, the *inside plant*. The outside plant has the responsibility to aggregate inbound traffic for

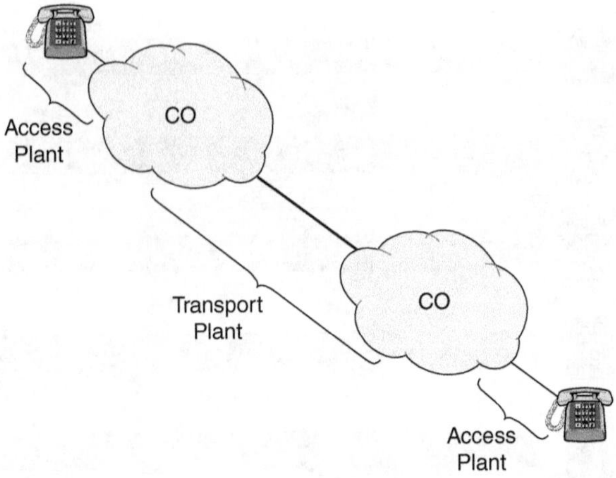

FIGURE 4.64 The layout of a typical telecommunications network.

switching and transport across the long-haul, as well as to terminate traffic at a particular destination. The inside plant, on the other hand, has the responsibility to multiplex incoming traffic streams, switch the streams, and select an outbound path for ultimate delivery to the next central office in the chain or the final destination. Switching, therefore, was centrally important to the development of the modern network.

Multiplexing

Equally important as the development of the central office switch was the concept of multiplexing, which allowed multiple conversations to be carried simultaneously across a single shared physical circuit. The first such systems used frequency-division multiplexing, a technique made possible by the development of the vacuum tube, in which the range of available frequencies is divided into *chunks*, which are then parceled out to subscribers. For example, Fig. 4.65 illustrates that subscriber no. 1 might be assigned the range of frequencies between 0 and 4,000 Hz, while subscriber no. 2 is assigned 4,000 to 8,000 Hz, no. 3 is assigned 8,000 to 12,000 Hz, and so on, up to the maximum range of frequencies available in the channelized system. In frequency-division multiplexing, we often observe that users are given *some of the frequency all of the time*, meaning that they are free to use their assigned frequency allocation at any time, but may *not* step outside the bounds given to them. Early FDM systems were capable of transporting 24 two KHz channels, for an overall system bandwidth of 96 KHz. Frequency-division multiplexing, while largely replaced today by more efficient systems that will be discussed later, served the industry well for many years.

FIGURE 4.65 Frequency assignments in a multichannel analog system.

This model worked well in early telephone systems. Because the lower regions of the 300 to 3,300 Hz voice band carry the frequency components that allow for recognizability and intelligibility, telephony engineers concluded that while the higher frequencies enrich the transmitted voice, they are not necessary for calling parties to recognize and understand each other. This understanding of the makeup of the human voice helped them create a network that was capable of faithfully reproducing the sounds of a conversation while keeping the cost of consumed bandwidth to a minimum. Instead of assigning the full complement of 10 KHz to each end of a conversation, they employed filters to bandwidth-limit each user to approximately 4,000 Hz, a resource savings of some 60%. Within the network, subscribers were frequency-division multiplexed across shared physical facilities, thus allowing the telephone company to efficiently conserve network bandwidth.

Time, of course, changes everything. As with any technology, there were downsides to frequency division multiplexing. It is an analog technology, and therefore suffers from the shortcomings that have historically plagued all transmission systems. The wire over which information is transmitted behaves like a long-wire antenna, picking up noise along the length of the transmission path and very effectively homogenizing it with the voice signal. Additionally, the power of the transmitted signal diminishes over distance, and if the distance is far enough the signal will have to be amplified to make it intelligible at the receiving end. Unfortunately, the amplifiers used in the network are not particularly discriminating: they have no way of separating the voice noise; the result is that they convert a weak, noisy signal into a loud noisy signal—better, but far from ideal. A better solution was needed.

The better solution came about with the development of time-division multiplexing, which became possible because of the transistor and integrated circuit electronics that arrived in the late 1950s and early 1960s. TDM is a digital transmission scheme, which implies a small number of discrete signal states, rather than the essentially infinite range of values employed in analog systems. While digital systems are as susceptible to noise impairment as their analog counterparts, the discrete nature of their binary signaling makes it relatively easy to separate the noise from the transmitted signal. In digital carrier systems, there are only three valid signal values, one positive, one negative, and zero; anything else is construed to be noise. It is therefore a trivial exercise for digital repeaters to discern what is desirable and what is not, thus eliminating the problem of cumulative noise. The role of the regenerator, as shown is Fig. 4.66, is to receive a weak, noisy digital signal, remove the noise, reconstruct the original signal, and amplify it before transmitting the signal onto the next segment of the transmission facility. For this reason, repeaters are also called *regenerators*, because that is precisely the function they perform.

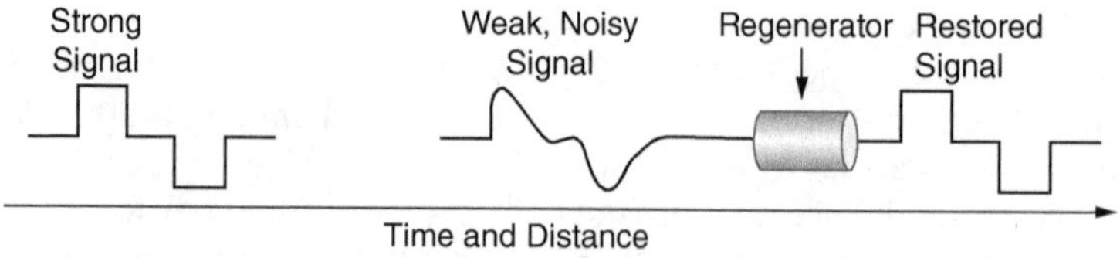

FIGURE 4.66 The role of the regenerator in transmission systems.

One observation: it is estimated that as much as 60% of the cost of building a transmission facility lies in the regenerator sections of the span. For this reason, optical networking, discussed a bit later, has various benefits, not the least of which is the ability to reduce the number of regenerators required on long transmission spans. In a typical network, these regenerators must be placed approximately every 6,000 ft along a span, which means that there is considerable expense involved when providing regeneration along a long-haul network.

Digital signals, often called *square waves*, comprise a very rich mixture of signal frequencies, as we discussed earlier in the book. Not to bring too much physics into the discussion, we must at least revisit the Fourier series, which describes the makeup of a digital signal. The Fourier series is a mathematical representation of the behavior of waveforms. Among other things, it notes the following fact: If we start with a fundamental signal and mathematically add to it its odd harmonics (a harmonic is defined as a wave whose frequency is a whole-number multiple of another wave), we see a rather remarkable thing happening: the waveform gets steeper on the sides and flatter on top. As we add more and more of the odd harmonics (there is, after all, an infinite series of them), the wave begins to look like the typical square wave. Now of course, there is no such thing as a true square wave; for our purposes, though, we'll accept the fact.

It should now be intuitive to the reader that digital signals comprise a mixture of low-, medium-, and high-frequency components, which means that they cannot be transmitted across the bandwidth-limited 4-KHz channels of the traditional telephone network. In digital carrier facilities, the equipment that restricts the individual transmission channels to 4 kHz *chunks* is eliminated, thus giving each user access to the full breadth of available spectrum across the shared physical medium. In frequency division systems, we observed that we give users *some of the frequency all of the time;* in time division systems, we turn that around and give users *all* of the frequency *some* of the time. As a result, high-frequency digital signals can be transmitted without restriction.

Digitization brings with it a cadre of advantages, including improved voice and data-transmission quality; better maintenance and troubleshooting capability, and therefore reliability; and dramatic improvements in configuration flexibility. In digital carrier systems, the time division multiplexer is known as a channel bank; under normal circumstances, it allows either 24 or 30 circuits to share a single, four-wire facility. The 24-channel system is called T-carrier; the 30-channel system, used in most of the world, is called E-carrier. Originally designed in 1962, as a way to transport multiple channels of voice over expensive transmission facilities, they soon became useful as data transmission networks as well. That, however, came later. For now, we focus on voice.

Voice Digitization

The process of converting analog voice to a digital representation in the modern network is a logical and straightforward process. It comprises four distinct steps: pulse amplitude modulation (PAM) sampling, in which the amplitude of the incoming analog wave is sampled every 125 ms (microseconds); companding, during which the values are weighted toward those most receptive to the human ear; quantization, in which the weighted samples are given values on a nonlinear scale; and finally encoding, during which each value is assigned a distinct binary value. Each of these stages of pulse code modulation (PCM) will now be discussed in detail.

Pulse Code Modulation

Thanks to the work performed by Harry Nyquist at Bell Laboratories in the 1920s, we know that to optimally represent an analog signal as a digitally encoded bitstream, the analog signal must be sampled at a rate that is equal to twice the bandwidth of the channel over which the signal is to be transmitted. Since each analog voice channel is allocated 4 KHz of bandwidth, it follows that each voice signal must be sampled at twice that rate, or 8,000 samples per second. In fact, that is precisely what happens in T-carrier systems, which we now use to illustrate our example. The standard T-carrier multiplexer accepts inputs from 24 analog channels as shown in Fig. 4.67. Each channel is sampled in turn, every one eight-thousandth of a second in round-robin fashion, resulting in the generation of 8,000 pulse amplitude samples from each channel every second. The sampling rate is important. If the sampling rate is too high, too much information is transmitted, and bandwidth is wasted; if the sampling rate is too low, then we run the risk of aliasing. Aliasing is the interpretation of the sample points as a false waveform, due to the paucity of samples.

This PAM process represents the first stage of PCM, the process by which an analog baseband signal is converted to a digital signal for transmission across the T-carrier network. This first step is shown in Fig. 4.68.

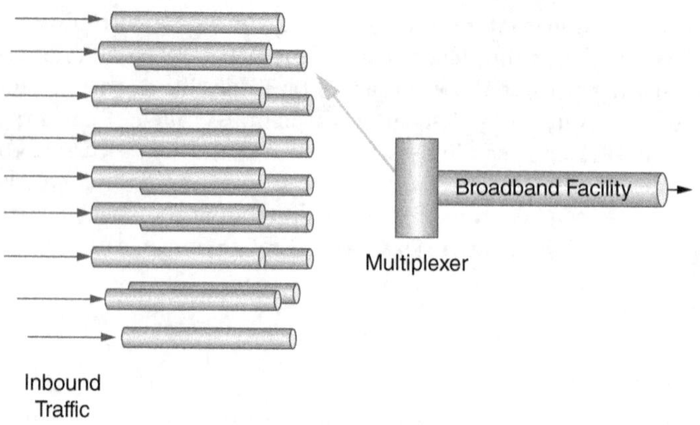

FIGURE 4.67 The channelization of a typical T-carrier.

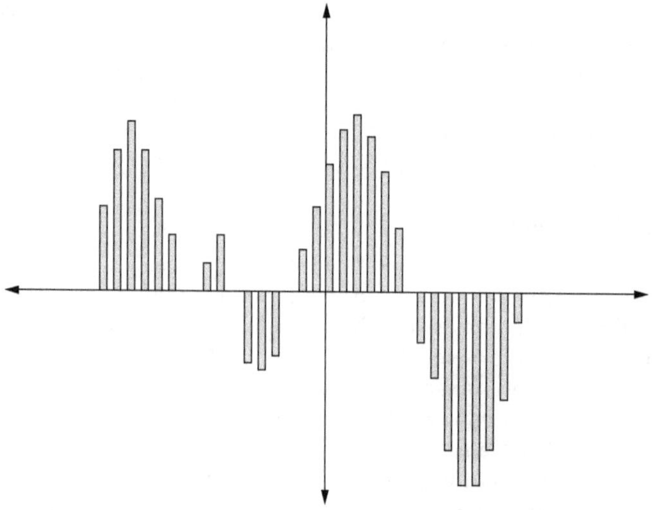

FIGURE 4.68 Pulse amplitude modulation, the first stage of voice digitization.

The second stage of PCM, shown in Fig. 4.69, is called quantization. In quantization, we assign values to each sample within a constrained range. For illustration purposes, imagine what we now have before us. We have *replaced* the continuous analog waveform of the signal with a series of amplitude samples which are close enough together that we can discern the shape of the original wave from their collective amplitudes. Imagine also that we have graphed these samples in such a way that the *wave* of sample points meanders above and below an established zero point on the x axis, so that some of the samples have positive values and others are negative, as shown.

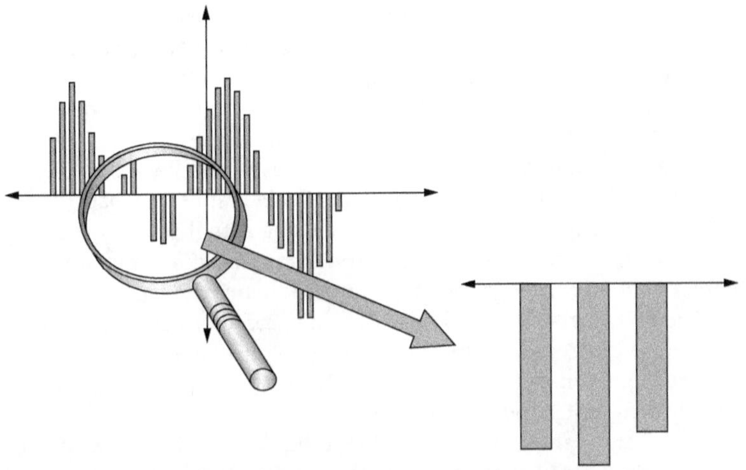

FIGURE 4.69 Quantization, the second stage of voice digitization.

The amplitude levels allow us to assign values to each of the PAM samples, although a glaring problem with this technique should be obvious to the careful reader. Very few of the samples actually line up *exactly* with the amplitudes delineated by the graphing process. In fact, most of them fall *between* the values, as shown in the illustration. It doesn't take much of an intuitive leap to see that several of the samples will be assigned the same digital value by the coder-decoder that performs this function, yet they are clearly *not* the same amplitude. This inaccuracy in the measurement method results in a problem known as *quantizing noise,* and is inevitable when linear measurement systems, such as the one suggested by the drawing, are employed in coder-decoders (CODECs).

Needless to say, design engineers recognized this problem rather quickly, and equally quickly came up with an adequate solution. It is a fairly well-known fact among psycholinguists and speech therapists that the human ear is far more sensitive to discrete changes in amplitude at low-volume levels than it is at high-volume levels, a fact not missed by the network designers tasked with optimizing the performance of digital carrier systems intended for voice transport. Instead of using a linear scale for digitally encoding the PAM samples, they designed and employed a nonlinear scale that is weighted with much more granularity at low volume levels—that is, close to the zero line—than at the higher amplitude levels. In other words, the values are extremely close together near the x axis, and get farther and farther apart as they travel up and down the y axis. This nonlinear approach keeps the quantizing noise to a minimum at the low-amplitude levels where hearing sensitivity is the highest, and allows it to creep up at the higher amplitudes, where the human ear is less sensitive to its presence. It turns out that this is not a problem, because the inherent shortcomings of the mechanical equipment (microphones, speakers, the circuit itself) introduce slight distortions at high-amplitude levels that hide the effect of the nonlinear quantizing scale.

This technique of *compressing* the values of the PAM samples to make them fit the nonlinear quantizing scale results in a bandwidth savings of more than 30 percent. In fact, the actual process is called *companding*, because the sample is first compressed for transmission, then expanded for reception at the far end, hence the term.

The actual graph scale is divided into 255 distinct values above and below the zero line. In North America and Japan, the encoding scheme is known as μ-law (mu-law); the rest of the world relies on a slightly different standard known as A-law.

There are eight segments above the line and eight below (one of which is the shared zero point); each segment, in turn, is subdivided into 16 steps. A bit of binary mathematics now allows us to convert the quantized amplitude samples into an 8-bit value for transmission. For the sake of demonstration, let's consider a negative sample that falls into the 13th step in segment 5. The conversion would take on the following representation:

1 101 1101

Where the initial '0' indicates a negative sample, '101' indicates the fifth segment, and '1101' indicates the thirteenth step in the segment. We now have an 8-bit representation of an analog amplitude sample that can be transmitted across a digital network, then reconstructed with its many counterparts as an accurate representation of the original analog waveform at the receiving end. This entire process is known as PCM, and the result of its efforts is often referred to as *toll-quality voice.*

Alternative Digitization Techniques

While PCM is perhaps the best-known, high-quality voice digitization process, it is by no means the only one. Advances in coding schemes and improvements in the overall quality of the telephone network have made it possible for encoding schemes to be developed that use far less bandwidth than traditional PCM. In this next section, we will consider some of these techniques.

Adaptive Differential Pulse Code Modulation

Adaptive differential pulse code modulation (ADPCM) is a technique that allows toll-quality voice signals to be encoded at half-rate (32 Kbps) for transmission. ADPCM relies on the predictability that is inherent in human speech to reduce the amount of information required. The technique still relies on PCM encoding, but adds an additional step to carry out its task. The 64-Kbps PCM-encoded signal is fed into an ADPCM transcoder, which considers the *prior* behavior of the incoming stream to create a prediction of the behavior of the *next* sample. Here's where the magic happens: instead of transmitting the actual value of the predicted sample, it encodes in four bits and transmits the *difference* between the actual and predicted samples. Since the difference from sample to sample is typically quite small, the results are generally considered to be very close to toll-quality. This 4-bit transcoding process, which is based on the known behavior characteristics of human voice, allows the system to transmit 8,000 4-bit samples per second, thus reducing the overall bandwidth requirement from 64 to 32 Kbps. It should be noted that ADPCM works well for voice, because the encoding and predictive algorithms are based upon its behavior characteristics. It does not, however, work as well for higher bit rate data (above 4800 bps), which has an entirely different set of behavior characteristics.

Continuously Variable Slope Delta

Continuously variable slope delta (CVSD) is a unique form of voice encoding that relies on the values of individual bits to predict the behavior of the incoming signal. Instead of transmitting the volume (height or y-value) of PAM samples, CVSD transmits information that measures the changing slope of the waveform. So instead of transmitting the actual change itself, it transmits the *rate* of change.

To perform its task, CVSD uses a reference voltage to which it compares all incoming values. If the incoming signal value is less than the reference voltage, then the CVSD encoder reduces the slope of the curve to make its approximation better mirror the slope of the actual signal. If the incoming value is *more* than the reference value, then the encoder will increase the slope of the output signal, again causing it to approach and therefore mirror the slope of the actual signal. With each recurring sample and comparison, the step function can be increased or decreased as required. For example, if the signal is increasing rapidly, then the steps are increased one after the other in a form of step function by the encoding algorithm. Obviously, the reproduced signal is not a particularly exact representation of the input signal: in practice, it is pretty jagged. Filters, therefore, are used to smooth the transitions.

CVSD is typically implemented at 32 Kbps, although it can be implemented at rates as low as 9,600 bps. At 16 to 24 Kbps, recognizability is still possible; down to 9,600, recognizability is seriously affected although intelligibility is not.

Linear Predictive Coding

We mention linear predictive coding (LPC) here only because it has carved out a niche for itself in certain voice-related applications such as voice mail systems, automobiles, aviation, and electronic games that speak to children. LPC is a complex process, implemented completely in silicon, which allows for voice to be encoded at rates as low as 2,400 bps. The resulting quality is far from toll-quality, but it is certainly intelligible and its low bit rate capability gives it a distinct advantage over other systems.

Linear predictive coding relies on the fact that each sound created by the human voice has unique attributes, such as frequency range, resonance, and loudness, among others. When voice samples are created in LPC, these attributes are used to generate *prediction coefficients.* These predictive coefficients represent linear combinations of previous samples, hence the name, *linear predictive coding.*

Prediction coefficients are created by taking advantage of the known *formants* of speech, which are the resonant characteristics of the mouth and throat, which give speech its characteristic timbre and sound. This sound, referred to by speech pathologists as the *buzz,* can be described by both its pitch and its intensity. LPC, therefore, models the behavior of the vocal cords and the vocal tract itself.

To create the digitized voice samples, the buzz is passed through an inverse filter that is selected based upon the value of the coefficients. The remaining signal, after the buzz has been removed, is called the residue.

In the most common form of LPC, the residue is encoded as either a *voiced* or *unvoiced* sound. Voiced sounds are those that require vocal cord vibration, such as the *g* in *glare,* the *b* in *boy,* the *d* and *g* in *dog.* Unvoiced sounds require no vocal cord vibration, such as the *h* in *how,* the *sh* in *shoe,* and the *f* in *frog.* The transmitter creates and sends the prediction coefficients, which include measures of pitch, intensity and whatever voiced and unvoiced coefficients that are required. The receiver undoes the process: it converts the voice residue, pitch and intensity coefficients into a representation of the source signal, using a filter similar to the one used by the transmitter to synthesize the original signal.

Digital Speech Interpolation

Human speech has many measurable (and therefore predictable) characteristics, one of which is a tendency to have embedded pauses. As a rule, people do not spew out a series of uninterrupted sounds; they tend to pause for emphasis, to collect their thoughts, to reword a phrase while the other person listens quietly on the other end of the line. When speech technicians monitor these pauses, they discover that during considerably more than half of the total connect time, the line is silent.

Digital speech interpolation (DSI) takes advantage of this characteristic silence to drastically reduce the bandwidth required for a single channel. Whereas 24 channels can be transported over a typical T-1 facility, DSI allows for as many as 120 conversations to be carried over the same circuit. The format is proprietary, and requires the setting aside of a certain amount of bandwidth for overhead.

A form of *statistical multiplexing* lies at the heart of DSI's functionality. Standard T-carrier is a time division multiplexed scheme, in which channel ownership is assured: a user assigned to channel three will *always* own channel three, regardless of whether they are actually using the line. In DSI, channels are not owned. Instead, large numbers of users share a *pool* of available channels. When a user starts to talk, the DSI system assigns an available time slot to that user and notifies the receiving end of the assignment. This system works well when the number of users is large, because statistical probabilities are more accurate and indicative of behavior in larger populations than in smaller ones.

There is a downside to DSI, of course, and it comes in several forms. *Competitive clipping* occurs when more people start to talk than there are available channels, resulting in someone being unable to talk. *Connection clipping* occurs when the receiving end fails to learn what channel a conversation has been assigned within a reasonable amount of time, resulting in signal loss. Two approaches have been created to address these problems; in the case of competitive clipping, the system intentionally *clips* off the front end of the initial word of the second person who speaks. This technique is not optimal, but does prevent loss of the conversation and also obviates the problem of clipping out the middle of a conversation, which would be more difficult for the speakers to recover from. The loss of an initial syllable or two can be mentally reconstructed far more easily than sounds in the middle of a sentence.

A second technique used to recover from clipping problems is to temporarily reduce the encoding rate. The typical encoding rate for DSI is 32 Kbps; in certain situations, the encoding rate may be reduced to 24 Kbps, thus freeing up significant bandwidth for additional channels.

Both techniques are widely utilized in DSI systems.

Enter the Modern World: Voice-over IP

For the longest time, beginning in the late 1990s, telecom professionals viewed VoIP the same way that many IT professionals viewed the Macintosh in the 1980s: As a passing fad, a toy technology that would have its brief time in the sun and then fade away. Skype, for example, appeared on the scene in 2003, making it possible to make computer-to-computer phone calls (and soon thereafter, video calls) anywhere in the world, free of charge. Vonage followed with a commercial voice offering for home and business, branding themselves with the very recognizable music from Kill Bill and securing themselves a place among the pantheon of telecom gods.

VoIP has fundamentally changed the way people talk on a daily basis; in this section we'll describe what it is, how it works, and how it is being used by a wide array of both residential and enterprise customers.

VoIP versus Internet Telephony: An Important Distinction

Let's examine how a telephone call is carried across an IP network (we'll assume that this is a traditional phone call, not a call using Skype or other computer-to-computer services). Please refer to Fig. 4.70.

A customer begins the call using a traditional telephone. The call is carried across the PSTN to an IP telephony gateway, which is nothing more than a special purpose router designed to interface between the PSTN and a packet-based IP network. As soon as the gateway receives the call, it interrogates an associated gatekeeper device that provides information about billing, authorization, authentication, and call routing. As soon as the gatekeeper has delivered the information

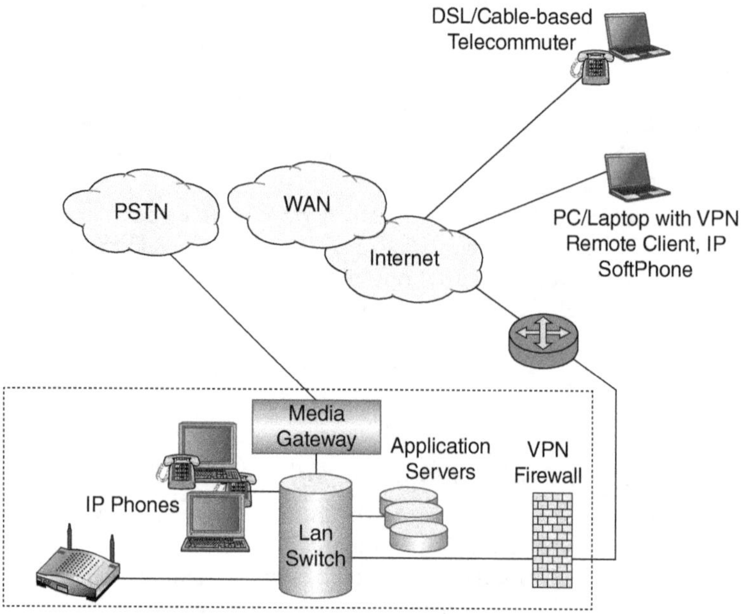

Figure 4.70 Layout of a typical VoIP network, showing interconnection with the Internet.

to the gateway, the gateway transmits the call across the IP network to another gateway, which in turn hands the call off to the local PSTN at the receiving end, completing the call.

Let's also address one important misconception. At the risk of sounding Aristotelian, "All Internet telephony is VoIP, but all VoIP is not Internet telephony." Got it? Let me explain. It is quite possible today to make telephone calls across the Internet, which by definition are IP-based simply because IP is the fundamental operational protocol of the Internet. However, IP-based calls can be made without involving the Internet at all. For example, a corporation may interconnect multiple locations using dedicated private line facilities or SIP trunks (we'll discuss this shortly in the section entitled, "IP-enabled Call Centers"), and they may transport IP-based phone calls across that dedicated network facility. This is clearly VoIP, but is *not* Internet telephony. There's a big difference.

VoIP Evolution

There was a time when VoIP was only for the brave: Those intrepid explorers among us who were willing to try Internet-based telephony just to prove that it could be done, with little regard for QoS, enhanced services and the like. In many ways it was reminiscent of the 70s, when we all drove around in our cars and talked to each other on CB radios (Remember? Oh, come on…"10-4, Rubber Duck. See you on the flip side." Don't deny it…). The quality was about that good, and it was experimental, and new, and somewhat exciting. It faded in and out of the public's awareness, but didn't really catch on in a big way until late 2002 or so, when the level of broadband penetration reached a point where adequate access bandwidth became available and CODEC technology far enough advanced to make VoIP not only possible but actually quite good. Everyone toyed with the technology; Cisco, Avaya, Lucent, and Nortel, not to mention Ericsson and Siemens, all built VoIP products to one degree or another because they knew that sooner or later the technology would become mainstream. But as always happens, it took *early introducers* (not early adopters) to make the transition from experiment to service. And while companies like TELUS in Canada played enormous roles in the ongoing development of VoIP (They were the first company to install a system-wide IP backbone, and one of the first innovators to implement an MPLS backbone and offer true VoIP services), it took one off-the-wall company to really push it to the front of the public consciousness. That company was Skype.

Skype

I knew it was over when I downloaded Skype. When the inventors of KaZaA are distributing for free a little program that you can use to talk to anybody else,

> *and the quality is fantastic, and it's free—it's over. The world will change now inevitably.*
>
> —Michael Powell, Chairman, FCC

> *The idea of charging for calls belongs to the last century. Skype software gives people new power to affordably stay in touch with their friends and family by taking advantage of their technology and connectivity investments.*
>
> —Niklas Zennström, CEO & Co-founder of Skype

Many readers will recall KaZaA, the file-sharing program that became well-known during the turbulent Napster controversy over the unauthorized sharing of commercial music. The same two programmers who created KaZaA, Niklas Zennström and Janus Friis, put their heads together and concluded that while peer-to-peer music works well, it is fraught with legal ramifications. So what else could be shared in a peer-to-peer fashion? *Voice.*

To date nearly 40 million people have downloaded the free Skype installer, and as I sit here writing this section in mid-December 2013 there are more than 50 million people online. I have used Skype for many years and the service, which now includes messaging, videoconferencing, and multipoint communications, is excellent.

Vonage and Skype are but two of a sizeable and growing herd of public VoIP providers that have redefined the world of voice services. In the enterprise space, VoIP is having an equally dramatic impact. Let's examine the various ways in which enterprise VoIP can be deployed to the benefit of customers.

Carrier Class IP Voice

There is no question that VoIP is a serious contender for voice services in both the residence and enterprise markets.

From a customer's perspective, the principal advantages of voice over IP include consolidated voice, data, and multimedia transport, elimination of parallel systems, the ability to exercise PSTN fallback in the event that the IP network becomes overly congested, and reduction of long-distance voice costs.

For a service provider, the advantages are different but no less dramatic, including efficient use of network investments due to traffic consolidation, new revenue sources from existing clients because of demand for service-oriented applications that benefit from being offered over an IP infrastructure, and the option of transaction-based billing. These can collectively be reduced to the general category of customer service, which service providers such as ISPs and CLECs should be focused on. The challenge they face will be to prove that the service quality they provide over their IP networks will be identical to that provided over traditional circuit switched technology.

Major carriers are voting on IP with their own wallets, a sure sign of confidence in the technology.

Winning with VoIP

IP is a powerful motivator for change, but that doesn't mean it isn't fraught with a certain degree of risk. One thing we do know is this: Companies will succeed in their evolution to an all-IP network if—and *only* if—they embrace the IP migration as a business transformation as much as it is a network transformation.

What IP and VoIP bring to the table more than anything else is the opportunity to modify the organizational operating and provisioning model, to simplify their infrastructure, and to focus on a stable of standardized products that meet the needs of the vast majority of their customers. This is a significant evolutionary decision that, if it is to take root and bear fruit, must emphasize three specific things. First, it must reduce the complexity of delivered services. Second, it must pave the road for the easy provisioning of standardized services. And finally, it must make it possible for the customer to self-provision.

An interesting fact about the telecom services industry is that roughly three-quarters of all customized services are self-imposed. In other words, they're not made available because a customer asked for them, but rather because the service provider believed that customers would want them. The thing is, customized services are costly, yet the cost is often overlooked. This doesn't mean that service providers should stop offering customized services. If they are to offer them, however, they should only offer those that can be done in software so that the customer can self-provision them, and should be provisioned from the edge of the network rather than from the core. The closer the point of service origin is to the customer, the more easily the service can be customized and delivered at a low cost. Along the same lines, service providers should rely on a limited set of hardware, network, and service agreement options to reduce cost.

Of course, there is a downside, and that is that sales teams are traditionally paid on the basis of the sale of products. To get the most out of VoIP as a product strategy, sales teams must be retrained to sell on the basis of *return on solution,* on their ability to demonstrate the strategic value of IP and VoIP.

The key to IP's success in the voice provisioning arena lies with its *invisibility.* If done correctly, service providers can add IP to their networks, maintaining service quality while dramatically improving their overall efficiency. IP voice (not to be confused with Internet voice) will be implemented by carriers and corporations as a way to reduce costs and move to a multiservice network fabric. Virtually every major equipment manufacturer has added SS7 and IP voice capability to their router products and access devices, recognizing that their primary customers are looking for IP solutions.

IP-Enabled Call Centers

Ultimately, well-run call centers are nothing more than enormous routers. They receive incoming data delivered using a variety of media (phone calls, e-mail messages, facsimile transmissions, and social media) and make decisions about handling them based on information contained in each message. One challenge that has always faced call center management is the ability to integrate message types and route them to a particular agent based on specified decision-making criteria such as name, address, telephone number, e-mail address, automatic number identification (ANI) triggers, product purchase history, the customer's geographic location, or language preference. This has resulted in the development of a technical philosophy (and ultimately product launch) called *unified messaging*. With unified messaging, all incoming messages for a particular agent, regardless of the media over which they are delivered, are housed centrally and clustered under a single account identifier. When the agent logs into the network, their PC lists all of the messages that have been received for them, giving them the ability to much more effectively manage the information contained in those messages.

Today, unified messaging systems also support road warriors. A traveling employee can dial into a message gateway and download all messages—voice, fax, e-mail—from that one central location, thus dramatically simplifying the process of staying connected while away from the office.

Call centers are undergoing tremendous change as the IP juggernaut hits them. The first of these is a redefinition of the market they serve. For the longest time, call centers have primarily served large corporations because they are expensive to deploy. With the arrival of IP, however, the cost is dropping precipitously, and all major corporations are now moving toward an IP-enabled call center model because of the ability to create unified application sets and to introduce enormous flexibility into their calling models.

The beauty of this is that the evolution supports one of the most important changes taking place in the domain of the contact center. The marketing budget dedicated to customer loyalty and retention in most companies is growing rapidly. Peer evaluation sites (think Yelp) have exploded in the last few years, and the universal customer history record has become the number one customer service trend under the banner of customer experience. One manifestation of this is that customers are demanding a level of experience quality that surpasses anything they have experienced in the past.

Integrating the PBX

There is an enormous installed base of legacy PBX equipment, and vendors did not initially enter the IP game enthusiastically. Early entrants arrived with enhancements to existing equipment that were proprietary and expensive, and did very little to raise customer awareness or engender trust in the vendor's migration strategy. Over time, however, PBX manufacturers began to embrace the concept of convergence as their customers' demands for IP-based systems grew, and soon products began to appear. Most have heard the message delivered by the customer: preserve the embedded base to the degree possible as a way of saving the existing investment; create a migration strategy that is seamless and as transparent as possible; and preserve the features and functions that customers are already familiar with to minimize churn.

Major vendors have responded with products that do exactly that. They allow voice calls and faxes to be carried over LAN, WANs, the Internet, and corporate intranets and function as both a gateway and gatekeeper, providing circuit-to-packet conversion, security, and access to a wide variety of applications including enhanced call features such as multiple line appearances, hunt groups, multiparty conferencing, call forward, hold, call transfer, and speed dial. They also provide access to voice mail, CTI applications, wireless interfaces, and call center features.

An Important Aside: Billing as a Critical Service

One area that is often overlooked when companies look to improve the quality of the services they provide to their customers is billing. And while it is not typically viewed as a strategic competitive advantage, studies have shown that customers view it as one of the top considerations when assessing capability in a service provider.

Billing offers the potential to strengthen customer relationships, improve long-term business health, cement customer loyalty, and generally make businesses more competitive. However, for billing to achieve its maximum benefit and strategic value it must be fully integrated with a company's other operations support systems including network and service-provisioning systems, installation support, repair, network management, and sales and marketing. If done properly the billing system becomes an integral component of a service suite that allows the service provider to quickly and efficiently introduce new and improved services in logical bundles; improve business indicators such as service timeliness, billing accuracy, and cost; offer custom service programs to individual customers based on individual service profiles; and transparently migrate from legacy service platforms to the so-called next generation network.

In order for billing as a strategic service to work successfully, service providers must build a business plan and migration strategy that takes into account integration with existing operations support systems; business process interaction; the role of IT personnel and processes; and post-implementation testing to ensure compliance with strategic goals stipulated at the beginning of the project.

VoIP Supporting Protocols

In addition to IP, VoIP relies on a superset of additional protocols to guarantee the level of rich functionality that users have come to expect. The relative importance of the protocols discussed in the sections that follow wax and wane like the stages of the moon, but all are important, and readers should be familiar with them.

H.323

H.323 started as H.320 in 1996, an ITU-T standard for the transmission of multimedia content over ISDN. Its original goal was to connect LAN-based multimedia systems to network-based multimedia systems. It defined a network architecture that included gatekeepers, which performed zone management and address conversion; endpoints, which were terminals and gateway devices; and multimedia control units, which served as bridges between multimedia types.

H.323 has been rolled out in four phases. Phase one defined a three-stage call setup process: a pre-call step, which performed user registration, connection admission, and exchange of status messages required for call setup; the actual call setup process, which used messages similar to ISDN's Q.931; and finally, a capability exchange stage, which established a logical communications channel between the communicating devices and identified conference management details.

Phase two allowed for the use of real-time transport protocol (RTP) over ATM, which eliminated the added redundancy of IP and also provided for privacy and authentication as well as greatly demanded telephony features such as call transfer and call forwarding. RTP has an added advantage: when errors result in packet loss, RTP does not request resends of those packets, thus providing for real-time processing of application-related content. No delays result from errors.

Phase three added the ability to transmit real-time fax after establishing a voice connection, and phase four, released in May 1999, added call connection over UDP, which significantly reduced call setup time; inter-zone communications; call hold, park, pickup; and call and message waiting features. This last phase bridged the considerable gap between IP voice and IP telephony.

Several Internet telephony interoperability concerns are addressed by H.323. These include gateway-to-gateway interoperability, which ensures that telephony can be accomplished between different vendors' gateways; gatekeeper-to-gatekeeper interoperability, which does the same thing for different vendors' gatekeeper devices; and finally, gateway-to-gatekeeper interoperability, which completes the interoperability picture.

Session Initialization Protocol

While H.323 has its share of supporters, it is being edged out by the IETF's session initialization protocol (SIP). SIP supporters claim that H.323 is far too complex and rigid to serve as a standard for basic telephony setup requirements, arguing that SIP, which is architecturally simpler and imminently extensible, is a better choice. In reality H.323 is an *umbrella standard* that includes (among others) H.225 for call handling; H.245 for call control; G.711 and G.721 for CODEC definitions; and T.120 for data conferencing. Originally created as a technique for transporting multimedia traffic over a local area network, gatekeeper functions have been added that allow LAN traffic and LAN capacity to be monitored so that calls are established only if adequate capacity is available on the network. Later, the gatekeeper routed model was added, which allowed the gatekeeper to play an active role in the actual call setup process. This meant that H.323 had migrated from being purely a peer-to-peer protocol to having a more traditional, hierarchical design.

The greatest advantage that H.323 offers is maturity. It has been available for some time now, and while robust and full-featured, it was never designed to serve as a peer-to-peer protocol. Its maturity, therefore, is not enough to carry it. It currently lacks a network-to-network interface, nor does it adequately support congestion control. This is not generally a problem for private networks, but it does become problematic for service providers that wish to interconnect PSTNs and provide national service among a cluster of providers. As a result many service providers have chosen to deploy SIP instead of H.323 in their national networks.

Why SIP?

SIP is now considered a mature protocol and has found a *home* in the growing IP services pantheon. Part of the reason for its success is a radical shift in market demand and a shift in the architecture of the services delivery model.

One shift that has taken place, thanks to the growing capabilities of cloud services, is the migration of content and applications from the user's device where they have traditionally resided to a centralized data store in the cloud, a trend that is sometimes called *Dumb Terminal 2.0.* Those of you who have been in the industry for a while will recall the days of the dumb terminal, an intelligence-free display device (hence the name, dumb terminal) connected to a centralized mainframe computer (for all intents and purposes, a cloud resource) over a low-speed facility. All of the intelligence in those early systems was concentrated in the mainframe, leaving the device to serve as a display.

Today, a similar thing is happening in the mobile world of laptops, tablets, and phones. Applications are moving into the cloud where they execute; a small client replaces them on the mobile device. The point is that this phenomenon frees up "real estate" on the mobile device that can be used for other functions—expanded memory, a more complex camera (or in some cases, multiple cameras), a larger battery, more radio technology.

A second piece of this evolutionary puzzle that is catalyzing the role of SIP is unified communications (UC). UC assumes a common IP protocol platform and includes unified voicemail, fax, and e-mail; IP telephony; presence; Web conferencing; document sharing; and cross-platform application execution. UC is growing at a rapid pace; according to consultancy Frost & Sullivan, three-quarters of all telephony ports shipped in 2009 were intended for IP telephony systems. That number is expected to increase to 90% by the end of 2016.

SIP extends the capabilities of IP to include signaling for both wireline and wireless networks, thus bridging the gap between different networks, PBXs, and a host of end-user devices.

SIP Trunks

SIP trunks are an evolutionary benefit of SIP's growing influence, as it inexorably edges SS7 out of the pole position of signaling protocols. There's nothing magical about a SIP trunk; they're virtual circuits that are provisioned across a private IP backbone into the Internet. They are most commonly used as a replacement technology for the ISDN PRIs that most commonly interconnect an IP-enabled PBX (sometimes called B-VoIP) to the wider area global network. They have another advantage: they can provide tremendous reliability and fallback capability in the event of a network failure.

SIP is designed to establish peer-to-peer sessions between Internet routers. The protocol defines a variety of server types, including feature servers, registration servers, and redirect servers. SIP supports fully distributed services, which reside in the actual user devices, and because these are based on existing IETF protocols, they provide a seamless integration path for voice/data integration.

Ultimately, telecommunications, like any industry, revolves around profitability. Any protocol that allows new services to be deployed inexpensively and quickly immediately catches the eye of service providers. Like TCP/IP, SIP provides an open architecture that can be used by any vendor to develop products, thus ensuring multivendor interoperability. And because SIP has been adopted by all of the major equipment manufacturers and is designed for use in large carrier networks with potentially millions of ports, its success is assured.

Originally, H.323 was to be the protocol of choice to make this possible. And while H.323 is clearly a capable suite of protocols and is indeed quite good for voice over IP services that derive from ISDN implementations, it is still incomplete and is quite complex. As a result it has been relegated to use as a video control protocol and for some gatekeeper-to-gatekeeper communications functions.

The intense interest in moving voice to an IP infrastructure is driven by simple and understandable factors: cost of service and enhanced flexibility. However, in keeping with the "Jurassic Park Effect" (Just because you *can*, doesn't necessarily mean you *should.*), it is critical to understand the differences that exist between simple voice and full-blown telephony with its many enhanced features. It is the feature set that gives voice its range of capability; a typical local switch such as Lucent Technologies' 5ESS offers more than 3,000 features, and more will certainly follow. Of course, the features and services are possible because of the protocols that have been developed to provide them across an IP infrastructure.

Media Gateway Control Protocol—and Friends

Many of the protocols that are guiding the successful development of VoIP efforts today stem from work performed early on by Level 3 and Telcordia, which together founded an organization called the International SoftSwitch Consortium. In 1998, Level 3 brought together a collection of vendors that collaboratively developed and released the Internet protocol device control (IPDC). At the same time, Telcordia created and released the simple gateway control protocol (SGCP). The two were later merged to form the media gateway control protocol (MGCP), discussed in detail in RFC 2705.

MGCP allows a network device responsible for establishing calls to control the devices that actually perform IP voice streaming. It permits software call agents and media gateway controllers to control streaming media gateways at the edge of the network. These gateways can be cable modems, set top boxes, PBXs, VTOA gateways, and VoIP gateways. Under this design, the gateways manage the circuit-switch-to-IP voice conversion, while the agents manage signaling and call processing.

MGCP makes the assumption that call control in the network is software based, resident in external intelligent devices that perform all call control functions. It also makes the assumption that these devices will communicate with one another in a primary-secondary arrangement, under which the call agents send instructions to the gateways for execution.

Application	Required Bandwidth	Sensitivity to Delay
Voice	Low	High
Video	High	Medium to high
Medical imaging	Medium to high	Low
Web surfing	Medium	Medium
LAN interconnection	Low to high	Low
Electronic mail	Low	Low

Meanwhile, Lucent created a new protocol called the media device control protocol (MDCP). The best features of the original three were combined to create a full-featured protocol called the MeGaCo, also defined as H.248. In March 1999, the IETF and ITU met collaboratively and created a formal technical agreement between the two organizations, which resulted in a single protocol with two names. The IETF calls it MeGaCo, the ITU calls it H.GCP.

MeGaCo/H.GCP operates under the assumption that network intelligence is housed in the central office, and therefore replaces the gatekeeper concept proposed by H.323. By managing multiple gateways within a single IP-equipped central office, MeGaCo minimizes the complexity of the telephone network. In other words, a corporation might be connected to an IP-capable central office, but because of the IP-capable switches in the CO that have the ability to convert between circuit switched and packet switched voice, full telephony features are possible. Thus the next generation switch converts between circuit and packet, while MeGaCo performs the signaling necessary to establish a call across an IP wide area network. It effectively bridges the gap between legacy SS7 signaling and the new requirements of IP, and supports both connection-oriented and connectionless services.

A Final Thought: Network Management for QoS

Because of the diverse audiences that require network performance information and the importance of service level agreements (SLAs), the data collected by network management systems must be malleable so that it can be formatted for different sets of corporate eyes. For the purposes of monitoring performance relative to SLAs, customers require information that details the performance of the network relative to the requirements of their applications. For network operations personnel, reports must be generated that detail network performance to ensure that the network is meeting the requirements of the SLAs that exist between the service provider and the customer. Finally, for the needs of sales and marketing organizations, reports must be available which allow them to properly represent the company's capabilities to customers, and to allow them to anticipate requirements for network augmentation and growth.

For the longest time, the telecommunications management network (TMN) has been considered the ideal model for network management. As the network profile has changed, however, with the steady migration to IP and a renewed focus on service rather than technology, the standard TMN philosophy has begun to appear somewhat tarnished.

Originally designed by the ITU-T, TMN is built around the OSI model and its attendant standards, which include the common management information protocol (CMIP) and the guidelines for the development of managed objects (GDMO).

TMN employs a model, shown in Fig. 4.71, comprising a *network element layer*, an *element management layer*, a *network management layer*, a *service management layer*, and a *business management layer*. Each has a specific set of responsibilities closely related to those of the layers that surround it.

The network element layer defines each manageable element in the network on a device by device basis. Thus, the manageable characteristics of each device in the network are defined at this functional layer.

The element management layer manages the characteristics of the elements defined by the network element layer. Information found here includes activity log data for each element. This layer houses the actual element management systems responsible for the management of each device or set of devices in the network.

The network management layer has the ability to monitor the entire network based upon information provided by the element management layer.

The services management layer responds to information provided by the network management layer to deliver such service

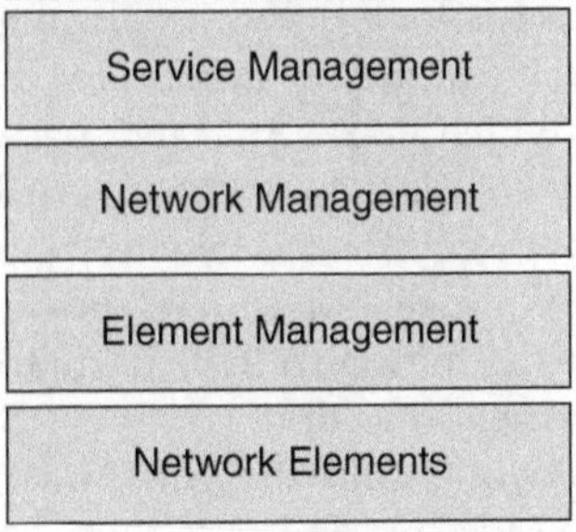

FIGURE 4.71 The telecommunications management network (TMN) model.

functions as accounting, provisioning, fault management, configuration, and security services.

Finally, the business management layer manages the applications and processes that provide strategic business planning and tracking of customer interaction vehicles such as SLAs.

OSI, while highly capable, has long been considered less efficient than IETF management standards, and in 1991, market forces began to effect a shift. That year, the Object Management Group was founded by a number of computer companies including Sun, Hewlett-Packard, and 3Com, and together they introduced the common object request broker architecture (CORBA). CORBA is designed to be vendor-independent and built around object-oriented applications. It allows disparate applications to communicate with each other, regardless of physical location or vendor. And while CORBA did not achieve immediate success, it has now been widely accepted, resulting in CORBA-based development efforts among network management system vendors. While this may seem to fly in the face of the OSI-centric TMN architecture, it really doesn't. TMN is more of a philosophical approach to network management, and does not specify technological implementation requirements. Thus, a conversion from CMIP to the simple network management protocol (SNMP), or the implementation of CORBA, does not affect the overall goal of the telecommunications management network.

Summary

IP is here to stay, and is profoundly changing the nature of telecommunications at its most fundamental levels. Applications for IP range from carrier-class voice that is indistinguishable from that provided by traditional circuit switching platforms to best effort services that carry no service quality guarantees. The interesting thing is that all of the various capability levels made possible by the incursion of IP have an application in modern telecommunications and are being implemented at rapid fire pace. There is still a tremendous amount of hype associated with IP services as they edge their way into the protected fiefdoms of legacy technologies, and implementers and customers alike must be wary of *brochureware* solutions and the downside of the Jurassic Park effect, which warns that just because you *can* implement an IP telephony solution doesn't necessarily mean you *should.* Hearkening back once again to the old telephone company adage that observes "If it ain't broke, don't fix it," buyers and implementers alike must be cautious as they plan their IP migration strategies. The technology offers tremendous opportunities to offer consolidated services, to make networks and the companies that operate them more efficient, and to save cost and pass those savings on to the customer. Ultimately, however, IP's promise lies in its ubiquity, and its ability to tie services together and make them part of a unified delivery strategy. The name of the game is service, and IP provides the bridge that allows service providers to jump from a technology-centric focus to a renewed focus on the services that customers care about.

Chapter Summary

A few years ago, while filming a video about telephony on location in Texas, I interviewed a small town sheriff about his use of telecommunications technology in his job. I wanted to know how it has changed the way he does his job. "Well sir," he began, puffing out his chest and sticking his thumbs into his waistband, "We use telecommunications all the time. We use our radios and cell phones in the cars, and our telephones there to talk here in town and for long distance. And now," he said, patting a fax machine on his desk, "we've got backup." I wasn't sure what he meant by that, so I asked. "We've always had trouble with the phones goin' out around here, and that was a bad thing. Can't very well do our job if we can't talk. But now we've got this here fax machine." I was puzzled—I was missing his point, so I probed a little deeper. "Don't you get it, son? Before, if we lost the phones, we were like fish outta water. Now, I know that if the phones go down, I can always send a fax."

Unfortunately, many peoples' understanding of the inner workings of the telephone network is at about the same level as the understanding of the sheriff (who was a wonderful guy, by the way. I set him straight on his understanding of the telephone network and he treated me to the best chicken fried steak I have ever eaten.). Hopefully this chapter, and those that follow, are helping to lift the haze a bit.

We will return to the central office several times as we examine the data transport side of telecommunications, but for now, let's introduce a necessarily painful component of telephony. Next we delve into the Byzantine (yet fascinating) world of telecom regulation.

Chapter Four Questions

1. Why was it called "The Bell System?"
2. What is the difference between a BOC, an RBOC, and an ILEC?
3. Who were the three principal inventors working on the earliest telephones?
4. What was the Kingsbury Commitment?
5. What was the Communications Act of 1934? Why was it important?
6. What was Computer Enquiry II?
7. The Divestiture of AT&T in 1984 was far-reaching. Was it a good idea? Have the results of it been positive?

8. What is a central office?

9. What is the difference between an NPA and an NXX?

10. How many telephone numbers are there in a single NXX? How many in an NPA? Why?

11. Explain the difference between a local switch and a tandem switch.

12. "SS7 is perhaps the single most important component of the modern telephone network." Explain.

13. Is twisted pair the only option for a local loop?

14. Explain the differences between D1, D4, and extended superframe (ESF).

15. What is SONET? How important is it in the modern network?

16. Explain the basic differences between a traditional circuit-switched network and a VoIP network.

CHAPTER 5

The Byzantine World of Regulation

I was to learn later in life that we tend to meet any new situation by reorganizing, and a wonderful method it can be for creating the illusion of progress while producing confusion, inefficiency, and demoralization.

—Petronius Arbiter, Emperor Nero's Minister of Entertainment

It had been a long and painful process, but starting today they would enjoy the fruits of their collective labors. Today, telecomm regulation has a whole new face, one that would redefine the way incumbent service providers, their competitors and the customers they serve interact. So radically different was it that it had taken three years to force the changes through Congress and another two to get the FCC to rally around the changes.

In his newly created cabinet-level role as Secretary of Telecommunications and Information Technology, Tom Vairetta had worked hard to bring about a sweeping rewrite of telecom law and a major philosophical shift on the part of the industry at large. Yes, today would be different. Instead of the contrived, artificially competitive model that drove the industry for so long, all incumbent local exchange carriers had agreed to subject themselves to structural separation, resulting in the creation of a wholesale arm and a retail arm for each company. The wholesale arm sells infrastructure; the retail arm sells services. All competitive carriers are free to buy from the wholesale division of any of the carriers at a cost that is the same for all. And to ensure that the services delivered by the infrastructure company are as good as they can be, a well-informed oversight committee ensures that the technology infrastructure is also as good as it can be. Of course, such an infrastructure is expensive; after all, the reason that the FCC and Congress were even willing to consider the concept was the fact that the cost of maintaining and upgrading the infrastructure to meet customer demands had become so onerously high that the service providers, faced with declining revenues on all service fronts, could not afford to make the necessary investment. So to offset the cost—this was a stroke of genius, if he did say so himself—Congress had passed a *chip and device tax*, which placed a surcharge on every manufactured semiconductor and optoelectronic device as well as on the finished product—a tax similar to the VAT used in Europe. The monies collected will pay for the Universal Service Fund and for guaranteed, scheduled network upgrades. The money also pays to guarantee quality of service (QoS), technological currency, and what his committee had come to call "quality of experience (QoE)." But now he had to go—he was late for his meeting with the global league of service providers.

"Wake up, Steve–it's time for your pill." You may laugh at my characterization of the regulatory environment of the future, but it isn't any more far-fetched than some of the other ideas that have come along in the last few years.

Today, writing a definitive section on regulation is like playing that wonderful arcade game from the 1970s, Whack-a-Mole: every time things seem like they're under control, another one pops up in an entirely new and unexpected place. In this section, we will attempt to create a snapshot of telecom regulation with enough detail to make it useful, but with the understanding that as I sit here typing, the environment is shape-shifting around me. I can't emphasize enough how critical it is for decision-makers in the tech sector to stay aware of changes in regulatory law and attuned to the impacts that regulation has on all facets of the technology industry.

The *reworking* of the telecommunications industry that began with the divestiture of AT&T in 1984 reaches far beyond the borders of the United States. It has become a global phenomenon as companies move to compete in what has become an enormously lucrative business. Customers, however, have been quick to point out that technology is a wonderful thing, but unless the provider positions it in a way that makes it useful to the customer it has very little value. Customers expect service provider representatives to have relevant technical knowledge as well as knowledge about them as customers: business issues, competition, major segment concerns, and the like. Furthermore, the level of customer satisfaction or dissatisfaction varies from industry segment to industry segment. Ninety-two percent of all enterprise customers want to work with an account team when purchasing services, yet according to both formal and informal surveys, account teams are not creating long-lasting relationships or loyalty in their day-to-day dealings with customers. Eighty-eight percent of those customers buy their services from the local provider, but only 53 percent say that they would continue to do so if presented with a capable alternative. According to numerous studies, the single most important factor that customers take into consideration when selecting a service provider is cost. Yet the most important factor identified for improvement is customer service. In some ways the telecom industry resembles the insurance industry. I had a conversation yesterday with a telecom professional who sells into the insurance world and he told me that 20 percent of insurance customers change carriers after going through the process of filing a claim—across the board. Telephone companies face similar churn numbers and are struggling to redefine themselves to ensure that they offer a top-quality experience when dealing with their customers.

The most daunting challenge that has historically faced service providers is the requirement to provide true universal service, in rural areas as well as in metropolitan and suburban areas. The American Communications Act of 1934, signed into law by President Roosevelt, mandated that telephony service be universal and affordable. To make this happen, AT&T agreed in the 1940s to offer low-cost service with the tacit approval of the Justice Department to charge a slightly higher price for long distance, some business line services, and value-added services, as a way to offset the cost of deploying high-cost rural service and to support low-income families. These subsidies made it possible for AT&T to offer true universal, low-cost service.

Today, the model—and the focus—has shifted drastically. Instead of concerning themselves with universal dial tone under the auspices of the universal services mandate, service providers in the United States today worry about the availability of universal broadband.

The task of regulating an industry as large and complex as telecommunications is daunting at best—an exercise in cat-herding of the first order. But when all of the various elements are boiled down to the "golden elixir" of regulation's ultimate responsibilities, we find two relatively simple and straight-forward goals: (1) to ensure that the regulated service, whether it be voice, television service, Internet access, or broadband, is as broadly available as it can possibly be within the service area covered by the governing tariff; and (2) to ensure that the customers using the regulated service have the best possible experience while doing so.

Many believe that the job of the regulator is to create marketplace competition. Not so: they can use competition as a way to achieve the first two, but in that case, competition is a means to an end, not the ultimate goal. If, for example, it is determined that customer service can be improved by introducing a competitor into a particular segment of the marketplace, then the regulator has the right to mandate competition in that geography and the creation of a secondary player to achieve that goal. As a consequence, regulators now find themselves concerned more with quality of experience measures (highly qualitative) rather than the historically significant quality of service (highly quantitative) measures. This shift makes a great deal of sense and comes on the heels of a number of service-related changes that are in fact best addressed by a qualitative rather than quantitative regulatory approach. TELUS, for example, perhaps the most enlightened service provider in the world, has adopted *likelihood to recommend* as its primary measure of customer satisfaction. "How likely is this customer to recommend us to someone they care about?"

Regulation Challenge No. 1: Net Neutrality

The first major regulatory challenge to face the industry is the rising challenge of what has come to be known as *Net neutrality.* Net neutrality is a governing principle that has been proposed initially for residential broadband networks but that will eventually apply to all networks. A broadband network that meets the criteria of net neutrality is a network that is free of all restrictions related to content, Web sites, access platforms, and the types of equipment that may be attached to the network.

Formally, net neutrality states that if a user pays for a certain level of Internet access quality, and another user pays for the same level of access, the two users should be able to connect to each other at that rate. Informally, the definition has been extended to take into account content types as well. Beginning in the early 2000s, net neutrality advocates began to raise concerns about the ability of broadband providers to use their *choke hold* on the last mile infrastructure to block Internet applications and content as they desired, particularly access to their competitors. For example, if a Comcast customer who uses Comcast.net to access the Internet does so to download free movies from the Web or to make free VoIP telephone calls via Skype, Comcast might reduce the bandwidth available to that customer as a way to punitively control the use of the Comcast network resource—a process known as *throttling*. Comcast's argument would be that they are protecting other subscribers from oversubscription issues, but customers who experience the throttling complain that they are suffering from the technological equivalent of a bait-and-switch tactic. This came to a head when the FCC issued their 2005 broadband policy statement, sometimes called the Internet policy statement. This document, which set the stage for decisions that followed, lists four principles that they believe should guide the model of an open Internet:

- To encourage broadband deployment and to preserve and promote the open and interconnected nature of the public Internet, consumers are entitled to access the lawful Internet content of their choice.
- To encourage broadband deployment and preserve and promote the open and interconnected nature of the public Internet, consumers are entitled to run applications and use services of their choice, subject to the needs of law enforcement.
- To encourage broadband deployment and preserve and promote the open and interconnected nature of the public Internet,

consumers are entitled to connect their choice of legal devices that do not harm the network.

- To encourage broadband deployment and preserve and promote the open and interconnected nature of the public Internet, consumers are entitled to competition among network providers, application and service providers, and content providers.

As we will see when we get to the second major regulatory shift that is underway, these four principles served as the basis for others that came later.

Regulation Challenge No. 2: Broadband Stimulus

As part of the Obama administration's economic recovery strategy as well as the long-term technology plan for the nation, a deployment plan known as Broadband Stimulus was created with the goal of expanding the degree to which broadband access is available throughout the country, particularly in rural areas. Two bodies were given massive funding and a mandate to carry out this expansion—the Rural Utilities Service (RUS) and the National Telecommunications and Information Administration (NTIA). The RUS created the Broadband Initiatives Program (BIP), exclusively targeting rural areas, while the NTIA created its Broadband Technology Opportunities Program (BTOP), which is largely (but not exclusively) targeted at rural deployment.

A Notice of Funds Availability (NOFA) was issued, making $2.4 billion available to the RUS's BIP, and $1.6 billion available to the NTIA's BTOP organization. The money is to be made available in a series of funding tranches, with 96% of it rolled out in the first tranche as a way to *jumpstart* the telecom economy and create massive change in rural areas. We won't go into the details of how the money is to be divided, nor will we discuss the details of the government's definition of a rural area.[1] What we will do, however, is note that both BIP and BTOP augmented the FCC's original four mandates related to net neutrality, as follows: In addition to the mandates of the original four rules, applicants for stimulus funding must also not favor, in any way, particular content or applications over others; must publicly explain any network management policies that they put into place (which could result in the *throttling* problem); must connect to the public Internet; and must offer interconnection at reasonable rates and terms. It is important to note that these rules are binding on

[1] For an *excellent* overview of Broadband Stimulus, go to *http://www.commlawblog.com/2009/07/articles/unlicensed-operations-and-emer/broadband-stimulus-101-the-print-version-is-now-available/*.

applicants as well as all *downstream* contractors that they may engage to carry out the work.

Technology and Regulation

The introduction of new technologies always conveys a temporary advantage to the first mover. Innovation involves significant risk; the temporary first mover advantage allows first movers to extract profits as a reward for taking the initial risk, thus allowing them to recover the cost of innovation. As technology advances, the first mover position usually erodes, so the advantage is fleeting. The relationship, however, fosters a zeal for ongoing entrepreneurial development.

Under regulatory rules in many countries today, incumbent providers are required to open their networks through element unbundling, and to lease their network resources—including new technologies—at a wholesale price to their competitors. In the minds of regulators this creates a competitive marketplace, but it is artificial. In fact, the opposite often happens: the incumbents lose their incentive to invest in new technology, and innovation progresses at "telco time." Under wholesale unbundling requirements, the rewards for innovative behavior are socialized, while the risks undertaken by the incumbents are privatized. Why should they invest and take substantial economic risk, they argue, when they are required by law to immediately share the rewards with their competitors?

Ultimately, the truth comes down to this: success in the access marketplace, translated as sustainable profits, relies on network ownership—period. During the heady bubble years, many would-be competitors such as E.spire and ICG bought switches and other network infrastructure components, built partial networks, and created business plans that consciously relied on the incumbent local exchange carriers (ILECs) for the remainder of their network infrastructure. Because of the behavior of the marketplace and fickle consumers, most competitive local exchange carriers (CLECs) failed to attract a viable customer base and quickly learned that this model could not succeed. Most of them have now disappeared.

Current Issues in Regulation

Today, global regulators are being called upon to oversee a swath of functions. Their responsibilities are driven largely by convergence, the rise of IP, and broadband penetration. These organizations have enjoyed growing influence since the end of 2012.

There were 158 national regulators with authority over telecom and IT functions in their respective countries. Those functions include:

- *Competition:* The extent to which competition should be used to enhance the delivery of a superior customer service experience in a given area, or the extent to which that same competition hinders delivery of QoE. It is absolutely clear that those markets that are *not* dominated by a single large player are more healthy, more vibrant, and more innovative, making it equally clear that competition is crucial for economic development. By the end of 2012, according to the ITU, more than 90% of all member countries were at least partially competitive. Oddly enough, basic telephony is still monopolized in many countries, yet it is increasingly the least relevant of many services!

- *Privatization:* In those countries where the service provider (PTT) is still a state-run entity, how best to privatize it as a way to foster innovation, economic growth, and healthy competition?
- *License and spectrum management:* The extent to which wireless licenses should be centrally controlled and the manner in which they should be fairly disseminated.
- *Infrastructure sharing:* The infrastructure required to properly operate a telecommunications or IT service delivery platform is breathtakingly expensive. Is it within the realm of the regulator to mandate infrastructure sharing as a way to reduce cost to both providers and consumers? Two directions for managing this challenge have emerged. The first is to simplify the process by consolidating licenses so that a carrier doesn't have to apply for different licenses for different services. The second is to simplify market entry procedures by eliminating bureaucracy. Deregulation is also working in many countries; in some cases we have seen this take so far as to effect a total elimination of all license and concession requirements.

As far as spectrum management is concerned, the recognized goal of the regulator is to manage a finite resource for all concerned. One of the ongoing trends is a competitive auction process during which players bid for control of blocks of available spectrum. To ensure proper management of a scarce national resource, most countries still retain control of the overall process.

- *Impact of IP/VoIP:* Regulators are not responsible for the regulation of technology: they are responsible for the regulation of service delivery. IP and VoIP are not services; they are technology options.
- *Universal broadband:* A global mandate that has far-reaching implications for not only service delivery but also healthcare, education, economic growth, and transparent government.

Life, Liberty, and Broadband

It began in Estonia and was soon followed by legislators in Finland, the UK, France, and a host of other countries. Access to the Internet has become so inextricably important that these countries have written law that makes access to broadband an inalienable human right. Some have gone so far as to recommend that the law be written into their national constitutions.

Should broadband be held in the same esteem as access to fair government, due process of law, the right to religious and political freedom, and the like? Some say yes. What matters here is that the impact of telecom and IT on the people of the world has been recognized as a powerful, life-affecting force, and cannot be trivialized. The role of the regulator has never been more important.

Chapter Five Questions

1. What are the primary goals of telecom regulation?

2. How would you compare the Communications Act of 1934 with the Telecommunications Act of 1996?

3. How did the ILECs basically control the regulatory process when it came to the introduction of new services and technologies?

4. What are the 14 requirements under TA96 that the ILECs were required to meet as a prerequisite for entry into long distance?

5. Why did the long-distance carriers not have a similar list of requirements?

6. What is UNE-P?

7. Explain the concept of "structural separation." Is it a good idea?

8. Should VoIP be regulated the same way as traditional telephony? Why or why not?

9. Should telecom be regulated exclusively at the federal level, or should there be state-level regulatory oversight as well?

10. "The local loop is a natural monopoly." Explain.

CHAPTER 6

Premises Technologies

Everybody is a genius. But if you judge a fish by its ability to climb a tree, it will live its whole life believing that it is stupid.

—Albert Einstein

In the previous chapters we laid the foundation for connectivity by examining the basics of telecommunications: protocols, standards, the inner workings of telephony, and of course, the regulatory decisions that govern the overall industry. In Chap. 6, we're going to change directions a bit by turning our attention to the devices that connect to the network—essentially the premises devices that make the network useful by creating and terminating media-based content that will be transported over the network and that will most likely be stored, analyzed, and monetized by an adjacent IT infrastructure, which we'll discuss in more detail in a later chapter. We begin with an examination of a typical computer.

The computer is the ultimate premises technology device. In one form or another it appears in essentially every device used by a customer to access the network, regardless of whether that person is a residential customer or one of thousands working for a large enterprise. Furthermore, the computer takes on different appearances depending on the application to which it will address itself. Smart phones contain sophisticated, self-contained computers; other examples include tablets, netbooks, laptops, desktop PCs, servers, and mainframes.

The Computer

For all its complexity—and capability—a typical computer has a small number of functional components, shown in Fig. 6.1. These are the central processing unit (CPU); main memory; secondary memory; input-output devices, sometimes called I/O; and a parallel bus that ties all of these elements together. It also has two types of software that make the computer useful to a human. The first is application

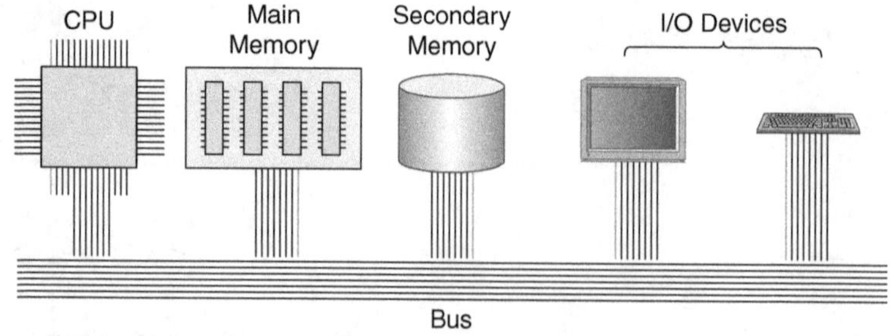

FIGURE 6.1 Functional components of a typical computer.

software such as word processors, spreadsheet applications, presentation software, and MP3 encoders. This is the software that most people are aware of, because it is the software with which they interact on a daily basis.

The second is the operating system that manages the goings on within the computer. Examples of operating systems include the various manifestations of Windows, Apple's IOS, Linux, UNIX, and, in the world of industrial computing, MVS, VMS, and VM. The functions of the operating system are analogous to those of an office manager, including internal component inventory, awareness of the existence of peripherals such as printers, scanners, monitors, mice, and keyboards, knowledge of available applications including the compute resources they each need, and the location of data files on the computer's hard drive(s). In a sense, the OS is the executive assistant to the computer itself; in fact, some mainframe environments refer to their operating system as the EXEC.

The concept of building modular computers came about in the 1940s when Hungarian-born mathematician John Von Neumann applied the work he had done in logic and game theory to the challenge of building large electronic computers (Fig. 6.2). As one of the primary contributors to the design of the Electronic Numerical Integrator and Computer (ENIAC), Von Neumann introduced the concept of *stored program control* and modular computing, the design under which all modern computers are built today. Stored program control implies that some form of internal memory houses the instruction set that the computer executes. This is important because the first computers had to be physically rewired with patch cords every time the instruction set changed.

The internals of a typical modern computer are shown in Fig. 6.3. Let's look at them.

The CPU

The CPU is literally the brain of the computer. Its job is to receive data input from the I/O devices (keyboard, mouse, modem, and the like),

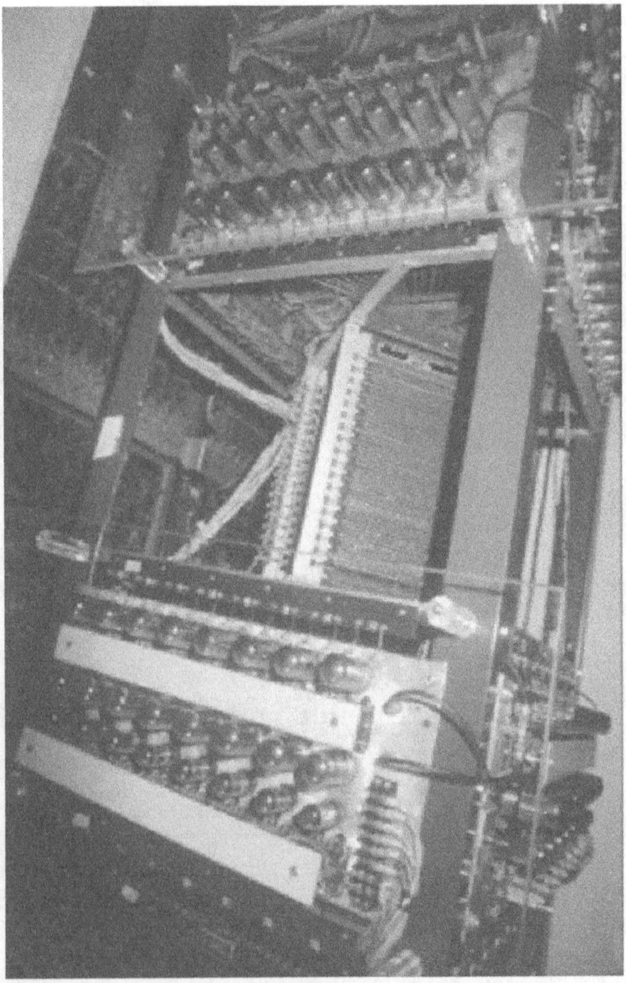

FIGURE 6.2 An example of a very large computer.

manipulate the data in some way based on a set of commands from a resident application, and package the newly gerrymandered data for presentation to a human user at another I/O device (monitor). Let's be clear here. Keep in mind from our studies of data communications earlier in the book that data is represented as string of zeroes and ones—in effect, a complex binary number. But complex or not, a binary number can be manipulated (added, subtracted, multiplied, divided) and can thus be made to represent the information it contains in a variety of ways. This is how computers work: they manipulate the numbers that represent the information in the data to create something that can be displayed or played.

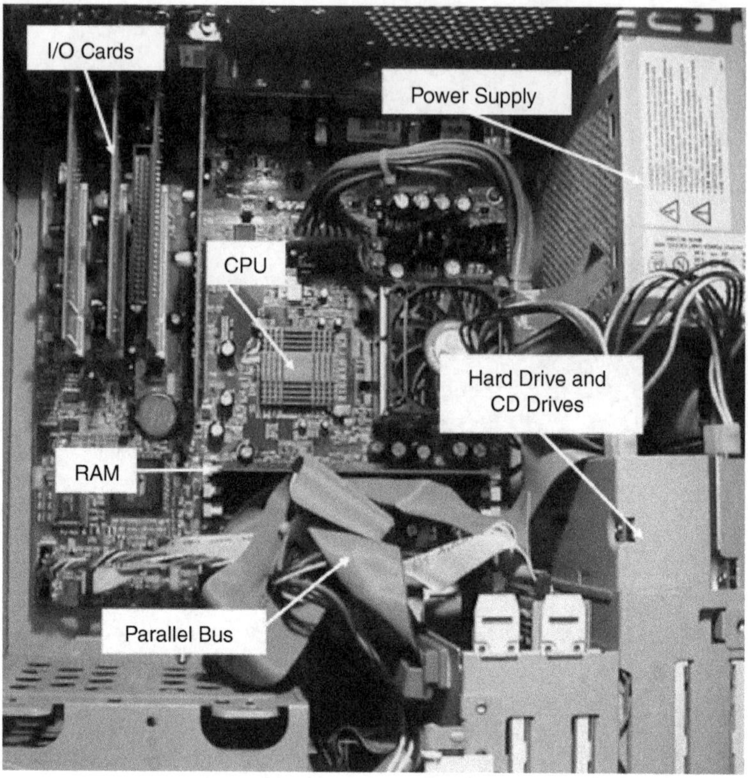

FIGURE 6.3 Internals of a computer.

For the purposes of discussion, let me give you an example of the complexity of the math that often lies behind application execution—and the work that the computer must do to make it acceptable (and useful) to a human user. As many of you know, I am a photographer. I do commercial nature and landscape photography, and rely pretty heavily on Adobe's Lightroom and Photoshop programs to sharpen my images, remove dust spots, and the like.

One of the challenges that photographers face in digital photography is that the camera is not capable of seeing the range of brightness that the human eye can see. As a result, we sometimes have to do things to *compress* the range of light into a range that both the camera and the software can work with.

One of the capabilities that Photoshop offers is called *Blend Mode.* Blend Mode allows a photographer to blend together two layers to achieve an effect in the finished photograph. For example, if the sky is completely washed out because of the camera's limited ability to capture a wide range of exposure values, two layers can be blended to ensure that the sky is properly exposed and the darker foreground is as well. Seems simple on the surface, right? All I have to do to blend two layers is create and open them in Photoshop, click on a drop-down menu, and select the Blend Mode option that I want. That's it. But what goes on behind the scenes? Well, I looked into it, and thanks to Wikipedia,[1] I found a gentle example of the math going on behind the scenes. Here's the Wikipedia explanation:

There are a variety of different methods of applying a soft light blend. All the flavors produce the same result when the top layer is pure black; same for when the top layer is pure neutral gray. The Photoshop and illusions.hu flavors also produce the same result when the top layer is pure white (the differences between these two are in how one interpolates between these 3 results).

These three results coincide with gamma correction of the bottom layer with $\gamma = 2$ (for top black), unchanged bottom layer (or, what is the same, $\gamma = 1$) (for top neutral gray), and $\gamma = 0.5$ (for top white).

The formula used by Photoshop has a discontinuity of local contrast and other formulas correct it. Photoshop's formula is

$$f_{photoshop}(a,b) = \begin{cases} 2ab + a^2(1-2b), & \text{if } b < 0.5 \\ 2a(1-b) + \sqrt{a}(2b-1), & \text{otherwise} \end{cases}$$

where a is the base layer value and b is the top layer value. Depending on b, one gets a linear interpolation between three gamma corrections: $\gamma = 2$ (for $b = 0$), $\gamma = 1$ (for $b = 0.5$), and $\gamma = 0.5$ (for $b = 1$).

$$f_{pegtop}(a,b) = (1-2b)a^2 + 2ba$$

Pegtop's formula is smoother and corrects the discontinuity at $b = 0.5$:

This is a linear interpolation between gamma correction with $\gamma = 2$ (for $b = 0$), and a certain tonal curve (for $b = 1$). (The latter curve is equivalent to applying $\gamma = 2$ to the negative of image.)

A third formula defined by illusions.hu corrects the discontinuity in a different way, doing gamma correction with γ depending on b:

For $b = 0$, one still gets $\gamma = 2$, for $b = 0.5$ one gets $\gamma = 1$, for $b = 1$ one gets $\gamma = 0.5$, but it is not a linear interpolation between these three images.

The formula specified by recent W3C drafts for SVG and Canvas is mathematically equivalent to the Photoshop formula with a small variation where $b \geq 0.5$ and $a \leq 0.25$:

$$f_{w3c}(a,b) = \begin{cases} a - (1-2b)\cdot a\cdot(1-a) & \text{if } b \leq 0.5 \\ a + (2b-1)\cdot(g_{w3c}(a) - a) & \text{otherwise} \end{cases}$$

[1]The full explanation can be found here: *http://en.wikipedia.org/wiki/Blend_modes.*

where

$$g_{w3c}(a) = \begin{cases} [(16a-12)\cdot a+4]\cdot a & \text{if } a<0.25 \\ \sqrt{a} & \text{otherwise} \end{cases}$$

It is still a linear interpolation between 3 images for $b = 0, 0.5, 1$. But now the image for $b = 1$ is not $\gamma = 0.5$, but the result of a tonal curve which differs from the curve of $\gamma = 0.5$ for small values of a; while gamma correction with $\gamma = 0.5$ may increase the value of a many times, this new curve limits the increase of a by coefficient 4.

I'm not trying to give you nightmares, but I want you to understand that (1) the job of the CPU is purely mathematical, and (2) it's math that makes it all work. Incredible job that CPU has, wouldn't you say?

CPU Sub-Components

The CPU has its own set of sub-components. These include a clock, an arithmetic-logic unit (ALU), and an array of registers. The clock is the device that provides synchronization and timing to all devices in the computer and is characterized by the number of clock cycles per second it is capable of generating. These cycles are called Hertz; modern systems today operate at in the range of 3-4 *billion* cycles per second, or GHz. The faster the clock, the faster the machine can perform computing operations.

The ALU is the specialized silicon intelligence in the CPU that performs the mathematical permutations that make the CPU useful. All functions performed by a computer—word processing, spreadsheets, multimedia, videoconferencing, editing an image, as described earlier—*all functions*—are viewed by the computer as mathematical functions and are therefore executed as such. It is the job of the ALU to carry out these mathematical permutations.

Registers are nothing more than an array of very fast memory that is located physically close to the ALU for rapid input/output functions during execution cycles.

Main Memory

Main memory, sometimes called random access memory (RAM), is another measure of the *goodness* of a computer today. RAM is the segment of memory in a computer used for execution space and as a place to store the operating system, user data, and application files that are in current use. RAM is silicon-based, solid-state memory and is extremely fast in terms of access speed. It is, however, volatile: when the PC is turned off, or power is lost, whatever information is stored in RAM disappears. When basic computer courses tell students to "save often," this is the reason. When a computer user is writing a document in a word processor, or populating a spreadsheet,

Figure 6.4 My array of computers: a MacBook Air (left); my 27″ iMac (center) and my original Mac.

or manipulating a digital photograph, the file lives in RAM until the person saves, at which time it is written to main memory, which is non-volatile as we'll see in a moment. Modern systems typically have a minimum of 2 to 3 gigabytes (GB) of RAM.

It amazes me to look across my desk at my first Mac, which I purchased in 1986 (see Fig. 6.4). It has a 40-MB hard drive, 640 KB of RAM, and a processor that doesn't run—it dribbles. Yet it runs the same Office applications I use today (although much reduced in terms of bells and whistles) and boots in seconds, compared to the very powerful iMac that I'm writing on today. This machine has 16 GB of RAM, a 500-GB solid-state non-volatile hard drive (discussed in a moment) and a 3 TB (terabyte) hard drive. The amazing thing? This machine cost only slightly more than my little Mac from 1986.

Secondary Memory

Secondary memory has become a very popular line of business in the evolving PC market, not to mention in the world of digital photography. It provides a mechanism for long-term storage of data files and is non-volatile—when the power goes away the information it stores does not. There are lots of examples to choose from including mechanical devices such as hard drives, and solid-state devices such as USB memory sticks (flash drives), CF cards, SD cards, Micro SD cards, and a number of others.

Secondary memory tends to be a much slower medium in terms of access time than main memory because of both architecture and location—it isn't adjacent to the CPU/ALU. Solid-state memory such as flash drives and memory cards tend to be faster than a traditional hard drive because they are all-electronic—there are no moving parts. Any time a mechanical element is introduced, speed goes down.

Let's examine the inner workings of a typical hard drive such as the one shown in Fig. 6.5. This particular drive chose a hot July day to fail, right in the middle of writing the last edition of this book (and yes, I had backed up often). That's why the disk platters are exposed. Notice that the drive comprises three platters and an armature that sweeps back and forth over the surface of the platters. Mounted on the armature are multiple read-write heads, similar to those used in old tape decks. The heads write data to and from the surfaces of the platters.

Figure 6.6 is a schematic diagram that more clearly illustrates how hard drives actually work. The platters, typically made of aluminum and cast to extremely exacting standards, are coated with iron oxide similar to what was used on recording tape. Remember when we were discussing the importance of saving often when working on a computer? The *save* process reads information stored temporarily in RAM, transports the information to the disk controller, which in turn, under the guidance of the operating system, transmits the information to the write heads, which in turn writes the bits to the disk

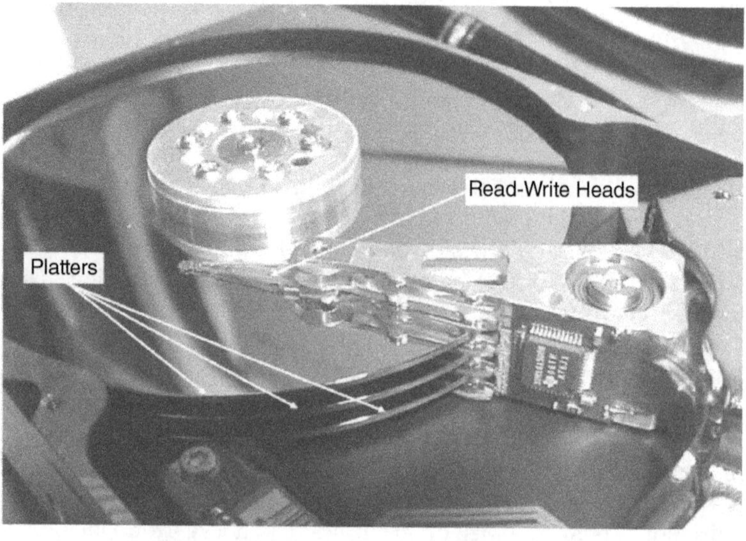

FIGURE 6.5 A hard drive, showing the platters upon which data is written.

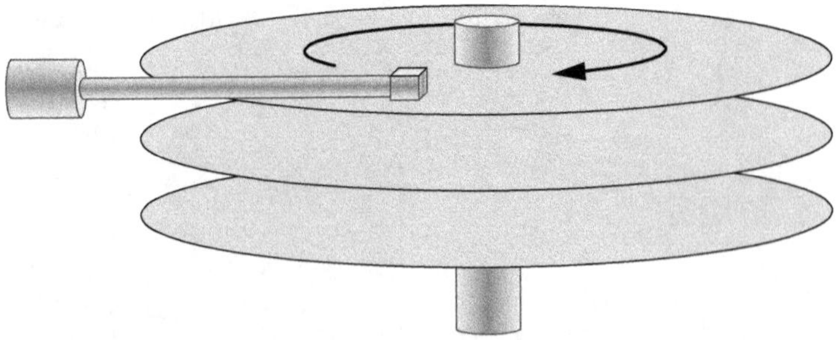

FIGURE 6.6 Illustration of how hard drives work.

surface by magnetizing the iron oxide surface. Of course, once the bits have been written to the disk, it's important to keep track of where the information is actually stored. This is a function of the operating system. Have you ever connected a new hard drive to a computer and had the computer ask you if you're sure you want it to format the drive? In order to keep track of where files are stored on a disk, the operating system *maps* the disk with a collection of road markers that help it find the *beginning* of the disk so that it can use that as a reference. Obviously there is no beginning on a circle, by design; by creating an arbitrary start point, however, the operating system can then find the files it stores on the drive. This start point is called the file allocation table (FAT); if the FAT goes away for some reason, the operating system will not be able to find files stored on the drive. When my hard drive failed, my initial indication that something bad was about to happen was an ominous message that appeared during the boot cycle that said 'UNABLE TO FIND FAT TABLE.'

When the operating system formats a disk, it logically divides the disk into what are called cylinders, tracks, and sectors as shown in Fig. 6.7. It then uses those delimiters as a way to store and recall files on a drive using those three dimensions. When the operating system (OS) writes the file to the disk surface, there is every possibility (in fact it is likely) that the file will not be written to the surface in a contiguous stream of bits. Instead, the file may be broken into pieces, each of which may be written on a different part of the disk array. The pieces are linked together by the operating system using a series of pointers that tell the OS where to go to get the next piece of the file when the time comes to retrieve it. The pointer, of course, is a cylinder : track : sector indication.

Cylinders, tracks, and sectors are relatively easy to understand. A track is a single writeable path on one platter of the disk array. A cylinder is a *stack of tracks* on multiple platters. And a sector is a "pie slice" of the disk array. With a little imagination it is easy to see how a file can be stored or located on a disk by tracking those three indicators.

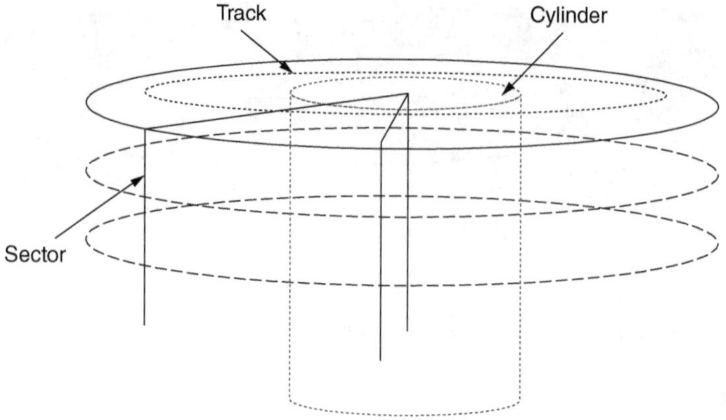

Figure 6.7 Cylinders, tracks, and sectors on a hard drive.

It should also be easy to see now how file-recovery applications work. When you direct your computer to erase a file, it doesn't really erase it; it just removes the pointers so that it is no longer a *registered* file and can no longer be found on the hard drive. What file restore utilities do is remember where the pointers were when you told the computer to erase the file. That way, when you beg the computer to find the file after you've done something stupid (this is where many people kill the chicken and light candles), the utility has a trivial task: simply restore the pointers. As long as there hasn't been too much disk activity since the deletion and the file hasn't been overwritten, it can be recovered.

Another useful tool is the disk optimization utility, which today is in most cases, part of the operating system. It keeps track of the files and applications that you access most often while at the same time keeping track of the degree of fragmentation that the files stored on the disk are experiencing. Think about it: when disks start to get full there is less of a chance that the OS will find a single contiguous piece of writeable disk space and will therefore have to fragment the file before writing it to the disk. Disk optimization utilities perform two tasks. First, they rearrange files on the disk surface so that those accessed most frequently are written to new locations close to the spindle, which spins faster than the outer edge of the disk and is therefore accessible more quickly. Second, they rearrange the file segments and attempt to reassemble files or at least move them closer to each other so that they can be more efficiently managed.

A Hybrid: Solid-State Drives

Solid-state drives (SSDs) are interesting beasts. They perform the same function as a traditional spinning hard drive but have no moving parts—they are entirely solid-state, meaning that all of the storage functionality is performed in silicon-based memory. They are much faster than traditional drives in terms of read-write speed and are also more robust: they are significantly more resistant to physical shock. They are also completely silent. The one downside of SSDs is their cost: They are still several times more expensive than a comparable traditional hard drive.

A variation on the solid-state theme is the hybrid drive such as the Fusion drive used in Apple computers. Hybrid drives combine the best of both solid-state and traditional drives, offering the cost-effective storage of a traditional spinning drive in combination with a large SSD cache (often as large as 512 GB) to improve performance and offer access to frequently used caches of data.

Input/Output (I/O) Devices

I/O devices are elements of a computer that provide the interface between the user and the computer. They include mice, keyboards, monitors, printers, scanners, modems, speakers, digital editing tablets, and increasingly such things as wireless health monitors—essentially any device that takes data into or spits data out of the computer.

The Bus

Interconnecting the components of a computer is a cable known as a *parallel bus*. It is called *parallel* because the bits that make up one or more 8-bit bytes travel down the bus side-by-side on individual conductors rather than one after the next as occurs in a serial cable on a single conductor. Both are shown schematically in Fig. 6.8.

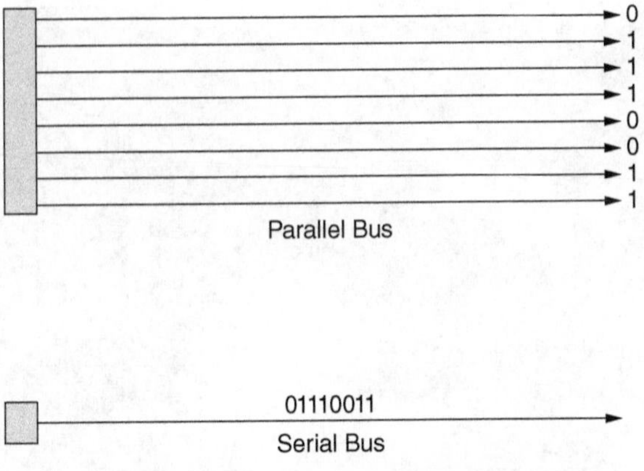

FIGURE 6.8 Illustration of serial and parallel buses, both used in modern computers.

The advantage of a parallel bus is speed: by pumping multiple bits into a device in the computer simultaneously, the CPU can process them faster. Obviously the more leads there are in the bus the more bits can be transported. It should come as no surprise, then, that another differentiator of computers today is the width of the bus. A 32-bit bus is four times faster than an 8-bit bus; as long as the internal device to which the bus is transporting data has as many input/output leads as the bus, it can handle the higher volume of traffic. The parallel bus does not have to be the flexible gray ribbon cable; for those devices that are physically attached to the main printed circuit board (often called the *motherboard*), the parallel bus is extended as wire leads that are etched into the surface of the motherboard. Devices such as I/O cards attach to the board via edge connectors.

Enough about computer internals. A brief word about computer history and evolution is now in order. The work performed by Von Neumann and his contemporaries led to the development of the modern mainframe computer (Fig. 6.9) in the 1960s and its many succeeding machine design generations. Targeted primarily at corporate applications, the mainframe continues to provide computing services required by most corporations: access to enormous databases, security, and support for hundreds if not thousands of simultaneous users. These systems host enormous disk pools (Fig. 6.10) and tape libraries (Fig. 6.11) and are housed in totally self-contained windowless computer centers. Their components are interconnected by large cables that require that they be installed on raised floor (Fig. 6.12). The space beneath the floor provides space for power and data cables, as well as the pipes required for the cooling system. These machines produce so much heat that they are fed enormous amounts of chilled

Figure 6.9 A modern mainframe computer.

FIGURE 6.10 An array of large disk drives.

FIGURE 6.11 A tape library.

air and water and require constant attention and monitoring. It makes sense: these are very powerful creatures.

These machines also had a tendency to mirror the organizations they served from an architectural perspective. Think about it: corporations in the 1960s were hierarchical, managed from the top-down, highly centralized in their decision-making practices, and tightly controlling in their demeanor. Mainframe computer systems such as the one shown in Fig. 6.13 were the same. The typical architecture included a mainframe computer where all the power in the system was concentrated, just like the corporation that owned it (all the power was at the top of the house). Just below the mainframe was a device called a front-end controller (FEP). The job of the FEP was to

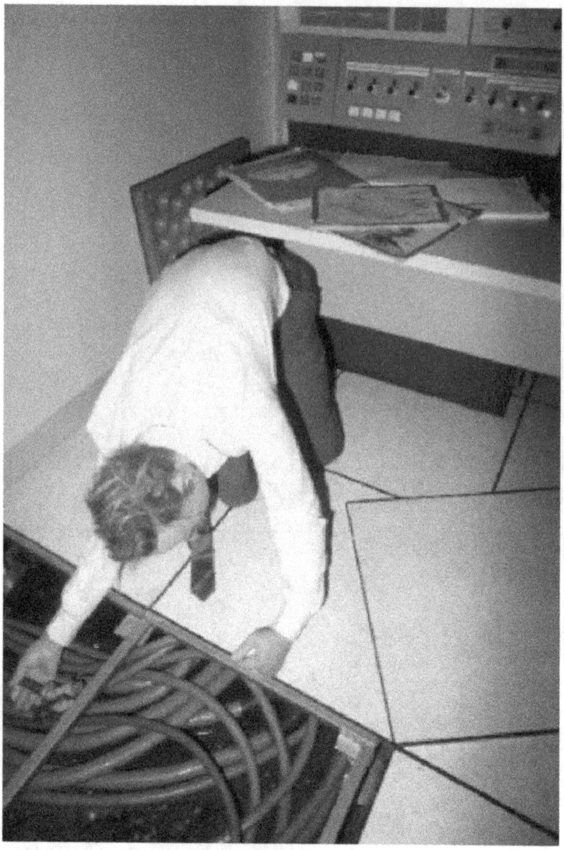

Figure 6.12 Underneath the raised floor.

Figure 6.13 Another example of a mainframe.

serve as traffic control between the mainframe and the scads of devices that were attempting to access it for compute time.

Immediately below the front end were one or more cluster controllers that were responsible for managing the pool of end-user devices that sometimes numbered in the hundreds. In effect, the cluster controllers served as traffic-management devices for the FEP, which in turn provided traffic management services to the mainframe. Redundancy, anyone?

Over time, computer technology advanced and soon a new need arose. Mainframes were fine for the computing requirements of large, homogeneous user communities, but as the technology became cheaper and more ubiquitous, the applications for which computers could be used became more diverse. At the same time, corporations became flatter as they began to eliminate redundant levels of management—and cost. I remember very well that when I joined the telephone company in 1981, there were 13 levels of management between the non-management ranks and the CEO. Today there are about eight. As the corporation flattened, so too did their compute requirements—and the networks that connected them. Soon a need arose for smaller machines that could be used in more specialized departmental applications, and in the 1970s, thanks to companies like Xerox, Digital Equipment Corporation (DEC), and Data General, the minicomputer was born (Fig. 6.14). The minicomputer made it possible for individual departments in a corporation or even small corporations to take charge of their own computer destinies and not be shackled to the centralized data centers of yore. The transition carried with it a price, of course: not only did these companies or organizations lose their dependency on the data center; they also lost the centralized support that came with it. So there was a downside to this evolution.

FIGURE 6.14 A minicomputer.

Enter the PC

The real evolution, of course, came with the birth of the personal computer. Thanks to Bill Gates and his concept of a simple operating system (DOS), and Steve Jobs with his vision of "computing for the masses," truly ubiquitous computing became a reality. From the perspective of the individual user, this evolution was unparalleled. The revolution began in January 1975 with the announcement of the MITS Altair (Fig. 6.15). Built by Micro Instrumentation and Telemetry Systems (MITS) in Albuquerque, New Mexico, the Altair was designed around the Intel 8080 microprocessor and a specially designed 100-pin connector. The machine ran a BASIC operating system developed by Bill Gates and Paul Allen; in effect, MITS was Microsoft's first customer.

The Altair was really a hobbyist's machine, but soon the market shifted to a small business focus. Machines like the Apple Macintosh, the Osborne, and the Kaypro were smaller and offered integrated keyboards and video displays. In 1981, of course, IBM entered the market with the DOS-based personal PC, and soon the world was a very different place. Apple followed with the Macintosh and the first commercially available graphical user interface (GUI), and the rest, as they say, is history. Soon corporations embraced the PC; "a chicken in every pot, a computer on every desk" seemed to be the rallying cry for IT departments everywhere.

There was a downside to this evolution, of course. The arrival of the PC heralded the arrival of a new era in computing that allowed every individual to have their own applications, their own file

FIGURE 6.15 The MITS Altair, Microsoft's first customer. (*Photo courtesy Jim Willing, The Computer Garage.*)

structures, and their own data. The good news was that each person now controlled their own individual computer resources; the bad news was that each person now controlled their own individual computer resources. Suddenly any semblance of centralized control was gone; instead of having *a copy* of the database, there were now as many copies as there were users. This led to huge problems. Furthermore, PC proliferation led to another challenge: connectivity, or lack of it. Whereas before every user had electronic access to every other user via the mainframe or minicomputer-based network that hooked everyone together, the PC did not offer that advantage (yet). Furthermore, PCs eliminated the efficiency with which expensive network resources such as printers could be shared. Some of you may remember the days when being the person in the office with *the laser printer* attached to your machine was akin to approaching a state of Nirvana. You were able to do your work and print anytime you wanted—except, of course, for the disruption caused by all of those people standing behind you with floppy disks in their hands promising you that "It's only one page—it'll just take a second." Thus was born the term *Sneakernet;* in order to print something, you had to put the document on a diskette (probably a 5-1/4 inch floppy, back when floppies really were floppy), walk over to the machine with the directly-attached printer, and beg and wheedle for permission to print the file—not the most efficient technique for sharing a printer. Something else was needed. That *something* was the local area network (LAN).

LAN Basics

A local area network (LAN) is exactly what the name implies: A physically small network, typically characterized by high-speed transport, low error rates, and private ownership, that serves the data transport needs of a geographically small community of users. Most LANs provide connectivity, resource sharing, and transport services within a single building, although they can operate within the confines of multiple buildings on a campus. Initially, there were exclusively wired, but over time, as we'll see, they have evolved to include wireless options as well.

When LANs were first created the idea was to design a network option that would provide a low-cost solution for transport. Up until their arrival, the only transport option available was a dedicated private line from the telephone company or X.25 packet switching. Both were expensive and X.25 was less reliable than was desired. Furthermore, the devices being connected together were relatively low-cost devices; it simply didn't make sense to interconnect them with expensive network resources. That would defeat the purpose of the LAN philosophy.

LAN Characteristics

All LANs, regardless of the access mechanism, share certain characteristics. All rely on some form of transmission medium that is shared among all the users on the LAN; all use some kind of interrupt and contention protocol to ensure that all devices get an equal opportunity to use the shared medium; and all have some kind of software called a network operating system (NOS) that controls the environment within which they operate.

All local area networks have the same basic components as shown in Fig. 6.16. A collection of devices such as PCs, servers and printers serve as the interface between the human user and the shared medium. Each of these machines, if connected to a wired LAN, has a built-in device known as a network interface card (NIC), which provides the physical connectivity between the user device and the shared medium.

The NIC also implements the access protocol that the devices wishing to access the shared medium use on their particular LAN. These access schemes, which include such protocols as ethernet, token ring and Wi-Fi, will be discussed shortly. The NIC also provides the connectivity required to attach a user device to the shared network.

Topologically, LANs differ greatly. The earliest LANs used a bus architecture, shown in Fig. 6.17, so-called because they were literally

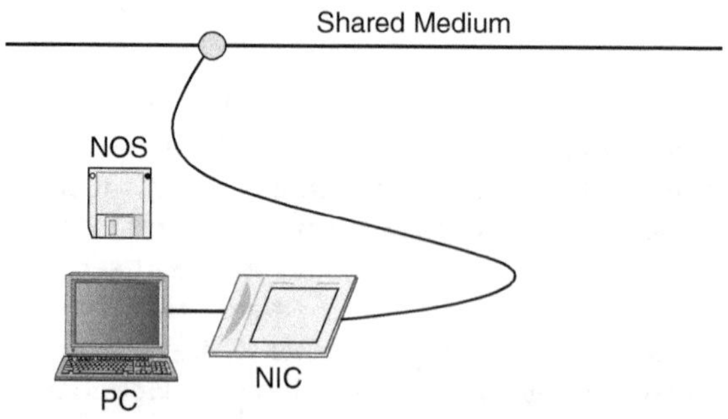

FIGURE 6.16 A typical local area network (LAN).

FIGURE 6.17 A bus-based LAN.

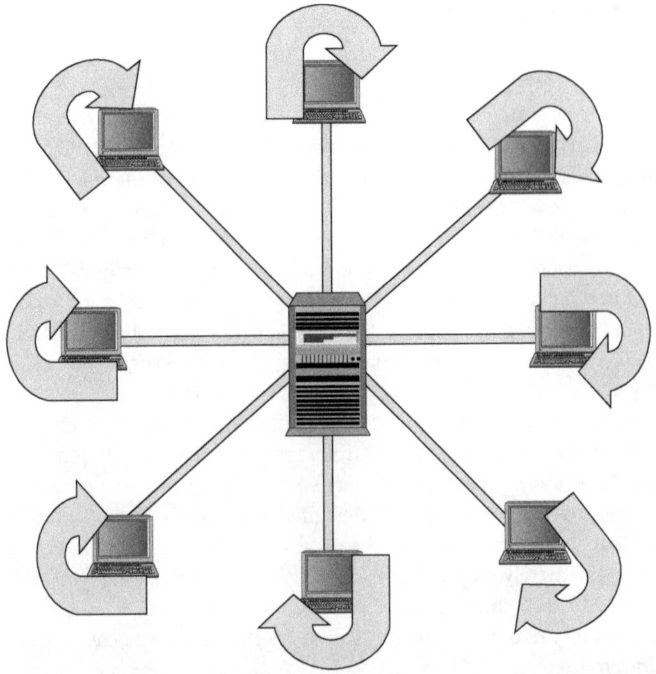

FIGURE 6.18 A token-passing ring LAN.

a long run of twisted pair wire or coaxial cable to which stations were periodically attached. Attachment was easy; in fact, early coax systems relied on a device called a *vampire tap*, which poked a hole in the insulation surrounding the center conductor in order to suck the digital blood from the shared medium. Later designs such as IBM's token ring used a contiguous ring architecture such as that shown in Fig. 6.18. Both had their advantages, and even though IBM's token ring was technologically better than ethernet, it lost the battle and has for the most part disappeared today. We'll discuss its rise and fall later in the chapter.

Local Area Network Access Schemes

Local area networks have traditionally fallen into two primary categories characterized by the manner in which they access the shared transmission medium (shared among all the devices on the LAN). The first, and most common, is called *contention*, and the second group is called *distributed polling*. I tend to refer to contention-based LANs as the Berkeley method, while I view distributed polling LANs as users of the Harvard method. I'll explain in a moment.

Contention-Based LANs

Dear Mr. Metcalf:

We're not sure how to break this to you, but we have discovered that your claim of a patent for the invention of Ethernet must be denied after the fact due to its existence prior to the date of your claim of invention. There is a small freshwater fish, Gymnarchus Niloticus, that uses an interesting technique for locating mates, food, and simply communicating with peers. The fish's body is polarized, with a "cathode" on its head and an "anode" on its tail. Using special electric cells in its body similar to those employed by the electric eel or the California electric ray, Gymnarchus emits nominal 300 Hz, 10 volt pulses, which reflect back and inform the fish about its immediate environment.

In the event that two Gymnarchus are in the same area, their emissions interfere with one another (intersymbol interference?), rendering their detection mechanisms ineffective. But, being the clever creatures that they are, Gymnarchus has designed a technique to deal with this problem. When the two fish "hear" each other's transmissions, they both immediately stop pulsing. Each fish waits a measurably random period of time while continuing to swim, after which they begin to transmit again, but this time at slightly different frequencies to avoid interference.

We hope that you understand that under the circumstances we cannot in good conscience grant this patent.

Sincerely yours,
U.S. Patent Office

I don't think a fish can hold a patent, but if it could, *Niloticus* would hold the patent for a widely used technique called "carrier sense, multiple access with collision detection (CSMA/CD)." Please read on.

Perhaps the best-known contention-based medium access scheme is ethernet, a product developed by 3Com founder and Xerox PARC veteran Bob Metcalfe. In contention-based LANs, devices attached to the network vie for access using the technological equivalent of gladiatorial combat. "If it feels good, do it" is a good way to describe the manner in which they share access (hence the "Berkeley method"). If a station wants to transmit, it simply does so, knowing that there exists the possibility that the transmitted signal may collide with the signal generated by another station that transmits at the same time. Even though the transmissions are electrical and are occurring on a local area network, there is still some delay between the time that both stations transmit and the time that they both realize that someone else has transmitted. This realization is called a collision and it results in the destruction of both transmitted messages as shown in Fig. 6.19. In the event that a collision occurs as the result of simultaneous transmission, both stations immediately back off: they stop their

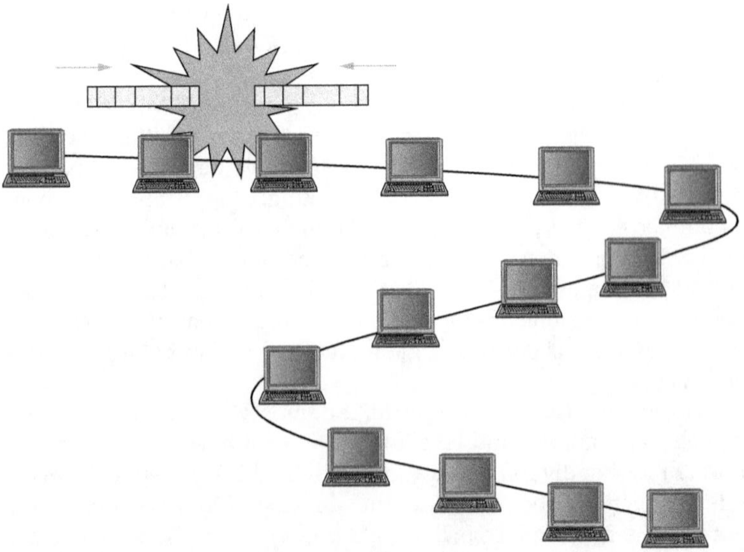

FIGURE 6.19 A collision on the LAN.

transmissions, wait a random amount of time, and try again. This technique has the *wonderful* name of *truncated binary exponential back-off*. It's one of those phrases you just *have* to commit to memory because it sounds *so good* when you casually let it roll off the tongue in conversation.

Ultimately, each station *will* get a turn to transmit, although how long they may have to wait is based on how busy the LAN is. Contention-based systems are characterized by what is known as *unbounded delay,* because there is no upward limit on how much delay a station can incur as it waits to use the shared medium. As the LAN gets busier and traffic increases, the number of stations vying for access to the shared medium—which only allows a single station at a time to use it, by the way—also goes up, which naturally results in more collisions. Collisions translate into wasted bandwidth, so contention-based LANs do everything they can to avoid them. We will discuss how a bit later in this chapter.

The protocol that contention-based LANs employ is called carrier sense, multiple access with collision detection (CSMA/CD). In CSMA/CD, a station observes the following guidelines when attempting to use the shared network. First, it listens to the shared medium to determine whether it is in use or not—that's the carrier sense part of the name. If the LAN is available (not in use), it begins to transmit, but continues to listen while it is transmitting, knowing that another station could also choose to transmit at the same time—that's the multiple access part. In the event that a collision is detected, usually indicated by a dramatic increase in the signal power measured on the shared LAN, both stations back off and try again. That's the collision detection part.

Ethernet is the most common example of a CSMA/CD LAN. Originally released as a 10 Mbps product based on IEEE standard 802.3, ethernet rapidly became the most widely deployed LAN technology in the world. As bandwidth-hungry applications such as e-commerce, enterprise resource planning (ERP), and Web access evolved, transport technologies advanced and as bandwidth availability (and capability) grew, 10 Mbps ethernet began to show its age. Today, new versions of ethernet have emerged that offer 100 Mbps (fast ethernet), 1,000 Mbps (gigabit ethernet) transport, and 10 Gbps ethernet.

Gigabit ethernet has become a fundamentally important technology as its popularity and level of deployment have climbed. It supports a range of diverse applications including LAN telephony, server interconnection, and video to the desktop. Many vendors have entered the marketplace including Alcatel-Lucent, Ciena, and Cisco Systems.

Gigabit ethernet is designed to operate over a range of media types including both twisted pair and optical fiber. A collection of physical layer standards have emerged over the years including gigabit ethernet over optical fiber (1000BASE-X), gigabit ethernet over twisted pair cable (1000BASE-T), gigabit ethernet over shielded balanced copper cable (1000BASE-CX, basically obsolete today), gigabit ethernet over multimode fiber (1000BASE-SX), and gigabit ethernet over single-mode fiber (1000BASE-LX).

The other aspect of the LAN environment that began to show weaknesses was the overall topology of the network itself. LANs are broadcast environments, which means that when a station transmits, every station on the LAN segment hears the message (Fig. 6.20). While this is a simple implementation scheme it is also wasteful of bandwidth since stations hear broadcasts that they have no reason to hear. In response to this, a technological evolution occurred. It was obvious to LAN implementers that the traffic on most LANs was somewhat domain-oriented, that is, it tended to cluster into communities of interest based on the work groups using the LAN. For example, if employees in sales shared a LAN with shipping and order

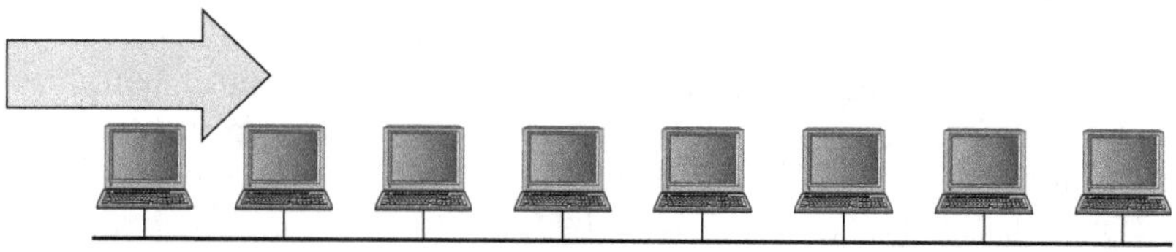

Figure 6.20 Listening for a transmitted message before sending.

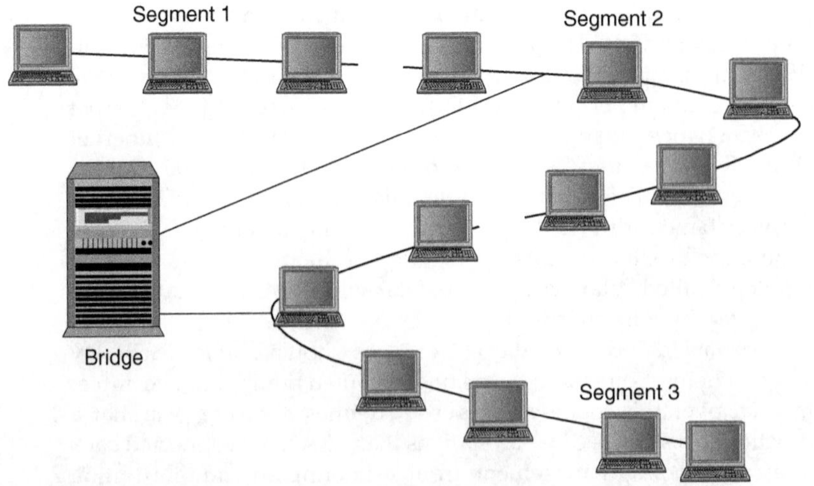

Figure 6.21 A bridged LAN.

processing, three discernible traffic groupings emerged according to what network architects call "the 80:20 rule." The 80:20 rule simply states that 80% of the traffic that originates in a particular work group tends to stay in that work group, an observation that makes network design distinctly simpler. If the traffic naturally tends to segregate itself into groupings, then the topology of the network should change to reflect those groupings. Thus was born the bridge.

Bridges are devices with two responsibilities: they filter traffic that does not have to propagate in the forward direction, and forward traffic that does. For example, if the network described above were to have a bridge inserted in it (Fig. 6.21), all of the employees in each of the three work groups would share a LAN segment, and each segment would be attached to a port on the bridge. When an employee in sales transmits a message to another employee in sales, the bridge is intelligent enough to know that the traffic does not have to be forwarded to the other ports. Similarly, if the sales employee now sends a message to someone in shipping, the bridge recognizes that the sender and receiver are on different segments and thus forwards the message to the appropriate port, using address information in a table that it maintains (the filter/forward database). Bridges operate at layer two of the OSI model, and as such are frame switches.

LAN Switching

Following close on the heels of bridging is *LAN switching*. LAN switching qualifies as *bridging on steroids*. In LAN switching, the filter/forward database is distributed—that is, a copy of it exists at each port, which implies that different ports can make simultaneous traffic handling decisions. This allows the LAN switch to implement full-duplex transmission, reduce overall throughput delay, and in some cases implement per-port rate adjustment. The first 10 Mbps ethernet LAN Switches emerged in 1993 followed closely by fast ethernet (100 Mbps) versions in 1995 and gigabit ethernet (1,000 Mbps) switches in 1997. Fast ethernet immediately stepped up to the marketplace bandwidth challenge and was quickly accepted as the next generation of ethernet. LAN switching also helped to propagate the topology called "star wiring." In a star-wired LAN all stations are connected by wire runs back to the LAN switch or a hub that sits in the *geographical center* of the network, as shown in Fig. 6.22. Any access scheme (contention-based or distributed polling) can be implemented over this topology because it defines a wiring plan, not a functional design. Because all stations in the LAN are connected back to a center point, management, troubleshooting and administration of the network is simplified. Today, high-speed LAN switches have become the architectural norm for network deployment in multitenant office buildings and corporate facilities because of low cost, high utility, and ease of management.

LAN switching is a variation of packet switching that is implemented across LANs. It is essentially a next-generation of bridging, designed to address some of the limitations of bridging as networks grew larger, more complex in terms of the traffic they were carrying and more demanding. Like bridging, LAN switching allows traffic to be directed very discretely to its intended destination using hardware-based decision-making capabilities, thus limiting bandwidth consumption.

There are a number of thematic variations of LAN switching as we'll see in this brief section. Remember our discussion of the OSI model back in Chap. 3? Time to dust it off—we're going to talk protocols for a bit.

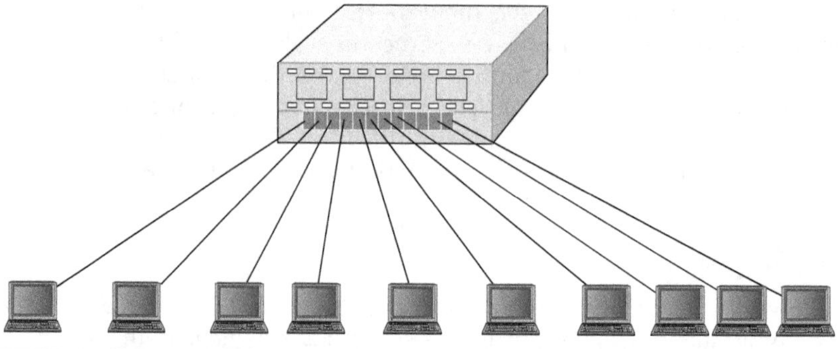

Figure 6.22 A LAN switch.

Layer 2 Switching

LANs rely on a physical layer address for each communicating device known as a media access control (MAC) address. Layer 2 switching uses the MAC address to make data frame forwarding decisions. This technique is hardware-based, meaning that much of the decision-making process is performed in silicon, typically in a device known as an application-specific integrated circuit (ASIC). And because the filter-forward tables are implemented in silicon, decisions tend to be made extremely fast—so fast, in fact, that these devices tend to operate at what is called wire-speed—very little latency, since no software interrupts need to be performed to make filtering and forwarding decisions. A layer two switch, then, is more capable than a traditional bridge; in effect it behaves like a multiport bridge. As a result it is used effectively for workgroup connectivity and for dividing networks into segments according to the 80:20 rule historically known as *collision domains.*

Layer 3 Switching

Layer 3 switching is really routing by a different name—with a few significant differences. For example, LAN switches perform their routing decisions in hardware, whereas routers rely on more complex (and slower) microprocessors to make the same decisions. The advantage that LAN switching offers over a routed environment is that LAN switches are fully capable of handling high-volume LAN traffic and can therefore be used as a replacement for the (typically) more expensive routers.

Layer 4 Switching

Layer 4 switching is something of an intentional misnomer because in reality it is just another layer 3 switching option. It provides a sophisticated routing capability above layer 3 by relying on the port numbers that the transport layer uses to make end-to-end routing decisions. Remember, the role of the transport layer is to ensure end-to-end integrity.

The most important advantage that layer 4 switching offers is the ability to prioritize traffic based on the application it is intended for. Voice and video, for example, can be given a higher priority than e-mail, for example, thus guaranteeing individual QoS.

Contention-based LANs are the most commonly deployed LAN topologies. Distributed polling environments, however, do have their place.

Distributed Polling LANs

In addition to the gladiatorial combat approach to sharing access to a transmission facility, there is a more civilized technique known as distributed polling, or as it is more commonly known, *token passing*. IBM's token-passing ring is perhaps the best-known of these products, followed closely by FDDI, a 100-Mbps version occasionally seen in campus and metropolitan area networks (although the sun seems to be setting on FDDI).

In token-passing LANs, stations take turns with the shared medium, passing the right to use it from station-to-station by handing off a *token* that gives the bearer the one-time right to transmit while all other stations remain quiescent. Thus, the Harvard approach. This is a much fairer way to share access to the transmission medium than CSMA/CD, because while every station has to wait for its turn, it is absolutely guaranteed that it will get that turn. These systems are therefore characterized by bounded delay, because there is a maximum amount of time that any station will ever have to wait for the token.

Token-passing rings work as shown in Fig. 6.23. When a station wishes to transmit a file to another station on the LAN, it must first wait for the *token*—a small and unique piece of code that must be *held* by a station to validate the frame of data that is created and transmitted. Let's assume for a moment that a station has secured the token because a prior station has released it. The station places the token in the appropriate field of the frame it builds (actually, the medium

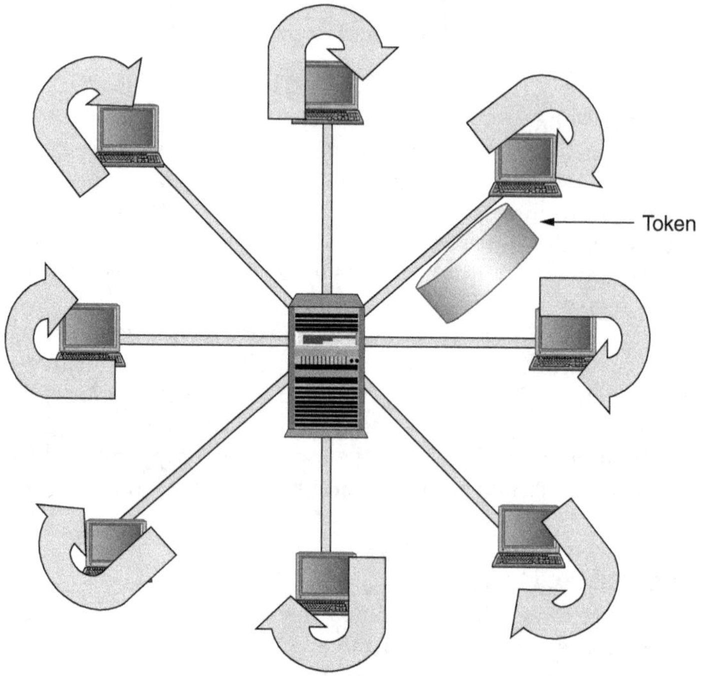

FIGURE 6.23 The architecture and function of a token-passing ring.

access control scheme, called a MAC scheme and implemented on the NIC card builds the frame), adds the data and address, and transmits the frame to the next station on the ring. The next station, which also has a frame it wishes to send, receives the frame, notes that it is not the intended recipient, and also notes that the token is busy. It does not transmit, but instead passes the frame of data from the first station on to the next station. This process continues, station by station, until the frame arrives at the intended recipient on the ring. The recipient validates that it is the intended recipient at which time it makes a copy of the received frame, sets a bit in the frame to indicate that it has been successfully received, *leaves the token set as busy,* and transmits the frame on to the next station on the ring. Because the token is still shown as busy, no other station can transmit. Ultimately the frame returns to the originator, at which time it is recognized as having been received correctly. The station therefore removes the frame from the ring, frees the token, and passes it on to the next station (it is not allowed to send again just because it is in possession of a free token).

This is where the overall fairness scheme of this technique shines through. *The very next station to receive a free token* is the station that first indicated a need for it. It will transmit its traffic, after which it will pass the token on to the next station on the ring, followed by the next station, and so on.

This technique works very well in situations where high traffic congestion on the LAN is the norm. Stations will always have to wait for what is called *maximum token rotation time,* that is, the amount of time it takes for the token to be passed completely around the ring, but they will *always* get a turn. Thus for high-congestion situations a token passing environment may be better.

The Decline of Token-Passing

For the most part, token passing rings have fallen out of favor. They're still out there in some large, legacy installations, but for the most part they have lost the race to ethernet. The interesting thing about this outcome is that by most measures token-passing is a superior technology to contention-based LANs in almost every way. First, they don't suffer from the unbounded delay that occurs in ethernet LANs when traffic gets heavy. Second, token-passing environments allow for the transport of larger packets than ethernet. Third, the early versions of ethernet and token ring both offered plenty of bandwidth, but the maximum speed of an ethernet LAN was 10 Mbps, whereas token ring offered 16 Mbps.

So why did token ring lose? A number of reasons, one of which may have been a combination of arrogance and greed. Token ring was always more expensive than ethernet, and the economics were not helped by the fact that IBM charged third-party vendors that wanted to manufacture token ring-compatible cards a hefty royalty for the right to do so. Meanwhile, ethernet evolved to support faster and faster levels of bandwidth. Then LAN switches arrived on the scene, eliminating the problem of contention and collisions.

IBM and its partner companies tried to play catch up with token ring by increasing the bandwidth and playing games with the price, but by the time they did anything meaningful it was too late. Ethernet had won.

Logical LAN Design

One other topic that should be covered before we conclude our discussion of local area networks is logical design. Two designs have emerged over the years for LAN data management. The first is called *peer-to-peer*. In a peer-to-peer LAN, shown in Fig. 6.24, all stations on the network operate at the same protocol layer, and all have equal access *at any time* to the shared medium and other resources on the network. They do not have to wait for any kind of permission to transmit; they simply do so. Traditional CSMA/CD is an example of this model. It is simple, easy to implement, and does not require a complex operating system to operate. It does, however, result in a free-for-all approach to networking, and in large networks can result in security and performance problems.

The alternative and far more commonly seen technique is called *client-server*. In a client-server LAN, all data and application resources are archived on a designated server that is attached to the local area network and accessible by all stations (user PCs) with appropriate permissions, as illustrated in Fig. 6.25. Because the server houses all of the data and application resources, client PCs do not have to be particularly robust. When a user wishes to execute a program such as a word processor, they go through the same keystrokes they would on a stand-alone PC. In a client-server environment, however, the application actually executes on the server, giving the user the

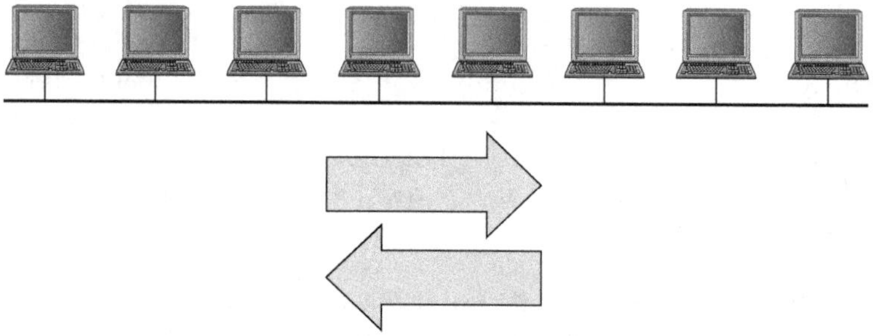

FIGURE 6.24 A peer-to-peer LAN.

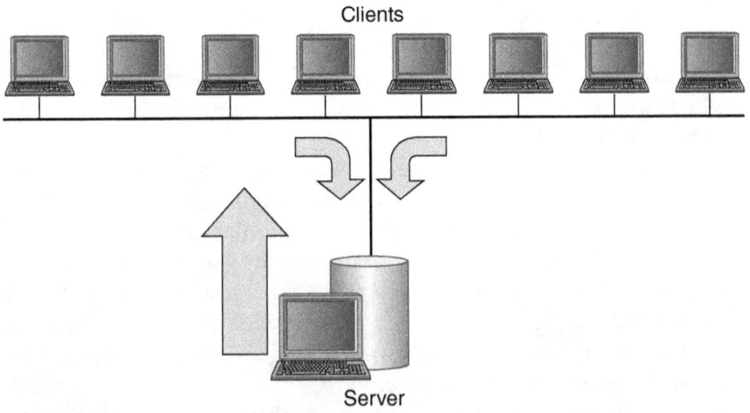

Figure 6.25 A client-server architecture.

appearance of local execution. Data files modified by the user are also stored on the server, resulting in a significant improvement in data management, cost control, security, and *software harmonization* compared to the peer-to-peer design. This also means that client devices can be relatively inexpensive, because they need very little in the way of onboard computing resources. The server, on the other hand, is really a PC with additional disk, memory, and processor capacity so that it can handle the requests it receives from all the users that depend on it. Needless to say, client-server architectures are more common today than peer-to-peer in corporate environments. In fact, they are more common than most people realize: The vast majority of the applications that you enjoy on your smartphone or tablet are client-server based. You don't actually own the software; its creator grants you permission to use it, and when you download it you are actually downloading a client that executes locally but accesses the server via the wireless network. We'll see more about this when we discuss cloud solutions in Chap. 11.

Other Relevant Technologies

Before we wrap this chapter we should take a few minutes to discuss a number of other technologies that are both important and relevant.

Following a Goat Trail: Wi-Fi

In keeping with the technologically schizophrenic nature of this book we begin our discussion of Wi-Fi by talking about goat trails.

Go to any reasonably sized business park or college campus and you will see that the designers of the place have gone to great lengths to make the environment pleasing. They have tastefully positioned the buildings in such a way that they are optimally collocated, sometimes even exhibiting the sublime characteristics of *feng shui*. Interconnecting the buildings are carefully poured concrete sidewalks, angling up and over and up and over, providing walkways from one building to the next.

The careful observer, however, will also notice the warren of "goat trails" worn into the grass that run from building to building (Fig. 6.26), demonstrating that while the angular sidewalks are aesthetically pleasing, they violate the rule which observes correctly that the shortest distance between two points is a straight line. In fact, I maintain that architects should dispense with the sidewalks altogether, wait 6 months, and then pave the goat trails. It makes a lot more sense and saves on concrete.

This is essentially how Wi-Fi came about, and it's important because other *competing* standards did not follow this path—and are experiencing very different market acceptance profiles. The technology was introduced in 1992 when wireless LAN companies began to make use of the spectrum component where Wi-Fi operates today. It is important to note that the technology was released before the standards were complete and manufacturers, seeing the promise of a new technological innovation, jumped on the bandwagon and began building Wi-Fi devices. Standards development efforts followed and

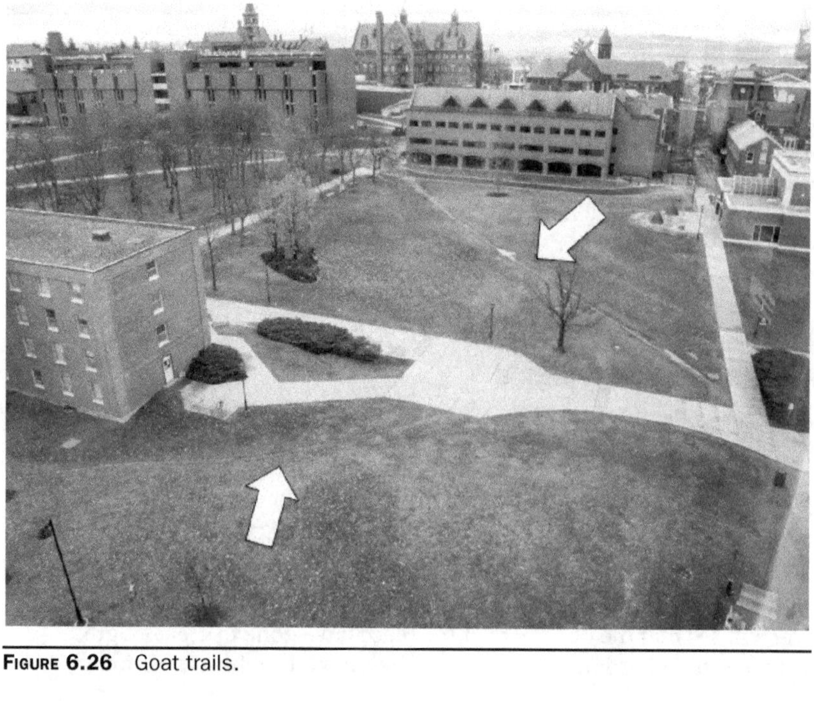

FIGURE 6.26 Goat trails.

the result was what you might expect: a collection of incompatible standards that created problems in the industry that were reminiscent of the *modem wars* of the 90s, when two competing and incompatible 56-Kbps modem standards were released, creating implementation paralysis throughout the industry. Hence the goat trail comparison: Wi-Fi was created as the result of a *path of least resistance* progression model in which developers rushed to create products. And while this resulted in rapid introduction of Wi-Fi products and services, it created a variety of problems—including, at least initially, the release of dysfunctional standards.

The technology waddled around through varying degrees of commercial success for the intervening five years until June 1997 when the IEEE released the initial 802.11 standard for high-speed transmission using infrared and two forms of radio transmission within the unlicensed 2.4-GHz frequency band. The first of these was frequency hopping spread spectrum (FHSS); the second was direct sequence spread spectrum (DSSS). It is useful to understand the difference between these two techniques; they are described in the following sections and will have relevance when we get to the chapter on wireless mobility. We begin with spread spectrum.

Spread Spectrum

The idea behind spread spectrum transmission is a simple concept: disguise a narrowband transmission within a broadband transmission by *spreading* or *smearing* the narrowband component within the broadband carrier. This is done using either of two basic techniques: *frequency hopping,* in which a control signal directs the two communicating devices to *hop* randomly from frequency to frequency to avoid eavesdropping, and *direct sequence,* in which the signal is combined with a random *noise* signal to disguise its contents under the command of a control signal that knows how to separate the signal *wheat* from the noise *chaff.*

Frequency Hopping Spread Spectrum

The development of frequency hopping spread spectrum is one of those stories that is worth knowing about—simply because it is so much fun to tell and so remarkable in the telling. During World War II there was considerable angst among Allied forces over the Axis Powers' ability to interfere with radio-controlled torpedoes launched from submarines by jamming the radio signals that guided them to their targets. To that end a pair of industrious entrepreneurs filed patent number 2,292,387 titled, "Secret Communications System." The inventors were orchestra composer George Antheil and electrical engineer Hedy K. Markey, better known as early film star Hedy Lamarr. The technique described in their patent application is remarkably simple: a central *authority* (the base station, shown schematically in Fig. 6.27) communicates with the two communicating endpoint

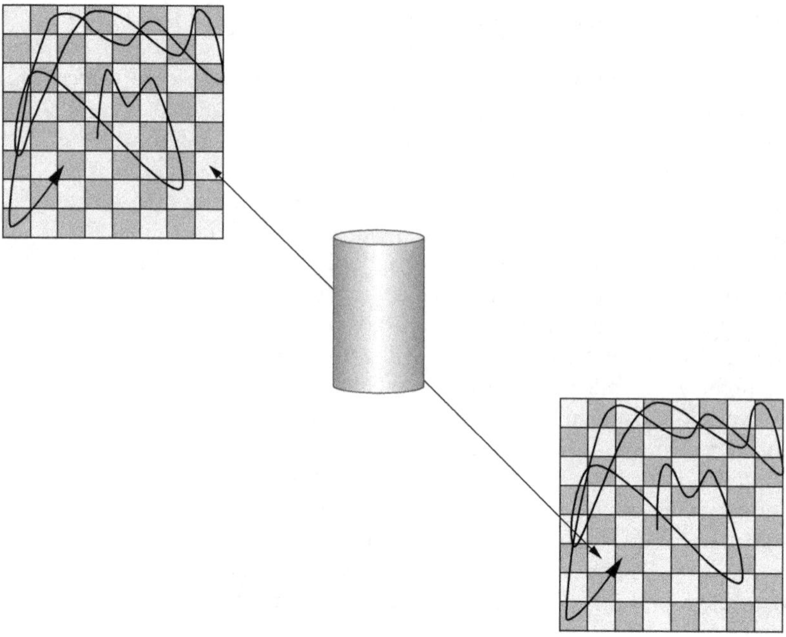

FIGURE 6.27 A schematic illustration of frequency-hopping.

devices, instructing them to *hop* randomly from frequency to frequency at randomly selected times and for random intervals, typically on the order of 5 ms (milliseconds) or less. Only the base station (think of an orchestra conductor) and the two devices know when and where to jump—and for how long. To an outsider looking in, that is, any device wishing to eavesdrop on the conversation, the hopping process appears completely random. It isn't, however: the base station knows precisely what it is doing and when it is doing it, so the hopping behavior is actually *pseudorandom*, and the technique is often referred to as a *pseudorandom hopping code.* This technique is most commonly used in code division multiple access (CDMA) cellular systems, and is (as you might expect) used extensively in secure military communications systems.

Direct Sequence Spread Spectrum

Direct sequence spread spectrum, sometimes called *noise-modulated spread spectrum,* is a very different technique than its FHSS cousin. In frequency hopping, a *conversation* jumps from frequency to frequency on a seemingly random basis. In noise modulation, the actual signal is combined with a carefully crafted *noise* signal which disguises it. The otherwise narrowband signal, shown in Fig. 6.28, is spread across a much wider channel—typically on the order of 1.25 MHz. The bits in the data stream are combined with *noise bits* to create a much

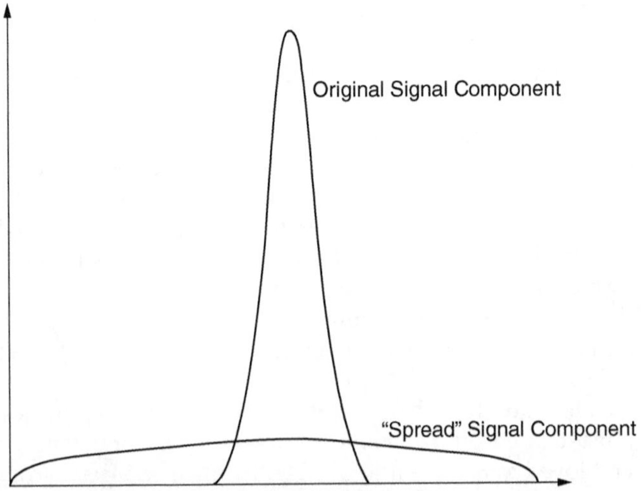

FIGURE 6.28 Direct Sequence Spread Spectrum.

broader signal, and as before, only the base station and the communicating devices know the code that must be used to extract the original signal from the noise. The code is sometimes called a chipping code, and the technique is referred to as producing a *chipped signal.*

One way to think about this technique is as follows: Imagine that you have just been appointed ambassador to South Africa (congratulations!). As part of your pre-departure indoctrination you have learned that the country has 11 official languages: Afrikaans, IsiNdebele, IsiXhosa, IsiZulu, Northern Sotho, Sesotho, Setswana, SiSwati, Tshivenda, Xitsonga, and English. Upon arriving in the country, you attend a reception in your honor, and in attendance are dignitaries from the many regions of South Africa, all speaking their local language over glasses of *mampoer* and South African wine. The room is filled with people and the cacophony of 11 distinct languages is simply noise to your ears. Suddenly, however, from across the room, emerging from the general din, you hear someone calling to you. From across the ballroom your newly appointed aide de camp is talking to you in English at a normal level, yet you understand her perfectly.

If you understand this description, then you understand noise modulated spread spectrum. These two techniques, FHSS and DSSS, formed the functional basis of the original wireless LAN standards that ultimately coalesced into the phenomenon known as Wi-Fi.

Wi-Fi Today

Today several 802.11 standards exist. 802.11b and 802.11g devices transmit in the unlicensed 2.4-GHz range, while 802.11a devices transmit at 5 GHz. Needless to say this presents a radio compatibility problem that had to be overcome. As far as implementation history is concerned, 802.11b was the first version of the Wi-Fi standard to reach the market. It is the least costly and offers the lowest bandwidth of the three, transmitting in the 2.4-GHz spectral band at speeds up to 11 Mbps.

Next in the developmental lineup was 802.11a. Operating at 5 GHz, it can handle transmission speeds up to 54 Mbps. Its biggest challenge, however, was its incompatibility with the widely accepted 802.11b. To remedy this, 802.11g was introduced, which operates in the same 2.4-GHz band as 802.11b but offers transmission rates at speeds equivalent to 802.11a—upward of 54 Mbps. As a result it is not only compatible with 802.11b but also benefits from the lower cost, unlicensed spectrum at 2.4 GHz.

To achieve the high bandwidth levels that these systems offer, innovative data-encoding techniques are used. One of the most common, employed in 802.11a and 802.11g, is called *orthogonal frequency division multiplexing (OFDM).* OFDM divides the radio signal into multiple sub-signals that are then transmitted simultaneously across a range of different frequencies to the receiver.

802.11b, on the other hand, uses a different encoding technique called *complementary code keying (CCK).* As one knowledge base defines it, complementary codes are *binary complementary sequences with the mathematical property that their periodic auto-correlative vector sum is zero except at the zero shift.* And while that definition may set the engineers in the audience all aquiver, it makes me feel like I'm reading the bridge column in the newspaper. So in English, CCK is simply a technique that creates a set of 64 8-bit code words that encode data for 11-Mbps data transmission in the 2.4-GHz band (802.11b). These code words are mathematically unique so that they can be distinguished from one another by the receiver, even when there is substantial noise and multipath interference. As a result, CCK-based systems can achieve substantially higher data rates by eliminating the impact of such effect that would otherwise limit its ability to achieve them.

It should also be noted that complementary code keying only works with DSSS systems specified in the original 802.11 standard. It does not work with FHSS.

Other Wi-Fi standards have followed, as shown in the following list:

802.11-2007

In 2003, one of the Wi-Fi working groups was given the go-ahead to combine a cluster of amendments put forward in the 1999 version of 802.11, resulting in a 802.11ma, made up of the 802.11 standard combined with amendments a, b, d, e, g, h, i, and j. It was approved on March 8, 2007 and quickly renamed to IEEE 802.11-2007. Are these people imaginative, or what?

802.11n

802.11n adds support for multiple-input multiple-output antennas, a technology known as MIMO.

MIMO uses an array of antennas at both the transmitter and receiver to improve both spectral efficiency and link reliability. Often referred to as *smart antenna technology,* MIMO offers a lot of promise in terms of both range and throughput. As a consequence, it has been incorporated into a number of wireless standards including Wi-Fi, HSPA, and LTE.

802.11n operates on both the 2.4-GHz and the 5-GHz bands, although support for 5-GHz operation is considered to be an optional feature. Here's where the magic comes into play: 802.11n operates at speeds ranging from 54 to 600 Mbps! Ratified and published in 2009, it has enjoyed huge popularity—for all-too-obvious reasons.

802.11-2012

Building on the work done in 2003, further consolidation of amendments was undertaken and accomplished beginning in 2007. The group combined amendments k, r, y, n, w, p, z, v, u, and s with the existing standard to create IEEE 802.11-2012.

802.11ac

IEEE 802.11ac was developed between 2011 and 2013 and has enjoyed significant attention since its release. It delivers massive amounts of bandwidth in the 5-GHz band; the standard allows for throughput of 1 Gbps for multi-station WLANs and single-link throughput of nominally 500 Mbps.

802.11ad

IEEE 802.11ad, sometimes called *WiGig,* has been published but as of this writing has not yet been implemented. Operating in the 60-GHz range, it will offer a maximum theoretical bandwidth of 7 Gbps—respectable by anyone's measure.

802.11ah

There are standards, and there are standards—and then there are the really odd, albeit intriguing, standards. 802.11ah is one of those. Designed to meet the low-power, low-bandwidth requirements of machine-to-machine communications and near-field communications, it shows promise as a potential solution for the future.

802.11 Physical Layer

All 802 standards address themselves to both the physical (PHY) and media access control (MAC) layers. At the PHY layer, IEEE 802.11 identifies three options for wireless local area networks: diffused infrared, DSSS, and FHSS.

While the infrared PHY operates at a baseband level, the other two operate at 2.4 GHz, part of the industrial, scientific, and medical (ISM) band. It can be used for operating wireless LAN devices and does not require an end-user license. All three PHYs specify support for 1-Mbps and 2-Mbps data rate.

802.11 Media Access Control Layer

The 802.11 MAC layer, like CSMA/CD and token-passing, presents the rules used to access the wireless medium. The primary services provided by the MAC layer are as follows:

- *Data transfer:* Based on a carrier sense multiple access with collision avoidance (CSMA/CA) algorithm as the media access scheme.
- *Association:* The establishment of wireless links between wireless clients and access points (APs).
- *Authentication:* The process of conclusively verifying a client's identity prior to a wireless client associating with an AP. 802.11 devices operate under an open system where any wireless client can associate with any AP without verifying credentials. True authentication is possible with the use of the wired equivalent privacy protocol (WEP), which uses a shared key validation protocol similar to that used in public key infrastructures (PKI). Only those devices with a valid shared key can be associated with an AP.
- *Privacy:* By default, data is transferred in the clear; any 802.11-compliant device can potentially eaves drop like-PHY 802.11 traffic that is within range. WEP encrypts the data before it is transmitted using a 40-bit encryption algorithm known as RC4. The same shared key used in authentication is used to encrypt or decrypt the data; only clients with the correct shared key can decipher the data.
- *Power management:* 802.11 defines an *active mode,* where a wireless client is powered at a level adequate to transmit and receive, and a *power save mode,* under which a client is not able to transmit or receive, but consumes less power while in a standby mode of sorts.

Wi-Fi in Action

Wi-Fi is a remarkably simple technology. In fact, most new laptops come equipped with a built-in Wi-Fi card. The card, which is simply a radio, serves to connect the computer or PDA in which it is installed to a base station known as a hotspot. The hotspot is simply the central point of connection for a computer to the network; typically it is a small device that is connected to the Internet, often via a DSL facility. The hotspot typically contains an 802.11-based radio that has the ability to engage in simultaneous conversations with multiple 802.11 cards. Today it is common to see them deployed in coffee shops, airports, hotels, and conference centers. The technology has also gained an enormous following in the residence and home office markets because of the degree to which it simplifies home or small office networking. It has been slower to catch on, however, in the enterpriser space—read on.

Wi-Fi Security

Anytime data is transmitted over a wireless facility there are concerns about the confidentiality of the information being transmitted. The broad appeal of Wi-Fi has resulted in the incorporation into Wi-Fi an encryption technique known as *wired equivalent privacy*. WEP is an encryption technique with two variants: a 64-bit encryption technique and a 128-bit encryption scheme. 128-bit encryption is naturally more secure and is the most widely used technique.

Gaining access to a WEP-enabled system requires the use of a WEP key, which is a decryption code required for the system to recognize and allow access to your system. One of the reasons Wi-Fi has been slow to deploy in the enterprise environment is its inherent vulnerability. Consider the following scenario: A well-meaning employee purchases a hotspot at a local office supply store. The employee installs it at the office as a favor to his or her colleagues, but fails to enable a WEP key. The hotspot is now wide open—and anyone driving by with a Wi-Fi-enabled laptop can gain access to the hotspot—and to the enterprise network that lies behind it. It is critical, therefore, that all hotspots be WEP-enabled.

An Aside: Mobility versus Ubiquity

As a telecom industry analyst and author of numerous books about the business and technology sides of the telecom marketplace, I tend to focus on the words used by the people and companies that comprise the industry, because the careless use of those words often leads to misinterpretation, which in turn leads to inappropriate investment and infrastructure decisions. To that end, I realized recently that two important words in our technology lexicon are used synonymously. The words, *mobility* and *ubiquity*, are *not* synonyms. However, applications, network designs, investment decisions, strategic planning efforts, and end-user device engineering are being done as if they were one and the same. At worst, this will lead to another spate of ill-placed investments and the requisite marketplace blowback; at best it will lead to confusion and the resulting slowdown of effective deployment. Hopefully the former won't occur; the latter already has. So before this gets too far down a dead-end road, let me clarify the difference between the two, beginning with definitions.

According to the American Heritage Dictionary, *mobility* means "the quality or state of being mobile," while *ubiquity* is defined as "existence or apparent existence everywhere at the same time." From a communications perspective, mobility means being able to move freely while staying connected, as when engaging in the increasingly socially unacceptable practice of using a cell phone while driving. Ubiquity, on the other hand, means universal connectivity, that is, the ability to count on the presence of a connection of one kind or another from the bottom of the Grand Canyon to the top of Mount McKinley and everywhere in between.

From a development point-of-view, these two concepts are being used interchangeably, creating confusion, and developmental paralysis. Consider, for example, Wi-Fi. Wi-Fi is a high-speed wireless technology that provides 11 Mbps (or more—or less) of access bandwidth to roaming laptops and PDAs. Its installation in restaurants and coffee shops by the likes of McDonald's and Starbucks is being advertised as a great step forward in mobility. But this isn't mobility: People aren't walking around Starbucks with a Venti Latte in one hand and a laptop in the other, surfing for MP3s. They are in fact sitting in a booth or at a table, enjoying the merits of *ubiquitous access.* Yes, the connection happened to be wireless, and these people are not in their homes or offices, but the presence of Wi-Fi in this case is not an addition to the pantheon of mobility applications. It is, however, an application of ubiquitous connectivity. So in this particular situation, what value does the wireless loop add? It seems to me that Starbucks, McDonald's, and others could provide equal or better service to their customers who have a need to be connected by providing an RJ-45 cable or two at each table. It would be cheaper, more secure, and better in terms of service quality. Yet there is a perceived sexiness associated with wireless that somehow precludes wired connectivity as an access option in ad hoc situations.

Wi-Fi, of course, provides a convenience factor that clearly has value to the user, however intangible that value may be. Being able to connect anywhere in the coffee shop without being physically tethered is an advantage, but how much of an advantage is it? Bluetooth, an alternative wireless technology, promised to eliminate dependence on wires between computer peripherals, yet it remains a largely stillborn solution. Does it work? Yes. Does it provide a level of value that overcomes the price? Apparently not, because its usage levels are near zero.

Consider the following scenario: McDonald's, Starbucks, Barnes & Noble, Borders, major airports, and large department stores all purchase DSL lines. They terminate the lines on a low-cost router, which in turn connects to a multi-port hub. They then install convenient RJ-45 connections at each table or at some easily recognized spot in the store. Laptop users can either provide their own cable or borrow/purchase one from the host store. The service is offered at no cost and provides differentiation for the business, or is available via subscription in much the same way that T-Mobile and Wayport offer connectivity in hotels and airports. Note that the connectivity technology is wired, and while the businesses listed could also add a wireless access point to their network, providing connectivity for Wi-Fi users, they still offer near-ubiquitous connectivity using a wired option.

There is another issue that seems to have been lost in this discussion, and that's the issue of power. Wireless connectivity sucks the life out of laptop batteries at a dizzying rate, which means that even with a *wireless* network loop a user will need a *wired* power loop in short order. In other words, they will have to be tethered anyway if they use the connection more than 10 or 15 minutes, and this negates the sexiness of the wireless connection to a large extent.

So what is the point of this discussion? Both mobility and ubiquity are important in the emerging world of network usage, but they are not necessarily the same. Mobility implies the ability to connect to the network via a wireless local loop. Ideally it offers predictable high bandwidth, easy, dependable connections, and secure transmission. Ubiquity, on the other hand, implies the ability to connect to the network anywhere, anytime, regardless of the characteristics of the loop. Ubiquitous access may include wireless as an option, but may also include wired solutions such as Ethernet, T-1, and DSL.

Another issue is the usage of the two words in terms of parts of speech. Mobility, used as it is in our lexicon, is a noun, because in the minds of many it defines (incorrectly) an application. Mobility is no more an application than DSL is an application. It is a technology option, no more and no less.

Ubiquity, on the other hand, is used as an adjectival modifier to qualify the nature of an individual's access to the network. Ubiquitous access implies the delivery of something that is superior because it is—everywhere. But isn't wireline access universally available, and therefore ubiquitous as well? And isn't it significantly more secure than a wireless connection, particularly in an enterprise application? After all, if a business decides to implement wireless (802.11) as its connectivity option of choice within a workplace, how can it possibly guarantee that (1) all users implement a secure wireless protocol over the local loop, and (2) signals do not leak out of the building? Wireless is a great technology, offering freedom and mobility to users, but there is a price. And that price can be steep if it is implemented without forethought.

I have observed in this book that customers are not really looking for the next great killer application, but rather for a *killer* way to access existing applications because those applications offer solutions to most of the challenges that users encounter. Consider the typical business user. As long as they are in their home-office environment, wired connectivity is perfectly acceptable for both voice and data. When they leave the office and get in the car, *mobile telephony* becomes important. Mobile data has no application (thankfully!) in the car other than for those applications optimized for that environment—OnStar service, for example, or GPS-based guidance systems, or specific applications related to public safety. If, however, the user stops at Starbucks for coffee before going home and decides to check e-mail one last time, *ubiquitous* connectivity, whether wired or wireless, provides the value to the user from a data perspective, while *mobile* telephony remains valuable for voice. An RJ-45 connection on the tabletop is just as serviceable as a Wi-Fi connection, not to mention far more secure and predictable.

The bottom line is this. *Mobility* defines the characteristics of a *lifestyle choice* that involves networking, whether personal or work-related, while *ubiquity* defines the *characteristics of the technology infrastructure* required to support the mobile lifestyle. Anywhere, anytime connectivity has become the chanted mantra of the mobile user, and while wireless (Wi-Fi) is the most loudly proclaimed option, it is *not* the *only* option. This, I believe, is part of the reason that revenues associated with Wi-Fi remain elusive. It is sexy, cool, and functional. But of those three characteristics the only one that has revenue potentially associated with it is *functional*, and there are too many alternatives to wireless that offer lower cost, greater security, and more predictable connections. Until a service provider comes up with a compelling argument for Wi-Fi's performance superiority, the only companies that will make money on it will be those building wireless access points and routers.

ZigBee

ZigBee—what kind of name is that?!!

More than a 100 years ago scientists began to take notice of the complex dance, known as the waggle dance, that honeybees use to communicate vast amounts of complex information to other members of the hive such as the location of food, predators, nesting sites, and so on. Because the bees tend to zig and zag while performing the dance and communicating with each other, the network professionals who developed this low-power, low-cost mesh network decided to credit them by choosing a name in their honor.

As odd as it may be, ZigBee offers some serious advantages. Because it is a low-power solution, it is easy on batteries. And because it costs very little, it can be widely deployed in applications that require low-cost networking solutions over a short distance, such as mobile payments systems and machine-to-machine communications applications.

Finally, because it utilizes a mesh architecture, it is highly survivable.

ZigBee operates in the industrial, scientific, and medical (ISM) spectrum and is based on the 802.15.4 IEEE standard. In the United States and Australia, ISM occupies the 915-MHz range; 868 MHz in Europe; and 2.4 GHz in most other parts of the world. This is not a barn-burner as far as bandwidth is concerned: on a good day it offers anywhere from 20 to 250 Kbps.

Applications for ZigBee are interesting and include wireless alarm and sensor systems, home automation, medical monitoring, embedded sensing (in conjunction with RFID), and industrial process control. This is not a protocol that we hear a lot about but trust me, as the phenomenon known as the Internet of Things takes root and begins to show promise, this technology will leap to the fore in its relevance for cost-effectively interconnecting vast numbers of sensors.

The Final Three: Firewire, Thunderbolt, and USB

By rights these three don't really belong here, but truthfully I can't think of a better place to put them in the book. They are important for the interconnection of user devices, so for lack of a better home I'm going to place them here.

Firewire

Once again Apple leaps into the fray, this time with a high-speed interface for data transfer.

FireWire is what Apple calls the standard known as IEEE 1394, which defines a high-speed serial bus technology. And while Apple initially kicked off development of the standard in concert with the IEEE, FireWire also had developmental help from TI, DEC, SONY, STMicroelectronics, and IBM.

FireWire is a serial bus architecture designed to replace the parallel SCSI bus used in higher-end audio and video systems. It supports both isochronous (occurring at regular intervals, like video frames) and asynchronous applications and delivers data rates in excess of 800 Mbps.

Thunderbolt

Contrary to popular belief, Thunderbolt was not developed by Apple—it was actually developed by Intel, although it first appeared commercially on Apple's MacBook Pro in 2011 and relying on the Mini DisplayPort that *was* developed by Apple.

Thunderbolt ports multiplex several *data lanes* together to achieve high levels of throughput—up to 10 Gbps when connected to a Thunderbolt-capable device.

One interesting fact about Thunderbolt is that it was originally designed to run exclusively over an optical interface because of the bandwidth it was designed to offer. During the development phase, however, it became clear that copper could deliver the same bandwidth at a lower price, so the design shifted to a metallic facility.

Universal Serial Bus

Universal serial bus (USB) was developed in the mid-90s and released in 1996 to serve as a low-cost, high-performance, power-passing bus for the interconnection of computers and peripherals (Fig. 6.29). One of the primary goals of its designers was to standardize the interface for an array of peripheral devices, thus lowering the manufacturing cost for both computers and the devices that connect to them.

Three connectors emerged from the standardization process. The most common is the interface seen on most PCs and laptops. Second is the USB-Mini connector, found on many digital cameras. And the third is the USB-Micro, found on most cell phones today.

USB supports four different data transfer speeds: low speed, full speed, high speed (USB 2.0), and super speed (USB 3.0). Low speed and full speed were supported in the first generation of USB and supported 1.5 Mbps and 12 Mbps, respectively. USB 2.0, which came out in April of 2000, upped the ante to 280 Mbps. USB 3.0 appeared on the scene in 2008, offering a whopping 5 Gbps of data transfer speed. USB 3.1, currently under development, advances USB into the realm of Thunderbolt, offering speeds of 10 Gbps.

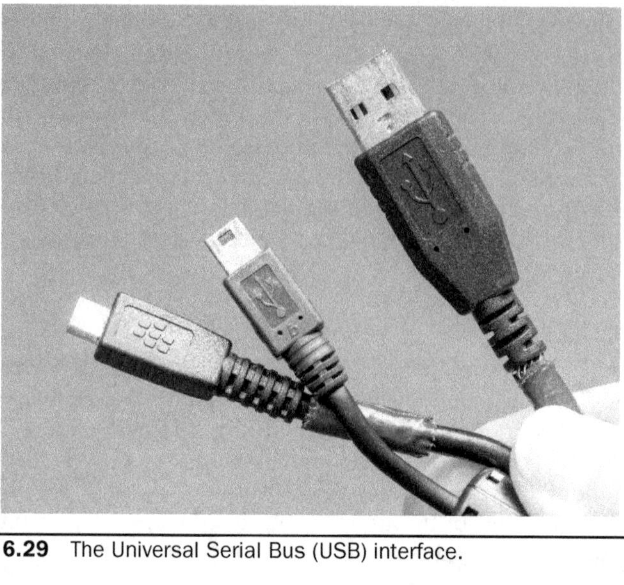

FIGURE 6.29 The Universal Serial Bus (USB) interface.

USB PD

At the time of this writing a new USB standard is emerging called USB power delivery (USB PD). Today, USB can trickle up to 10 W of charging power to a connected device, enough to fill the batteries of cell phones, MP3 players or to power small devices such as portable hard drives. USB PD goes well beyond that, increasing the power delivery to a much as 100 W, and to do so in a bidirectional manner—in other words, power could be delivered from the connected device to the computer, if need be. This is a fascinating development and one that will change the way in which peripherals and main devices both communicate and provide power to one another.

Summary

Premises technologies, including local area networks, wireless solutions, and the content that rides on both of them are the most visible components of the overall network to a customer. In the next chapter we dive into the world of content and multimedia.

Chapter Six Questions

1. What are the five principal components of a computer? Describe the functions of each.
2. What factors affect a CPU's processing speed?
3. What does "4.0 GHz" actually mean?
4. What is a bit? A byte? A megabyte?
5. What are the functions of the ALU?
6. Explain the differences between main and secondary memory. Give examples of each.
7. Explain the differences between a token-passing and a contention-based LAN.
8. Why is ethernet the preferred LAN scheme?
9. What is the difference between 802.3 and ethernet?
10. What is the difference between a peer-to-peer network and a client-server network?

CHAPTER 7

Content and Media

Competencies owned and nurtured by a company represent its critical resource and competitive advantage, and the company should create a portfolio of services that contribute to and extract value from those competencies.

—G. Hamel & C. Prahalad

As the third leg of our telecom, media, and technology (TMT) stool, content—sometimes referred to as media—is the most visible element of the services triumvirate because ultimately, it's the thing that people actually want to pay money for. The network that transports the content tends to be invisible (except, of course, when it breaks and becomes *glaringly* visible, and the IT or technology part of the equation is often vilified because it generates invoices or is perceived as intrusive. The content, however, is everybody's friend: Whether we're talking about music, a YouTube video, NetFlix, personal banking records, a videoconference, a digital TV signal, a stream of data from the Mars Rover, a remote teleradiology consult, or a downloadable app, the response is the same: it's the content that matters. How it gets to me, how it gets tracked and monetized, isn't my problem. And therein lies the growing challenge for network and IT providers. They must struggle to define their relevance to both the consumer and business markets with the same argument we introduced at the beginning of the book: I don't care how good your content is, how compelling your app, how exciting your movie, how detailed your digital tomographic image. Without a good network over which to transport, and a solid and responsive IT infrastructure to track and store it, it has little value.

The section that follows explores multimedia and all its many flavors. Multimedia is the primary driving force for bandwidth today; an understanding of what it is and where it is going is important. Besides, it's fascinating stuff.

The World of Multimedia

In the last decade, a revolution has taken place in visual applications. Starting with simple, still image-based applications such as grayscale facsimile, the technology diverged into a collection of visually oriented applications that include video and virtual reality. Driven by aggressive demands from sophisticated, applications-hungry users and fueled by network and computer technologies capable of delivering such bandwidth-and-processor-intensive services, the telecommunications industry has undergone a remarkable metamorphosis as industry players battle for the pole position.

Why this rapid growth? Curt Carlson, Vice President of Information Systems at the David Sarnoff Research Institute in Princeton, New Jersey, observes that more than half of the human brain is devoted to vision-related functions—an indication that vision is our single most important sense. He believes that this rapid evolution in image-based systems is occurring because those are the systems that people actually *need.*

"First we invented radio," he observes, "then we invented television. Now we are entering what we call the age of interactivity, in which we will take and merge all of those technologies and add the element of user interaction. Vision is one of the key elements that allow us to create these exciting new applications." Indeed, many new applications depend on the interactive component of image-based technologies. Medical imaging, interactive customer service applications, and multimedia education are but a few.

Setting the Stage

A revolution of sorts has taken place over the course of the last decade or so, a revolution that has gotten us to this place where media, IT, and telecom occupy seats of equal power in the technology game.

For the purposes of the conversation that follows please refer to Fig. 7.1. We begin in the upper left corner.

The last decade-and-a-half saw exciting advances in the technology behind digital imaging, specifically compression standards that preserve the quality of the compressed image and hardware-based coding schemes, typically called CODECs. These advances in audio, video and imaging software led ultimately to user-generated content (UGC), which most of us know as YouTube (although to be fair, voice was the first form of user-generated content). UGC, in turn, opened up new avenues for the delivery of highly targeted and effective advertising, something that the advertising industry salivated over as they saw the negative impact that digital video recorders (DVRs) had on their declining ability to keep commercial messaging in front of a viewer. That new advertising mechanism drove revenue, depicted on the right side of the diagram.

Another motive force behind this brave new world of content was the development of the application-programmer interface (API).

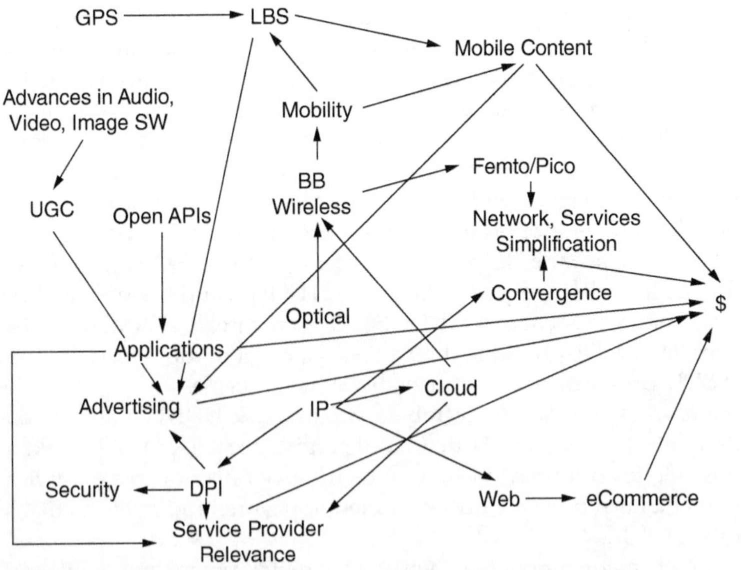

FIGURE 7.1 Diagram showing the complex relationship between different elements of the T, M, and T sectors.

Application-Programmer Interface

An API consists of standards, protocols, development tools, and subroutines that are used to build applications in a standardized manner or to give an application the ability to communicate with a remote application, typically using the Internet as the internet-working medium. This communications process is accomplished via a series of calls, which are requests for information from one communicating entity to the other. A call may ask for such things as data that is specific to a particular organization, metadata that defines the characteristics of a particular database record, the ability to update or change the data in a record, or it may perform some kind of database maintenance function. Whatever the case, the calls that take place between the processes (applications) running on the local and remote machines are implemented via a process called *web services*, which are in turn usually written in extensible markup language (XML), which facilitates large swaths of Internet communications.

An API, for example, could be written to schedule and set up a corporate videoconference. The API for this particular responsibility would necessarily include XML-encoded scripts that define the name and contact information of the person establishing the conference, the invitees, the IP addresses of the endpoint devices, the time of the conference, the expected duration, access to downloadable documents, and so on.

APIs are designed to be operating system-agnostic, meaning that they will run on any OS or hardware platform in the same way that Web resources are universally viewable. They're usually packaged in what is known as a software development kit (SDK); the SDK includes the API itself, a set of development tools, and operating instructions.

APIs are coded in XML (which is very similar to HTML, the markup language in which most Web pages are written). Three additional elements support the development and operation of functional APIs. They are the simple object access protocol (SOAP), universal description, discovery, and integration (UDDI), and the Web services description language (WSDL). SOAP generates the XML-encoded content; UDDI provides directory and collaboration services, and WSDL provides a mechanism by which businesses can describe themselves and their deliverable services to the UDDI database. Similarly, it is through APIs that Apple and Google (Android) make it possible for hundreds of thousands of developers to create applications that can be easily hosted at their app stores and made available to the world.

APIs made it possible, then, for the development and delivery of a whole new generation of applications that could be downloaded at very little cost, in turn making it possible for each individual with a mobile device to customize that device to their personal requirements. The device became the receptacle for the content and apps; the apps became the truly desirable thing. The apps created two outcomes. First was the opportunity for additional revenue; the other was an opportunity for service providers to become more relevant in the mind of their customers. As a result many of the larger telephone companies have opened their own app stores, not so much as a revenue-generation strategy, but rather a strategy to create mindshare.

Don't be mistaken: This is big business. The Apple and Google apps stores now host close to 800,000 each, and revenue from app sales in 2013 topped $25 billion. They may only cost 99 cents, but when you multiply that number by the number of people downloading and paying for them, sooner or later you're talking about real money. But it's a fickle market: advertising is the most common mechanism for attracting new customers and the cost is going up. Mobile game developers are seeing the cost of acquisition rise by double digits year-over-year with no end in sight. So developers are finding themselves forced to spend more on analytics to better understand the customer; they're also spreading their wings to develop apps and games for a broader array of devices including tablets and television.

Move with me now, if you would, to the bottom center of the diagram where you find the acronym IP. IP, of course, has been around for a very long time (details to follow in Chap. 11) and is responsible for a complete reinvention of the concept of networking.

Clearly it lies at the functional heart of the World Wide Web and the Internet upon which it lives, and is a requirement for e-commerce and all that that implies. It is also the fundamental basis for convergence, the coming-together of companies, technologies, industries, and media, which in turn led to the growing simplification of the network-at-large. Consider this: there was a time when a typical telephone company operated a voice network, a frame relay network, an ATM network, a wireless network, an ISDN network, an Internet cloud, and in many cases an X.25 network. Each network required different expertise, which implied different people; different technology, which implied different vendors and spares inventories; and a massively complicated network management operation. Imagine being able to replace all of that with a single IP *cloud* (network) that has a variety of interfaces for each of the different network types. Now, instead of running seven networks we run one, and we only have to maintain a spares inventory for a single network infrastructure. Frodo and company had their one ring to bind them all; the world of technology has its one technology that binds it all.

IP also is the basis for cloud technologies, providing as it does the connectivity between the user and a vast, virtual resource that is global in scope.

The other thing that IP makes possible is the most discrete understanding of the customer we have ever had, thanks in no small part to protocols like deep packet inspection (DPI). DPI, together with its sister deep content inspection (DCI), make it possible to examine them and analyze the contents of an IP packet. This analysis, conducted discretely and anonymously, allows a service provider—not necessarily a telephone company but really any entity involved in the delivery of digital products—to develop a rich and highly nuanced profile of the customer so that customized and highly targeted advertising content can be directed to that customer. Think about how pleased we are today when Amazon makes book recommendations to us based on our purchase history. Imagine now what life could be like when all aspects of commercial messaging were that on-target, especially if that messaging is delivered on an opt-in basis only.

Let me take you for just a moment across a back road on our diagram to the upper right corner where it says Femto-Pico. "Femto-Pico" refers to femtocells and picocells, a relatively new wireless technology that is helping service providers improve the quality of service experience they deliver.

Many of you probably live in an area where the wireless signal at your home is not ideal. Here in northern Vermont I have to stand in the northwest corner of my office with a frying pan against my head to collect enough signal to get more than a couple of bars of service on my mobile, because I live at the fringe of the nearest cell.

Femtocells eliminate this problem. They are small devices, about the size of a small home router, and they cost about the same. When a customer purchases a femtocells they are directed to take it home and plug it into their broadband connection, whether a cable modem or DSL router. They then register the device at their service provider's Web site, a process that identifies the nearest cellular tower and therefore the transmit and receive frequency pairs that the femtocell should use to avoid any problems of interference. Once registered the device wakes up and starts to work. Congratulations: You now have your own private little cell tower in your home and you can now enjoy five bars of cellular service on your mobile device.

But let's now take this conversation to the next level (Fig. 7.2). The electronics that provide femtocell functionality are nothing more than a small circuit board, so let's assume we can make the plastic housing a little bit bigger so that it has a bit more real estate inside. The first thing we'll include is another circuit board that adds a router to the box, which means that we now have to add RJ-45 LAN connectors on the back of the box. While we're at it let's attach a daughter board to the router that implements WiFi, and since we have the room we'll add another daughter board that implements Bluetooth. And since we still have some headroom inside the case, let's go ahead and add one more circuit board, which adds a digital tuner to the mix of capabilities that the box is capable of delivering.

Now we'll add one final thing: a small piece of silicon wafer that implements DPI for content inspection inside the box. So what have we created? Well, think about it: 100% of all the packets that go into this house or office and out of this house or office … flow through this box. Which means that 100% of those very same packets are readable

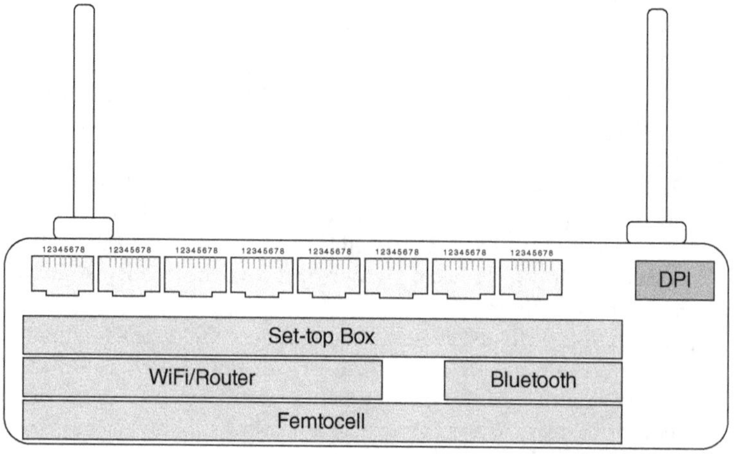

FIGURE 7.2 The inner workings of the femtocell-of-the-future.

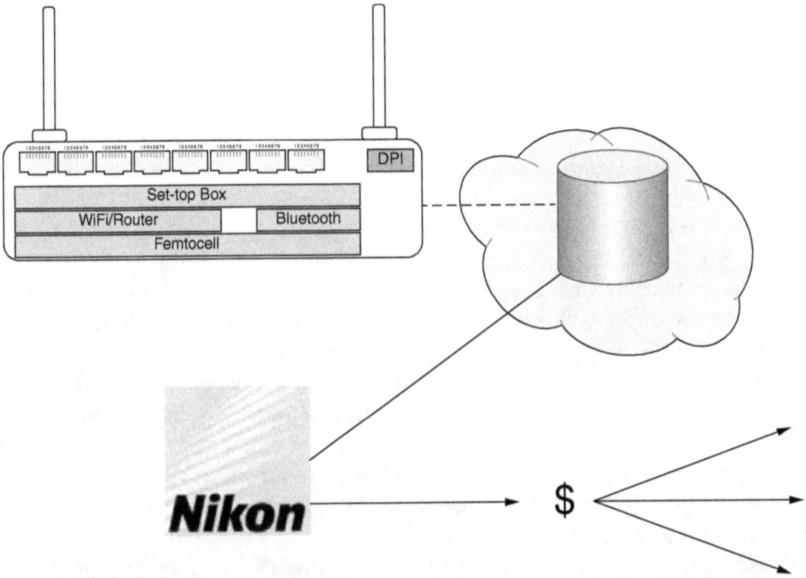

FIGURE 7.3 The implications of a multifunction femtocell with analytics capabilities.

and analyzable by the DPI client that lives in the box. And what are the implications of this? Well, take a look at Fig. 7.3.

Here we have our *augmented femtocell* in the upper left corner. It is connected to the customer's broadband connection, and somewhere out in the cloud is a database where the customer's information is analyzed: What they watch, when they watch, how often they watch, where they surf, whether they use their tablet while they're watching TV, what books they purchase and read, which commercials they actually watch all the way through rather than change channels, which programs they record on their DVR for future viewing, and so on. All of this information is a gold mine of intelligence for someone who really wants to understand their customer's likes, dislikes, preferences, biases, and emotional attachments.

Let me go on record here with one very, very large caveat. The model of experience delivery that I am describing is based on one fundamental and very important—as in deal-breaker—assumption: That this is an *opt-in option only*. In other words, this is not like the *do-not-call* list where customers have to deliberately go to a Web site and inform the keepers of the site that they do not want to be called by solicitors. This model is the opposite: In this case the subscriber must consciously and deliberately go to the service provider and ask to be included—that is, inform the service provider that they would like their information to be analyzed. Then and only then will their information be analyzed. This model avoids privacy and confidentiality concerns;

involvement in the analytics process is 100 percent under the control of the customer.

So for the purposes of this example let's assume that I am the person being profiled by my service provider. I have agreed to have my data analyzed anonymously with the understanding that it will not be shared and there will be immediate and ongoing benefit to me as a subscriber.

Before we get into the process let me give you a bit of background to help you understand how the analysis of seemingly unrelated bits of data can lead to a particular desired outcome. As an undergraduate student I studied two widely different fields: Romance Philology (the study of the origins of Romance languages) and biology. Because I grew up overseas I developed a strong interest in language and culture, hence the language degree; and I am fascinated by the natural world, which explains my work in biology.

Here's something else you need to know. In addition to working in the technology world I am also a professional nature, landscape, and travel photographer.

When the system analyzes the various streams of data that go into and out of my home and office here's what they will learn:

- I spend quite a bit of time watching National Geographic and Discovery Channel programs about nature, wildlife, national parks, exotic places, and unique destinations.
- I buy a lot of books at Amazon about those very same topics as well as books about digital photography, various aspects of biology, and biographies of famous scientists in the various fields of natural science.
- I spend way too much time (and money) at the Web site of B&H Photography in New York. If I'm buying cameras and lenses and flashes, they're made by Nikon; printers, Epson; tripods and camera supports, Gitzo.
- I pay Adobe $10 every month for the cloud-based service they offer that gives me access to Photoshop, Lightroom, and a collection of other related applications.
- I visit my Facebook page about twice a week.
- I enjoy novels by Michael Connelly, Lee Child and Douglas Preston, and Lincoln Child.
- I watch the Big Bang Theory, Two-and-a-Half Men, NCIS, 60 Minutes, and Criminal Minds every week and record them on my DVR.
- Every once in a great while I watch a show on NetFlix.
- I buy a lot of music from iTunes, especially progressive rock from the 70s: Yes, Emerson, Lake and Palmer, Pink Floyd, Starship. I also love 1960s progressive music such as King Crimson.

- I listen to six different Podcasts every week, four of them about photography, one about storytelling, one about history.

So with that raw data in-hand the analytics engine kicks in and, over a period of time, begins to develop an accurate and insightful profile that describes me pretty accurately. Over time it becomes increasingly more accurate until a moment arrives when it is considered statistically accurate. At that point the agency (service provider, etc.) conducting the analysis looks at the results and concludes that they have *got* to get my profile in front of Nikon. So they take my profile, now carefully redacted so that it does not have any personal information in it, and they share it with Nikon. The profile clearly indicates that this is a person who buys a great deal of Nikon equipment and could probably be incented to buy more, based on their demographic, their perceived income and other indicators captured during the analytics phase. Nikon begins to salivate, but as they reach for the profile the service provider pulls back, saying, "Not so fast—let's talk," rubbing their index finger and thumb together in the universal sign that means, "show me the money." In other words, Nikon, how much would you be willing to pay for access to this customer for advertising purposes? The service provider agree on a price, at which point the service provider makes it clear that they will not give the customer's information to Nikon; instead, they will act on behalf of both Nikon and the customer (in this case, me) as a proxy for the exchange of information. In effect the service provider acts as an information firewall between the customer's private information and any commercial entity that wants access to it. They agree to take commercial messaging from Nikon and put it in front of me, in exchange for a mutually agreed fee.

Let's stop for a moment and analyze what just happened. Here's what we know:

- The service provider has done a good job identifying a potentially lucrative target for Nikon marketing.
- Because of the good job they did there is a high likelihood that I (the customer) will respond positively to whatever commercial message they put in front of me, dramatically increasing the overall effectiveness of the advertising effort.
- As a consequence, Nikon is willing to pay a premium for the placement of the ad, because the target is well known.

Nikon pays the service provider the agreed fee. But where does the money actually go? How is it disbursed, once it's in the hands of the service provider? This is a fundamentally central question. Some of it, of course, stays with the service provider as payment for the analytical services they performed. Some of it goes to the network that will ultimately host the commercial message. And *some* of it goes to—can you guess? The customer! That's right. In exchange for agreeing to be profiled by the system, the customer is compensated, not necessarily monetarily, but via some other means. For example, the customer might be bumped to a higher broadband tier. Or they might be given a higher tier of channel selections on their TV service. Or perhaps they are given additional minutes of use or broadband access on their mobile plan. One option might be that the customer receives a *menu* of selections from which they may choose, as their reward for taking part in the process.

So what's the outcome of all this? Well, there are several, and have merit. First, the customer sees commercial advertising that is topical and personally relevant. Second, the advertiser (Nikon, in this case) sees a potential upside on sales. And third, the service provider enjoys a higher degree of relevance, enters a new line of business, and can now take advantage of a new revenue stream in the form of analytics and advertising. It also means that the service provider now finds itself moving inexorably toward reinvention as an IT provider, something we'll talk more about in Chap. 11.

Let's go back to Fig. 7.1 one more time. Directly above the IP acronym in the center of the page is the word *optical,* which as you'll see has an arrow pointing up to broadband wireless. This relationship may seem counterintuitive, but it isn't, for the following reason. If you're selling third- or fourth-generation wireless HSPA or LTE service to your customers and your backhaul from the tower into the network is a 1.544 Mbps T1 facility, you're in trouble. For the levels of bandwidth that are being demanded by customers today to feed their never-ending demand for more and more content, a much bigger pipe is required and that pipe is optical. Broadband wireless in turn makes mobility far easier since it facilitates the always-on lifestyle and makes access to personal content a universal reality. This brings up an important observation. The terms *wireless* and *mobile* tend to be used interchangeably, but they are profoundly different. Wireless is a technology option; mobility is a lifestyle choice.

Mobility leads logically to mobile content, including music, video, images, banking data, voice, and a host of applications. It also leads to the accelerated deployment of location-based services (LBS), a growing area of interest that in combination with machine-to-machine communications and near-field communications is redefining a number of industries. Finally, GPS weighs in. In partnership with location-based services it adds another layer of accuracy to the overall system.

It's All about the Ecosystem

I apologize for dragging you through that hairball of a diagram, but hopefully you understand how important the exercise was. At the beginning of this book, I made the case that it is nigh-on impossible to study any one of the three TMT sectors without also paying close

Technology	TMT Sector
Audio, video, imaging compression and CODECs	Media, Tech
User-Generated Content	Media
Advertising	Media
Open APIs	Tech, Telecom
Applications	Tech, Media
Security	Tech
Deep Packet Inspection	Tech, Telecom
IP	Tech, Telecom, Media
Cloud	Tech, Telecom
WWW/eCommerce	Tech, Telecom, Media
Convergence	Tech, Telecom, Media
Femtocell	Tech, Telecom
Mobile Content	Telecom, Media
Optical	Tech, Telecom
Broadband Wireless	Telecom
Location-Based Services	Tech, Telecom, Media
GPS	Tech, Telecom, Media

Table 7.1 The Responsibilities of the T, M, and T Sectors

attention to the other two. Figure 7.1 and the relationships it describes illustrate this observation pretty clearly.

Take a look at Table 7.1, which presents the information in a slightly different form. Here I have prepared a two-column table. The left column lists the technologies we just discussed; the right column shows which of the three T, M, or T sectors the technology is found in. You'll notice that relatively few of them are found in a single area, and quite a few encompass all three. Again, these three areas are inextricably intertwined—and for good reason.

For the most part, the media sector concerns itself with content, which is a big territory. It includes still images, video, television, music, and a host of other content types. Let's now turn our attention to the manner in which content is encoded, interpreted, and displayed. Some of this technology we have already touched on briefly, but we're going to provide more detail here. We begin with still images.

Still Images

Still image applications have been in widespread use for quite some time. Initially there were photocopy machines. They did not provide storage of documents nor did they offer the ability to electronically transport them from one place to another. That capability arrived on a limited basis with the fax machine.

While a fax transmission allows a document to be moved from one location to another, what actually moves is not document content but rather an *image* of the document's content. This is important, because there is no element of flexibility inherent in this system that allows the receiver to make immediate and easy changes to the document. The image must be converted into machine-readable data, a capability that is just now becoming reasonably good with optical character recognition (OCR) software. Companies like Amazon and Barnes & Noble, for example, that convert thousands of paper books into digital documents that can be displayed on a book reader like the Kindle or Nook, rely on OCR technology to scan the pages, convert the scanned images into machine-readable text, and output a readable document. This is also why you occasionally run across odd errors in the text of a digital book: Some character was misinterpreted by the OCR software and the "human in the loop" who was supposed to catch the glitch failed to do so.

As imaging technology advanced and networks grew more capable, other technological variations emerged. The marriage of the copy machine and the modem yielded the scanner, which allows high-quality images to be incorporated into documents or stored on a machine-accessible medium such as a hard drive.

Other advances followed. The emergence of high-quality television (HQTV) coupled with high-bandwidth, high-quality networks led to the development and professional acceptance of medical imaging applications with which diagnosis-quality x-ray images can be used for remote *teleradiology* applications. This made possible the delivery of highly specialized diagnostic capabilities to rural areas, a significant advancement and extension of medicine. Suddenly, diagnostic capability could travel to the patient, rather than the patient having to travel to a diagnostic center. This put a completely new face on healthcare and is one force that is redefining the nature of service delivery in medicine.

Equally important are imaging applications that have emerged for the banking, insurance, design, and publishing industries. Images convey enormous amounts of information. By digitizing them, storing them online, and making them available simultaneously to large numbers of users, the applications for which the original image was intended are enhanced. Distance ceases to be an issue, transcription errors are eliminated, and the availability of expertise becomes a non-problem.

There are downsides, of course. Image-based applications require expensive end-user equipment. Image files tend to be large, so storage requirements are significant. Furthermore, because of the bandwidth-intensive nature of transmitted image files, network infrastructures must be re-examined. Of course, thanks to our TMT triumvirate and the fact that all three sectors operate in lockstep, they tend to evolve together.

The Arrival of Compression

To deal with the storage and transmission issues associated with image-based applications, corollary technologies such as digital compression have emerged. One of the main techniques used today for still image compression is JPEG, developed by the Joint Photographic Experts Group, a cooperative effort between the International Organization for Standardization (ISO), the ITU-T, and the International Electrotechnical Commission (IEC).

JPEG, discussed earlier, works as follows. Digital images, composed of thousands of picture elements (pixels), are *dissolved* into a mosaic of 16-pixel × 16-pixel blocks. These blocks are then reduced to 8 × 8 blocks by removing every other pixel value. JPEG software then calculates an average brightness and color value for each block that is stored and used to reconstruct the original image during decompression.

Today, still-image technologies are widely used and will continue to play a key role in the application of visual technologies. But others have emerged as well. One of the most promising is video. But what is also very interesting to anyone who works in or adjacent to the TMT industry is the various places in which video and still images touch and become a new hybrid form of media. One example of this is Google Glass.

Google Glass

Google Glass (Fig. 7.4) is one of those mysterious projects that everyone knows about and can talk about, but precious few have ever actually seen one or used one. So what is it, and why all the hype around it?

In essence, Google Glass (or "Glass," as the cognoscenti have taken to calling it) is a wearable computer that looks like a pair of glasses that Data would wear on Star Trek. It has a heads-up display in the form of a lens that sits in front of the wearer's right eye. It communicates via natural voice commands and communicates wirelessly with the Internet using either Wi-Fi or Bluetooth. Testing of the device began in early 2012, and while Glass is not yet available to the general public, Google does run an early adopter program called the Glass Explorer Program. If accepted into the program, users may buy the device (current cost is about $1,500).

To communicate with the headset, users simply say, "ok, Glass," which readies the device for the next command. Follow-on commands include "Take a picture," "Record a video," "Start a Google Plus hangout with Kenn Sato," or natural language questions such as,

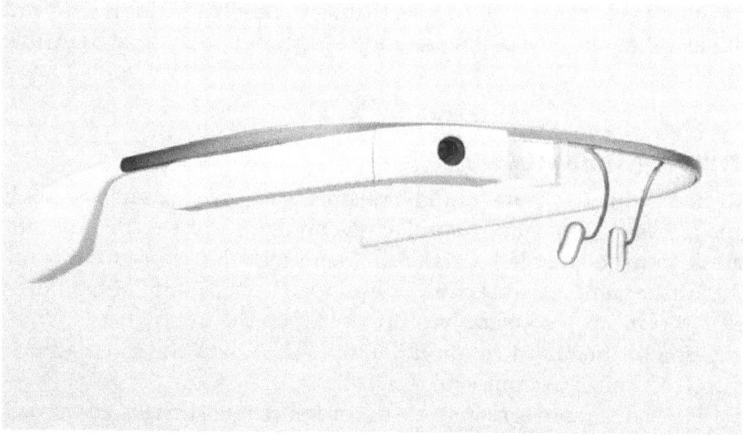

Figure 7.4 Google Glass. (*Image courtesy of Google.*)

"Where is the closest Mexican restaurant?" They can also scroll through a series of commands by touching the side of the headset to activate a scrolling list on the screen. Audio response from Glass is done via a bone conduction speaker mounted in the earpiece.

Glass also has the ability to take pictures and record 720p HD video on-command, which is one of the things that causes twitchiness in some people. Concerns about privacy invasion abound, although the reality is that (my opinion, now) anyone who thinks their daily routines are still a closely guarded secret needs a new drug. The camera makes a shutter-click sound when it takes a picture that cannot be silenced, and a small LED is illuminated when video is being recorded. Proponents are equally vocal in their support of Google Glass. Imagine walking down the street and having the heads-up display tell you that your best friend is 100 meters ahead on the left. Or that the person approaching you and smiling, who you haven't seen in over a year and whose name you can't remember, is Dave Heckman who works for the University of Pennsylvania.

The application possibilities are unbounded—and speaking of applications, Google Glass is a third-party world. Apps are being developed for Glass that do a variety of impressive things including facial recognition, real-time language translation, streaming video, access to Google Maps, image editing, social media access, location-based services, and a host of others.

Why This Matters

You may see Google Glass and other edge-of-the-envelope technologies as intrusive, too much like science fiction, or irrelevant to your day-to-day life—and you may be right. However, don't dismiss the long-term implications of this and other innovations that push the limits of human creativity and innovation, no matter how outlandish they may appear at first blush. Consider this: In June 2013 Pedro Guillén, a surgeon and Chief of Trauma Medicine at Madrid's Clínica CEMTA, used Google Glass to broadcast a knee surgery to Stanford University so that Dr. Homero Rivas, a specialist in telemedicine, could consult on the procedure.

Apple's innovations have equally important impact. While you may not like using Siri on your iPhone or iPad, the implications of Siri's voice response capabilities are profound. At first glance, Siri seems like a novelty—one of those things that Apple added in a whimsical moment to make their devices more fun to use (for example, try asking Siri, "What is the movie 'Inception' about?"). But Siri's influence goes way beyond entertainment. For example, in the world of search and search-engine optimization (SEO), Siri stands poised to redefine the practice of local SEO because of the extraordinary depth of the information "she" can access and the various responses she can make. For example, "Find me a good Mexican restaurant" is easy for Siri because location-based services automatically narrow the search to the area from which the request is being made. At the same time, advertising can be customized to the local area, the user's preferences, and other personal indicators of relevance.

Siri has clearly broken the mold when it comes to voice recognition. I recently had a conversation with a friend, and as an experiment decided to use one of the online transcription services to convert the conversation to text. My actual words went something like this:

Hey Chris, the outlook is pretty bright, although a few things have cropped up. I'm looking for a visual that describes the history of fiber. Particularly want to see information about ring architectures and the number of homes they pass, alright? We're getting errors on the copper out by the airport, but the airport has access to fiber.

The conversation went on for about 30 seconds. Here's what the online service returned as a transcription of the conversation:

Hey Chris, This. Yvonne alright. All that propped up, I guess I'm looking for a visual have to get out the Velcro fiber. That's right, the fiber. Lou. But there ring. And then I. Victor, over away. Maybe on that, or somehow however you baby. And most visually He If you have We're looking for where the holes are, and Since, where the homes alright. Where there is any so bye bye. Our Now, or an E how we could answer those other error is register conversation now. Sure. Piper or wireless Janet, Matt. I hope I'm making myself clear. Someone is. Where we have, by right now, and at what also include where. Airport has Piper, where, the tell. Sprint bill. Eric alright. I guess I don't know the, and I don't want right now, fine man currently. And the project and share our plans, plane it which is you were the Yeah, hey. All right now, project test that or fine for me company phone. After that is Donna me yet. All of that. We have any under serve terriers, black. And com ex-wife, okay white be able to. Ciao. Now I understand. Of course, fish, C fiber, project basically no longer others. Anne currents demonstration pilot And I don't know if our 3. That's my current on Thursday. Thank you all be in my car for about another 15 or 25th. If you wanna call me And I'm available, After, hey.

Not quite ready for primetime.

Another innovation with far-reaching implications—global, in this case—is Apple's decision to add a fingerprint reader to its mobile devices. On the surface, this is good for device and information security. But consider this: fingerprints are unique to every individual, worldwide. Mobile phones are becoming one of the most democratized devices in the world; by 2015 there will be more mobile devices on the planet than people. The addition of the reader, in combination with a secure application, makes it possible for every individual on the planet to take part in fair, corruption-free elections in their country, without fear of going to a voting station where violence might erupt. With the addition of something as simple as a thumbprint reader, Apple may have unwittingly (or wittingly) democratized ... democracy, the result of which is the most transparent government model ever created.

We'll talk more about reinvention and innovation when we get into the IT conversation in Chap. 11. For now, though, let's look at video—beginning with its colorful history.

Video

The video story begins in 1951 at RCA's David Sarnoff Research Institute. During a celebration dinner, Brigadier General David Sarnoff, the Chairman of RCA and founder of NBC, requested that the institute work on three new inventions, one of them a device that he called a *videograph*. In his mind, a videograph was a device capable of capturing television signals on some form of inexpensive medium, such as magnetic tape.

Remember the time frame that we're talking about. Thanks to Philo T. Farnsworth who invented the predecessor of today's cathode ray tube (CRT—and yes, that's his real name), electronic television became a reality in the 1920s. By the early 1950s, black and white television was widespread in America. The gap between the arrival of television and the demand for video, therefore, was fairly narrow.

Work on the videograph began almost immediately and a powerful cast of characters was assembled. There was one unlikely member of this cast who served as a catalyst: Bing Crosby. Keenly interested in broadcast technologies, Crosby wanted to be able to record his weekly shows for later transmission. The Bing Crosby Laboratories played a key role in the development and testing of video technology.

The Sarnoff Institute called upon the capabilities of several companies to reach its goal of creating what Sarnoff dubbed the "hear-see machine." One of them was Ampex, developer of the first commercial audio tape recorder. The Sarnoff team believed that audiotape technology could be applied to video recording.

To a certain extent they were correct. Marvin Camras, a Sarnoff team member and scientist who developed the ability to record audio signals on steel wire used during WWII, soon discovered that the video signal was dramatically broader than the relatively narrow spectrum of the audio signal. Early audio tape machines typically moved the tape along at a stately 15 inches per second (in/s or IPS). To meet the bandwidth requirements of the video signal, the tape had to be accelerated to somewhere between 300 and 400 IPS—roughly 25 mi/h (miles per hour).

To put this into perspective, a tape that would accommodate an hour's worth of audio in those days would hold *1 minute* of video—which did not take into account the length of the leader that had to be in place to allow the recorder to reach its ridiculously high tape transport speed. To hold 15 minutes of video, a reel of ¼-in tape would have had to be 3 ft in diameter—not exactly portable. Put another way, a 1-hour show would require 25 mi of tape!

To get around this problem, Camras invented the spinning record head. Instead of moving the tape rapidly past the recording head, he moved the tape slowly, and rapidly spun the head. By attaching the head to a 20,000-rpm Hoover vacuum cleaner motor (stolen, by the way, from his wife's vacuum cleaner), he was able to use 2-in tape and reduce the tape transport speed to 30 to 40 IPS—a dramatic improvement.

Is That a Horse ... or a Tree?

The first video demonstrations were admirable but rather funny. First, the resolution of the television was only 40 lines-per-inch compared to more than 250 on modern systems. The images were so poor that audiences required a narrator to tell them what they were seeing on the screen.

Luckily, things got better. The original video systems rendered black and white images, but soon a color system was developed. It recorded five tracks (red, blue, green, synch, and audio) on ½-in tape and ran at the original speed of 360 IPS. The system was a bit unwieldy: It required a *mile* of color tape to capture a 4-minute recording.

Obviously, mile-long tapes were unacceptable, especially if they would only yield 4-minute programs. As a result, the Sarnoff/Ampex team re-examined the design of the recording mechanism. Three scientists—Charles Ginsburg, Alex Maxey, and Ray Dolby (later to be known for his work in audio)—redesigned the rotating record head, rotating it about 90 degrees so that the video signal was now written on the tape in a zig-zag design. This redesign, combined with FM instead of AM signal modulation, allowed the team to reduce the tape speed to a remarkable 17½ in/s. For comparison's sake, modern machines consume tape at about 2½ in/s. This, by the way, is why the record head in your home VCR sits at a funny angle (it's that big silver cylinder you see when you open the slot where the tape is inserted).

By 1956, Sarnoff and Ampex had created a commercially viable product. They demonstrated "The Mark IV" to 200 CBS affiliate managers in April of that year. When David Sarnoff walked out on the stage and stood next to his own pre-recorded image playing on the television next to him, the room went berserk. In four days, Ampex took $5M in video machine orders.

Modern Video Technology

Today's palm-size videotape recorders—or for that matter, mobile phones—are a far cry from the washer-dryer-size Mark IV of 1956. But physical dimensions are only a piece of the video story.

The first video systems were analog, and relied on a technique called *composite video*. In composite video, all of the signal components, such as color, brightness, audio, and so on, are combined into a single multiplexed signal. Because of the interleaved nature of this technique, a composite signal is not particularly good, and suffers from such impairments as clarity loss between generations (in much the same way an analog audio signal suffers over distance), and color bleeding. Unfortunately, bandwidth at the time was extremely expensive and the cost to transport five distinct high-bandwidth channels was inordinately high. Composite video, therefore, was a reasonable alternative.

As the cost of bandwidth dropped in concert with advances in transmission technology, *component video* emerged, in which the signal components are transported separately, each in its own channel. This eliminated many of the impairments that plagued composite systems.

Component Formats

Several distinct component formats have emerged including RGB (for red-green-blue); YUV (for luminance [Y], hue [U], and saturation [V]); YIQ; and a number of others. An interesting aside: luminance is analogous to brightness of the video signal, whereas hue and saturation make up the chrominance (color) component.

All of these techniques accomplish the task of representing the red, green, and blue components needed to create a color video signal. In fact, they are mathematical permutations of one another.

One final observation is that RGB, YUV, and the others previously mentioned are *signal formats*, a very different beast from *tape formats*, such as D1, D2, Betacam, VHS, and S-VHS. Signal formats describe the manner in which the information that represents the captured image is created and transported; tape formats define how the information is encoded on the storage medium.

Of course, these signal formats are analog, and therefore still suffer from analog impairments such as quality loss, as downstream generations are created. Something better is needed.

Digital Video

In the same way that digital data transmission was viewed as a way to eliminate analog signal impairments, digital video formats were created to do the same for video. Professional formats such as D1, D2, and digital betacam (the latter often discounted because it incorporates a form of compression), and even High-8 and MiniDV, virtually eliminate the problem of generational loss.

One downside is that even though these formats are digital, they still record their images sequentially on videotape. Today, video editors rely on nonlinear systems. Because the video signal can be digitized ("bits is bits"), it can be stored on a hard drive just as easily as a text or image file can be. Of course, these files tend to be large and therefore require significant amounts of hard drive space. For example, a full-motion, full color, TV-screen-size, uncompressed 2-hour movie requires 26 MB/s storage capacity—a total of nearly 24 GB storage. Of course, video is compressed for streaming and viewing, the result of which is that a high-quality movie file is typically on the order of 4 to 5 GB.

The Dichotomy of Quality of Experience

This brings up an interesting dichotomy—or perhaps an irony. Remember that the focus of this book—and my own personal philosophy—is about the implications of the technologies we discuss rather than the technologies themselves. There's plenty of technology in this book as you've already seen, but what really matter is what we do with it.

Whenever I talk about the impact of mobility I always ask the audience to tell me what they believe is the most important thing that mobile technology has given the world. And they always say the same things: spontaneity, freedom, safety and security, true mobility, and flexibility. I always nod in agreement and then inform them that those are all good answers. They're *wrong* answers, but good answers. I believe that the most important thing that mobility brought to the world is a willingness on the part of the market to accept crappy service (that's a technical term). Think about it: It is because of mobility that we have the phrase in the popular lexicon, "Can you hear me now?" It is because of mobility that we now start conversations, not with "How are you?" but with, "If I lose you I'll call you back." The service quality just isn't there. However, *I'm in my car.* Or, I'm in the bush in South Africa photographing cheetahs for a magazine, and the Food Bank in Vermont calls to thank me and my wife for our donation this year, not realizing that I'm halfway around the world. *I'm in South Africa and my phone rings.*

The reason I'm waxing eloquent about this is because the same holds true for modern video delivery. When video was delivered to the customer on a VHS tape, the quality was extremely good and looked fine on a TV screen. When the industry left VHS behind and migrated to DVD, the image got even better. Blue-Ray, even better. Yet the vast majority of people today use streaming services such as NetFlix with its 40 million subscribers, YouTube which receives 60 hours of new video *every minute,* and cable television alternatives such as HULU, iTunes, and Redbox to gain access to entertainment content. See Table 7.2 for a comparison of common viewing and purchase options.

My point is that the *quality of the service* doesn't have to be perfect, as long as the *quality of the experience* is what the customer is looking for. Here's an example. About a year ago I flew into a very small town in New Mexico to deliver a keynote the next morning to an industry group. As sometimes happens, my suitcase failed to appear, but the agent located the bag and told me that it would arrive on the 8 p.m. flight. They offered to deliver it but I told them that I'd just swing by and pick it up after dinner since I was staying near the airport and the airport was roughly the size of a dinner plate.

Service	Cost
HULU	Free for basic service; $8 per month for HULU+
Netflix	$8 per month
iTunes	$1 per viewed TV episode, $2 to buy; $4 to rent movies, $15 to buy
Amazon	$1–2 per viewed TV episode; $4 to rent movies, $15 to buy
CinemaNow	$2 per viewed TV episode; $4 to rent movies, $16 to buy
RedBox	$1 per DVD, $1.50 for Blu-Ray
VuDu	$2 per viewed TV episode; $4 to rent movies, $15 to buy

TABLE 7.2 Typical Costs of On-demand Services

I went to eat at a recommended restaurant in the vicinity of my hotel around 6 p.m. because I was starving. I ate slowly, had dessert, had a cup of coffee, had another cup of coffee, but it was still only 7 PM by the time I got to the airport. So I settled down in the baggage claim area with an hour to go before my bag arrived. Then I had an epiphany—I could catch up on at least one episode of the Walking Dead on NetFlix! Logging into the NetFlix app with my iPhone, I enjoyed two episodes before the plane arrived with my bag. Was it fantastic viewing quality? Of course not. Did it occasionally freeze because of bandwidth exhaustion? Twice during the two episodes for about 5 seconds each time. Would it have been a better experience on the large screen in my hotel room? Yep—*but I wasn't in my hotel room. I was in the airport.* And my iPhone was delivering precisely the experience I wanted at that moment in time.

I cannot emphasize enough the importance *and relevance* of this evolving relationship between quality-of-service and quality-of-experience.

Video Capture and Encoding Technologies

Today, the market demands an inexpensive, high-quality solution to the storage of video. CDs or DVDs are one option. Before we go any farther with this discussion, though, let's explore what it takes to put video on a CD-ROM.

The Video Process

The signal that historically emerges from a commercial video camera, while it can be either analog or digital, is typically analog. The signal is laid down on either analog tape (Betacam, VHS, S-VHS), digital tape (D1, D2, digital betacam), or in some cases a built-in hard drive. Digital tape is expensive, but it eliminates generational loss from copy-to-copy and is therefore common in commercial video.

To create a digital representation of analog video, the analog signal is typically sampled at 3 to 4 times the bandwidth of the video channel (unlike audio, which typically relies on a 2× sample). As a result, the output bandwidth for digital tape machines is quite high—411 Mbps for D1, 74 Mbps for D2.

A single CD-ROM drive holds approximately 650 MB data; a DVD, 4 GB and change. The best numbers available today, including optimal sampling and compression rates (discussed a bit later) indicate that VHS-quality recording requires roughly 5 MB of storage per second of recorded movie. That's 7,200 seconds for a 2-hour movie, or roughly 30 CDs (or 7 DVDs). Finally, the maximum transfer rate across a typical bus in a PC is 420 KB/s, somewhat less than that required for a full-motion movie.

What the Market Wants, the Market Gets: Compression

Growth in desktop video applications for videoconferencing, on-demand training, and gaming has fueled the growth in digital video technology, but the problems mentioned above still loom large. Recent advances have had an impact; for example, storage and transport limitations can often be overcome with compression.

The most widely used compression standard for video is MPEG, created by the Moving Pictures Expert Group, the joint ISO/IEC/ITU-T organization that oversees standards development for video. We discussed MPEG in detail earlier in the book. MPEG assumes (correctly) that a relatively small amount of the information in a series of video frames changes from frame to frame. The background, for example, stays relatively constant in most movies, for long periods of time. Therefore, only a small amount of the information in each frame needs to be re-captured, re-compressed, and re-stored.

Other techniques exist, but they are largely proprietary. These include PLV, Indeo, RTV, Compact Video, AVC, DV, DVCAM, and a few others.

Television Standards

It's interesting to note that in the drive toward digitization the ultimate goal is to create a video storage and transport technology that will yield as good a representation of the original image as analog transmission does.

Historically, there have been two primary governing organizations that dictate television standards. One is the National Television System Committee, or NTSC (sometimes said to stand for "Never Twice the Same Color" based on the sloppy color management that characterizes the standard); the other, used primarily in Europe, is the Phased Alternate Line (PAL) system. NTSC is built around a 525 line-per-frame, 30 frame-per-second standard, while PAL uses 625 lines-per-frame and 25 frames-per-second. While technically different, both address the same concerns and rely on the same characteristics to guarantee image quality.

From NTSC to ATSC

In 2009, the United States made the conversion from analog television to digital, leaving NTSC behind. The Advanced Television Systems Committee (ATSC) developed digital television standards in the early 90s to address the emerging requirements of digital broadcast over fixed, cable, and satellite networks. This group of manufacturers and service providers created the standards that ultimately became the specifications for high-definition television (HDTV).

These ATSC standards define a letterbox, 16:9 display format with resolution up to 1920 × 1080. This format is about six times the resolution of prior standards.

On June 12, 2009, the United States replaced most of the analog NTSC television system with ATSC; Canada followed in 2011, South Korea in 2012, and Mexico will follow in 2015.

The audio signal specified in ATSC is Dolby AC-3, an innovative audio standard that supports five sound channels and an additional low-frequency channel for special audio effects.

Video Quality Factors

Four factors influence the "richness" of the video signal. They are frame rate, color resolution, image quality, and spatial resolution.

Frame rate is a measure of the refresh rate of the actual image painted on the screen. The NTSC video standard is 30 frames per second, meaning that the image is updated 30 times every second. Each frame consists of odd and even fields. The odd field contains the odd-numbered screen lines, while the even field contains the even numbered screen lines that make up the picture.

Television sets paint the screen by first painting the odd field, then the even. they repeat this process at the rate of 60 fields—or 30 frames—per second. The number 60 is chosen to coincide with the frequency of electricity in the United States. By the same token, PAL relies on a scan rate that is very close to 50 Hz, the standard in Europe. This odd-even alternation of fields is called *interlaced video*.

Many monitors, on the other hand, use a technique called *progressive scan*, in which the entire screen is painted 30 times per second, from top to bottom. This is referred to as *non-interlaced video*. Non-interlaced systems tend to demonstrate less flicker than their interlaced counterparts.

Computers often rely on VGA (variable graphics array) monitors, which are much sharper and clearer than television screens. This is due to the density of the phosphor dots on the inside of the screen face that yield color when struck by the deflected electron beams, as well as a number of other factors. The scan rate of VGA is much higher than that of traditional television, and can therefore be non-interlaced to reduce screen flicker.

Another quality factor is *color resolution*. Most systems resolve color images using a technique called RGB—for the red, green, and blue primary colors. While video does rely on RGB, it also uses a variety of other resolution techniques, including YUV and YIQ. YUV is a color scheme used in both PAL and NTSC. Y represents the luminance component; U and V, hue and saturation, respectively, make up the color component. Varying the hue and saturation components changes the color.

Image quality plays a critical role in the final outcome, and the actual resolution varies by application. For example, the user of a slow scan, desktop videoconferencing application might be perfectly happy with a half-screen, 15 frame-per-second, 8-bit image, whereas a physician using a medical application might require a full frame, 768 × 484 (the NTSC standard screen size) pixel image, with 24-bit color for perfect accuracy. Both frame rate (frames per second) and color density (bits per pixel) play a key role.

Finally, *spatial resolution* comes into the equation. Many PCs have displays that measure 640 × 480 pixels. This is considerably smaller than the NTSC standard of 768 × 484, or even the slightly different European PAL system. In modern systems, the user has great control over the resolution of the image, because he or she can vary the number of pixels on the screen. The pixels seen on the screen are simply memory representations of the information displayed. By selecting more pixels, and therefore better resolution, the graininess of the screen image is reduced. Some VGA adapters, for example, have resolutions as dense as 1024 × 768 pixels.

The converse, of course, is also true. By selecting less pixels, and therefore increasing the graininess of the image, special effects can be created, such as *pixelization* or *tiling*.

To ensure long-term relevance and operation, ATSC supports a variety of frame rates, aspect ratios, and resolution, some of which are shown in Table 7.3.

Ultra High-Definition TV

Also known as ultra HD (UHD or UHDTV), this evolving standard supports unparalleled screen resolution. If you've been to the Consumer Electronics show (CES) in Las Vegas in the last couple of years and had your eyes open at the time, you most likely saw examples of ultra high-definition television. At least one of the screens I stood in front of was about 6-ft wide and from a distance of 6 in, I could see absolutely no pixels whatsoever.

Resolution				
Vertical	**Horizontal**	**Aspect Ratio**	**Scanning**	**Frame Rate**
1080	1920	16:9	Progressive Interlaced	24, 30 30, 60
720	1280	16:9	Progressive	24, 30, 60
480	704 640	4:3, 16:9 4:3	Progressive Progressive Interlaced	24, 30, 60 24, 30, 60 30

Table 7.3 ATSC Service Characteristics

Developed at NHK Science and Technology Research Labs in Japan, UHD supports a 16:9 screen format and two impressive resolutions: 4K UHDTV, which is an 8.3 megapixel progressively-scanned screen that is 3840 pixels by 2160 pixels; and 8K UHDTV, a 33.2 megapixel progressively scanned screen that measures 7680 pixels by 4320 pixels, delivering resolution that is 16 times the resolution of current HDTV standards.

3DTV

3D television has finally become a reality, although artificial 3D vision has been around for a long time—in fact, it was originally demonstrated by Charles Wheatstone, whom you'll remember from our telephony discussion earlier in the book. When the newest Blu-Ray standard emerged it included specifications for 3D, and a number of vendors including Samsung, Panasonic, and Sony were quick to jump on the bandwagon with viable products. Through a combination of 3D glasses along with 3D-capable video players and TVs, these vendors were able to deliver true 3D viewing to customers.

3D works as follows. To see 3D, separate images from each eye must be blended by the brain into one. The physical separation between the eyes gives each eye a slightly different view, and when blended they become a single three-dimensional view of the environment. To create 3D television, the player transmits and the TV displays twice as many signals as an ordinary TV, alternating between the left and right eye. In synchrony the glasses open and close a *shutter* over each eye, so fast that the person doesn't perceive it. Typically the TV's refresh rate must be at least 120 frames per second, which means that the TV must contain a signal converter to create the alternating images. A wireless interface with the glasses synchs the glasses to the signal emanating from the TV.

Video Summary

Let's review the factors and choices involved in creating video.

The actual image is captured by a camera that is either analog or digital. The resulting signal is then encoded on either analog or digital tape. If a VHS tape is to be the end result, it will be created directly from the original tape.

If desktop video or CD-ROM video is the end result, then additional factors come into play. Because of the massive size of the original *file*, it must be modified in some way to make it transportable to the new medium. It might be compressed, in which case a 200:1 reduction in size (or greater) could occur, with a resulting loss of quality (rain, for example, because it is a random event, tends to disappear in MPEG-compressed movies); it might be sampled less frequently, thus lowering the total number of frames, but causing the animation-like jerkiness that results from lower sampling rates; or, the screen size could be reduced, thus reducing the total number of pixels that need to be encoded.

During the last decade, video has achieved a significant role in a wide variety of industries. It has found a home in medicine, travel, engineering, and education, and it provides not only a medium for the presentation of information, but combined with telecommunications technology, it makes possible such applications as distance learning, video teleconferencing, and desktop video. The ability to digitize video signals has brought about a fundamental change in the way video is created, edited, transported, and stored.

Its emergence has also changed the players in the game. Once the exclusive domain of filmmakers and television studios, video is now fought over by creative companies and individuals who want to control its content, cable and telephone companies who want to control its delivery, and a powerful market that wants it to be ubiquitous, richly featured and cheap. Regulators are in the mix as well, trying to make sense of a telecommunications industry that, once designed to transport voice, now carries a broad mix of fundamentally indistinguishable data types.

Before we close this chapter, let's spend a few minutes talking about music.

The World of Music

I recently had the opportunity to spend a few days with a music industry executive. This person has worked for a variety of music industry players including Warner/Chappell, Disney, and iTunes. At one point in our conversation I asked him to describe how music distribution specifically and the music industry at large will change over the next few years and he told me a fascinating story. As I recount the major points of our conversation here, think about how the TMT convergence relates to the story.

Music has gone through a series of phases over the years. Music distribution is a very old business—about 200 years—and it has a rich history behind it. Currently, however, it's facing unprecedented change that began with Steve Jobs and the iTunes revolution and continues today.

The first phase in the industry was all about music ownership, during which we bought wax cylinders, 78s, 45s, 33-1/3 LPs, reel-to-reel, 8-tracks, cassettes, and CDs. Now we're entering the second phase, which is all about downloading the music from a source like iTunes. Wal-Mart was the largest music distributor in the world for the longest time; Apple iTunes replaced them in 2005 or so.

Now we're entering the post-download phase of access in which customers buy a subscription to a streaming service like Spotify, Pandora, Apple Radio, Amazon, or Google Play. These models are interesting because they can be bundled with mobile plans, creating a revenue opportunity for both the music service and the carrier and extending the relevance message of the mobile service provider. In fact, one of the most interesting companies that has come onto the scene is Move Music, a distributor of music from new artists that is forging relationships with pre-paid carriers that target lower income customers.

What is particularly interesting here is that the music industry, like so many industries that have managed to consolidate market power during their lifetimes, is facing an unsettling reality which simply observes that the seat of power no longer lies in the hands of these large labels, but rather with smaller players—and most notably, with the market itself. This, combined with a byzantine contract model, has caused a revolt among the ranks that is redefining the nature of music distribution.

Consider this example. In the past, contracts were often created such that multiple artists or companies shared ownership of a song. For example, let's say that three musicians create a song. Musician number one writes the lyrics and the advertising hook; the other two perform background. Artist number one gets 50% of the revenues, while the other two get 25% each.

This is where things get interesting. Artist number one is signed with Sony, artist number two with Warner/Chappell, and artist number three with Universal. When the time comes to market the song, none of the three companies are willing to do so because of the partial ownership issue: in effect, by marketing the song, they spend money to market a song that also belongs to the competition. Instead, music executives often gave this order: "Don't pitch any songs that are jointly owned; only pitch the wholly owned catalog." As a result, many very talented artists never get a chance to share their music.

So what other options are there? In the new world of music distribution there are quite a few, thanks largely to the Internet. An artist can go directly to YouTube, or sign with Warner Music, or get into the Fox pipeline to have their work be used in a movie. But this requires submitting ones self to the bureaucracy: why would an artist want to subject himself or herself to that?

What Google has managed to do somewhat subversively is create a new model of delivery targeted at the creative class that gives them the ability to go directly to the end consumer, bypassing the label and the traditional music industry. This is precisely what Radiohead did with their music: They put it on their Web site, announced it to the public, and told them that the price is "whatever you think is fair." They made *plenty* of money. [Note: This is an example of the phenomenon I described at the beginning of the book when I said that anyone wishing to gain influence over a market must give up control of that market].

Consider the following example. An Artists and Repertoire (A&R) person at one of the big labels is trying to sign a potentially big new artist. Somewhere along the way the licensing department licenses one of the artist's songs to a major brewery. But this creates a major conflict with the artist because he or she is a recovering alcoholic, and now their song is the theme song for a major alcohol producer. Did the label have the right to license the song to the brewery? Yes. Did doing so ruin the relationship with the artist and the A&R person? Yes. But take note: This had nothing to do with contracts, and begs the question, who has the power?

Until relatively recently, the music industry was largely unaware that kids in dorm rooms at colleges and universities will often create a YouTube video playlist, connect a pair of good speakers to their laptop or tablet, and run it at a party. The video is playing but no one is watching—it's the music they care about. This opens a discussion about licensing. Under the domain of streamed content there are two types of licenses: interactive streaming and non-interactive streaming. Interactive means that I have the ability to click on a track and it plays exactly what I want. Services like Pandora, however, are examples of a noninteractive streaming license. In Pandora I want to hear songs that are *like* the artist I want to hear.

This is where we uncover a potential dark side of the YouTube channel. YouTube earns about $10 billion annually much of it driven by music. YouTube, it seems, has a video license that covers music distribution, a license that results in payments from them to the music industry of tens of millions of dollars annually. Here's how the model works: Somebody uploads a video to YouTube which comprises a digital photo of an album cover and a sound track, which is a song from the album. YouTube's music detection algorithms identify the songs, and YouTube, using its advertising insight, places ads on those videos. Revenue from the ads is distributed to the labels that own the artist, and the artists is—hopefully—paid for the use of their song.

The problem is that much of the user-generated content that appears on YouTube isn't picked up by the algorithms that find and identify music—and therefore artists. As a result, 100% of the revenue that comes from the placement of those ads goes to Google, with nothing going to the artist.

Artists claim that they have a legitimate complaint. The YouTube advertising banner at the top of the home page costs somewhere on the order of $100,000 per day, with a commitment to $250,000 in future advertising. That one banner generates about $172 million in a single year, much of which is driven by the music business—yet pays out nothing to artists.

So how should the revenue model work? The same way it works at services like Spotify, where artists are paid a pro rata share of what people are listening to.

Without music, YouTube would not exist. Many music industry executives believe that the industry as a whole must take a stand to demand payment from online services or deny the use of the intellectual property. Artists need to get paid—just like software engineers, album artists, and sales representatives.

"If you want to gain influence over a market, you must give up control of that market." See the corollary?

Steve Jobs' Influence

Many credit Steve Jobs and the iTunes team with the reinvention of the music industry, and there's no question that iTunes "made a dent in the music universe," to steal a phrase from Jobs.

Truthfully, Apple's influence had four key impacts on the music industry. The first was a redefinition of online sales and distribution. In 2003, Apple announced iTunes and a business model under which a single downloadable song cost 99 cents, an album about $10. Jobs struck quickly and decisively, signing distribution deals with the four major labels at the time: Universal, Sony, EMI, and Warner Music Group. The involvement of these companies painted iTunes and the world of digital distribution with a patina of legitimacy.

The second major impact was the manner in which the market consumed media. Thanks to the iPod, a consumer could now walk around with 1,000 songs in his or her pocket. They could hear any song they wanted, when they wanted. Playlist creation allowed them to build their own "albums" of favorites, deliberately shifting the control of the music experience from the label, which previously required them to buy the entire album, to the consumer, who now had the right to buy a single song if they desired.

The third impact was the democratization of music production. Thanks to apps like Garage Band, the extraordinarily expensive gap between creation and production disappeared. Suddenly, music became democratized, and the music industry gauntlet that previously determined who made it and who didn't, disappeared. Instead, artists created music, produced it themselves, put it on their Web site, and let the market decide whether they had a future. Let me say that in a slightly different way: They turned control over to the market, gaining in return a degree of influence—and relevance.

Cracking the Code

Whether we are talking about music, video, still images, television, e-mail, corporate data records, or any other form of consumable content, the story is the same—in the digital domain, it's what people want, need and desire. Consider the information in Fig. 7.5.

Personal Data: Address, Name, Profile, Gender, Resource Preferences	Invoicing and Service: Payment, Collection, Spend, Balance	Contacts, Groups, Applications: News Feeds, Cookies, Social Networks, Content Profile
Device Profile: Device Capabilities, OS, Memory, Installed Apps	**Carrier**	Communications: Messages, Voicemail, Text, Video, Calls, Call Log, Click-through Data
Contract Data: SIM Information, Options, Service Details, Contract Terms, Service Package Details	User Context: Location, Presence, Roaming, Service Profile	Personal Content: Documents, Images, Audio, Video

Figure 7.5 The information contained in a typical service provider's databases is quite detailed and comprehensive.

This diagram details *some* of the customer information contained in the vast databases of a typical mobile service provider. Conspiracy theorists, be forewarned: They *have* to have this data to do their jobs. Without it they would not be able to deliver content, services, software upgrades, or invoices, nor would they be able to respond to customer requests for information or to handle troubleshooting procedures. As a customer, now, think about how the fact that the service provider has access to this information could serve you, and you'll understand why the federation of information—that is, the ability of a trusted third party to have access to your information—can be a very good thing:

- The service provider monitors your wireless data usage in real-time and notifies you before you exceed the limit, beyond which you will incur additional charges.
- Your carrier determines that you are in another country that is outside of your normal coverage area and automatically signs you up for the lowest cost plan to a avoid an unpleasant billing surprise at the end of the month as well as a interruption of your service.
- Based on the types of queries you make with your phone, the service provider sends you a text that recommends certain applications that will make the searches easier for you.
- Using the location-based capabilities that are built into your phone, you receive a text message warning you that you are entering an unsafe area.

- Based on the ability of the phone (in concert with a cloud-based service) to analyze the content of the photographs you take with the phone's digital camera, you receive recommendations of places to visit in the city you're visiting.
- You receive advertisements in real-time through your phone for products that you are interested in and that are available close-by.

The thing to keep in mind is this. Do you remember when radios looked like the one shown in Fig. 7.6? Did you ever wonder why all car radios had (and continue to have, by the way) five buttons? No, it has nothing to do with the number of fingers on each hand. It has to do with a well-known neuropsychological observation, which is that the human brain is capable of remember five, plus or minus two, simultaneous things that are somewhat related. In other words, I could give you a radio that has 100 buttons, and you'll gleefully spend the afternoon programming every one of them. But tomorrow you'll remember … about five of them. Take this simple test: How many cable or satellite TV channels do you have at home? 150-plus? Now, ask yourself this: How many do you actually watch on a regular basis? For most people, the answer is about five.

So what do we learn from this? The answer is interesting. Conventional wisdom tells us that if we don't have enough content to deliver to the customer, we're dead where we stand. Current wisdom, however, tell us that *too much* content is even more deadly,

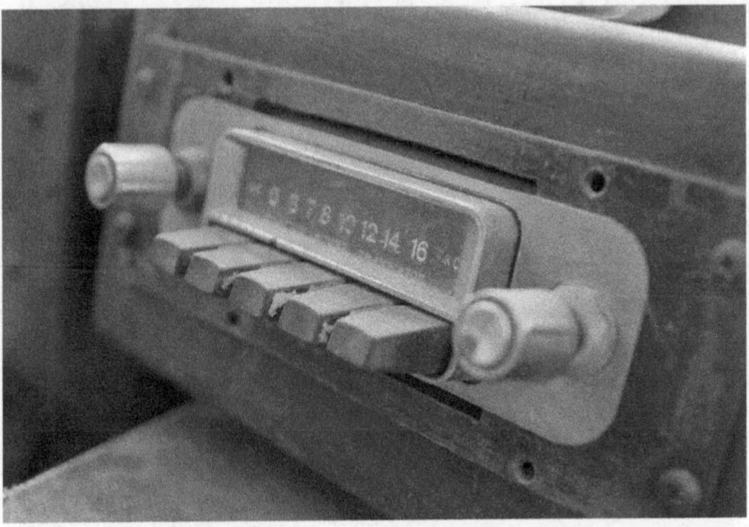

Figure 7.6 Remember when car radios looked like this? Did you ever wonder why they all had five buttons?

because it overwhelms the ability of the customer to find what they are looking for. The thing we learn from this is that to a large extent, the content itself doesn't really matter—nor does the volume of the content or even the quality of the content. What *does* matter is the ability to connect the right person to the right content, where *right* is determined by the customer, not the provider.

So what is needed to make this happen? Four things. First we need the ability to aggregate—not content, but a market. In other words we need the ability to reach a large, dispersed, and perhaps unknown population that is big enough to be statistically valid. Second, we need navigation capability so that we can easily and seamlessly connect the customer to the available content. Third, we must ensure digital awareness—that is, ensure that the customer is aware of the availability of the digital content and has the skill and wherewithal to reach it. And finally, we must have a monetization mechanism.

In October 2004, author Chris Anderson published an article[1] in *Wired Magazine* entitled "The Long Tail". In it he describes an economic model (the Long Tail) in which a business can be successful in a very different way than it ever has before, as long as it follows the four observations made earlier. If the business can aggregate a very large population of customers and connect them with a very large array of products, even if they only sell a few of each product, they will make money because of the diverse volumes they are moving. Here's an example. All of of my books have enjoyed success, although some of them I wonder about. For example, *SONET/SDH Demystified* is terrific if you have a need to understand the inner workings of this optical technology that works like a T1 or E1 on steroids—or, if you have a serious sleeping disorder. I joke about the fact that all four people who read it, loved it. Let's face it, we're not talking Tom Clancy here. But truth be told, several thousand copies of the book have sold. Why? Because of the Internet's ability to aggregate a large population of dispersed customers, and the ability of companies like Amazon to connect those people to the book. Suddenly, my market is global—and I'm very thankful for it.

Final Thoughts

Media is the third leg of the TMT stool, and it is a crucial one. The network provides a conduit for the delivery of the content to the user, including both access to the network and transport across it. Technology, or IT, provides the means to house the media, track its use, analyze it, and monetize its consumption. For the most part, however, the telecom and IT elements of the TMT triumvirate are invisible to

[1]Anderson, Chris. "The Long Tail" *Wired*, October 2004.

the customer, yet they are every bit as important as the content they so desperately want. In fact, without them, the content cannot reach the customer, nor can the customer search for it, pay for it, or even request it.

In the next chapter, we turn our attention to the technologies that give us the ability to access the content that we all want.

Chapter Seven Questions

1. "Voice is a form of user-generated content." Explain.
2. In what way could APIs be used to enhance your own business?
3. Do you see DPI as a boon to society or the facilitator of invasive scrutiny?
4. Explain the differences between QoS and QoE. Which is more important? Why?

CHAPTER 8

Access Technologies

Man's greatest invention is not the TV
Nor is it the radio, it seems to me
It isn't the airplane, it isn't the car
As wonderful as these inventions all are.
It's not the computer or cellular phone
Or any device or machine that is known.
No nuclear weapon of war ever shook
The world like man's greatest invention—
the book.

—Charles Osgood

For the longest time, *access* described the manner in which customers reached the network for the transport of voice services. In the last 20 years, however, that definition has changed dramatically. In 1981, just over 30 years ago, IBM changed the world when it introduced the PC, and in 1984 the Macintosh arrived, bringing well-designed and logically organized—not to mention approachable—computing power to the proverbial masses. Shortly thereafter, hobbyists began to take advantage of emergent modem technology and created online databases—the first bulletin board systems that allowed people to send simple text messages to each other. This accelerated the modem market dramatically and before long data became a common component of local loop traffic. At that time there was no concept of instant messenger or of the degree to which e-mail would fundamentally change the way people communicate and do business. At the same time the business world found more and more applications for data, and the need to move that data from place to place became a major contributor to the growth in data traffic on the world's telephone networks.

In those heady, early days, data did not represent a problem for the bandwidth-limited local loop. The digital information created by a computer and intended for transmission through the telephone network was received by a modem, converted into a modulated analog waveform that fell within the 4-KHz voice band, and fed to the network without incident. As we mentioned in an earlier chapter, the modem's job was (and is) quite simple: Invoke the Wizard of Oz protocol. When a computer is doing the talking, the modem must make the network think it is talking to a telephone. "Pay no attention to that man behind the curtain!"

Over time modem technology advanced, allowing the local loop to provide higher and higher bandwidth. This increasing bandwidth was made possible through clever signaling schemes that allowed a single signaling event to transport more than a single bit. These modern modems, often called Shannon-Busting modems because they defy the limits of signaling defined by the work done by Claude Shannon in the early twentieth century, are commonplace today. They allow baud levels to reach unheard-of extremes and permit the creation of very high bit-per-signal rates. DSL, for example, uses modems at either end of the circuit to achieve extremely high bit rates—over an analog facility.

The analog local loop is used today for various voice and data applications in both business and residence markets. The new lease on life it enjoys thanks to advanced modem technology as well as a focus by installation personnel on the need to build clean, reliable outside plant has resulted in the development of faster access technologies designed to operate across the analog local loop, including traditional high-speed modem access and such options as DSL.

Marketplace Realities

According to recent demographic studies there are about 3.3 million people in the United States today—just over 3 percent of the workforce—who consider their home to be their primary workplace, and the number is growing rapidly. These numbers include telecommuters but do not include those who are self-employed and happen to work out of their homes. They require the ability to connect to remote LANs and corporate databases, retrieve e-mail, access the Web, and conduct videoconferences with colleagues and customers. The traditional bandwidth-limited local loop is not capable of satisfying these requirements with traditional modem technology. Dedicated private line service, which would solve the problem, is far too expensive as an option, and because it is dedicated it is not particularly efficient. Other solutions are required and these have emerged in the form of access technologies that take advantage of expanded capabilities of the traditional analog local loop such as DSL. In some cases a whole new architectural approach is causing excitement in the industry; cable access, for example, has become a popular option as the cable infrastructure has evolved to a largely optical, all-digital system with high-bandwidth, two-way capability. We will discuss each of these options in the pages that follow.

Integrated Services Digital Network

Prior to the arrival of DSL, the integrated services digital network (ISDN) was considered to be a major step forward for digital access. Often described as the technology that took 15 years to become an overnight success, ISDN's relevance as an access technology has seriously declined with the exception of the primary rate interface, which is still very popular for PBX environments.

The Basic Rate Interface

There are two well-known implementations of ISDN. The most common (and the one intended primarily for residence and small business applications) is called the basic rate interface (BRI). In BRI, the two-wire local loop supports a pair of 64-Kbps digital channels known as B-channels and a 16-Kbps D-channel which is primarily used for signaling but can also be used by the customer for low-speed (up to 9.6 Kbps) packet data. The B-channels can be used for voice and data and in some implementations can be bonded together to create a single 128-Kbps channel for videoconferencing or other higher bandwidth applications. Figure 8.1 shows the layout of a typical ISDN BRI implementation.

The reference points mentioned earlier identify circuit components between the functional devices just described. The U reference point is the local loop; the S/T reference point sits between the NT1 and the TEs; the R reference point is found between the TA and the TE2.

The Primary Rate Interface

The other major implementation of ISDN is called the primary rate interface (PRI). The PRI is really nothing more than a T-carrier in that it is a four-wire local loop, uses AMI and B8ZS for ones-density control and signaling, and provides 24 to 64 Kbps channels that can be distributed among a collection of users as the customer sees fit (see Fig. 8.2). In PRI, the signaling channel operates at 64 Kbps (unlike the

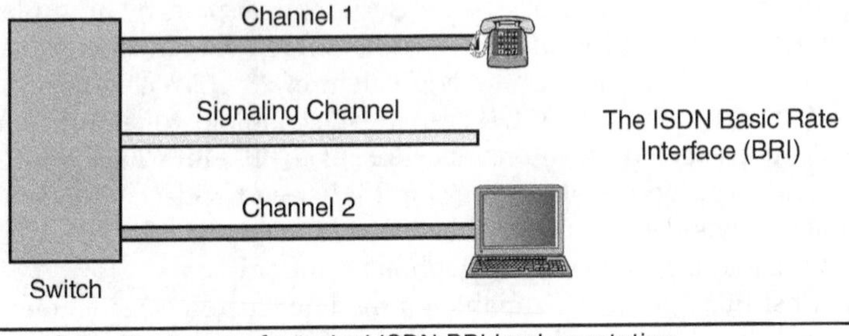

FIGURE 8.1 The layout of a typical ISDN BRI implementation.

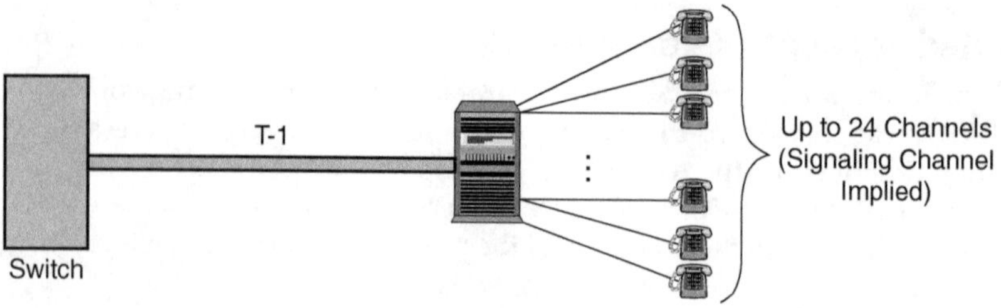

Figure 8.2 Channel distribution in an ISDN PRI.

Channel	Bandwidth
H0	384 Kbps (6B)
H10	1.472 Mbps (23B)
H11	1.536 Mbps (24B)
H12	1.920 Mbps (30B)

Table 8.1 Bandwidth Options in an ISDN PRI

16 Kbps D-channel in the BRI) and is not accessible by the user. It is used solely for signaling purposes, that is, it cannot carry user data. The primary reason for this is service protection: in the PRI, the D-channel is used to control the 23 B-channels and therefore requires significantly more bandwidth than the BRI D-channel. Furthermore, the PRI standards allow multiple PRIs to share a single D-channel, which makes the D-channel's operational consistency all the more critical.

A PRI can be provisioned as a collection of super-rate channels to satisfy the bandwidth requirements of higher bit rate services. These are called *H-channels,* and are provisioned as shown in Table 8.1.

PBX Applications

The PRI's marketplace is the business community and its primary advantage is pair gain, that is, to conserve copper pairs by multiplexing the traffic from multiple user channels onto a shared facility. Inasmuch as a PRI can deliver the equivalent of 23 voice channels to a location over a single circuit, it is an ideal technology for a number of applications including interconnection of a PBX to a local switch, dynamic bandwidth allocation for higher-end videoconferencing applications, and interconnection between an Internet service provider's network and that of the local telephone company.

Most PBXs are ISDN-capable on the line (customer) side, meaning that they have the ability to deliver ISDN services to users that emulate the services that would be provided over a direct connection to an ISDN-provisioned local loop. On the trunk (switch) side, the PBX is connected to the local switch via one or more PRIs, which in turn provide access to the telephone network. This arrangement results in significant savings, faster call setup, more flexible administration of trunk resources, and the ability to offer a diversity of services through the granular allocation of bandwidth as required.

As service providers move inexorably toward the all-IP environment, however, TDM becomes an expensive alternative and can now be replaced with far more cost-effective (and flexible) SIP trunks. The deployment of SIP requires the simultaneous deployment of a VoIP gateway to handle the TDM-to-IP conversion. The advantage of the migration to SIP via a media gateway protects the in-place investment, allowing the analog and digital payloads to be handled simultaneously.

ISDN has had its day—but for the most part it can now safely retire. A new technology is now on the scene. Meet DSL.

Digital Subscriber Line

The access technology that enjoys the greatest amount of attention today is digital subscriber line (DSL). It provides a good solution for residential data requirements as well as remote LAN access, Internet surfing, and access for telecommuters to corporate data resources.

DSL came about largely as a direct result of the Internet's success. Prior to its arrival the average telephone call lasted approximately four minutes, a number that central office personnel used to engineer switching systems to handle anticipated call volumes. This number was arrived at after nearly 125 years of experience designing networks for voice customers. Central office switching engineers knew about Erlang theory, hundred call seconds (CCS) loading objectives, peak-calling days/weeks/seasons, and had decades of trended data to help them anticipate load problems. So in 1993, the unspeakable happened. When the Internet—and with it, the World Wide Web—arrived, network performance became unpredictable as callers began to surf, often for hours on end. The average four-minute call became a thing of the past as service providers such as AOL and Prodigy began to offer flat rate plans that did not penalize customers for long connect times. Suddenly, switches in major metropolitan areas, faced with unpredictably high call volumes and hold times, began to block—that is, deny dial tone to customers—during normal business hours, a phenomenon that had only occurred in the past during disasters or on Mother's Day.

A number of solutions were considered, including charging different rates for data calls than for voice calls, but none of these proved feasible until a technological solution was proposed. That solution was DSL.

It is a common belief that the local loop is incapable of carrying more than the frequencies required to support the voice band. This is a misconception. Even basic rate ISDN, for example, requires significantly high bandwidth than 64 Kbps to support its digital traffic. When ISDN is deployed, the loop must be modified to eliminate its designed-in bandwidth limitations. You will recall from our telephony discussion that the local loop is tuned to the voice band through the placement of load coils that narrow the accessible channel width as a way to conserve precious bandwidth. They make the transmission of frequencies above the voice band impossible but allow the relatively low-frequency voice band components to be carried across a local loop. High frequency signals tend to deteriorate faster than low-frequency signal components, so the elimination of the high frequencies extends the transmission distance and reduces transmission impairments that would result from the uneven deterioration of a rich, multi-frequency signal, which is the nature of a digital transmission (remember our square wave discussion and the Fourier series?). These load coils, therefore, must be removed if digital services are to be deployed.

A local loop is only incapable of transporting high-frequency signal components if it is *designed* not to carry them. The capacity is still there; the network design, through load coil deployment, simply makes that additional bandwidth unavailable. DSL services, especially ADSL, take advantage of this *disguised* bandwidth.

DSL Technology

In spite of the name, DSL is an analog technology. The devices installed on each end of the circuit are sophisticated high-speed modems that rely on complex encoding schemes to achieve the high bit rates that DSL offers. Furthermore, several of the DSL services, specifically ADSL, G.lite, VDSL, and RADSL, all discussed later in this section, are designed to operate in conjunction with voice across the same local loop. ADSL is the most commonly deployed service and offers a great deal to both business and residence subscribers.

DSL Services

DSL comes in a variety of flavors designed to provide flexible, efficient, high-speed service across the existing telephony infrastructure. From a consumer point-of-view, DSL, especially ADSL, offers a remarkable leap forward in terms of available bandwidth for broadband access to the Web. As content has steadily moved away from being largely text-based and has become intensely visual, the demand for faster delivery services has grown in lockstep. DSL may provide the solution at a reasonable cost to both the service provider and the consumer.

Businesses also benefit from DSL. Remote workers, for example, rely on DSL for LAN and Internet access; they implement a virtual private network (VPN) security solution across the interface and enjoy access to enterprise content and applications as if they were in the office. DSL is available in a variety of both symmetric and asymmetric services and therefore offers a high-bandwidth access solution for a variety of applications. The most common DSL services are ADSL, ADSL2+, HDSL, HDSL-2, RADSL, and VDSL. The special case of G.lite, a form of ADSL, will also be discussed.

Asymmetric Digital Subscriber Line

When the World Wide Web and flat rate access charges arrived, the typical consumer phone call went from roughly four minutes in duration to several times that. All the engineering that led to the overall design of the network based on an average four-minute hold time went out the window as the switches staggered under the added load. Never was the expression, "In its success lie the seeds of its own destruction," more true. When ADSL arrived, it provided the offload required to save the network.

The typical ADSL installation is shown in Fig. 8.3. No change is required to the two-wire local loop; minor equipment changes, however, are necessary. First, the customer must have an ADSL modem at their premise. This device allows both the telephone service and a data access device, such as a router, to be connected to the line.

The ADSL modem is more than a simple modem in that it also provides the frequency division multiplexing process required to separate the voice and data traffic for transport across the loop. The device that actually does this is called a splitter because it splits the voice traffic away from the data. It is usually bundled as part of the ADSL modem although it can also be installed as a card in the PC, as a stand-alone device at the demarcation point, or on each phone at the premises. The most *common* implementation is to integrate the splitter as part of the DSL modem; this, however, is the least *desirable* implementation because this design can occasionally lead to crosstalk between the voice and data circuitry inside the device. When voice traffic reaches the ADSL modem it is immediately encoded in the traditional voice band and handed off to the local switch when it arrives at the central office. The modem is often referred to as an ADSL transmission unit

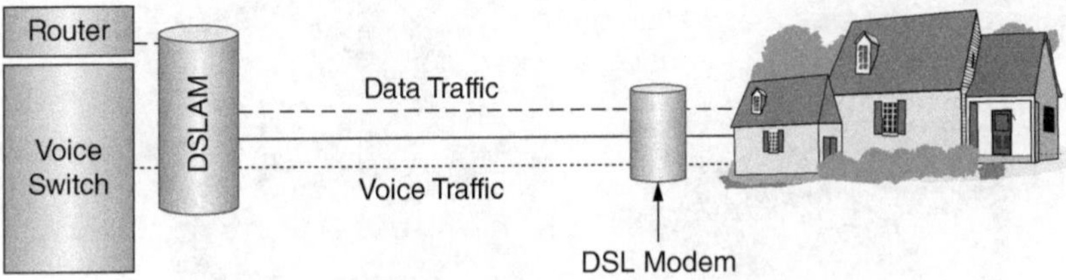

FIGURE 8.3 A typical ADSL installation.

for remote use (ATU-R). Similarly, the device in the central office often called an ATU-C (for central office).

When a PC wishes to transmit data across the local loop the traffic is encoded in the higher-frequency band reserved for data traffic. The ADSL modem knows to do this because the traffic is arriving on a logical port that is reserved for data devices. Upon arrival at the central office the data traffic does not travel to the local switch; instead, it stops at the ADSL modem that has been installed at the central office-end of the circuit. In this case, the device is actually a bank of DSL modems that serves a large number of subscribers, and is known as a digital subscriber line access multiplexer (DSLAM) (pronounced 'dee-slam'). A DSLAM is shown in Fig. 8.4.

FIGURE 8.4 A DSLAM.

Instead of traveling on to the local switch, the data traffic is now passed around the switch to an IP router that is in turn connected to the Internet. This process is known as a *line-side redirect*.

The advantages of this architecture are fairly obvious. First, the redirect offloads the data traffic from the local switch so that the switch can go back to doing what it does best—switching voice traffic. Second, it creates a new line of business for the service provider. As a result of adding the router and connecting the router to the Internet, the service provider instantly becomes an ISP. This is a near-ideal combination, because it allows the service provider to take a step toward becoming a *true service provider* by offering much more than simple access and transport.

As the name implies, ADSL provides two-wire asymmetric service—that is, the upstream bandwidth is different from the downstream. In the upstream direction, data rates vary from 1 to 5 Mbps, while the downstream bandwidth varies from 8 to 50 Mbps. Because most applications today are asymmetric in nature, this disparity poses no problem for the average consumer of the service. What do I mean by this? Imagine the following scenario. You log onto the NetFlix Web site, search for a bit and decide to watch an episode from the National Geographic Channel. You click on the link and the program begins to play. The upstream transmission that you requested ("Please play this show for me") is *enormously* smaller than the downstream channel (the streamed program). This kind of relationship is quite typical for most application usage; asymmetric architectures like ADSL, then, lend themselves perfectly to this kind of usage.

A Word about the DSLAM

This device gets a lot of attention because of the central role that it plays in the deployment of broadband access services. Obviously, the DSLAM must interface with the local switch so that it can pass voice calls on to the PSTN. However, it often interfaces with a number of other devices as well. For example, on the customer side, the DSLAM may connect to a standard ATU-C, directly to a PC with a built-in network interface card (NIC), to a variety of DSL services, or to an integrated access device of some kind. On the trunk side (facing the switch), the DSLAM may connect to IP routers as described before, to an ATM switch, or to some other broadband service provider. It therefore becomes the focal point for the provisioning of a wide variety of access methods and service types.

The physical location of the DSLAM can vary as well. On the one hand, it can be collocated with the switch in the central office; on the other hand, it can be installed remotely. For example, it might be installed in a neighborhood so that it is closer to the customer, thus reducing the length of the local loops that are attached to it as way to improve the bandwidth available to the customer. In the case of this

Version	Standard Name	Common Name	Downstream	Upstream
ADSL	ANSI T1.413-1998 Issue 2	ADSL	8.0 Mbps	1.0 Mbps
ADSL	ITU G.992.1	ADSL (G.DMT)	8.0 Mbps	1.3 Mbps
ADSL	ITU G.992.1 Annex A	ADSL over POTS	12.0 Mbps	1.3 Mbps
ADSL	ITU G.992.1 Annex B	ADSL over ISDN	12.0 Mbps	1.8 Mbps
ADSL	ITU G.992.2	ADSL Lite (G.Lite)	1.5 Mbps	0.5 Mbps
ADSL2	ITU G.992.3	ADSL2	12.0 Mbps	1.3 Mbps
ADSL2	ITU G.992.3 Annex J	ADSL2	12.0 Mbps	3.5 Mbps
ADSL2	ITU G.992.3 Annex L	RE-ADSL2	5.0 Mbps	0.8 Mbps
ADSL2	ITU G.992.4	splitterless ADSL2	1.5 Mbps	0.5 Mbps
ADSL2+	ITU G.992.5	ADSL2+	20.0 Mbps	1.1 Mbps
ADSL2+	ITU G.992.5 Annex M	ADSL2+M	24.0 Mbps	3.3 Mbps
ADSL2++		ADSL4	52.0 Mbps	5.0 Mbps

TABLE 8.2 DSL Service Options

remote DSLAM, the device is connected to the central office via a fiber backhaul facility.

There are many different versions of DSL technology; they are described on the following pages. Please refer to Table 8.2.

ADSL2

ADSL2 superseded pre-existing ADSL standards. It is interoperable with pre-existing ADSL deployments and is critical for service providers and customers alike; it means that they can continue to use their installed base of equipment with minimal disruption.

ADSL2+

ADSL2+, on the other hand, is an extension of ADSL2. It is capable of doubling the transmission speed of typical ADSL installations to over 50 Mbps downstream on loops shorter than 8,000 ft. A newer version, ADSL2++, commonly known as ASDL4, is under development and will offer downstream speeds of 52 Mbps, upstream 5 Mbps.

High Bit Rate Digital Subscriber Line

The greatest promise of *high bit rate digital subscriber line* (HDSL) is that it provides a mechanism for the deployment of four-wire

T1 and E1 circuits without the need for span repeaters, which can add significantly to the cost of deploying data services. It also means that service can be deployed in a matter of days rather than weeks, something customers certainly applaud.

DSL technologies in general allow for repeaterless facilities as far as 12,000 ft, while traditional four-wire data circuits such as T1 and E1 require repeaters every 6,000 ft. Consequently, many telephone companies are now using HDSL "behind the scenes" as a way to deploy these traditional services. Customers do not realize that the T1 facility they are plugging their equipment into is being delivered using HDSL technology. The important thing is that they don't *need* to know. All the customer should have to care about is that there is now a SmartJack installed in the basement, and through that jack they have access to 1.544 Mbps or 2.048 Mbps of bandwidth—period.

HDSL2

HDSL2 offers the same service that HDSL offers, with one added (and significant) advantage: it does so over a single pair of wires, rather than two. It also provides other advantages. First, it was designed to improve vendor interoperability by requiring less equipment at either end of the span (transceivers, repeaters). Second, it was designed to work within the confines of standard telephone company carrier serving area (CSA) guidelines by offering a 12,000-ft wire-run capability that matches the requirements of CSA deployment strategies.

A number of companies have deployed T-1 access over HDSL2 at rates 40 percent lower than typical T-carrier prices. Furthermore, a number of vendors including 3Com, Lucent, Nortel Networks, and Alcatel have announced their intent to work together to achieve interoperability among DSL modems.

Rate-Adaptive Digital Subscriber Line

Rate-adaptive digital subscriber line (RADSL, pronounced "Rad-zel") is a variation of ADSL designed to accommodate changing line conditions that can affect the overall performance of the circuit. Like ADSL, it relies on DMT encoding, which selectively *populates* sub-carriers with transported data, thus allowing for granular rate-setting.

Very High-Speed Digital Subscriber Line

Very high-speed digital subscriber line (VDSL) is the newest DSL entrant in the bandwidth game and shows promise as a provider of extremely high levels of access bandwidth—as much as 52 Mbps over a short local loop. VDSL requires fiber-to-the-curb (FTTC) architecture and recommends ATM as a switching protocol; from a fiber hub, copper tail circuits deliver the signal to the business or residential premises. Bandwidth available through VDSL ranges from 1.5 to 6 Mbps on the upstream side, and from 13 to 52 Mbps on the downstream side. Obviously, the service is distance-sensitive and actual achievable bandwidth drops as a function of distance. Nevertheless, even a short loop is respectable when such high bandwidth levels can be achieved. With VDSL, 52 Mbps can be reached over a loop length of up to 1,000 ft—a not unreasonable distance by any means. Cable providers are also deploying this technology over their coaxial networks; they are able to offer bandwidth levels as high as 85 Mbps, symmetrically.

G.lite

Because the installation of splitters has proven to be a contentious and problematic issue, the need arose for a version of ADSL that did not require them. That version, known as either *ADSL Lite* or *G.lite* (After the ITU-T G-series standards that govern much of the ADSL technology). In 1997, Microsoft, Compaq, and Intel created the Universal ADSL Working Group (UAWG),[1] an organization that grew to nearly 50 members dedicated to the task of simplifying the rollout of ADSL. In effect, the organization had four stated goals:

- To ensure that analog telephone service will work over the G.lite deployment without remote splitters, in spite of the fact that the quality of the voice may suffer slightly due to the potential for impedance mismatch.
- To maximize the length of deployed local loops by limiting the maximum bandwidth provided. Research indicates that customers are far more likely to notice a performance improvement when migrating from 64 Kbps to 1.5 Mbps than when going from 1.5 Mbps to higher speeds. Perception is clearly important in the marketplace, so the UADSL Working Group chose 1.5 Mbps as their downstream speed.
- To simplify the installation and use of ADSL technology by making the process as "plug-and-play" as possible.
- To reduce the cost of the service to a perceived reasonable level.

Of course, G.lite is not without its detractors. A number of vendors have pointed out that if G.lite requires the installation of microfilters at the premises on a regular basis, then true splitterless DSL is a myth, since microfilters are in effect a form of splitter. They contend that if the filters are required anyway, then they might as well be used in full-service ADSL deployments to guarantee high-quality service delivery. Unfortunately, this flies in the face of one of the key tenets of G.lite, which is to simplify and reduce the cost of DSL deployment by

[1]The group "self-dissolved" in the summer of 1999 after completing what they believed their charter to be.

eliminating the need for an installation dispatch (a "truck roll" in the industry's parlance). The key to G.lite's success in the eyes of the implementers was to eliminate the dispatch, minimize the impact on traditional POTS telephones, reduce costs, and extend the achievable drop length. Unfortunately, customers still have to be burdened with the installation of microfilters, and coupled noise on POTS is higher than expected. Many vendors argue that these problems largely disappear with full-feature ADSL using splitters; a truck dispatch is still required, but again, it is often required to install the microfilters anyway, so there is no net loss. Furthermore, a number of major semiconductor manufacturers support both G.lite and ADSL on the same chipset, so the decision to migrate from one to the other is a simple one that does not necessarily involve a major replacement of internal electronics.

Wild Cards

A handful of de facto implementations of DSL have also emerged in the last few years in response to specific demands from the market. These include etherloop, sometimes called ethernet local loop, a half-duplex technology that operates at a modest 6 Mbps over copper; Internet protocol subscriber line (IPSL), which supports 40 Mbps service over a mile-long twisted pair loop (the company that created it has now disappeared, but discussions of the technology occasionally pop up); and gigabit DSL, although so far support for this technology has been lukewarm at best.

Naked DSL

Yes, you read correctly. Don't get too excited, though. Naked DSL is a technique that is used to provision DSL service without traditional analog voice over the same local loop. Scenarios might include a customer who subscribes to VoIP service and doesn't need analog voice, but wants broadband access. This model has historically been called unbundled network element (UNE) service. Some carriers actively market the service as a way to preserve their declining local loop business.

DSL Market Issues

DSL technology offers advantages to both the service provider and the customer. The service provider benefits from successful DSL deployment because it serves not only as a cost-effective technique for satisfying the bandwidth demands of customers in a timely fashion, but also because it provides a Trojan horse approach to the delivery of certain pre-existing services. As we noted earlier, many providers today implement T-1 and E-1 services over HDSL because it offers a cost effective way to do so. Customers are blissfully unaware of the fact; in this case, it is the service provider rather than the customer who benefits most from the deployment of the technology. From a customer point-of-view, DSL provides a cost-effective way to buy medium-to-high levels of bandwidth, and in some cases embedded access to content.

Provider Challenges

The greatest challenges facing those companies looking to deploy DSL are competition from cable and wireless; pent-up customer demand; installation issues; and plant quality.

Competition from cable and wireless companies represents a threat to wireline service providers on several fronts. First, cable modems enjoy a significant amount of press and are therefore gaining well-deserved marketshare—in fact, the majority of the broadband installed base today is delivered via cable modems, *not* via DSL. The service they provide is for the most part well-received and offers high-quality, high-speed access. Wireless, on the other hand, is a slumbering beast. It has largely been ignored as a serious competitor for data transport, but the wild success of WiFi, HSPA, and LTE (all discussed in detail in the next chapter) have pushed broadband wireless into the limelight.

The second challenge is unmet customer demand. If DSL is to satisfy the broadband access requirements of the marketplace, it must be made available throughout ILEC service areas. This means that incumbent providers must equip their central offices with DSLAMs that will provide the line-side redirect required as the initial stage of DSL deployment. The law of primacy is evident here: the ILECs must get to market first with broadband offerings if they are to achieve and keep a place in the burgeoning broadband access marketplace. The refocusing that has occurred in the last few years on broadband has changed all the rules and created opportunities for the traditional carriers as well as the cable providers. The call for universal dial tone has been replaced by a call for universal broadband; carriers and cable companies, not to mention fixed wireless providers, have an opportunity to come into the market swinging with new offerings.

The third and fourth challenges to rapid and ubiquitous DSL deployment are installation issues and plant quality. A significant number of impairments have proven to be vexing for would-be deployers of widespread DSL. These challenges fall into two categories: electrical disturbances and physical impairments. And while solutions have been crafted for most of these, they still pop up occasionally as vexing problems.

Electrical Disturbances

The primary cause of electrical disturbance in DSL is crosstalk, caused when the electrical energy carried on one pair of wires "bleeds" over to another pair and causes interference (noise) there. Crosstalk exists in several flavors. Near-end crosstalk (NEXT) occurs when the transmitter at one end of the link interferes with the signal received by the receiver at the same end of the link, while far-end crosstalk (FEXT) occurs when the transmitter at one end of the circuit causes problems for the signal received by a receiver at the far end of the circuit. Similarly, problems can occur when multiple DSL services of the same type exist in the same cable and interfere with one another. This is referred to as self-NEXT or self–FEXT. When different flavors of DSL interfere with one another, the phenomenon is called Foreign-NEXT or foreign–FEXT. Other problems that can cause errors in DSL and therefore a limitation in the maximum achievable bandwidth of the system are simple radio frequency interference (RFI) that can find its way into the system, and impulse, Gaussian, and random noise that exist in the background but can affect signal quality even at extremely low levels.

Physical Impairments

The physical impairments that can have an impact on the performance of a newly deployed DSL circuit tend to be characteristics of the voice network that typically have minimal effect on simple voice and low-speed data services. These include load coils, bridged taps, splices, mixed gauge loops, and weather conditions.

Load Coils and Bridged Taps

We have already mentioned the problems that load coils can cause for digital transmission, because they limit the frequency range that is allowable across a local loop they can seriously impair transmission. Bridged taps are equally problematic. When a multipair telephone cable is installed in a newly built-up area, it is generally some time before the assignment of each pair to a home or business is actually made. To simplify the process of installation when the time comes, the cable's wire pairs are periodically terminated in terminal boxes (sometimes called B-boxes) installed every block or so. As we discussed in Chap. 4, there may be multiple appearances of each pair on a city street, waiting for assignment to a customer. When the time comes to assign a particular pair, the installation technician will simply go to the terminal box where that customer's drop appears and cross-connect the loop to the appearance of the cable pair (a set of lugs) to which that customer has been assigned. This eliminates the need for the time-consuming process of splicing the customer directly into the actual cable.

Unfortunately, this efficiency process also creates a problem. While the customer's service has been installed in record time, there are now unterminated appearances of the customer's cable pair in each of the terminal boxes along the street. While these so-called bridged taps present no problem for analog voice, they can be a catastrophic source of noise due to signal reflections that occur at the copper-air interface of each bridged tap. If DSL is to be deployed over the loop, the bridged taps must be removed. And while the specifications indicate that bridged taps do not cause problems for DSL, actual deployment says otherwise. To achieve the bandwidth that DSL promises, the taps must be terminated.

Splices and Gauge Changes

Splices can result in service impairments due to cold solder joints, corrosion, or weakening effects caused by repeated bending of wind-driven aerial cable.

Gauge changes tend to have the same effects that plague circuits with unterminated bridged taps: when a signal traveling down a wire of one gauge jumps to a piece of wire of another gauge, the signal is reflected, resulting in an impairment known as intersymbol interference. The use of multiple gauges is common in loop deployment strategy because it allows outside plant engineers to use lower cost small gauge wire where appropriate, cross-connecting it to larger gauge, lower resistance wire where necessary.

Weather

"Weather" is perhaps a bit of a misnomer: the real issue is moisture. One of the greatest challenges facing DSL deployment is the age of the outside plant. Much of the older distribution cable uses inadequate insulation between the conductors (paper in some cases). In some cases the outer sheath has cracked, allowing moisture to seep into the cable itself. The moisture causes crosstalk between wire pairs that can last until the water evaporates. Unfortunately, this can take a considerable amount of time and result in extended outages.

Solutions

All of these factors have solutions. The real question is whether they can be remedied at a reasonable cost. Given the growing demand for broadband access, there seems to be little doubt that the elimination of these factors would be worthwhile at all reasonable costs, particularly considering how competitive the market for the local loop customer has become.

The electrical effects, largely caused by various forms of crosstalk, can be reduced or eliminated in a variety of ways. Physical cable deployment standards are already in place that, when followed, help to control the amount of near- and far-end crosstalk that can occur within binder groups in a given cable. Furthermore, filters have been designed that eliminate the background noise that can creep into a DSL circuit.

Physical impairments can be controlled to a point, although to a certain extend the service provider is at the mercy of their installed network plant. Obviously, older cable will present more physical impairments than newer cable, but there are steps that the service provider can take to maximize their success rate when entering the DSL marketplace. The first step that they can take is to pre-qualify local loops for DSL service to the extent possible. This means running a series of tests using Mechanized Loop Testing (MLT) to determine whether each loop's transmission performance falls within the bounds established by the service provider and existing standards and industry support organizations.

For DSL pre-qualification, MLT tests the following performance indicators (and others as required):

- Cable architecture
- Loop length
- Crosstalk and background noise

Cable-Based Access Technologies

In 1950, Ed Parsons placed an antenna on a hillside above his home in Washington State, attached it to a coaxial cable distribution network, and began to offer television service to his friends and neighbors. Prior to his efforts, the residents of his town were unable to pick up broadcast channels because of the blocking effects of the surrounding mountains. Thanks to Parsons, community antenna television (CATV) was born; from its roots came cable television.

Since that time the U.S. cable industry has become massive. Today it's a $200 billion industry that accounts for two million jobs. 7,000 headends in 1,100 companies deliver content to more than 60 million subscribers over more than a million miles of coaxial and fiber-optic cable. And it's not just entertainment content: 27 million subscribers buy their phone service from the cable company as well.

As the industry's network has grown, so too have the aspirations of those deploying it. Their goal has been to make it much more than a one-way medium for the delivery of television and pay-per-view; they want to provide a broad spectrum of interactive, two-way services that will allow them to compete head-to-head with the telephony industry. To a large degree, they have succeeded.

Playing in the Broadband Game

Unlike the telephone industry that began its colorful life under the scrutiny of a small number of like-minded individuals (Alexander Graham Bell and Theodore Vail, among others), the cable industry came about thanks to the combined efforts of hundreds of innovators, each building on Parson's original concept. As a consequence, the industry, while enormous, is in many ways fragmented. Powerful industry leaders like John Malone of TCI and Gerald Levine of Time-Warner were able to exert "Tito-like" powers to unite the many companies, turning a loosely cobbled-together collection of players into cohesive, powerful corporations with a shared vision of what they wanted to accomplish.

Today, the cable industry is a force to be reckoned with and upgrades to the original network are ongoing. This is a crucial activity that will ensure the success of the industry's ambitious business plan, and provide a competitive balance for the traditional telcos.

The Cable Network

The traditional cable network is an analog system based on a tree-like architecture. The head end, which serves as the signal origination point, serves as the signal aggregation facility. It collects programming information from a variety of sources including satellite and terrestrial feeds. Headend facilities often look like a mushroom farm; they are typically surrounded by a variety of satellite dishes (Fig. 8.5).

The headend is connected to the downstream distribution network by 1-in diameter rigid coaxial cable, as shown in Fig. 8.6. That cable delivers the signal to a neighborhood, where splitters divide the signal and send it down ½-in diameter semi-rigid coax that typically runs down a residential street. At each house, another splitter (Fig. 8.7) pulls off the signal and feeds it to the set-top box in the house over the drop wire, a *local loop* of flexible ¼-in coaxial cable.

FIGURE 8.5 A typical cable headend facility showing the downlink satellite dish array.

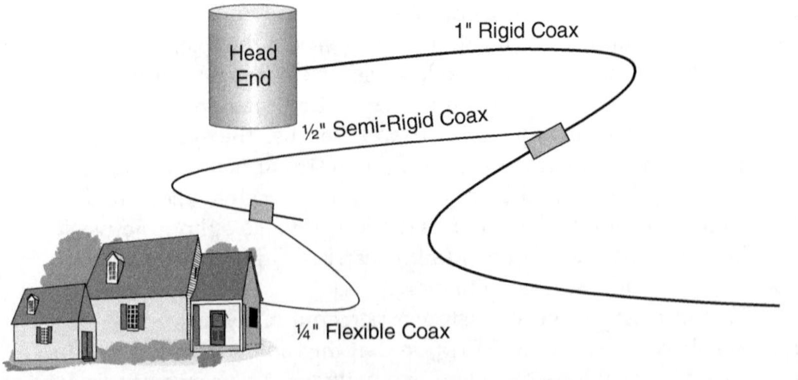

FIGURE 8.6 Cable distribution network.

FIGURE 8.7 A cable splitter used for signal distribution.

While this architecture is perfectly adequate for the delivery of one-way television signals, its shortcomings for other services should be fairly obvious to the reader. First of all, it is, by design, a broadcast system. It does not typically have the ability to support upstream traffic (from the customer toward the headend) and is therefore not suited for interactive applications. Second, because of its design, the network is prone to significant failures that have the potential to affect large numbers of customers. The tree structure, for example, means that if a failure occurs along any *branch* in the tree, every customer from that point downward loses service. Contrast this with the telephone network where customers have a dedicated local loop over which their service is delivered. Second, because the system is analog, it relies on amplifiers to keep the signal strong as it is propagated downstream. These amplifiers are powered locally—they do not have access to *central office power* as devices in the telephone network do. Consequently, a local power failure can bring down the network's ability to distribute service in that area.

The third issue is one of customer perception. For any number of reasons, there is a general perception that the cable network is not as capable or as reliable as the telephone network. As a consequence of this perception the cable industry is faced with the daunting challenge of convincing potential voice and data customers that they are in fact capable of delivering high-quality service. Some of the concerns are justified. In the first place, the telephone network has been in existence for almost 125 years, during which time its operators have learned how to optimally design, manage, and operate it in order to provide the best possible service. The cable industry, on the other hand, came about 50 years ago, and didn't benefit from the rigorously administered, centralized management philosophy that characterized the telephone industry. Additionally, the typical 450-MHz cable system did not have adequate bandwidth to support the bi-directional transport requirements of new services.

Furthermore, the architecture of the legacy cable network, with its distributed power delivery and tree-like distribution design, does not lend itself to the same high degree of redundancy and survivability that the telephone network offers. Consequently, cable providers have been somewhat hard-pressed to convert customers who are vigorously protective of their telecommunications services.

The Ever-Changing Cable Network

Faced with these harsh realities and the realization that the existing cable plant could not compete with the telephone network in its original analog incarnation, cable engineers began a major rework of the network in the early 1990s. Beginning with the headend and working their way outward, they progressively redesigned the network to the extent that in many areas of the country their coaxial *local loop* is capable of competing on equal footing with the telco's twisted pair—and in some cases beating it.

A Four-Phase Conversion

The process they have used in their evolution consists of four phases. In the first phase, they converted the headend from analog to digital.

This allowed them to digitally compress the content, resulting in far more efficient utilization of available bandwidth. Second, they undertook an ambitious physical upgrade of the coaxial plant, replacing the 1-in trunk and ½-in distribution cable with optical fiber. This brought about several desirable results. First, by using a fiber feeder, network designers were able to eliminate a significant number of the amplifiers responsible for the failures the network experienced due to power problems in the field. Second, the fiber makes it possible to provision significantly more bandwidth than coaxial systems allow. Third, because the system is digital, it suffers less from noise-related errors than its analog predecessor did. Finally, an upstream return channel was provisioned as shown in Fig. 8.8, which makes possible the delivery of true interactive services such as voice, Web surfing, and videoconferencing.

The third phase of the conversion had to do with the equipment provisioned at the user's premises. The analog set-top box has now been replaced with a digital device that has the ability to take advantage of the capabilities of the network, including access to the upstream channel. It decompresses digital content, performs content ("stuff") separation, and provides the network interface point for data and voice devices.

The final phase is business conversion. Cable providers look forward to the day when their networks will compete on equal footing with the twisted pair networks of the telephone company, and customers will see them as viable competitors. In order for this to happen they must demonstrate that their network is capable of delivering a broad variety of competitive services; that the network is robust; that they have operations support systems (OSSs) that will guarantee

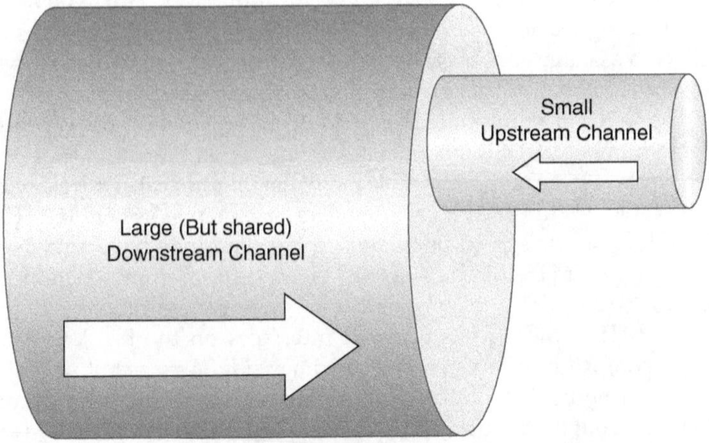

FIGURE 8.8 Upstream return channel for two-way data communications.

the robustness of the infrastructure; and that they are cost competitive with incumbent providers. They must also continue to create a presence for themselves in the business centers of the world.

Cable Modems

As cable providers have progressively upgraded their networks to include more fiber in the backbone, their plan to offer two-way access to the Internet has become a reality. Cable modems offer access speeds of up to 30 Mbps, and the market uptake has been spectacular.

Cable modems provide an affordable option to achieve high-speed access to the Web. They offer asymmetric access, that is, a much higher downstream speed than upstream, but for the majority of users this does not represent a problem since the bulk of their use will be for Web surfing during which the bulk of the traffic travels in the downstream direction anyway.

While cable modems do speed up access to the Web and other online services by several orders of magnitude, they do have a number of downsides that must be considered. The greatest concern that has been voiced about cable modems is security. Because cable modems are *always on,* they represent an easier entry point for hackers looking to break into machines. It is therefore *critical* that cable subscribers use some form of firewall software or a router that has the ability to perform filtering.

Data over Cable Service Interface Specification

As interest grew in the late 1990s for broadband access to data services over cable television networks, CableLabs®, working closely with the ITU and major hardware vendors, crafted a standard known as the data over cable service interface specification (DOCSIS). The standard is designed to ensure interoperability among cable modems as well as to assuage concerns about data security over shared cable systems, DOCSIS has done a great deal to resolve marketplace issues.

The DOCSIS architecture looks a lot like DSL architecture. Whereas a large number of DSL modems communicate with a DSLAM in the telco world, a large number of cable modems communicate with a device in the cable world called a cable modem termination system (CMTS). Depending on the served population, a headend may have anywhere from 1 to 12 CMTSs, and a single CMTS can support anywhere from 4,000 to 150,000 cable modems depending on its configuration.

A CMTS typically has ethernet interfaces on the headend side and coaxial RF interfaces on the customer side. As a result, Internet traffic can be bridged through the ethernet interfaces, through the CMTS, and out to the customer over the coaxial RF connections. CMTSs are typically IP devices; they transport IP packets that are encapsulated for transport as stipulated by the DOCSIS standard.

DOCSIS 1.0

Under the standards, CableLabs® crafted an original cable modem certification standard called DOCSIS 1.0, which guarantees that modems carrying the certification will interoperate with any headend equipment, are ready to be sold in the retail market, and will interoperate with other certified cable modems. Engineers from Askey, Broadcom, Cisco Systems, Ericsson, General Instrument, Motorola, Philips, 3Com, Panasonic, Digital Furnace, Thomson, Terayon, Toshiba, and Com21 participated in the development effort.

The DOCSIS 1.1 specification was released in April 1999, and included two additional functional descriptions which began to be implemented in 2000. The first specification details procedures for guaranteed bandwidth, as well as a specification for quality of service guarantees. The second specification is called baseline privacy interface plus (BPI+); it enhances the current security capability of the DOCSIS standards through the addition of digital certificate-based authentication and support for multicast services to customers.

DOCSIS 2.0

In December 2002, the ITU-T announced the approval of a standard defining the second-generation data over cable system, known as DOCSIS 2.0.

ITU Recommendation J.122 gives cable operators the ability to offer speeds up to 600 times faster than a standard dial-up telephone modem provides. The new standard can be used as the foundation upon which IP-based telephony services can be offered.

The enhancements that J.122 provides over its predecessor are primarily focused on the upstream transmission path from the customer to the network. Changes include increased capacity and improved robustness over the upstream path.

While the DOCSIS name is in widespread use, CableLabs now refers to the overall effort as the CableLabs certified cable modem project.

DOCSIS 3.0

DOCSIS 2.0 served the market well for quite some time, but in August 2006 it was upgraded to DOCSIS 3.0. Among other things the 3.0 version added support for IPv6; like its earlier versions, DOCSIS 3.0 provides backward compatibility with all previous versions of the standard.

DOCSIS 3.1 is in the design phase and will support a minimum of 10 Gbps downstream and 1 Gbps upstream, using a 4,096 QAM encoding scheme. It also eliminates the standard 6- and 8-MHz-wide channels used in earlier cable systems, opting instead for 20- to 50-KHz channels.

OK, enough terrestrial technology. Let's get into the air.

Wireless Access Technologies

To understand wireless communications it is necessary to examine both radio and telephone technologies, because the two are inextricably intertwined. In 1876, Alexander Graham Bell, a part-time inventor and a teacher of hearing-impaired students, invented the telephone while attempting to resolve the challenge of transmitting multiple telegraph signals over a shared pair of wires. His invention changed the world forever.

In 1896, a mere 20 years later, Italian engineer and inventor Guglielmo Marconi developed the spark gap radio transmitter which eventually allowed him to transmit long-wave radio signals across the Atlantic Ocean as early as 1901. Like Bell, his invention changed the world; for his contributions, he was awarded the Nobel Prize in 1909.

It wasn't until the 1920s, though, when these two technologies began to dovetail, that their true promise was realized. Telephony provided interpersonal, two-way, high-quality voice communications, but required the user to be stationary. Radio, on the other hand, provided mobile communications, but was limited by distance, environmentally induced signal degradation, and spectrum availability. Whereas telephony was advertised as a universally available service, radio was more of a catch-as-catch-can offering that was subject to severe blocking. If a system could be developed that combined the signal quality and ubiquity of telephony with the mobility of radio, however, a truly promising new service offering could be made available.

Today, cellular telephony (and other services like it) provides high-quality, ubiquitous, wireless telephone service. Thanks to advances in digital technology, wireless telephony also offers services identical to those provided by the wired network. And pricing for wired and wireless services are now reaching parity: flat rate nationwide pricing models are commonplace today that have no roaming or long-distance restrictions.

A Bit of History

The road to the modern wireless communications network was fraught with technical, bureaucratic, and social speed bumps. In 1905, listeners in New York City, accustomed to the staccato beat of Morse Code broadcasts spewing from their radios, were astounded to hear music and voice as Reginald Fessenden, a local radio commentator, broadcast the results of the annual New York City yacht races. Listeners could hear the results immediately, instead of waiting for the late edition of the paper. Public perception of the possibilities of radio suddenly began to grow.

At roughly the same time, the U.S. military, recognizing the potentially strategic value of radio communications in the armed forces, helped to create the Radio Corporation of America (RCA), ushering wireless technology into the technological forefront.

The first wireless applications were developed in 1915 and 1916 for ship-to-ship and ship-to-shore communications. In 1921, the Detroit Police Department inaugurated the first land-based mobile application for radio communications, when they installed one-way radios in patrol cars—the original "calling all cars" application (Fig. 8.9). The system required police officers to stop at a phone box to place a call to their precinct when a call came in, but the ability to quickly and easily communicate with roving officers represented a giant step forward for the police and their ability to respond quickly to calls.

Unfortunately, the system was not without its problems. Even though the transmitter was powerful, Detroit's deep concrete canyons prevented the signal from reaching many areas. Equally vexing was the problem of motion: the delicate vacuum tubes, subjected to the endless vibrations stemming from life in the trunk of a moving vehicle, caused both physical and electronic problems that were so bad that in 1927 the Detroit P.D. shut down the station out of sheer frustration. It wasn't until 1928, when Purdue University student Robert Batts developed a super heterodyne radio that was somewhat resistant to buffeting and vibration, that Detroit re-inaugurated their radio system.

Other cities followed in Detroit's footsteps. By 1934, there were nearly 200 radio-equipped city police departments communicating with nearly 5,000 roving vehicles. Radio had become commonplace,

FIGURE 8.9 A telephone box, reserved for police use.

and in 1937 the FCC stepped in and allocated 19 additional radio channels to the 11 that they had already made available for police department use. No one realized it, but the spectrum battles had begun.

Early radio systems were based on amplitude modulation (AM) technology. One drawback of AM radio is the fact that the strength of the transmitted signal diminishes according to the square of the distance. A signal measured at 20 W 1 mi (mile) from the transmitter will be *one-quarter* that strength, or 5 W, 2 mi from the transmitter. When Edwin Armstrong (see Chap. 3) announced and demonstrated his invention of frequency-modulated radio in 1935, he turned the broadcast industry on its ear. He offered a technology that not only eliminated the random noise and static that plagued AM radio, but also lowered transmitter power requirements and improved the ability of receivers to lock onto a weak signal.

World War II provided an ideal testing ground for FM. AM radio, used universally by the Axis powers during the war, was easily jammed by powerful Allied transmitters. FM, on the other hand, used only by Allied forces, was unaffected by AM jamming techniques. As we mentioned earlier, many historians believe that FM radio was crucially important to the Allies' success in World War II.

Radio's Evolution

Radio is often viewed as having evolved in two principal phases. The first was the *pioneer phase*, during which the fundamental technologies were developed, initial technical bugs were ironed out, applications were created (or at least conceived), and bureaucratic and governmental haranguing over the roles of various agencies initiated. This phase was characterized by the development and acceptance of FM transmission, the dominance of military and police usage of radio, and the challenge of building a radio that could withstand the rigors of mobile life. The pioneer phase lasted until the early 40s when the *commercial phase* began.

During the 1940s, radio technology advanced to the point that FM-based car phones were commercially available. Initially AT&T provided mobile telephone service under the umbrella of its nationwide monopoly. In 1949, however, the Justice Department filed suit against AT&T, settling the case with the Consent Decree of 1956. Under this Consent Decree, AT&T kept its telephone monopoly, but was forced to give up a number of other lines of business, including the manufacture of mobile radiotelephones. This allowed companies like Motorola to enter the fray and paved the way for the creation of radio common carriers (RCCs), much to AT&T's chagrin. These RCCs were typically small businesses that offered wireless service to several hundred people in a fixed geographical area.

A Touch of Technology

Before proceeding further, let's introduce some fundamentals of radio transmission.

For radio waves to carry information such as voice, data, music, or images, certain characteristics of the waves are changed or *modulated.* Those characteristics include the amplitude, or loudness, of the wave; the frequency, or pitch, of the wave; and the phase, or angle of inflection, of the wave. By modulating these characteristics, a *carrier wave* can be made to transport information, whether it is a radio broadcast, a data stream or a telephone conversation. (Please review this topic in Chap. 1 for more information.)

Transmission *space* is allocated based on ranges of available frequency. This range is known as a spectrum, a word that means a range of values. The radio spectrum represents a broad span of electromagnetic frequencies between 1 kHz (1,000 cycles per second) and ten *quintillion* (that's ten followed by 19 zeroes) cycles per second. Most radio-based applications operate at frequencies between 1 kHz and 300 GHz.

One concern that has surfaced repeatedly over the years since the inception of radio is spectrum availability. Different services require differing amounts of spectrum for proper operation. For example, a modern FM radio channel requires 30 kHz of bandwidth, while an AM channel only requires 3 kHz. Television, on the other hand, requires 6 MHz. As you might imagine, battles between the providers of these disparate services over spectrum allocation became quite heated in the years that followed.

Improving the Model

The first FM radiotelephone systems were massively inefficient. They required 120 kHz of channel bandwidth to transport a 3 kHz voice signal. As the technology advanced, though, this requirement was reduced. In the 1950s, the FCC mandated that channels be halved to 60 kHz, and by the 1960s things had advanced to the point that the spectrum could be further reduced to 30 kHz, where it remains today in analog systems. Realize that this is *still* a 10:1 ratio.

AT&T introduced the first commercial mobile telephone service in St. Louis in 1946. It relied on a single, high-power transmitter placed atop one of the city's tallest buildings and had an effective service range of about 50 miles. This FM system was fully interconnected with AT&T's wireline telephone network.

The service was not cheap. There was a basic $15.00 monthly charge, and usage was billed at a flat 15¢ per minute. In spite of the cost the service was quite popular, a fact that led to its undoing.

System engineers based their design on existing radio systems which were primarily designed for dispatch applications. The traffic patterns of radio dispatches are different than telephony: a dispatch occupies several seconds of airtime, while a telephone call typically lasts several minutes. Almost immediately, blocking became a serious problem. In fact, in New York City, where the service quickly became popular, AT&T had 543 active subscribers in 1976 and a waiting list of nearly 4,000 anxious-to-be-connected customers that the system's limited capacity could not accommodate.

Radio design played a role in improving spectrum utilization. The first systems, such as the one in St. Louis, used a scheme called *non-trunked radio.* In non-trunked radio systems the available spectrum was divided into channels, and groups of users were assigned to those channels. The advantage of this technique is that non-trunked radios were relatively inexpensive. The downside was that certain channels could become severely overloaded, while others remained virtually unused.

The invention of *trunked radio* relieved this problem immensely. Trunked radios were *frequency agile,* meaning that all radios could access all channels. When a user placed a call, the radio unit would search for an available channel and use it. With the arrival of this capability blocking became a non-issue. The downside, of course, was that frequency-agile radios, because of their more complex circuitry, were significantly more expensive than their non-trunked predecessors.

Most of this work was conducted during the turbulent 1960s and led to Bell Labs' introduction of improved mobile telephone service (IMTS). IMTS used two 30-kHz, narrowband FM channels for each conversation, and provided full-duplex talk paths, direct dialing, and automatic trunking. Introduced commercially in 1965, IMTS is considered to be the predecessor of modern cellular telephony.

The Spectrum Battles Heat Up

Starting in the mid-1940s, mobile telephony went head-to-head against the television industry in a pitched battle for spectrum. In 1947, the Bell System proposed to the FCC a plan for a large-scale mobile telephone system, asking them to allocate 150 two-way, 100-kHz channels. The proposal was not accepted.

In 1949, while the FCC wrestled with the assignment of spectrum in the 470- to 890-MHz range that would become UHF television, the telephone industry argued that they should be granted a piece of the electromagnetic pie. Their issue was that the 6 MHz of bandwidth required for a *single* TV channel was more than had *ever* been allocated for mobile telephony. Unfortunately, television had penetrated the American household, Captain Kangaroo, Roy Rogers and Winky-Dink captivated viewers, and consumers were hungry for additional programming. Wireless telephony wasn't even on their radar screens; it would have to wait.

In the 20 years that followed, mobile telephony continued to take a backseat to television. UHF usage grew slowly; in 1957 the Bell System petitioned the FCC to give them a piece of the earmarked spectrum in the 800-MHz range that was virtually unused. But television was still the darling of the nation, and the FCC was ferociously dedicated to expanding the deployment of UHF television. In 1962, the government signed into law the All Channels Receiver Act that mandated that all new televisions must have both VHF and UHF receivers. Remember when televisions had two tuner knobs on them? TV won again.

Between 1967 and 1968, the FCC and the House of Representatives, under pressure from the telephone industry, studied the issue of spectrum allocation once again. In 1968, the FCC convened the *Cellular Docket,* a contentious and highly visible collection of lawmakers who eventually ruled that mobile telephony's concerns could only be solved by giving them spectrum. In 1970, the FCC reallocated UHF channels 70 to 83 from television to mobile services. From the resulting 115-MHz electromagnetic chunk, 75 MHZ was allocated to mobile telephony, with 40 MHz available immediately and 35 MHz held in reserve.

Once spectrum was allocated, the political games began. It was roundly assumed that AT&T would design, install, and operate the wireless network as an extension of its own universal wireline network. After all, the spectrum allocation was "deeded" to the wireline telephone companies, and since AT&T provided service to roughly 85 percent of the market, this assumption was somewhat natural. Their advanced mobile phone system (AMPS) architecture, based on the same design philosophy as the wireline network, relied on a massively expensive switching infrastructure, hundreds of cells, and centralized control of all functions.

AT&T's only competitors in this market were the RCCs, mentioned earlier. Among them were Motorola, even then a large player in the industry. Initially Motorola sided with AT&T, but when AT&T chose other vendors to manufacture the equipment required to establish the network, Motorola changed its spots and sided with the other 500 some-odd RCCs in the country to sue the FCC for unfair treatment. Their suit called for the commission to deny AT&T's application for development of AMPS and to re-examine the spectrum allocation. In 1970, Motorola, in partnership with another large RCC, applied for permission to offer wireless service in Washington, D.C. After examining the petition, the courts decided that the answer to the industry's woes lay in a competitive market. In 1980, they began another cellular rule-making effort to determine the regulatory structure of the market they were attempting to create.

Three options emerged from their discussions. The first was preservation of the monopolistic, single-operator concept; the second advocated an open market, in which competition was opened to all comers and the market would sort itself out; and the third involved a duopoly approach, in which two systems would be allowed in each major market area. After long debate, regulators and lawmakers decided on the duopoly concept, with two 20-MHz systems ('A' frequencies and 'B' frequencies) allocated in each market. The 'A' frequencies were allocated to the non-wireline company, while the 'B' frequencies went to the wireline carrier.

The first commercial cellular telephone system became operational in October of 1983. By 1985, there were 340,000 cellular subscribers; today there are over 6 billion worldwide, with annual revenues of nearly $1.5 trillion. It's interesting to note that in 1982, AT&T proudly predicted that there would be over 100,000 cellular subscribers by 1990.

Cellular Telephony

As modern as cellular telephony is considered to be, it was originally conceived in the 1940s at Bell Labs, as part of a plan to overcome the congestion and poor service issues associated with mobile telephone systems.

There are four key design principles of cellular systems that are the same today as they were in the 1940s. They are the use many of low-power, small coverage area transmitters instead of a single, powerful, monolithic transmitter to cover a wide area; frequency reuse; the concept of cell splitting; and central control and cell-to-cell handoff of calls. These concepts are fairly straightforward. The first relies on the philosophy that the whole is greater than the sum of its parts. By using a large number of low-power transmitters scattered across a broad coverage zone, each capable of handling multiple simultaneous calls, more users can be supported than with a single, monolithic transmitter.

The second, frequency reuse, takes into account the fact that cellular telephony, like all radio-based services, has been allocated a limited number of frequencies by the FCC. By using geographically small, low-power cells, frequencies can be reused by non-adjacent cells.

When usage areas become saturated by excessive usage, cells can be split. When traffic engineers observe that the number of callers refused service because of congestion reaches a critical level, they can split the original cell into two cells, by installing a second site within the same geographical area that uses a different set of non-adjacent frequencies. This has an extended impact; as the cells become smaller, the total number of calls that the aggregate can support climbs dramatically. Because of cellular geometry, if the radius of the cell is halved, the number of supported calls is quadrupled. The smaller the cell, therefore, the larger the total number of simultaneous callers that can be accommodated. Of course this also causes the cost of deployment to climb dramatically, and while the architectural goal of most cellular providers is to create a mosaic of thousands of very small cells, called microcells or picocells, that's an expensive proposition and will not happen immediately.

Finally, cellular systems rely on a technique called cell-to-cell handoff, which simply means that the cellular network has the ability to track a call (using relative signal strength as an indicator) as the user moves across a cell. When the signal strength detected by the current cell is perceived by the system to be weaker than that detected by the cell the user is approaching, the call is handed off to the second cell. Ideally, the user hears nothing. Cell handoff and other cellular system capabilities are under the central control of a mobile telephone switching office (MTSO, sometimes pronounced "Mitso").

OK, But How Does It Work?

When a user turns a cellular phone on, several things happen. First, the phone identifies itself to anyone willing to listen (hopefully, a local cell) by transmitting a unique identification code on a frequency that is designated as a *control channel*. The control channel is used for the transmission of operations and maintenance messages between cellular telephones and cell sites. If the phone is within the operating area of a cell site, the site registers the presence of the phone within its operating area, notifies the MTSO to which all the cells in an area are connected, and tracks its position, based on signal strength, as it moves around the cell.

When the user wants to place a call, he or she simply pushes the right buttons, which create simulated touch-tone sounds. Once dialing is complete, the user pushes the *send* button, which causes the handset to transmit the buffered digits across the control channel to the local cell. The local cell hands the call off to the MTSO. The MTSO analyzes the digits, instructs the handset and the cell to use a particular set of frequencies for the call, and routes the call to the appropriate destination. MTSOs are interconnected to the wireline network, and can therefore terminate calls at any location, including to another cellular user.

If driving while talking, the user may approach a cell boundary. The MTSO, which tracks the relative signal strength of each user as they move among the various cells within its domain, will effect handoff of a call from one cell to another if the user's movement (based on signal strength) indicates that they are approaching a cell boundary.

Access Methods

The original analog cellular telephony systems relied on a technique called *frequency division multiple access (FDMA)* as the access- and frequency-sharing scheme between mobile users and the cellular network. In FDMA systems, the available spectrum is divided into channels which are assigned to users on demand. One or more of the channels are reserved and set aside as control channels, used to transmit maintenance and operations information between the mobile phone and the network.

Each conversation requires two 30-kHz channels—one for the *forward,* or base-to-mobile direction, and one for the *reverse,* or mobile-to-base station direction. This pairing of channels permits true, full-duplex telephony. Most analog systems have since gone by the wayside in favor of the far more popular, efficient and battery-friendly digital systems. During the developmental battles two techniques emerged.

The first, *time-division multiple access* (TDMA), resembles FDMA in that it divides the available frequency spectrum into channels. That, however, is where the resemblance ends. In TDMA, each of the analog channels carries telephone calls that are time-division multiplexed—that is, they share access to the channel. As in FDMA, a control channel is reserved for communication between the network and mobile users.

The biggest advantage that TDMA systems have over FDMA systems is that they support significantly more users. If each channel is divided into four time slots, then the system capacity is quadrupled. And while TDMA electronics are significantly more complex, the fact that they are digital means that they can easily evolve as technology advances.

The second digital access technique is called *code-division multiple access* (CDMA). CDMA systems are dramatically different from FDMA and TDMA systems in that they do not channelize the available bandwidth—instead, they allow all users to access and use the available spectrum simultaneously. This technique is called *spread spectrum transmission.* Spread spectrum techniques are described in detail later in the section entitled, "Access Evolution."

Not only are digital systems more secure than narrowband technologies, they also support significantly larger numbers of simultaneous users. In fact, whereas FDMA systems support a single user per frequency slot, CDMA systems can support hundreds.

Radio-based telephony has enjoyed a wild, tumultuous ride along the way to its position today as a mainstream, foundation-level technology. Starting in the late 1800s with the parallel work of Marconi and Bell, radio and telephony wandered down different paths until fairly recently, when they converged and joined forces, leading to the development of cellular telephony.

The story doesn't end with cellular telephony, however. Today, mobile users are clamoring for the ability to extend the reach of LANs, videoconferencing systems, medical image devices, and database access, without having to deal with the restrictions of a copper tether. Developing nations have realized that with cellular technology, telephony infrastructures can be installed in countries in a fraction of the time it takes to install a wired network. In fact, my work in Africa has shown me the degree to which humans demonstrate their ability to innovate in the face of technological challenge.

Access Evolution

Wireless access technologies have undergone an evolution comprising three generations. First generation systems, which originated in the late 1970s and continued to be developed and deployed throughout the 1980s, were analog and designed to support voice. They were characterized by the use of *frequency division multiple access* (FDMA) technology. In FDMA systems, users are assigned analog frequency pairs (and access to a control channel) over which they send and receive. One frequency serves as a voice transmit channel, the other as a receive channel. The control channel is used for signaling.

Second-generation systems (2G), which came about in the early 1990s, were all-digital and still primarily voice oriented, although rudimentary data transport was part of the service package. In 2G systems, digital access became the norm through such technologies as TDMA. In TDMA systems, the available frequency is broken into channels as it is in FDMA, but here the similarity between the two ends. Each channel is shared among a group of users based on a time division technique that ensures fair and equal access to the available channel pairs. This increases the so-called packing ratio, that is, the number of simultaneous users that the system can support.

An offshoot of 2G TDMA systems is 2.5G, the most common of which is *global system for mobile communications* (GSM). GSM is widely deployed in Europe and other parts of the world (well over 750 million subscribers globally) and goes beyond 2G by offering a suite of user services that enhances the value of the technology. These services include not only voice but two-way radio, paging, short messaging service (SMS, similar to instant messaging), and the incorporation of smart card technology that permits the customization of the handset and its features. These services are both popular and lucrative: In Europe and Africa, more people send and receive SMS/IM text messages via their mobile phones than those who use the Internet for similar functions. Three-quarters of all adult mobile users employ the texting function; virtually 100 percent of younger subscribers (Millennials) do. In fact, the average younger user sends about 40 text messages per day.

GSM is now present in the United States through AT&T Mobility and T-Mobile, the only all-GSM network providers in the country.

Data Enhancements to GSM

Related to GSM is an enhancement called *enhanced data rate for global evolution* (EDGE). Originally developed by Ericsson and called GSM384, EDGE allows a service provider to offer 384 Kbps service when eight GSM timeslots are used—a less-than-efficient use of available bandwidth. The conversion to EDGE requires that an additional transceiver be added to base stations and naturally requires the use of handsets that can handle the additional protocol.

EDGE is one technology that paves the way for the implementation of the *universal mobile telephony system* (UMTS). According to the original mandate of the European Telecommunications Standards Institute (ETSI), "UMTS will be a mobile communications system that can offer user benefits including high-quality wireless multimedia services to a convergent network of fixed, cellular and satellite components. It will deliver information directly to users and provide them with access to new and innovative services and applications. It will offer mobile personalized communications to the mass market regardless of location, network or terminal used." UMTS succeeded and continued to evolve.

Generalized Packet Radio Service

"Native-mode" GSM offers relatively low-speed data access. In response to demands for higher mobile bandwidth, a number of new technological add-ons have been created including the *generalized packet radio service* (GPRS) which can conceivably achieve higher data rates by *clustering* as many as eight 14.4-Kbps channels to offer as much as 115 Kbps to the mobile user. There are, of course, downsides to this model including the ability to support fewer simultaneous users because of channel clustering and the need to build an overlay on the existing network to support packet-mode data transport. Nevertheless, GPRS deployment is proceeding apace.

The third generation of wireless systems (3G) offers broadband access to wireless users over a high user count, digital network. One access protocol that will be deployed is CDMA. In CDMA networks, there is no channelization per se.

The Reality

There are effectively two evolutionary 3G paths underway in the United States, one rooted in GSM, the other in CDMA. Both have merit, both will undoubtedly survive for some time to come, but they are different and must be explained.

The GSM path begins with traditional 2.5G access. Over time, GPRS is added for data service, and in many cases EDGE is later added to the mix. This is often referred to in the industry as *wideband*

CDMA and leads to the ultimate end state, the *universal mobile telephone system (UMTS).*

The CDMA path, on the other hand, begins with CDMA2000 and adds high-speed data through an overlay service known as *1× evolution-data only (EV-DO).* 1×EV-DO technology, sometimes called high data rate (HDR), offers high-bandwidth wireless connectivity that is optimized for packet data over CDMA networks. EV-DO supports wireless Web access at download speeds ranging from about 384 Kbps to 2.4 Mbps.

Of course, things eventually *do* get better, as we'll see in the next section.

The Wireless Data Conundrum

As voice revenues decline through competitive commoditization, wireless carriers look for alternative revenue sources that they can turn to. One is wireless data, a service that has seen a remarkable roller coaster ride of success and failure over the years since demand for it began.

The first protocol that was announced for data transport over mobile environments was the *wireless application protocol* (WAP). Originally developed by Phone.com for mobile Internet access, it was largely a disappointment among users. Early on in its deployment, Germany's D2 network administrators announced that customers were using it less than a minute a day. The WAP acronym was soon redefined as the "wrong approach to portability" because of the complexity involved in its use (one article reported that it took 32 clicks and scrolls to access a simple stock quote). Many, however, believe that WAP failed because service providers that implemented it were unwilling to share revenues with the content providers that they supposedly partnered with. Historically, given the choice of protecting their existing service margins or allowing new markets to emerge based on new models, service providers choose what they are most comfortable with—protecting existing service margins. Now they are changing the way they view this because of the success of text messaging and other applications, but they still have a long way to go. For all intents and purposes WAP died a long time ago.

The Road to 3G—and Beyond

If we go back and look at the evolution of mobile broadband over the last 10 years or so, we find a very interesting continuum as shown in Fig. 8.10. The two primary (and incompatible) access standards for mobility, GSM and CDMA, ultimately converge to a service specification called UMTS. UMTS in turn leads us to the next phase of wireless evolution called high speed packet access; it, in turn, leads us to the

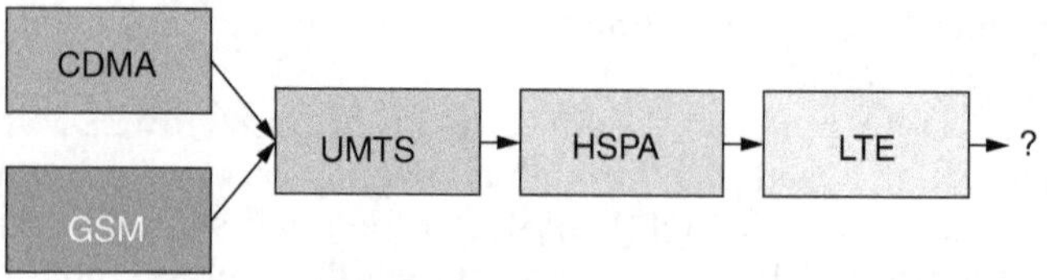

FIGURE 8.10 The evolution of mobile broadband.

most current standard for high-speed mobility, long-term evolution (LTE). We'll look at these in the next few pages.

Universal Mobile Telephone System

Universal mobile telephone system is the first true 3G wireless standard. And while it is based on GSM, it plays well with CDMA, affording the opportunity to bring the competing CDMA and GSM standards closer as we strive for a global universal access standard.

UMTS defines an entire technological ecosystem—not just an access standard. The UMTS environment comprises the UMTS terrestrial radio access network (UTRAN); the core network (sometimes called the mobile application part, or MAP) and the subscriber identity module, or SIM card.

From UMTS to HSPA

The next evolutionary stage from UMTS is the move to high speed packet access (HSPA). HSPA actually comprises two protocols, high-speed downline packet access (HSDPA) and high-speed upline packet access (HSUPA). Together they make up the HSPA offering.

HSPA builds on UMTS' Wideband CDMA foundation and improves the performance of third-generation networks. The standard has gone through a number of iterations, the most recent of which was released in 2008 and widely adopted by 2010. This standard is impressive, supporting downline speeds as high as 168 Mbps and uplink speeds in the range of 22 Mbps.

From HSPA to LTE

When I think about how much the technology world has changed between the last edition of the book and this one, it makes me smile and shake my head in wonder. There were no integrated smart phones; Android didn't exist; iPhones didn't exist; apps didn't exist; broadband wireless meant 9.6 Kbps on a good day with the wind at your back; and mobile access was spotty at best. Today, things are a bit ... different. Smart phones are the rule rather than the exception; access is essentially universal, with very few exceptions; Android and Apple rule the market; and apps number in hundreds of thousands.

And as the capability of the network to support such things grows, so does grows the demand for them. And that means the network has to be able to deliver, which is where LTE comes in.

Let's first talk about the future-proof network that LTE makes possible. LTE assumes that the underlying network is all-IP, which means that it integrates seamlessly with the direction of networking in general. It offers peak download speeds of 299.6 Mbps and upload speeds in excess of 75 Mbps. Think about that: LTE is faster than Wi-Fi!

It has other advantages as well. Data transfer latency is below 5 ms, and support is included for a variety of protocols including frequency-division duplex (FDD) and time-division duplex (TDD). LTE offers improved support for devices that are moving at a high rate of speed, with specific operating frequencies for vehicles moving in the 200- to 300-mph range.

LTE offers extremely good spectrum flexibility with 1.4 MHz, 3 MHz, 5 MHz, 10 MHz, 15 MHz, and 20 MHz wide cells standardized. And those cells can be pretty much any physical size, from tens of meters in diameter to 100 km across. Each 5-MHz cell can support as many as 200 active data clients, and the technology is designed to handle the delivery of television and other media-based services to a roaming device.

What really matters about LTE is that it is ambitious, aggressive, and real. It improves spectral efficiency, lowers operating costs for the carrier, supports a wide range of consumer and enterprise services, and offers seamless integration with pre-existing standards such as GSM, CDMAOne, W-CDMA, and CDMA2000.

TD-LTE

Be aware that there is an interesting variant of traditional LTE called TD-LTE. Instead of the dedicated channel pairs that traditional LTE mandates, TD-LTE dynamically assigns sub-frames on-demand for the uplink and downlink, which is much more efficient and is ideal for the demands of multimedia applications. It is also much more efficient from a power consumption perspective.

Voice over LTE

The idea for voice over LTE (VoLTE) originally came about as a fallback plan—that is, a system to support IP data with the option to *fall back* to 2G or 3G systems for voice. In the course of developing this all-IP capability, three options emerged: circuit-switched fallback (CSFB), simultaneous voice LTE (SV-LTE), and voice over LTE via generic access network (VoLGA).

Circuit-Switched Fallback

Circuit-switched fallback (CSFB) is a third-generation partnership project (3GPP) standard called 3GPP 23.272 which is designed to allow the connection to fall back to 2G or 3G, should it be required.

This option also transports SMS traffic, and while it is a viable option, it does require some changes to the EUTRAN network.

Simultaneous Voice LTE

Simultaneous voice LTE (SV-LTE) is perhaps the least desirable of the three options. It allows circuit-switched voice and packet-based data to run simultaneously, but it also requires that two handset radios be included in the handset and they must run simultaneously, which means that battery life will be affected.

Voice over LTE via Generic Access Network

Voice over LTE via generic access network (VoLGA) offers the most promise of the three. It is based on the 3GPP generic access network (GAN) standard and was created as a way for operators to have standardized voice and SMS services across all network types, including GSM, UMTS, and LTE. VoLGA lacks the dedicated circuit-switch voice channels that other options offer; instead it relies VoIP and end-to-end IP connections from the user device all the way through the network core. It also takes advantage of LTE's low delay and built-in QoS to guarantee better voice than can be delivered through the preexisting 2G and 3G services. This is a good direction, but it still has a ways to go. That said it is the only one of the three options that realistically enables mobile subscribers to receive consistent SMS, voice, and data services as they migrate between GSM, LTE, and UMTS networks.

Bluetooth

Bluetooth has had a tumultuous ride in the last few years. First heralded as the next great connectivity option, its future remains marginally questionable today. The Bluetooth standard, named for tenth-century Danish King Harald Bluetooth who united Denmark and part of Norway into a single kingdom, was developed by a group of manufacturers and allows any device, including computers, cell phones, and headphones, to automatically establish wireless device-to-device connections. The Bluetooth Special Interest Group, comprising more than 1,000 companies, works to support Bluetooth's role as a replacement technology for connectivity between peripherals, telephones, and computers. In effect it is a cable replacement solution and is often cited as the ideal way to simplify the rat's nest typically found behind most computer desks.

Bluetooth operates at 2.45 GHz, a spectrum allocation set aside for the use of industrial, scientific, and medical devices (ISM). The system is capable of transmitting data at 1 Mbps, although headers and handshaking information consume about 20 percent of this capacity. A number of other devices operate within this domain as well, including baby monitors, garage-door openers, and some cordless phones. The process of ensuring that Bluetooth and these other devices don't interfere with one another has been a crucial part of the design process, and today is one of the major detractors to the technology. Cordless phones of 2.5 GHz have become so popular that interference between them and Bluetooth-equipped devices has become a significant nuisance in some cases.

Applications suggested for Bluetooth are interesting, and while they are distance limited because of the technology's low operating power level, plenty of opportunities have presented themselves. The standard specifies operational distances of up to 100 m between Bluetooth-equipped devices, but because of power restrictions most devices will only be able to operate within a 10-m radius. Bluetooth is designed for the creation of *personal area networks* (PANs) that rely on *picocells* (very small service areas) for connectivity. To date, the most successful applications for Bluetooth are indeed cable replacement solutions on mobile phones, PCs (keyboards and mice), and headphones. The technology is also enjoying some success in the sensor and monitoring markets and is experiencing uptake in the manufacturing, security, and healthcare sectors.

Radio Frequency Identification

Radio frequency identification (RFID) has its origins in the 1940s, when the fundamental technology that underlies modern RFID systems was developed to discriminate between inbound friendly aircraft and inbound enemy aircraft. Allied aircraft carried transponders that broadcast a unique radio signal when interrogated by radar, identifying them as a *friendly*. This *Identify: Friend or Foe* (IFF) system was the basis for the development of the technology set we refer to today as RFID. Modern aircraft still use transponders to automatically and uniquely identify themselves to ground controllers (the well-known *squawk* function). Many believe that RFID is an extension of the common barcode, and while the two share some application overlap, RFID is much more than a barcode.

RFID Functionality

A typical RFID system is a remarkably simple collection of technology components. It comprises a collection of transponder tags, which form the heart of the system; a reader, which energizes the tags, collects information from them, and delivers the information to back-room analysis applications; and the application set that analyzes the data provided by the transponder tags to the reader. This application set is critically important, because RFID systems, by their very design, generate enormous amounts of data that must be analyzed and acted upon if the system is to have value.

Each transponder, an example of which is shown in Fig. 8.11, has a unique *serial number* (a card ID, or CID) that identifies the tag and

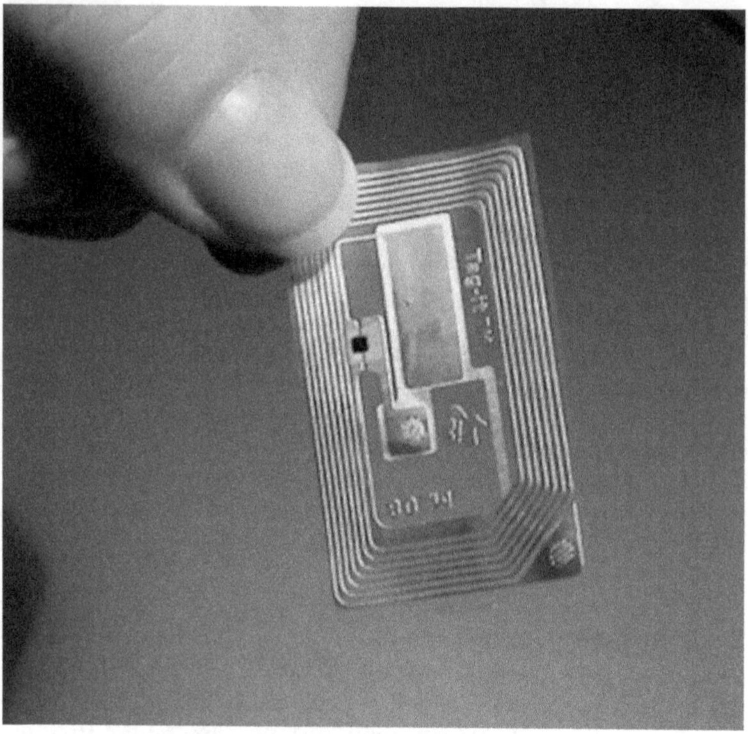

Figure 8.11 A typical RFID transponder, or tag.

therefore whatever it is attached to—a pallet of products, an identification card, or a beef cow in a herd. When the tag is within the operational range of a reader, the reader's magnetic field energizes the tag, causing it to go through a series of functions that culminate in the transmission of whatever piece of data is stored in the tag's memory. This information is programmable and might contain detailed product information, product perishability data, routing information, a sheep's bloodline, and so on. For the most part, the tags are passive, meaning that they have no battery but are in fact powered inductively by the RF signal emitted by the reader. These tags necessarily have a relatively short read range—as much as a foot, no more—but active tags, which do have internal power, can broadcast up to 20 ft under the right conditions. Passive tags are often used in applications where proximity to a reader is assured, such as in a warehouse or transportation-based supply chain environment. Active devices are commonly seen in such applications as automated toll-taking systems on major freeways. The EZ-Pass system deployed in the northeastern United States is a good example.

Readers are nothing more than RF emitters connected to back-end software analysis systems. The reader's role is to emit an RF signal that activates the tag or tags within its operational area; to transmit a series of simple commands to the tags; and to collect the data returned as a result of the transmitted commands. Because of the fact that multiple tags may be activated simultaneously as in a supply chain environment, most readers have anti-collision capability that allows them to control a multitag environment, forcing each tag to take a turn so that simultaneously activated devices can successfully transmit their information without interference from adjacent tags.

Tags come in a variety of forms as shown in Fig. 8.12. The wedge-shaped device (1) is designed to be attached in a number of ways to trackable items. The wedge shape makes it possible to guarantee the orientation of the tag in environments where signal strength is weak or impeded by metal in the transmission area. The paper tag (2) is often seen in bookstores, inserted between the pages of books as a theft deterrent. The tag contains a single bit of information. If the bit is set by the *deactivator* at checkout, the book was properly purchased. If not, it sets off the alarm upon exiting the store.

The card (3) is a typical contactless smart card, used in corporate access control environments, hotel room keyless entries, etc. The large disk (4) can be attached to large pallets or other containers via the convenient hole in the center of the disk, and is reusable. The thin sheet tag (5) is affixed to a sticky backing and can be affixed to the inside of computers, library books, and other small items.

The cylindrical tag (6) is an interesting transponder tag. The one shown is designed for high heat and corrosive environments, often used, for example, in automobile painting lines that are subject to extremes of temperature and caustic substances. The tag can be attached to an automobile body, for example, and as it passes through the line broadcasts the color that is to be applied and the chassis style so that the robotic painters apply the paint correctly. The cylinder

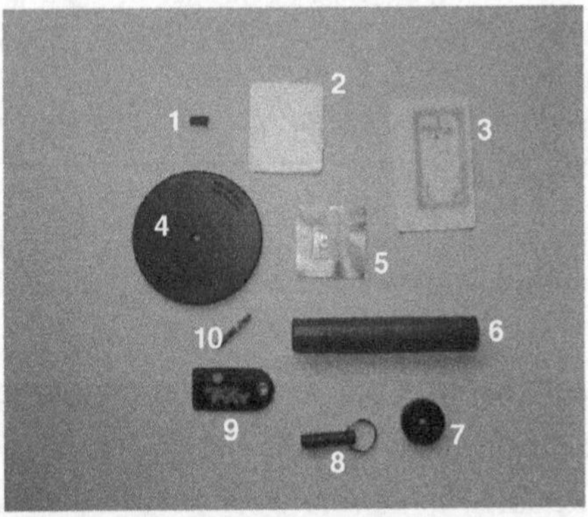

FIGURE 8.12 A collection of RFID tags. As you can see they are as varied as the applications for which they're intended.

style also comes in a ceramic form factor and is designed to be embedded in the stomach of cattle. The tag contains bloodline history, shipping information, and other data that is invaluable to veterinarians tracking disease. And because the device is ceramic, it is impervious to corrosive stomach acids.

The disk (7) is similar to the larger disk described earlier; it can be attached to smaller packages and like its larger cousin is reusable. In fact, these devices often remain attached to their shipping containers.

The key fobs shown in (8) and (9) are RFID tags designed to facilitate the purchase of fuel. The Speedpass (8) is commonly seen, and replaces the need for a credit card.

The glass transponder (10) is intriguing. These devices are designed for subcutaneous use, often injected under the skin of livestock, fish, and other wildlife that veterinarians and wildlife biologists wish to track.

Spectrum Considerations

RFID systems are generally divided into two categories: *passive or near-field systems,* and *active or far-field systems.* Passive systems typically operate at very short distances between the reader and the transponder and usually operate in the 13.56-MHz range of the electromagnetic spectrum. Active systems, which are more powerful and therefore capable of operating at significantly greater distances from the reader, operate in various regions of the spectrum between 800 MHz and 1 GHz, although there are a few other frequencies within which active RFID systems occasionally operate.

Applications for RFID

For the most part, RFID has been used as an extension of the well-known barcode system. It has numerous advantages over barcodes, however. Barcode labels can fall off, be torn or smudged, and must be properly oriented so that the laser reader can see the printed label. RFID devices do not suffer from these limitations: they are not subject to tearing or smudging and for the most part do not require specific orientation—as long as they are within the operational range of the reader, they can be activated and read. As a result, accuracy is increased and corporations see a reduction over time in both operating expense (OPEX) and capital (CAPEX) due to lowered personnel and equipment requirements.

Applications for RFID are wide ranging, and include:

- personnel identification
- access control
- security guard monitoring
- duty evasion

- food production control
- blood analysis identification
- water analysis
- refuse collection identification
- timber grade monitoring
- road construction material identification
- toxic waste monitoring
- vehicle parking monitoring
- valuable objects insurance identification
- gas bottle inventory control
- asset management
- stolen vehicle identification
- production line monitoring
- car body production
- parts identification
- barrel stock control
- machine tool management

This list represents a sample of possible applications; others emerge daily. One of the most intriguing new areas for RFID deployment lies within the realm of homeland security. There are clearly traditional applications for RFID—IFF, access control, package handling and identification, fire control, personnel movement, and production management—under the purview of defense and homeland security that are well understood and fully deployed. A new area that is enjoying a great deal of scrutiny today, however, is *port security*. Large ports are vulnerable environments because of the number of ships that come and go, and the far larger number of containers that enter the port system every hour of every day—over 90 percent of the world's volume of shipped goods. Until recently it has been next-to-impossible to examine every inbound container because of sheer volumes: Long Beach, shown in Fig. 8.13, handles the equivalent of eleven million containers a year, while Singapore and Hong Kong, arguably the largest roll-on, roll-off (RO-RO) ports in the world in terms of container volume, each move approximately 20 million units annually.

Shipboard container doors are closed and locked before the vessel leaves port. The doors are then sealed with anti-tamper protection to ensure that if they are opened in transit, the seal is broken and the fact that the container was opened will be evident to authorities upon inspection at the destination port. The concern, however, is that

FIGURE 8.13 The massive port of Long Beach and Los Angeles.

containers opened in transit may carry weapons or other destructive cargo, and by the time the tampered-with container is detected, the contents are already in port.

A number of firms build RFID-based electronic seals for containers that not only show that the container door has been opened, but also transmit the fact to a shipboard reader that then notifies authorities so that the ship can be intercepted and searched before entering port.

DoD and RFID

The Department of Defense is extremely interested in RFID applications. In a November 2003 summit on the technology, the DoD confirmed its commitment to RFID. The organization has mandated that all suppliers place passive RFID tags on products at the lowest possible level that is cost-effective by January 2005; placement may be at the individual product, pallet, or case level, and will vary somewhat by product type.

The DoD's primary interest in RFID is based on its move toward what it calls *knowledge-enabled logistics.* While the military is a unique *business,* it still relies on effective supply chain management to get the job done. Naturally, RFID lends itself to improved supply chain processes and faster deployment of resources to a forward theater. Frankly, the similarities between military and civilian requirements are far greater than the differences; consider the following quote:

In its ultimate form, the entire theater of operations will be networked. Sensors will reside on every piece of equipment and every person populating the field of operations, and information collected by those sensors will be processed in real time using artificial intelligence support to prioritize threats and challenges. In-charge personnel will be able to choose from a portfolio of response options to identify and select targets.

As a result, the time between sensing, processing, deciding, and acting will fall dramatically, allowing forces to target the opposition before they can respond.

—Retail executive, talking about his firm's
RFID product-tracking initiative

Network Impacts

Today, RFID's widespread implementation is somewhat limited by the cost of the tags—roughly 50 cents each, which means that no one is going to put them on a 30-cent toothbrush. However, it is widely believed that within a year or so the price per tag will drop to the sub-cent level, which means that they will be widespread. How widespread? According to ABI Research, RFID will be a $70 billion market by 2015. Consider, now, the impact on the network. Each tag can hold as many as 50 bytes of data. If each tag is activated by a reader once a day (although most will be activated more often that), and a significant percentage of that traffic finds its way into the network ...well, you do the math.

Final Thoughts

RFID currently falls into that wonderful Mark Twain-like space that is characterized by the quote, "The only thing worse than people talking about you is *nobody* talking about you." RFID is certainly getting its share of negative attention, largely due to concerns voiced by privacy advocates. Organizations like CASPIAN (the Committee against Privacy Invasion and Numbering) have risen up to fight the technology's widespread deployment because of concerns about the ability of various agencies to track an individual's movements and purchases without authorization. And while it is easy to assign these concerns with conspiracy theories, organizations like CASPIAN serve the same purpose in the technology world that Greenpeace serves in the oil industry: they force the industry to be at the top of its game by creating public awareness of potential hazards, real or not. RFID does need to be monitored, and while concerns over its abuse should be heard, the advantages that RFID brings in such applications as security, law enforcement, defense, health care, product manufacturing, veterinary medicine, food and water protection, and supply chain management far outweigh the risks associated with the potential for misuse of the technology.

RFID: What's Next?

So what's coming among the wireless ranks? Within the next 18 months expect significant segment buy-down and consolidation as the players close competitive ranks. I would expect to see a renewed focus on the applications that wireless technology can most effectively deliver to mobile users, particularly voice, e-mail, instant messaging, and perhaps some set of value-added services: games, downloadable ring-tones, photo sharing, multipoint IM communication, and the like. I would also expect to see the emergence of location-based services that rely on GPS technology to enhance intelligent purchasing applications and guarantee carrier-grade requirements such as E911. RFID will undoubtedly play a growing role as well; already it is showing up in retail, health care, and passport covers.

Satellite Technology

In October 1945, Arthur C. Clarke published a paper in *Wireless World* entitled, "Extra-Terrestrial-Relays: Can Rocket Stations Give World-Wide Radio Coverage?" In his paper Clarke proposed the concept of an orbiting platform that would serve as a relay facility for radio signals sent to it that could be turned around and retransmitted back to Earth with far greater coverage than was achievable through terrestrial transmission techniques. His platform would orbit at an altitude of 42,000 km (25,200 miles) above the equator where it would orbit at a speed identical to the rotation speed of the earth. As a consequence, the satellite would appear to be stationary to Earth-bound users.

Satellite technology may prove to be the primary communications gateway for regions of the world that do not yet have a terrestrial wired infrastructure, particularly given the fact that they are now capable of delivering broadband services. In addition to the United States, the largest markets for satellite coverage are Latin America and Asia, particularly Brazil and China.

Geosynchronous Satellites

Clarke's concept of a stationary platform in space forms the basis for today's geostationary or geosynchronous satellites. Ringing the equator like a string of pearls, these devices provide a variety of services including 64 Kbps voice, broadcast television, video-on-demand services, broadcast and interactive data, and point-of-sale applications, to name a few. And while satellites are viewed as technological marvels, the real magic lies more with what it takes to harden them for the environment in which they must operate and what it takes to get them there than it does their actual operational responsibilities. Satellites are, in effect, nothing more than a sophisticated collection of assignable, on-demand repeaters—in a sense, the world's longest local loop.

From a broadcast perspective, satellite technology has a number of advantages. First, its one-to-many capabilities are unequaled. Information from a central point can be transmitted to a satellite in geostationary orbit; the satellite can then rebroadcast the signal back to earth, covering an enormous service footprint.

Because the satellites appear to be stationary, the earth stations actually *can* be. One of the most common implementations of geosynchronous technology is seen in the very small aperture terminal (VSAT) dishes that have sprung up like mushrooms on a summer lawn. These dishes are used to provide both broadcast and interactive applications; the small DBS dishes used to receive TV signals are examples of broadcast applications, while the dishes seen on the roofs of large retail establishments, automobile dealerships and convenience stores are typically (although not always) used for interactive applications such as credit card verification, inventory queries, e-mail, and other corporate communications. Some of these applications use a satellite downlink, but rely on a telco return for the upstream traffic, that is, they must make a telephone call over a landline to offer two-way service.

One disadvantage of geosynchronous satellites has to do with their orbital altitude. On the one hand, because they are so high their service footprint is extremely large. On the other hand, because of the distance from the earth to the satellite, the typical transit time for the signal to propagate from the ground to the satellite (or back) is about half a second, which is a significant propagation delay for many services. Should an error occur in the transmission stream during transmission, the need to detect the error, ask for a retransmission, and wait for the second copy to arrive could be catastrophic for delay-sensitive services like voice and video. Consequently, many of these systems rely on forward error correction transmission techniques that allow the receiver to not only detect the error but correct it as well.

An interesting observation is that because the satellites orbit above the equator, dishes in the northern hemisphere always face south. The farther north a user's receiver dish is located, the lower it has to be oriented. Where I live in Vermont the satellite dishes are practically lying on the ground—they almost look as they are receiving signals from the depths of the mountains instead of a satellite orbiting 23,000 miles above the earth.

Low/Medium Earth Orbit Satellites

In addition to the geosynchronous satellite arrays, there are a variety of lower-orbit constellations deployed known as low and medium Earth orbit satellites (LEO/MEO). Unlike the geosynchronous satellites, these orbit at lower altitudes—400 to 600 miles, far lower than the 23,000-mile altitude of the typical GEO bird. As a result of their lower altitude, the transit delay between an Earth station and a LEO satellite is virtually nonexistent. However, another problem exists with low Earth orbit technology. Because the satellites orbit pole-to-pole, they do not appear to be stationary, which means that if they are to provide uninterrupted service they must be able to hand off a transmission from one satellite to another before the first bird disappears below the horizon. This has resulted in the development of sophisticated satellite-based technology that emulates the functionality of a cellular telephone network. The difference is that in this case, the user does not appear to move; the cell does!

Iridium

Perhaps the best-known example of LEOS technology is Motorola's somewhat ill-fated Iridium deployment. Comprising 66 satellites[2] in a polar array, Iridium was designed to provide voice service to any user, anywhere on the face of the Earth. The satellites would provide global coverage and would hand off calls from one to another as the need arose. Unfortunately, Iridium's marketing strategy was flawed; their prices were high, their phones large and cumbersome (one newspaper article referred to them as "manly phones"), and their market significantly overestimated. Additionally, their system was only capable of supporting 64-Kbps voice services, a puny bandwidth allocation in these days of customers with broadband desires. The company failed but was pulled from the ashes; today it is used extensively by the Department of Defense, but its hope for widespread deployment probably won't materialize.

Globalstar

Others have been successful, however. Globalstar's 48-satellite array offers voice, short messaging, roaming, global positioning, fax, and data transport up to 9600 bps. And while the data rates are miniscule by comparison to other services, the converged collection of services they provide is attractive to customers who wish to reduce the number of devices they must carry with them in order to stay connected.

[2]The system was named "Iridium" because in the original design, the system was to require 77 satellites, and 77 is the atomic number of that element. Shortly after naming it Iridium, however, the technologists in the company determined that they would only need 66 birds. They did not rename the system "Dysprosium."

ORBCOMM

ORBCOMM is an interesting company. Its satellites are nothing more than extraterrestrial routers which interconnect vehicles and earth stations to facilitate deployment of such packet-based applications as two-way short messaging, e-mail, machine-to-machine communications, and vehicle tracking. Their constellation has a total of 29 satellites that orbit at an altitude of 465 miles.

Satellite Services: What's the Catch?

As a service provisioning technology, satellites may seem so far out (no pun intended) that they may not appear to pose a threat to more traditional telecommunications solutions. At one time, that is, before the advent of low-Earth orbit technology, this was largely true. Geosynchronous satellites were extremely expensive, offered low bit rates, and suffered from serious latency that was unacceptable for many applications.

Today, this is no longer true. GEO satellites offer high-quality two-way transmission for certain applications. LEO technology has advanced to the point that it now offers low-latency, two-way communications at broadband speeds, is relatively inexpensive, and as a consequence poses a clear threat to terrestrial services. On the other hand, the best way to eliminate an enemy is to make the enemy a friend. Many traditional service providers have entered into alliances with satellite providers; consider the agreements between DirecTV and AT&T. By joining forces with satellite providers, service providers create a market block that will help them stave off the short-term incursion of cable. Between the minimal infrastructure required to receive satellite signals and the ubiquitous deployment of DSL over twisted pair, incumbent local telephone companies and their alliance partners are in a reasonably good position to counter the efforts of cable providers wishing to enter the local services marketplace. In the long term, however, wireless will win the access game.

Other Wireless Access Solutions

A number of wireless technologies have emerged in the last few years that are worth mentioning including WiMAX, Ultra-Wideband, Zigbee, and Bluetooth.

Worldwide Interoperability for Microwave Access

In April 2002, the IEEE published their 802.16 standard for broadband wireless access (BWA), also known as worldwide interoperability for microwave access (WiMAX). 802.16 specifies the details of the air interfaces for wireless metropolitan area networks (MANs). And while there are some similarities between WiFi and WiMAX, in other respects they could not be more different. First of all, WiMAX was not created as a *goat trail* technology. Instead, developers first quietly created standards which were socialized through other standards bodies such as the ITU-TSS. As a result of this strategy, spectrum was allocated globally for 802.16 implementation through a 2-year, open-consensus procedure that involved hundreds of engineers from major operators and vendors around the world. Consequently, 802.16, while still a nascent technology, enjoys global acceptance and what will be a relatively trivial implementation phase once it becomes more widely deployed. Furthermore, the capabilities of the standard are impressive. Whereas WiFi offers megabits of nominal bandwidth over service distances of 300 ft, WiMAX offers 100 Mbps over a service radius of several miles. And because it is orthogonal, it does not require line of sight for connectivity.

802.16 is initially targeted at the "first mile" challenge for metropolitan area networks. It operates between 10 and 66 GHz (the 2- to 11-GHz spectrum with point-to multipoint and optional mesh topologies), and defines a medium access control (MAC) layer that supports multiple physical layer specifications specific for each frequency band. The 10 to 66 GHz standard supports two-way transmission options at a variety of frequencies including 10.5 GHz, 25 GHz, 26 GHz, 31 GHz, 38 GHz, and 39 GHz. It also supports device interoperability so that carriers can use multiple vendors' products. Furthermore, the standard for the 2- to 11-GHz spectrum supports both unlicensed and licensed bands, a real boon for the entrepreneurial set that does so much to push the limits of any new innovation.

Over the last few years 802.16 has undergone a series of modifications, resulting in the existence of various flavors of the original standard including 802.16a and 802.16e. They are discussed in the following sections.

802.16a

IEEE 802.16a was ratified as an extension to the original 802.16 standard in January 2003. It enhances 802.16 and addresses radio systems that operate in the 2- to 11-GHz frequency ranges. It addresses the requirements of both licensed and unlicensed implementations, and supports point-to-multipoint networks as well as mesh topologies within the unlicensed region.

802.16e

IEEE 802.16e is an extension of 802.16a. Its potential impact is enormous in that it adds mobility to 802.16a systems. Given that the ultimate goal of 802.16 is to provide a technology solution that will bridge the gap between fixed and mobile wireless systems, the addition of mobility could support the end-to-end needs of a subscriber in both environments. Consider, however, what the addition of mobility means. I often hear people say that "WiMAX represents a real threat to WiFi!" Given that spectrum has already been allocated around the globe, mobile WiMAX represents more than a potential threat to WiFi—it represents a real threat to 3G! Consider the tens of billions of dollars that have been spent in the last few years on 3G spectrum, very little of which has actually been deployed. Then along comes WiMAX with its promise of wireless broadband connectivity with global reach, and suddenly 3G begins to appear a bit less attractive.

802.16e allows Wireless ISPs (WISPs) to enter and take over a market with minimal investment in infrastructure, then offer a complete package of services to subscribers. Wireless broadband, in the form of WiMAX, could also compete favorably with such options as cable modem and DSL (in fact, some analysts refer to WiMAX as wireless DSL).

Several industry players are leading WiMAX's implementation. The first of these is Intel, which is making heavy investments into WiMAX as part of a strategy to take the lead in WiMAX the same way they did in WiFi with Centrino. Their research shows that many people use their PDAs, broadband-equipped mobile phones, and laptops to access data networks while mobile, a phenomenon that is causing a significant number of communities to build metro-based broadband access areas to serve them.

As a founding member of the industry-led, non-profit WiMAX Forum, Intel led the charge to promote compatibility and interoperability among certified broadband wireless products. The forum's member companies support the industry-wide acceptance of 802.16 as well as the European ETSI HiperMAN wireless LAN standards. HiperMAN interworks with 802.16 and supports ATM as a high-speed switching fabric, but its primary focus is IP. Service characteristics are shown in Table 8.3.

And what of the continuing perspective that WiMAX competes with WiFi? Most informed players argue that WiMAX is really a complementary technology to WiFi, particularly in the metro arena. They see it as a broadband wireless alternative to cable or DSL, particularly in rural or greenfield areas.

The Future of WiMAX

According to the WiMAX Forum there were nearly 600 WiMAX networks deployed in about 150 countries by October 2010, offering service to more than 600 million subscribers. A year later the number had grown to 823 million people; today it is believed to be over a billion. WiMAX is a good technology, and it filled a specific niche for some time. Today, however, most carriers are moving to LTE instead.

Characteristic	802.16	802.16a	802.16e
Standard date	December 2001	January 2003	EOY 2004
Frequency range	10–66 GHz	< 11 GHz	< 6 GHz
Transmission limits	Line of sight	Non-line of sight	Non-line of sight
Bandwidth	32–134 Mbps in 28 MHz channels	Up to 75 Mbps in 20-MHz channels	Up to 15 Mbps in 5-MHz channels
Modulation scheme	QPSK, 16 QAM, 64 QAM	OFDM (256 subcarriers), QPSK, 16 QAM, 64 QAM	OFDM (256 subcarriers), QPSK, 16 QAM, 64 QAM
Mobility options	Fixed	Fixed and "portable"	Fixed and mobile
Channel bandwidth	20, 25, 28 MHz	Scalable from 1.5–20 MHz	Scalable from 1.5–20 MHz
Operating radius	2–5 Km	7–10 Km	2–5 Km

Table 8.3 WiMAX Service Characteristics

ZigBee

ZigBee, also known as IEEE 802.15.4, offers a cost-effective, standards-based wireless solution that specifically supports low data rates, low power consumption, security, and reliability. The name derives from a biological principal known as the *ZigBee principle,* seen in honeybee hives. The colonial honeybee lives with a queen, a few male drones, and thousands of worker bees. The survival of the colony depends on a process of continuous communication of vital information among all members of the hive. The technique that honeybees use to communicate with each other is called the *ZigBee principle,* which defines the zigzag dance pattern that the insects use to communicate critical information to each other. 802.15.4 was nicknamed ZigBee because it facilitates the ability of humans to emulate this survival behavior.

ZigBee is designed to support such applications as home and environmental controls, including lighting, automatic meter reading, telemetry for smoke and carbon monoxide detectors, HVAC, heating, security, drapery and shade controls, set top boxes, and specialized applications such as medical sensing and monitoring. To date, ZigBee is the only standard that supports the unique requirements of remote monitoring and control networks as well as sensory network applications. It is designed to support the broad deployment of

Characteristic	ZigBee	WiFi	Bluetooth
Application	Monitoring, control	E-mail, Web access, some video	Cable replacement
Memory requirements	4–32 KB	1 MB	256 KB
Battery life (days)	100–1,000	0.5–5	1–7
Network nodes	Unlimited	32	7
Bandwidth (Kbps)	20–250	11–54	720
Range (meters)	1–100	1–100	1–10

TABLE 8.4 ZigBee Service Characteristics

wireless networks that have low cost, low power requirements. In fact, ZigBee is designed to run for years on a single battery.

ZigBee products rely on the IEEE 802.15.4 physical radio standard, which operates globally in unlicensed bands at 2.4 GHz (global), 915 MHz (Americas) and 868 MHz (Europe). ZigBee supports data rates of 250 Kbps at 2.4 GHz (using 16 channels), 40 Kbs at 915 MHz (using 10 channels) and 20 Kbs at 868 MHz (using a single channel). Transmission distances range from 10 to 100 m, depending on power and environmental considerations.

In mid-December 2004 the ZigBee standard was ratified. Companies such as Figure 8 Wireless, Freescale, CompXs, Eaton, and Atmel, long involved in the development of the technology, have had ongoing involvement with the ZigBee alliance during the specification-writing phase. These companies already have ZigBee-based products.

The Table 8.4 compares ZigBee to the most common alternate technologies.

One Last Topic: Machine-to-Machine Communications

A few years ago, I was walking down the sidewalk near Shibuya Crossing in Tokyo, on my way to Electric Town—the nexus point for all of the major electronics stores in the city. This is a dangerous place—not because of any physical threat but because of the plethora of very cool things that the stores sell there. I can't tell you how many times I've walked into a store in Electronic Town, picked something off the shelf, and said to myself, "I don't even know what this does, but I need it."

As I walked down the sidewalk, trying not to look in the store windows, I noticed a large electronic billboard on the sidewalk in front of me. About 8-ft wide, it looked like Fig. 8.14.

I was gobsmacked. But here's what you need to know to really understand the implications of this scenario. As I mentioned earlier in the book, I am a dedicated Nikon photographer. The photograph

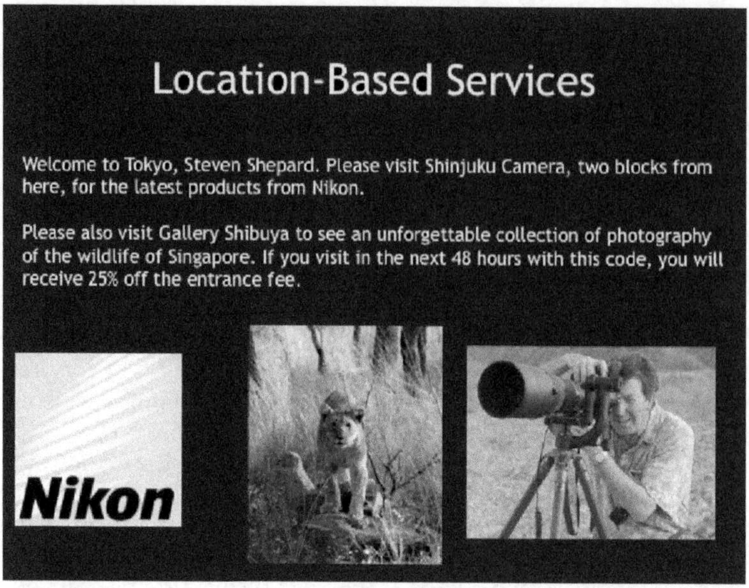

Location-Based Services
Welcome to Tokyo, Steven Shepard. Please visit Shinjuku Camera, two blocks from here, for the latest products from Nikon.
Please also visit Gallery Shibuya to see an unforgettable collection of photography of the wildlife of Singapore. If you visit in the next 48 hours with this code, you will receive 25% off the entrance fee.
Nikon

FIGURE 8.14 This sign appeared before me in Tokyo. The power of data and location!

of me with my 500-mm lens was taken a year earlier while I was in Namibia, shooting for a wildlife magazine. The lion cub was taken in 2009 in South Africa. And yes, I did visit Shinjuku Camera, three blocks away, where I had to buy a second camera bag to contain all of the things I bought while I was there. Luckily I had a 20-hour flight ahead of me so that I could come up with an explanation for my wife as to why I was coming home with a second camera bag.

So how did they do this?

I've never known for sure, but here's what I *think* went on. My phone, like most phones today, has Wi-Fi. I *suspect* that I walked through an open hotspot, and somehow in the process of doing that my e-mail address or my mobile number were captured. A quick Google search of either one will easily lead to www.ShepardImages.com, my photography Web site. Among other things the site has a page called "Gear" which is a thinly-veiled paean to Nikon, a collection of galleries, from which the lion was captured, and an "About the Photographer" page where the shot of me with Big Bertha is found. It's also clear from the galleries that I tend to focus my camera most often on the natural world, which explains the discounted invitation to visit Gallery Shibuya.

Now here's my question to you, my reader. Is this the coolest thing you've ever heard of, or is this kind of … creepy? If your answer is, "It depends," then you find yourself among the majority of people to whom I've presented this scenario. Let's face it: If instead of photography I was intensely interested in farm animals (and I don't mean in the agricultural sense), I probably wouldn't want that displayed on a billboard in Tokyo for all to see. For me it was a powerful demonstration of a relatively new technology family called *machine-to-machine communications* (M2M). It characterizes an environment where, based on shared personal information, the environment adapts to the user, not the other way around. Creepy? Perhaps. Powerful? Unquestionably.

The Internet of ...

This concept, which is closely related to RFID, has been around for quite some time. The Auto-ID Center at MIT calls it "the Internet of things," as does Cisco; Deloitte refers to it as "the Internet of Everything." In essence it is a technological ecosystem that connects devices, storage, analysis tools, and user, not to mention the power of mobility, managed services, and the cloud.

M2M comes in a variety of flavors that include person-to-person, a good example of which is the PSTN; device-to-device, which manifests as telemetry; data-to-person, which first showed up in the early Internet; and device-to-person, which is where we are today.

The technology is multifaceted and includes the Internet as the transport fabric; a collection of sensors which are embedded on people and objects; data collection and telemetry devices; access schemes, such as ZigBee, WiFi, and cellular; and a host of sophisticated analysis tools that generate insight and context.

M2M is enjoying widespread success as it creates new business models and competitive structures, improves supply chain and organizational processes, helps to reduce enterprise risk, and optimizes CAPEX and OPEX spend-through predictive analysis. Ultimately, through a collection of sensors, actuating events, analysis processes, and connectivity the physical world becomes a very sophisticated information system.

Machine-to-Machine Applications

Machine-to-machine applications fall into five general categories:

- Behavior tracking and response
- Situation awareness
- Sensor-driven analytics
- Process optimization
- Resource consumption optimization

Behavior Tracking and Response

This category includes such actions as environmental awareness and response, such as security system monitoring and event interpretation; enhanced decision-making and analysis, based on complex and seemingly unrelated inputs; control and process automation; optimized resource management, such as automated, environmentally-dependent cooling and lighting control to minimize energy consumption; real-time autonomous system response, such as automated shutdown of chemical plant processes or automated trading; location-based services, such as what I experienced in Tokyo; and behavior-driven supply chain enablement, an example of which is Amazon's decision to build more warehouses around the world for same-day order fulfillment, based on shoppers' tendencies to examine the product in the retail store but buy it online.

Situation Awareness

Situation Awareness speaks to the environment's ability to collect environmental data from a wide array of sensors (including people) as a way to make decisions about complex events such as weather, traffic congestion, or the spread of disease. I saw this firsthand while working on a healthcare project in Europe recently. This hospital happened to be equipped with state-of-the-art patient-monitoring equipment which included sensors attached to each patient and connected via a low-power wireless technology such as Zigbee to a centralized data collection facility. More important was the fact that the data collected from each patient was being analyzed in real-time. Here's how the scenario played out. The post-surgical patient in the room that was at the far right-hand end of the hallway spiked a mild fever, which was logged by the central monitoring system and which triggered a nurse alert to keep an eye on the patient's vitals. An hour later, the patient in the adjacent room also spiked a fever, and 45 minutes later the patient across the hall did as well. The centralized analysis detected that this was an anomaly and sent a major alert to the nursing station on that floor. They quickly determined that this was a methicillin-resistant staphylococcus aureus infection, the bane of hospitals everywhere because it tends to spread very rapidly, is difficult to treat, and is often spread by unwitting healthcare providers who are infected and move the infection from room-to-room. In this case they were able to stop the spread after the third room and quickly identify who the disease vector was. Think of the implications of this!

Situation awareness might also include crime prediction (think Tom Cruise in *Minority Report*); anticipation of financial discrepancies; and more accurate weather prediction. Many mobile weather apps today give their users the ability to post weather information. This isn't done because the developers want to be nice; they do it because by crowdsourcing information from a vast number of sources (in this case the people are the sensors) they develop far more accurate patterns.

Sensor-Driven Analytics

If you are a fan of the late Michael Crichton's novels, and you read his book *Swarm,* then you are familiar with the concept of Smart Dust. Smart Dust is a very real technology, a collection of hundreds or thousands of micro-electro-mechanical systems (MEMS) that are deployed in *swarms* and that communicate wirelessly with a centralized computer. Using RFID technology, they collect and share information about the environment in which they are deployed. The idea came about in 1992 thanks to work performed at the Rand Corporation and DARPA, where the original idea for the Internet was created. In one example, these tiny sensors might be dispersed above a battlefield from low-flying aircraft. They stick to the clothing of personnel in the field and to equipment, and as these battlefield assets move around they create an electronic movement signature that can be analyzed and responded to. Less dramatically, these *smart motes* could be affixed to products in a supply chain and tracked from source to destination.

Another example is consumer purchase behavior. Retail stores today are employing M2M communications as a way to analyze customer movement inside the retail store and provide them with a superior shopping experience, including real-time delivery of discount coupons based on the customer's physical location within the store. Another example is the idea of the smart mirror, which is now being put to use by at least one major retailer in the UK. A shopper goes into a changing room and tries on an outfit. They come out of the changing room and stand in front of the triptych mirror that gives them a near 360-degree view of the outfit. Unbeknownst to the customer, they are being photographed and filmed. An anatomically accurate wireframe of their body is generated, and thanks to the affinity band they wear (think Livestrong) or the information they give during checkout, the next time they log into the retailer's Web site to browse products, the clothing they look at is not worn by a model—it is worn by them.

Process Optimization

Anytime we are talking about complex industrial processes such as assembly lines, chemical plants, refineries, pharmaceutical manufacturing, power plant monitoring, automated trading, and supply chain management, we have a glowing opportunity to take advantage of M2M as a way to monitor and respond to event horizons that change far faster than a human could ever respond to. When things start to happen in a chemical plant that are not supposed to happen, for example, when a chemical reaction goes out of control as they are prone to do—time is of the essence. Placing sensor arrays at critical junctures that are connected to an automated analytics system can mean the difference between a routine plant shutdown and a major (and potentially unpleasant) event.

Resource Consumption Optimization

One place where M2M has a real opportunity to shine is in the area of resource management. Smart Grids and the automated home are good examples. If the power company has the ability to collect utilization data in real-time from thousands of sensors placed on the meters of homes and businesses, they can better anticipate usage spikes and turn them into non-events through intelligent resource management and redirection. By the same token, smart appliances can reduce their power consumption during low usage periods; for example, a refrigerator or freezer might turn itself off during the wee hours of the night when the chances of someone opening the door are nil.

Concerns and Issues

The big issues surrounding M2M communications are primarily concerned with data privacy, confidentiality, and security of personal information. These are valid concerns, although the benefits of M2M far exceed the potential risks. Another concern that has been voiced is one of process control: What happens when autonomous systems misbehave? As far as I'm concerned the people who voice these concerns have been watching too many Arnold Schwarzenegger movies and are worried that SkyNet might be real. However, we've seen what happens when automated trading systems "go rogue;" again, a degree of concern is merited—but only a degree. The idea that M2M is a new way for unmentionable agencies to get access to personal information is a bit off-base. No one complains when Amazon makes book recommendations based on your reading history, and anyone who has ever purchased anything with a credit card who believes that their purchase history is a secret needs a new drug. Regulation must be put into place to protect the public against intrusion, but those regulations must also recognize the extraordinary value that this technology brings to the world and must not get in the way.

The real issue is a far more practical one. These devices—the plethora of sensors that are being deployed all over the world—generate massive volumes of information, volumes the likes of which we have never seen before. Think about it: We're talking about a sensor array that includes both passive and active RFID tags, temperature monitors, medical devices, people, mobile phones, laptops, tablets, and many many more. The data that these devices generate is incalculably massive and must be housed somewhere if it is to have analytical value. So where will we put it? How will we analyze it, given that the volume far exceeds the ability of traditional analytical tools to handle? And most important, what do we do with the results of the analytics?

This is the domain of a new field called Big Data, which will be discussed in detail in Chap. 11. In my opinion it is the single biggest opportunity that exists for the enterprise today.

Chapter Summary

Access technologies, used to connect the customer to the network, come in a variety of forms and offer a broad variety of connectivity options and bandwidth levels. The key to success is to *not* be a bottleneck; access technologies that can evolve to meet the growing customer demands for bandwidth will be the winners in the game. DSL holds an advantage as long as it can overcome the availability challenge and the technology challenge of loop carrier restrictions. Wireless is hobbled by licensing and spectrum availability, both of which are regulatory and legal in nature rather than technology limitations.

In the next chapter, we will discuss transport technologies, including private line, frame relay, ATM, and optical networking.

Chapter Eight Questions

1. Is wireless an access or a transport technology?
2. Why is ISDN not a more popular technology in the United States?
3. Explain the differences between a BRI and a PRI.
4. "DSL is actually an analog technology." Explain.
5. Explain the purpose of a DSLAM.
6. What are the differences between FDMA, TDMA, and CDMA?
7. "RFID deployment represents an assault on privacy." Write a response that refutes this contention.
8. Why are satellites falling out of favor as an access solution?

CHAPTER 9

Transport Technologies

On the road again…
Just can't wait to get on the road again…

—Willie Nelson

We have now discussed the premises environment and a variety of wired and wireless access technologies. Remember, access refers to the collection of technologies that give the user (whether human or mechanized) the ability to access the transport domain of the network and gain access in turn to the content they seek. The next area we're going to explore, then, is transport.

Because businesses are rarely housed in a single building, and because their customers are typically scattered across a broad geographical area (particularly multinational customers), there is a growing need for high-speed, reliable wide-area transport. Wide-area can take on a variety of meanings. For example, a company with multiple offices scattered across the metropolitan expanse of a large city requires interoffice connectivity to do business properly. On the other hand, a large multinational with offices and clients in Madrid, San Francisco, Hamburg, and Singapore requires connectivity to ensure that the offices can exchange information on a 24-hour basis.

These requirements are satisfied through the proper deployment of wide-area transport technologies. These can be as simple as a dedicated private line circuit or as complex as a virtual installation that relies on asynchronous transfer mode (ATM) or Internet protocol (IP) for high-quality transport.

Dedicated facilities are excellent solutions because they are—well, dedicated. They provide fixed bandwidth that never varies and guarantees the quality of the transmission service. Because they are dedicated, however, they suffer from two disadvantages: First, they are expensive and only cost-effective when highly utilized. The pricing model for dedicated circuits includes two components: the mileage of the circuit and the bandwidth. The longer the circuit, and the faster it is, the more it costs. Second, because they are not switched and are often not redundant because of cost, dedicated facilities pose the potential threat of a prolonged service outage should they fail. Nevertheless, dedicated circuits are popular for certain applications and widely deployed. They include such solutions as T-1, which offers 1.544 Mbps (megabits per second) of bandwidth; DS-3, which offers 44.736 Mbps of bandwidth; and SONET or SDH, which offers a wide range of bandwidth from 51.84 Mbps to as much as 40 Gbps.

The alternative to a dedicated facility is a switched service such as frame relay, ATM, or in some cases gigabit ethernet—not to mention the various services that IP and multiprotocol label switching (MPLS) offer, which we'll discuss in the next chapter. These technologies provide *virtual circuits:* Instead of dedicating physical facilities, they dedicate logical timeslots to each customer who then share access to physical network resources. In the case of frame relay, the service can provide bandwidth as high as DS-3, thus providing a replacement technology for lower-speed dedicated circuits. ATM, on the other hand, operates hand-in-glove with SONET, and gigabit ethernet, when deployed in a switched configuration, offers speeds of 1 Gbps or higher. Finally, optical networking is carving out a large niche for itself as a bandwidth-rich solution with good quality of service (QoS).

We begin our discussion with dedicated private line, otherwise known as point-to-point.

Point-to-Point Technologies

Point-to-point technologies do exactly what the name implies: They connect one point directly with another. For example, it is common for two buildings in a downtown area to be connected by a point-to-point microwave or infrared circuit because the cost of establishing it is far lower than the cost to put in physical facilities through conduit in a crowded city—or worse, to get a permit to dig up streets to put in *new* conduit. Many businesses rely on dedicated, point-to-point optical facilities to interconnect locations, especially businesses that require dedicated bandwidth for high-speed applications. Of course, point-to-point does not necessarily imply high-bandwidth; many locations use 1.544 Mbps T-1 or 2.048 Mbps E-1 facilities for interconnection, and some rely on lower-speed circuits where higher bandwidth is not required. And increasingly, lower cost solutions are being used such as gigabit ethernet.

Dedicated facilities provide bandwidth from as low as 2,400 bps to as high as multiple gigabits per second. 2400 bps analog facilities are not commonly seen but are often used for alarm circuits and telemetry, while circuits operating at 4,800 and 9,600 bps are still used to access interactive, host-based data applications.

Higher-speed facilities are typically digital and are often channelized by dedicated multiplexers and shared among a collection of

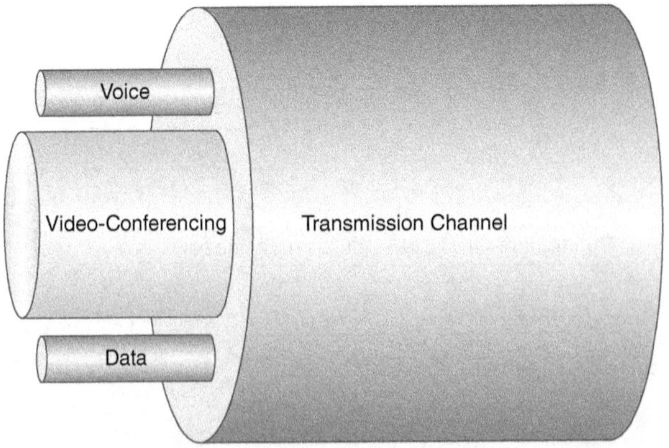

FIGURE 9.1 Channel distribution in a PBX environment.

users or by a variety of applications. For example, a high-bandwidth facility that interconnects two corporate locations might be dynamically subdivided into various-sized channels for use by a PBX for voice, a videoconferencing system, and data traffic, as shown in Fig. 9.1.

Dedicated facilities have the advantage of always being available to the subscriber. They also have the disadvantage of accumulating cost whether they are being used or not. For the longest time, dedicated circuits represented the only solution that provided guaranteed bandwidth—switched solutions simply weren't designed for the heavy service requirements of graphical and data-intensive traffic. Over time, however, that has changed. A number of switched solutions have emerged in the last few years that provide guaranteed bandwidth and only accumulate charges when they are being used (although some of them offer very reasonable fixed rate service). The two most common of these are frame relay and ATM. We begin with frame relay, although frame relay has fallen out of favor in the last few years as IP and MPLS have grown in popularity. Before we do, however, let's spend a few minutes discussing the hierarchy of switching. Part of this is review of prior material; most of it, though, is preparation for our discussion of so-called *fast packet* switching technologies.

The Switching Hierarchy

The switching hierarchy, shown in Fig. 9.2, has two major subheadings—circuit switching and store-and-forward switching. Circuit switching is something of an evolutionary dead end in that it lacks a future evolutionary path, other than to converge with packet switching.

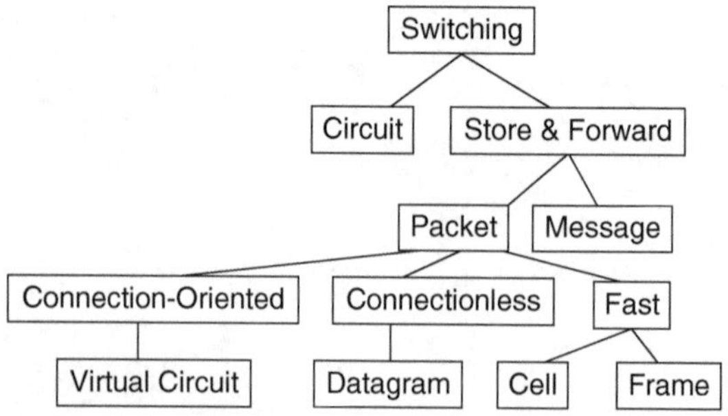

FIGURE 9.2 The switching hierarchy.

Circuit switching is the technique used in the traditional, legacy telephone network. A caller enters a destination address (the called telephone number) and the network establishes a temporary dedicated end-to-end path for the duration of the call. As we noted earlier, this technique works extremely well for voice and dial-up data calls but isn't particularly efficient—or dynamically survivable.

Packet switching evolved as one of the two descendants of store-and-forward technology. Message switching, the alternative and another evolutionary dead-end, is inefficient and not suited to the bursty nature of most data services today. Packet switching continues to hold sway and because of its many forms is a valid solution for most data applications.

In message switched environments, the entire message is transmitted *in toto* from end-to-end. As long as there are no transmission glitches, message switching works well. If, however, there is a problem during the transmission and the payload (the message) becomes corrupted, the entire message must be resent. Again, inefficiency reigns.

The difference between message and packet switching can best be described with the encyclopedia example. For those of you old enough to remember the bookshelf in your parent's home filled with the *World Book* or *Encyclopedia Britannica,* you know that it was an impressive array of 25 large, leather bound books (Preface, A to V, WXYZ, and an Index). If those books were instead published as a single gigantic book and we decided to transport it across a network, and we encountered a transmission problem, we'd have to retransmit the entire encyclopedia. If, however, we broke it into 25 books (packets) and one of the books encountered an error along the way, we'd only have to retransmit that one book. Make sense? That's the difference between message and packet switching. Of course, at the far end we have to have a mechanism to ensure that we have received all 25 books and that they are in order before putting them on the bookshelf, but that's why we have the transport layer. And you thought that OSI protocol stuff was silly.

Many Forms of Packet Switching

Packet switching has three major forms, two of which we discussed in earlier chapters. Connection-oriented packet switching manifests itself as virtual circuit service that offers the appearance and behavior of a dedicated circuit when in fact it is a switched service. It creates a path through the network that all packets from the same source (and going to the same destination) follow, the result of which is constant transmission latency for all packets in the flow. This is important: this technology doesn't eliminate delay; it simply ensures that all packets are delayed equally. Both frame relay and ATM, discussed later in this chapter, are virtual circuit technologies.

Connectionless packet switching does not establish a virtual dedicated path through the network. Instead, it hands the packets to the ingress switch and allows the switch to determine optimum routing on a packet-by-packet basis. This is extremely efficient from a network point-of-view but is less favorable for the client in that every packet is treated slightly differently, the result of which can be unpredictable delay and out-of-order delivery.

The third form of packet switching is called fast packet, two forms of which are frame relay and ATM. The technique is called *fast packet* because the processing efficiency is far greater than traditional packet switching because of reduced overhead.

Fast packet technologies are characterized by low error rates, significantly lower processing overhead, high transport speed, minimal delay, and relatively low cost. The switches accomplish this by making several assumptions about the network. First, they assume (correctly) that the network is digital and based largely on a fiber infrastructure, the result of which is a very low error rate. Second, they assume that unlike their predecessors, the end-user devices are intelligent, and therefore have the ability to detect and correct errors on an end-to-end basis (at a higher protocol layer) rather than stopping every packet at every node to check it for errors and correct them. These switches still check for errors, but if they find errored packets they simply discard them, knowing that the end devices will realize that there is a problem and take corrective measures on an end-to-end basis. Think back to the protocol chapter for a moment: This is the difference between layer-two error detection and layer-four error detection.

Frame Relay

Frame relay came about as a private line replacement technology and was originally intended as a data-only service. Today, it carries not only data but voice and video as well, and while it was originally crafted with a top speed of T-1/E-1, it now provides connectivity at much higher bandwidth levels.

One word of caution: Today, frame relay is considered a legacy technology that in many situations is beginning to show its age. And while it can still be found deployed in some industries such as banking where it is used to connect ATM machines to the host, for example, network providers are considering other options, including IP and MPLS. So we're going to spend a little bit of time on frame relay here because it is still found deployed in the network, but for the purposes of future planning readers would be wise to spend more time becoming familiar with the IP/MPLS solution discussed in the next chapter.

The Inner Workings of Frame Relay

In frame relay networks, the incoming data stream is packaged as a series of variable length frames that can transport any kind of data—LAN traffic, IP packets, SNA frames, voice, and video. In fact, it is a reasonably efficient transport mechanism for voice, allowing frame relay-capable PBXs to be connected to a frame relay permanent virtual circuit (PVC) which can cost effectively replace private line circuits. When voice is carried over frame relay it is usually compressed for transport efficiency and packaged in small frames to minimize the processing delay of the frames. Several hundred voice channels can be encoded over a single PVC, although the number is usually smaller when actually implemented.

Frame relay is a virtual circuit service. When a customer wishes to connect two locations using frame relay, they contact their service provider and tell the service representative the locations of the endpoints and the bandwidth they require. The service provider issues a service order to create the circuit. If at some point in the future the customer decides to change the circuit endpoints or upgrade the bandwidth, another service order must be issued. This service is called permanent virtual circuit (PVC), and is the most commonly deployed frame relay solution.

In frame relay, PVCs are identified using an address called a data link connection identifier (DLCI) (pronounced 'delsie'). At any given endpoint, the customer's router can support multiple DLCIs, and each DLCI can be assigned varying bandwidths based upon the requirements of the device/application on the router port associated with that DLCI. In Fig. 9.3, the customer has purchased a T-1 circuit to connect their router to the frame relay network. The router is connected to a videoconferencing unit at 384 Kbps, a frame relay-capable PBX at 768 Kbps, and a data circuit for Internet access at 512 Kbps. Note that the aggregate bandwidth assigned to these devices exceeds the actual bandwidth of the access line by 128 Kbps (1664–1536). Under normal circumstances this would not be possible, but frame relay assumes that the traffic that it will normally be transporting is bursty by nature. If the assumption is correct (and it usually is), there is very little likelihood that all three devices will burst at the same

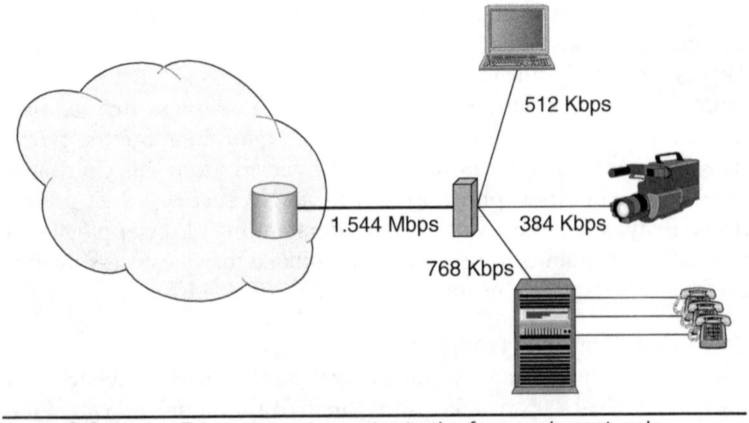

FIGURE 9.3 Using T-1 to connect a router to the frame relay network.

instant in time. As a consequence, the circuit's operating capacity can actually be *overbooked*, a process known as oversubscription. Most service providers allow as much as 200 percent oversubscription, something customers clearly benefit from provided the circuit is designed properly. This means that the frame relay salesperson must carefully assess the nature of the traffic that the customer will send over the link and ensure that enough bandwidth is allocated to support the requirements of the various devices that will be sharing access to the link. Failure to do so can result in an under-engineered facility that will not meet the customer's throughput requirements. This is a critical component of the service delivery formula.

Committed Information Rate

The throughput level, that is, the bandwidth that frame relay service providers absolutely guarantee on a PVC-by-PVC basis, is called the committed information rate (CIR). In addition to CIR, service providers will often support an excess information rate (EIR), which is the rate above the CIR they will attempt to carry, assuming the capacity is available within the network. However, all frames above the CIR are marked as eligible for discard, which simply means that the network will do its best to deliver them but makes no guarantees. If push comes to shove, and the network finds itself congested, the frames marked discard eligible (DE) are immediately discarded at their point of ingress. This CIR/EIR relationship is poorly understood by many customers because the CIR is taken to be an indicator of the absolute bandwidth of the circuit. Whereas bandwidth is typically measured in bits per second, CIR is a measure of *bits in 1 second.* In other words, the CIR is a measure of the average throughput that the network will guarantee. The actual transmission volume of a given CIR may be higher or lower than the CIR at any point in time because of the bursty nature of the data being sent, but in aggregate the network will maintain an average, guaranteed flow volume for each PVC. This is a selling point for frame relay. In most cases, customers get more than they actually pay for, and as long as the switch loading levels are properly engineered, the switch (and therefore the frame relay service offering) will not suffer adversely from this charitable bandwidth allocation philosophy. The key to success when selling frame relay is to have a very clear understanding of the applications the customer intends to use across the link so that the access facility can be properly sized for the anticipated traffic load.

Congestion Control in Frame Relay

Frame relay has two congestion control mechanisms. Embedded in the header of each frame relay frame are two additional bits called the forward explicit congestion notification bit (FECN) and the backward explicit congestion notification bit (BECN). Both are used to notify devices in the network of congestion situations that could affect throughput.

Consider the following scenario. A frame relay frame arrives at the second of three switches along the path to its intended destination, where it encounters severe local congestion (see Fig. 9.4). The congested switch sets the FECN bit to indicate the presence of congestion and transmits the frame to the next switch in the chain. When the frame arrives, the receiving switch takes note of the FECN bit, which tells it the following: "I just came from that switch back there, and it's extremely congested. You can transmit stuff back there if you want to, but there's a good chance that anything you send will be discarded, so you might want to wait awhile before transmitting." In other words, the switch has been notified of a congestion condition,

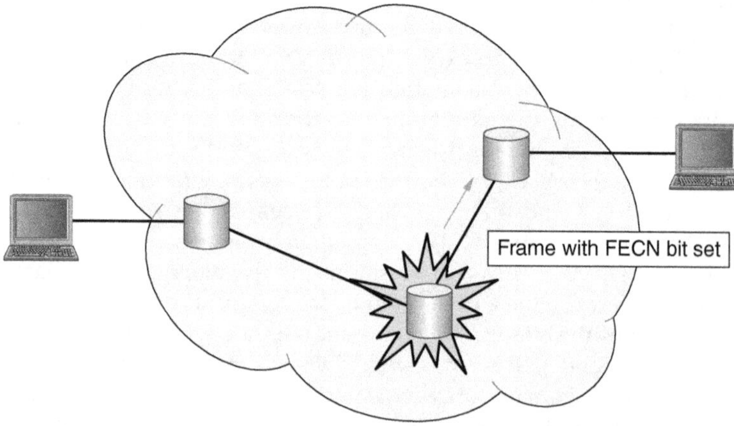

FIGURE 9.4 Using the FECN bit to report congestion in the network.

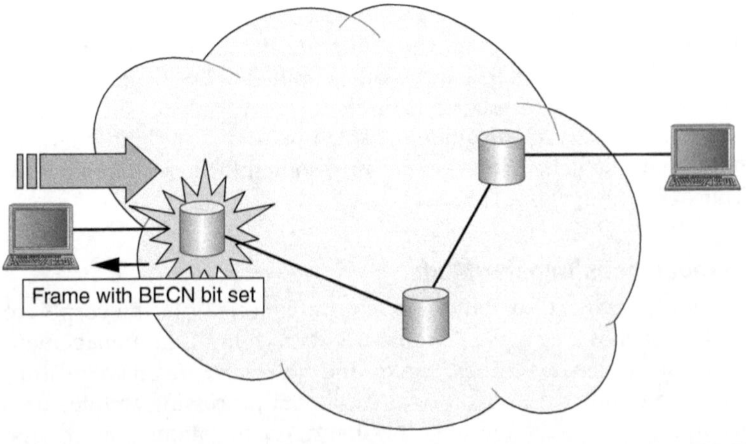

FIGURE 9.5 Using the BECN bit to report congestion in the network.

to which it may respond by throttling back its output to allow the affected switch time to recover.

On the other hand, the BECN bit is used to flow-control a device that is sending too much information into the network. Consider the situation shown in Fig. 9.5, where a particular device on the network is transmitting at a high volume level, routinely violating the CIR and perhaps the EIR level established by mutual consent. The ingress switch—that is, the first switch the traffic touches—has the ability to set the BECN bit on frames going toward the offending device, which carries the implicit message, "Cut it out or I'm going to hurt you." In effect, the BECN bit notifies the offending device that it is violating protocol, and continuing to do so will result in every frame from that device being discarded, without warning or notification. If this happens, it gives the ingress switch the opportunity to recover. However, it doesn't fix the problem— it merely forestalls the inevitable, because sooner or later the intended recipient will realize that frames are missing and will initiate recovery procedures, which will cause resends to occur. However, it may give the affected devices time to recover before the onslaught begins anew.

The problem with FECN and BECN lies in the fact that many devices choose not to implement them. They do not necessarily have the inherent ability to throttle back upon receipt of a congestion indicator, although devices that can are becoming more common. Nevertheless, proprietary solutions are in widespread use and will continue to be for some time to come.

Frame Relay Summary

Frame relay is clearly a Cinderella technology, evolving quickly from a data-only transport scheme to a multiservice technology with diverse capabilities. For data and some voice and video applications, it shines as a wide area network offering. In some areas, however, frame relay is lacking. Its bandwidth is limited to DS-3, and its ability to offer standards-based QoS is limited. Given the focus on QoS that is so much a part of customers' chanted mantra today, and the flexibility that a switched solution permits, something else was required. That something was ATM.

Asynchronous Transfer Mode

Network architectures often develop in concert with the corporate structures that they serve. Companies with centralized management authorities such as utilities, banks, and hospitals often have centralized and tightly controlled hierarchical data-processing architectures to protect their data. On the other hand, organizations that are distributed in nature, such as research and development facilities and universities, often have highly distributed data processing architectures. They tend to share information on a peer-to-peer basis and their corporate structures reflect the fact.

Asynchronous transfer mode came about not only because of the proliferation of diverse network architectures but also because of the evolution of traffic characteristics and transport requirements. To the well-known demands of voice we now add various flavors of data, video, MP3, an exponentially large variety of IP traffic, interactive real-time gaming, and a variety of other content types that place increasing demands on the network. Further, we have seen a requirement arise for a mechanism that can transparently and correctly transport the mix of various traffic types over a single network infrastructure, while at the same time delivering granular, controllable and measurable quality of service levels for each service type. In its original form, ATM was designed to do exactly that, working with SONET or SDH to deliver high-speed transport and switching throughout the network—in the wide area, the metropolitan area, the campus environment, and the local area network, right down to the desktop, seamlessly, accurately, and fast.

Today, because of competition from such technologies as QoS-aware IP transport, proprietary high-speed mesh networks, and fast and gigabit ethernet, ATM has for the most part lost the race to the desktop. ATM is a cell-based technology, which simply means that the fundamental unit of transport—a frame of data, if you will—is of a fixed size, which allows switch designers to build faster, simpler devices, since they can always count on their switched payload being the same size at all times. That cell comprises a 5-octet header and a 48-octet payload field as shown in Fig. 9.6. The payload contains user data; the header contains information that the network requires to both transport the payload correctly and ensure proper quality of service levels for the payload. ATM accomplishes this task well, but at a

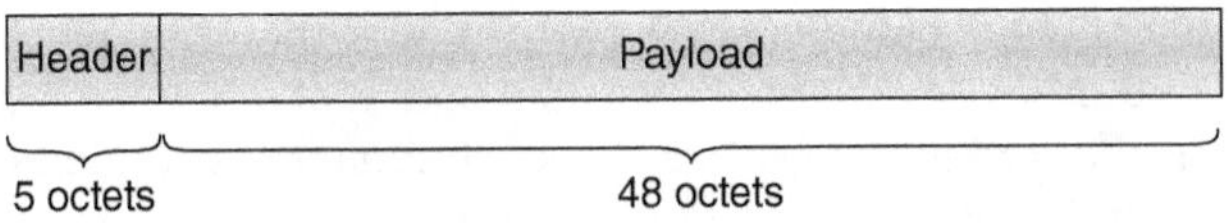

FIGURE 9.6 The structure of an ATM cell showing the five-byte header and 48-byte payload field.

cost. The five-octet header comprises nearly 10 percent of the cell, a rather significant price to pay, particularly when other technologies such as IP and SONET add their own significant percentages of overhead to the overall payload. This reality is part of the problem: ATM's original claims to fame, and the reasons it rocketed to the top of the technology hit parade, were its ability to switch cells at tremendous speed through the fabric of the wide-area network and the ease with which the technology could be scaled to fit any network situation. Today, however, given the availability of high-speed IP routers that routinely route packets at terabit rates, ATM's advantages have begun to pale to a certain degree.

ATM Evolution

ATM has, however, emerged from the flames in other ways. Today, many service providers see ATM as an ideal aggregation technology for diverse traffic streams that need to be combined for transport across a wide area network that will most likely be IP-based. ATM devices, then, will be placed at the edge of the network, where they will collect traffic for transport across the Internet or (more likely) a privately owned IP network. Furthermore, because it has the ability to be something of a chameleon by delivering diverse services across a common network fabric, it is further guaranteed a seat at the technology game.

It is interesting to note that the traditional, legacy telecommunications network comprises two principal *regions* that can be clearly distinguished from each other: the network itself, which provides switching, signaling, and transport for traffic generated by customer applications; and the access loop, which provides the connectivity between the customer's applications and the network. In this model, the network is considered to be a relatively intelligent medium, while the customer equipment is usually considered to be relatively stupid.

Not only is the intelligence seen as being concentrated within the confines of the network, so too is the bulk of the bandwidth since the legacy model indicates that traditional customer applications don't require much of it. Between central office switches, however, and between the offices themselves, enormous bandwidth is required.

Today, the world has changed. Customer equipment has become remarkably intelligent and many of the functions previously done within the network cloud are now performed at the edge. PBXs, computers, and other devices are now capable of making discriminatory decisions about required service levels, eliminating any need for the massive intelligence embedded in the core.

At the same time, the bandwidth is migrating from the core of the network toward the customer as applications evolve to require it. There is still massive bandwidth within the cloud, but the margins of the cloud are expanding toward the customer.

The result of this evolution is a redefinition of the network's regions. Instead of a low-speed, low-intelligence access area and a high-speed, highly intelligent core, the intelligence has migrated outward to the margins of the network, and the bandwidth, once exclusively a core resource, is now equally distributed at the edge. Thus we see something of a core and edge distinction evolving as customer requirements change.

One reason for this steady migration is the well-known fact within sales and marketing circles that products sell best when they are located close to the buying customer. They are also easier to customize for individual customers when they are physically closest to the situation for which the customer is buying them.

In 1997, Bell Labs researcher David Isenberg wrote *The Rise of the Stupid Network*, in which he observes:

> The Intelligent Network is a straight-line extension of four assumptions—scarcity, voice, circuit switching, and control. Its primary design impetus was not customer service. Rather, the Intelligent Network was a telephone company attempt to engineer vendor independence, more automatic operation, and some "intelligent" new services into existing network architecture. However, even as it rolls out and matures, the Intelligent Network is being superseded by a Stupid Network, with nothing but dumb transport in the middle, and intelligent user-controlled endpoints, whose design is guided by plenty, not scarcity, where transport is guided by the needs of the data, not the design assumptions of the network.

Isenberg continues:

> A new network "philosophy and architecture," is replacing the vision of an Intelligent Network. The vision is one in which the public communications network would be engineered for "always-on" use, not intermittence and scarcity. It would be engineered for intelligence at the end-user's device, not in the network. And the network would be engineered simply to "Deliver the Bits, Stupid," not for fancy network routing or "smart" number translation.

Interesting and somewhat prophetic, wouldn't you agree? Isenberg wrote that almost 20 years ago; it's still true today.

ATM Technology Overview

Because ATM plays such a major role in networks today it is important to develop at least a rudimentary understanding of its functions, architectures, and offered services.

ATM Protocols

Like all modern technologies, ATM has a well-developed protocol stack, shown in Fig. 9.7, which clearly delineates the functional breakdown of the service. The stack consists of four layers: the upper services layer, the ATM adaptation layer, the ATM layer, and the physical layer.

The upper services layer defines the nature of the actual services that ATM can provide. It identifies both constant and variable bit rate services. Voice is an example of a constant bit rate service, while signaling, IP, and frame relay are examples of both connectionless and connection-oriented variable bit rate services.

The ATM adaptation layer (AAL) has four general responsibilities:

- synchronization and recovery from errors
- error detection and correction
- segmentation and reassembly of the data stream
- multiplexing

The AAL comprises two functional sublayers. The convergence sublayer provides service-specific functions to the services layer so that the services layer can make the most efficient use of the underlying cell relay technology that ATM provides. Its functions include clock recovery for end-to-end timing management; a recovery mechanism for lost or out-of-order cells; and a timestamp capability for time-sensitive traffic such as voice and video.

The segmentation and reassembly sublayer (SAR) converts the user's data from its original incoming form into the 48-octet payload

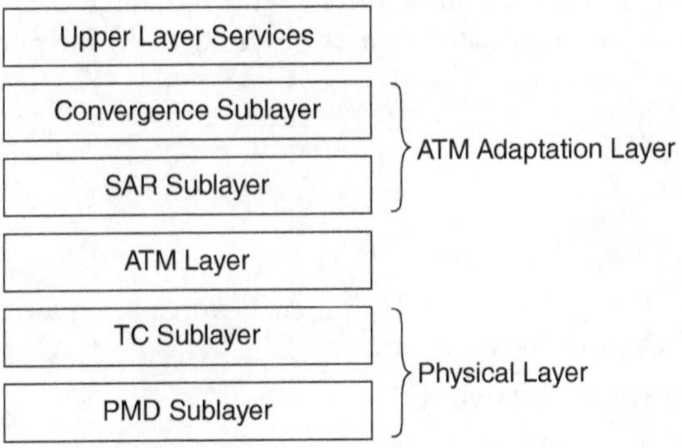

FIGURE 9.7 The ATM protocol stack.

chunks that will become cells. For example, if the user's data is arriving in the form of 64-kB IP packets, SAR chops them into 48-octet payload pieces. It also has the responsibility to detect lost or out-of-order cells that the convergence sublayer will recover from, and to detect single bit errors in the payload chunks.

The ATM Layer has five general responsibilities. These are

- cell multiplexing and de-multiplexing
- virtual path and virtual channel switching
- creation of the cell header
- generic flow control
- cell delineation

Because the ATM layer creates the cell header, it is responsible for all of the functions that the header manages. The process, then, is fairly straight-forward: the user's data passes from the services layer to the ATM adaptation layer, which segments the data stream into 48-octet pieces. The pieces are handed to the ATM layer, which creates the header and attaches it to the payload unit, thus creating a cell. The cells are then handed down to the physical layer.

The physical layer consists of two functional sub layers as well: the transmission convergence sublayer and the physical medium sublayer. The transmission convergence sublayer performs three primary functions. The first is called cell rate decoupling, which adapts the cell creation and transmission rate to the rate of the transmission facility by performing *cell stuffing,* similar to the bit stuffing process described earlier in the discussion of DS-3 frame creation. The second responsibility is cell delineation, which allows the receiver to delineate between one cell and the next. Finally, it generates the transmission frame in which the cells are to be carried.

The physical medium sublayer takes care of issues that are specific to the medium being used for transmission, such as line codes, electrical and optical concerns, timing, and signaling.

The physical layer can use a wide variety of transport options, including:

- DS1/DS2/DS3
- E1/E3
- 25.6 Mbps UNI over UTP-3
- 51 Mbps UNI over UTP-5 (TAXI: transparent asynchronous transmitter/receiver interface)
- 100 Mbps UNI over UTP-5
- OC-3/12/48c

Others, of course, will follow as transport technologies advance.

The ATM Cell Header

As we mentioned before, ATM is a cell-based technology that relies on a 48-octet payload field that contains actual user data, and a 5-byte header that contains information needed by the network to route the cell and provide proper levels of service.

The ATM cell header, shown in Fig. 9.8, is examined and updated by each switch it passes through, and comprises six distinct fields: the generic flow control field, the virtual path identifier, the virtual channel identifier, the payload type identifier, the cell loss priority field, and the header error control field.

Generic flow control (GFC) is a four-bit field used across the UNI for network-to-user flow control. It has not yet been completely defined in the ATM standards, but some companies have chosen to use it for very specific purposes. For example, Australia's Telstra Corporation uses it for flow control in the network-to-user direction, and as a traffic priority indicator in the user-to-network direction.

Virtual path identifier (VPI)—an 8-bit VPI identifies the virtual path over which the cells will be routed at the UNI. It should be noted that because of dedicated, internal flow control capabilities within the network, the GFC field is not needed across the network-to-network interface (NNI). It is therefore redeployed: the 4 bits are converted to additional VPI bits, thus extending the size of the virtual path *field.* This allows for the identification of more than 4,000 unique VPs. At the UNI, this number is excessive, but across the NNI it is necessary because of the number of potential paths that might exist between the switches that make up the fabric of the network.

Virtual channel identifier (VCI)—as the name implies, the 16-bit VCI identifies the unidirectional virtual channel over which the current cells will be routed.

Payload type identifier (PTI) is a 3-bit PTI field used to indicate network congestion and cell type, in addition to a number of other functions. The first bit indicates whether the cell was generated by

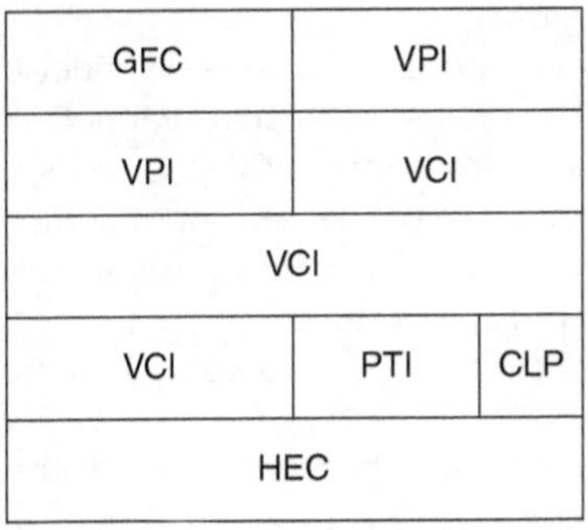

Figure 9.8 The ATM cell header.

the user or by the network, while the second indicates the presence or absence of congestion in user-generated cells, or flow-related operations, administration, and maintenance information in cells generated by the network. The third bit is used for service-specific, higher-layer functions in the user-to-network direction, such as to indicate that a cell is the last in a series of cells. From the network to the user, the third bit is used with the second bit to indicate whether the Operations, Administration, and Maintenance (OA&M) information refers to segment or end-to-end-related information flow.

Cell loss priority (CLP)—the single-bit cell loss priority field is a relatively primitive flow control mechanism by which the user can indicate to the network which cells to discard in the event of a condition that demands that some cells be eliminated, similar to the DE bit in frame relay. It can also be set by the network to indicate to downstream switches that certain cells in the stream are eligible for discard should that become necessary.

Header error control (HEC)—the 8-bit HEC field can be used for two purposes. First, it provides for the calculation of an 8-bit CRC that checks the integrity of the entire header. Second, it can be used for cell delineation.

Addressing in ATM

ATM is a connection-oriented, virtual circuit technology, meaning that communication paths are created through the network prior to actually sending traffic. Once established, the ATM cells are routed based upon a virtual circuit address. A virtual circuit is simply a connection that gives the user the appearance of being dedicated to that user, when in point of fact the only thing that is actually dedicated is a time slot. This technique is generically known as label-based switching, and is accomplished through the use of routing tables in the ATM switches that designate input ports, output ports, input addresses, output addresses, and quality of service parameters required for proper routing and service provisioning. As a result, cells do not contain explicit destination addresses, but rather contain timeslot identifiers.

There are two components to every virtual circuit address, as shown in Fig. 9.9. The first is the virtual channel (VC), which is a unidirectional conduit for the transmission of cells between two endpoints. For example, if two parties are conducting a videoconference, they will each have a VC for the transmission of outgoing cells that make up the video portion of the conference.

The second *level* of the ATM addressing scheme is called a virtual path (VP). A VP is a bundle of VCs that have the same endpoints and that, when considered together, make up a bi-directional transport facility. The combination of unidirectional channels that we need in our two-way videoconferencing example makes up a VP.

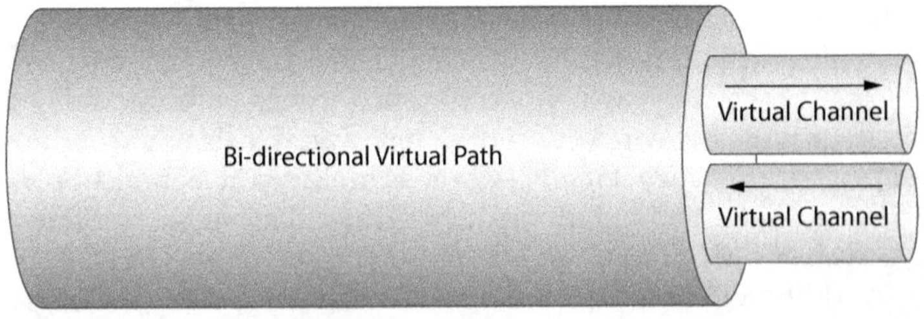

Figure 9.9 The components of an ATM virtual circuit address.

ATM Services

The basic services that ATM provides are based on three general characteristics: the nature of the connection between the communicating stations (connection-oriented vs. connectionless); the timing relationship between the sender and the receiver; and the bit rate required to ensure proper levels of service quality. Based on those generic requirements, both the ITU-T and the ATM Forum have created service classes that address the varying requirements of the most common forms of transmitted data.

ITU-T Service Classes

The ITU-T assigns service classes based on three characteristics: connection mode, bit rate, and the end-to-end timing relationship between the end stations. They have created four distinct service classes, based on the model shown in Fig. 9.10. Class A service, for example, defines a connection-oriented, constant bit rate, timing-based service that is ideal for the stringent requirements of voice service. Class B, on the other hand, is ideal for such services as variable

	Class A	Class B	Class C	Class D
AAL Type	1	2	5, 3/4	5, 3/4
Connection Mode	Connection-oriented	Connection-oriented	Connection-oriented	Connectionless
Bit Rate	Constant	Variable	Variable	Variable
Timing Relationship	Required	Required	Not required	Not Required
Service Types	Voice, Video	VBR Voice, Video	Frame Relay	IP

Figure 9.10 ITU-T service classes for ATM networks.

bit rate video, in that it defines a connection-oriented, variable bit rate, timing-based service.

Class C service was defined for such things as frame relay, in that it provides a connection-oriented, variable bit rate, timing-independent service. Finally, Class D delivers a connectionless, variable bit rate, timing independent service that is ideal for IP traffic as well as SMDS.

In addition to service classes, the ITU-T has defined ATM adaptation layer (AAL) service types, which align closely with the A, B, C, and D service types described previously. Whereas the service classes (A, B, C, D) describe the capabilities of the underlying network, the AAL types describe the cell format. They are AAL1, AAL2, AAL3/4, and AAL 5. However, only two of them have really *survived* in a big way.

AAL1 is defined for Class A service, which is a constant bit rate environment ideally suited for voice and voice-like applications. In AAL1 cells, the first octet of the payload serves as a payload header that contains cell sequence and synchronization information that is required to provision constant bit rate, fully sequenced service. AAL1 provides circuit-emulation service without dedicating a physical circuit, which explains the need for an end-to-end timing relationship between the transmitter and the receiver.

AAL5, on the other hand, is designed to provide both Class C and D services, and while it was originally proposed as a transport scheme for connection-oriented data services, it turns out to be more efficient than AAL3/4 and accommodates connectionless services quite well.

To guard against the possibility of errors, AAL5 has an 8-octet trailer appended to the user data which includes a variable size pad field used to align the payload on 48-octet boundaries, a 2-octet control field that is currently unused, a 2-octet length field that indicates the number of octets in the user data, and finally, a 4-octet CRC that can check the integrity of the entire payload. AAL5 is often referred to as the simple and easy adaptation layer (SEAL), and it may find an ideal application for itself in the burgeoning Internet arena. Recent studies indicate that TCP/IP transmissions produce comparatively large numbers of small packets that tend to be around 48 octets long. That being the case, AAL5 could well transport the bulk of them in its user data field. Furthermore, the maximum size of the user data field is 65,536 octets, coincidentally the same size as an IP packet.

ATM Forum Service Classes

The ATM Forum looks at service definitions slightly differently than the ITU-T, as shown in Fig. 9.11. Instead of the A-B-C-D services, the ATM forum categorizes them as real-time and non-real-time services. Under the real-time category, they define constant bit rate services that demand fixed resources with guaranteed availability. They also define real-time variable-bit rate service, which provides for statistical multiplexed, variable bandwidth service allocated on-demand.

Service	Descriptors	Loss	Delay	Bandwidth	Feedback
CBR	PCR, CDVT	Yes	Yes	Yes	No
VBR-RT	PCR, CDVT, SCR, MBS	Yes	Yes	Yes	No
VBR-NRT	PCR, CDVT, SCR, MBS	Yes	Yes	Yes	No
UBR	PCR, CDVT	No	No	No	No
ABR	PCR, CDVT, MCR	Yes	No	Yes	Yes

FIGURE 9.11 ATM forum service classes for ATM.

A further subset of real-time VBR is peak-allocated VBR, which guarantees constant loss and delay characteristics for all cells in that flow.

Under the non-real-time service class, unspecified bit rate (UBR) is the first service category. UBR is often compared to IP in that it is the best-effort delivery scheme in which the network provides whatever bandwidth it has available, with no guarantees made. All recovery functions from lost cells are the responsibility of the end-user devices.

UBR has two sub-categories of its own. The first, non-real-time VBR (NRT-VBR), improves the impacts of cell loss and delay by adding a network resource reservation capability. Available bit rate (ABR), UBR's other sub-category, makes use of feedback information from the far end to manage loss and ensure fair access to and transport across the network.

Each of the five classes makes certain guarantees with regard to cell loss, cell delay, and available bandwidth. Furthermore, each of them takes into account descriptors that are characteristic of each service described. These include peak cell rate (PCR), sustained cell rate (SCR), minimum cell rate (MCR), cell delay variation tolerance (CDVT), and burst tolerance (BT).

ATM Forum Specified Services

The ATM forum has identified a collection of services for which ATM is a suitable, perhaps even desirable, network technology. These include cell relay service (CRS), circuit emulation service (CES), voice and telephony over ATM (VTOA), frame relay bearer service (FRBS), LAN emulation (LANE), multiprotocol over ATM (MPOA), and a collection of others.

Cell relay service is the most basic of the ATM services. It delivers precisely what its name implies: a *raw pipe* transport mechanism for cell-based data. As such it does not provide any ATM bells and whistles, such as quality of service discrimination; nevertheless, it is the most commonly implemented ATM offering because of its lack of implementation complexity.

Circuit emulation service gives service providers the ability to offer a selection of bandwidth levels by varying both the number of cells transmitted per second and the number of bytes contained in each cell.

Voice and telephony over ATM is a service that has yet to be clearly defined. The ability to transport voice calls across an ATM network is a non-issue, given the availability of Class A service. What are not clearly defined, however, are corollary services such as 800/888 calls, 900 service, 911 call handling, enhanced services billing, SS7 signal interconnection, and so on. Until these issues are clearly resolved, ATM-based, feature-rich telephony will not become a mainstream service, but will instead be limited to simple voice—and there is a difference.

Frame relay bearer service refers to the ability of ATM to interwork with frame relay. Conceptually, the service implies that an interface standard allows an ATM switch to exchange data with a frame relay switch, thus allowing for interoperability between frame and cell-based services. Many manufacturers are taking a slightly different tack, however: they build switches with soft, chewy cell technology at the core, and surround the core with hard, crunchy interface cards to suit the needs of the customer.

For example, an ATM switch might have ATM cards on one side to interface with other ATM devices in the network, but frame relay cards on the other side to allow it to communicate with other frame relay switches, as shown in Fig. 9.12. Thus, a single piece of hardware

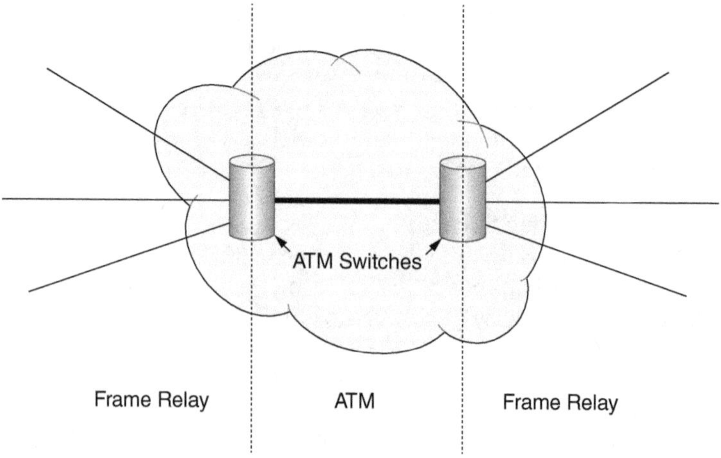

FIGURE 9.12 Interworking between ATM and other services, such as frame relay.

can logically serve as both a cell and frame relay switch. This design is becoming more and more common, because it helps to avoid a future rich with forklift upgrades.

LAN emulation (LANE) allows an ATM network to move traffic transparently between two similar LANs, but also allow ATM to transparently slip into the LAN arena. For example, two ethernet LANs could communicate across the fabric of an ATM network, as could two token ring LANs. In effect, LANE allows ATM to provide a bridging function between similar LAN environments. In LANE implementations, the ATM network does not handle MAC functions such as collision detection, token passing, or beaconing; it merely provides the connectivity between the two communicating endpoints. The MAC frames are simply transported inside AAL5 cells.

One clear concern about LANE is that LANs are connectionless, while ATM is a virtual circuit-based, connection-oriented technology. LANs routinely broadcast messages to all stations, while ATM allows point-to-point or multipoint circuits only. Thus, ATM must look like a LAN if it is to behave like one. To make this happen, LANE uses a collection of specialized LAN emulation clients and servers to provide the connectionless behavior expected from the ATM network.

On the other hand, multiprotocol over ATM (MPOA) provides the ATM equivalent of routing in LAN environments. In MPOA installations, routers are referred to as MPOA servers. When one station wants to transmit to another station, it queries its local MPOA server for the remote station's ATM address. The local server then queries its neighbor devices for information about the remote station's location. When a server finally responds, the originating station uses the information to establish a connection with the remote station, while the other servers cache the information for further use.

MPOA promises a great deal, but it is complex to implement and requires other ATM *components* such as the private NNI capability to work properly. Furthermore, it's being challenged by at least one alternative technology, known as IP switching.

IP switching is far less overhead-intensive than MPOA. Furthermore, it takes advantage of a known (but often ignored) reality in the LAN interconnection world: most routers today use IP as their core protocol, and the great majority of LANs are still ethernet. This means that a great deal of simplification can be done by crafting networks to operate around these two technological bases. And in fact, this is precisely what IP switching does. By using existing, low-overhead protocols, the IP switching software creates new ATM connections dynamically and quickly, updating switch tables on the fly. In IP switching environments, IP resides on top of ATM within a device as shown in Fig. 9.13, providing the best of both protocols. If two communicating devices wish to exchange information and they have done so before, an ATM mapping already exists and no layer three involvement (IP) is required—the ATM switch portion of the service

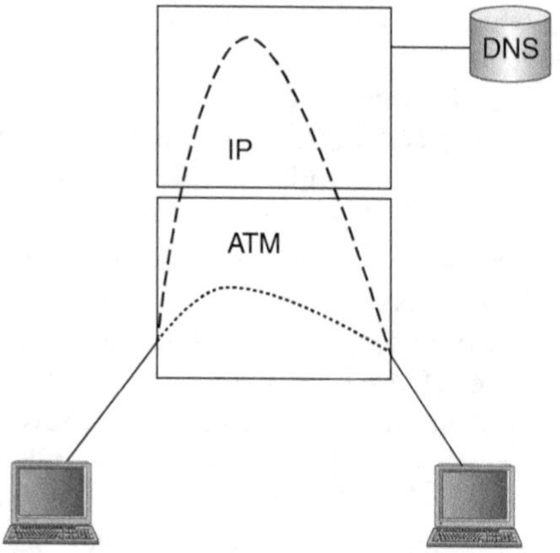

FIGURE 9.13 ATM and IP interworking in MPOA.

simply creates the connection at high speed. If an address lookup is required, then the *call* is handed up to IP, which takes whatever steps are required to perform the lookup (a DNS request, for example). Once it has the information, it hands it down to ATM, which proceeds to set up the call. The next time the two need to communicate, ATM will be able to handle the connection.

Other applications for ATM continue to emerge, including television and advanced medical imaging solutions. This leads to what I often refer to as *the great triumvirate:* ATM, SONET, or SDH, and broadband services. By combining the powerful switching and multiplexing fabric of ATM with the limitless transport capabilities of SONET or SDH, true broadband services can be achieved, and the idea of creating a network that can be all things to all services can finally be realized.

A major part of this all-things-to-all-services network is the ability to easily transport any kind of payload, anywhere in the world, without concern for bandwidth availability. That came with the arrival of modern optical networking technologies.

Optical Networking

Optical networking is often viewed as a point-to-point technology. It has achieved such a position of prominence in the last decade, however, that it qualifies as a transport option in its own right. Furthermore, optical switching adds to its relevance. In the next section, we discuss the development of optical networking, the various technologies that it employs, and the direction it is taking in this fast-paced market.

Early Optical Technology Breakthroughs

In 1878, two years after perfecting his speaking telegraph (which became the telephone), Alexander Graham Bell created a device that transmitted the human voice through the air for distances up to 200 m. That's right—he created the world's first wireless telephone. The device, which he called the Photophone, used carefully angled mirrors to reflect sunlight onto a diaphragm that was attached to a mouthpiece, as shown in Fig. 9.14. At the receiving end (Fig. 9.15), the light was concentrated by a parabolic mirror onto a selenium resistor, which was connected to a battery and speaker. The diaphragm vibrated when struck by the human voice, which in turn caused the intensity of the light striking the resistor to vary. The selenium resistor, in turn, caused the current flow to vary in concert with the varying sunlight, which caused the received sound to come out of the speaker with remarkable fidelity. This represented the birth of optical transmission.

Optical transmission in its early days was limited in terms of what it was capable of doing. Consider the following analogy: If you look through a 2-ft-square pane of window glass, it appears clear—if the glass is clean, it is virtually invisible. However, we all know that if you turn the pane on edge and look through it from edge to edge, the glass appears to be dark green. Very little light passes from one edge to the other. In this example, you are looking through a few inches of glass. Imagine trying to pass a high-bandwidth optical

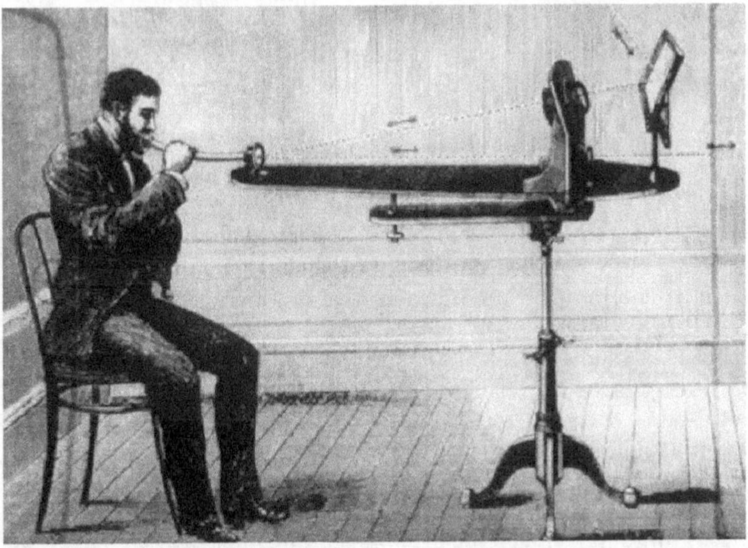

FIGURE 9.14 Photophone transmitter.

FIGURE 9.15 Photophone receiver.

signal through 40 or more kilometers of that same glass. Clearly (no pun intended) the signal would be attenuated to nothing in a matter of a few feet.

In 1966, Charles Kao (who was awarded the Nobel Prize in physics for his work in optical) and Charles Hockham at the UK's Standard Telecommunication Laboratory published their seminal work, *Dielectric-Fibre Surface Waveguides for Optical Frequencies,* in which they demonstrated that optical fiber could be used to carry information provided its end-to-end signal loss could be kept below 20 dB/km. Keeping in mind that the decibel scale is logarithmic, 20 dB of loss means that 99 percent of the light would be lost over each kilometer of distance. Only 1 percent would actually reach the receiver—and that's a 1-km run. Imagine the loss over today's fiber cables that are hundreds of kilometers long, if 20 dB was the modern performance criterion!

Kao and Hockham proved that metallic impurities in the glass such as chromium, vanadium, iron, and copper were the primary cause for such high levels of loss because the metal ions in the silica matrix unacceptably dispersed the signal. In response, glass manufacturers rose to the challenge and began to research the creation of ultra-pure products.

In 1970, Peter Schultz, Robert Maurer, and Donald Keck of Corning Glass Works (now Corning Corporation) announced the development of a glass fiber that offered better attenuation than the recognized 20-dB threshold. Today, fiber manufacturers offer fiber so incredibly pure that 10 percent of the light arrives at a receiver placed 50 km

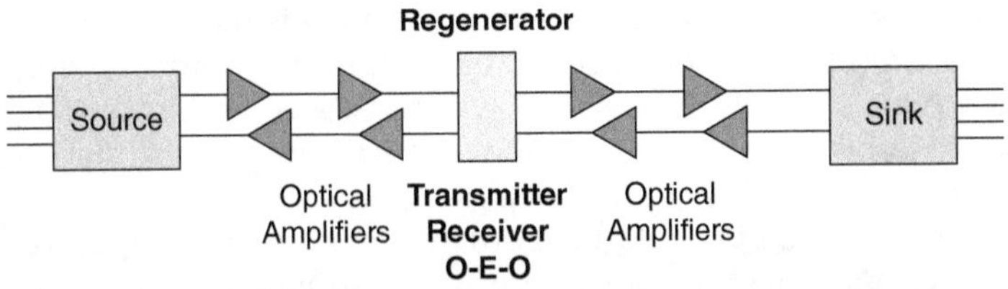

FIGURE 9.16 The three components of an optical system: a light sources, a transport medium, and a receive device.

away. Put another way, a fiber with 0.2 dB of measured loss delivers more than 60 percent of the transmitted light over a distance of 10 km. Remember the windowpane example? Imagine glass so pure that you could see clearly through a window 10-km thick!

Fundamentals of Optical Networking

At their most basic level, optical networks require three fundamental components as shown in Fig. 9.16: a source of light, a medium over which to transport it, and a receiver for the light. Additionally, there may be regenerators, optical amplifiers, and other pieces of equipment in the circuit. We will examine each of these generic components in turn.

Optical Sources

Today the most common sources of light for optical systems are either light-emitting diodes or laser diodes. Both are commonly used, although laser diodes have become the accepted standard for high-speed data applications because of their coherent signal. While lasers have gone through several iterations over the years including ruby rod and helium-neon, semiconductor lasers became the norm shortly after their introduction in the early 1960s because of their low cost and high stability.

Light-Emitting Diodes

Light-emitting diodes (LEDs) come in two varieties: surface-emitting LEDs and edge-emitting LEDs. Surface emitting LEDs emit a wide angle light pulse and therefore do not lend themselves to the more coherent requirements of optical data systems because of the difficulty involved in focusing their emitted light into the core of the receiving fiber. Instead, they are often used as indicators and signaling devices. They are, however, inexpensive, and are therefore commonly deployed in applications that do not require long distance transmission of a coherent light signal.

An alternative to the surface-emitting LED is the edge-emitting device. Edge emitters produce light at significantly narrower angles and have a smaller emitting area, which means that more of their emitted light can be focused into the core of the fiber where the actual signal transport takes place. They are typically faster devices than surface emitters but do have a downside: they are temperature-sensitive, and must therefore be installed in environmentally controlled devices to ensure the stability of the transmitted signal.

Laser Diodes

Laser diodes represent the alternative to LEDs. A laser diode has a very small emitting surface, usually no larger than a few microns in diameter, which means that a great deal of the emitted light can be directed into the fiber. Because they represent a coherent source, the emission angle of a laser diode is extremely narrow. It is the fastest of the three devices.

For the record, these are tiny devices, designed as they are to be installed on a silicon wafer for inclusion in a miniaturized electronic device. Figures 9.17 and 9.18 illustrate just how small these devices are.

The Origins of Optical Fiber

When Peter Schultz, Donald Keck, and Robert Maurer began their work at Corning to create a low-loss optical fiber, they did so using a newly crafted process called *inside vapor deposition* (IVD). Whereas

FIGURE 9.17 A laser diode device, showing its size relative to a U. S. penny.

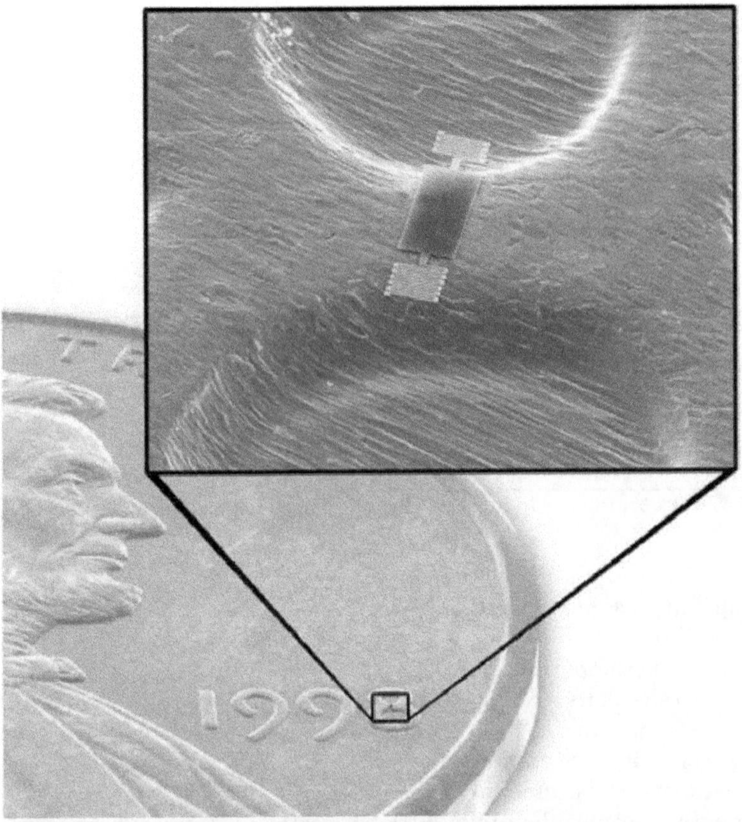

Figure 9.18 Actual size of the emitter inside the device–still compared to a penny. What continues to amaze me is that somebody manufactured and designed this tiny thing!

most glass is manufactured by melting and reshaping silica, IVD deposits various combinations of carefully selected compounds on the inside surface of a silica tube. The tube becomes the cladding, or outside layer, of the fiber; the vapor-deposited compounds become the core. The compounds are typically silicon chloride ($SiCl_4$) and oxygen (O_2), which are reacted under heat to form a soft, sooty deposit of silicon dioxide (SiO2) as shown in Fig. 9.19. In some cases, impurities such as germanium are added at this time to cause various effects in the finished product. In practice, the SiCl4 and O_2 are pumped into the fused silica tube as gases; the tube is heated in a high-temperature lathe, causing the sooty deposit to collect on the inside surface of the tube (Fig. 9.19). The continued heating of the tube causes the soot to fuse into a glasslike substance.

This process can be repeated as many times as required to create a graded refractive index. Ultimately, once the deposits are complete,

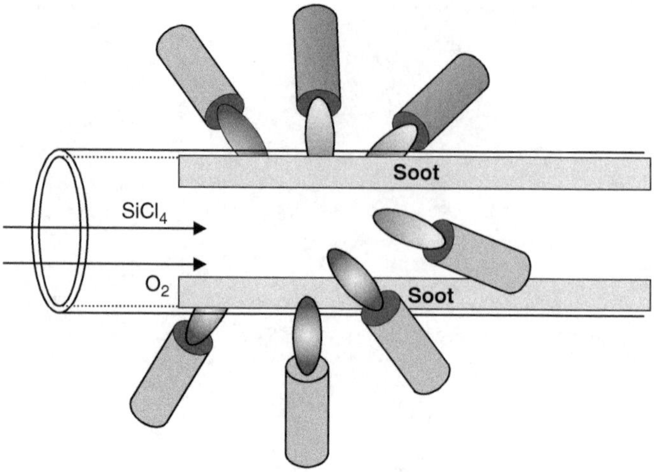

FIGURE 9.19 The inside vapor deposition process for manufacturing optical fiber.

the entire assembly is heated fiercely which causes the tube to collapse, creating what is known in the optical fiber industry as a preform. An example of a preform is shown in Fig. 9.20.

An alternative manufacturing process is called *outside vapor deposition* (OVD). In the OVD process, the soot is deposited on the surface of a rotating ceramic cylinder in two layers. The first layer is the soot that will become the core; the second layer becomes the cladding. Ultimately, the rod and soot are sintered to create a preform.

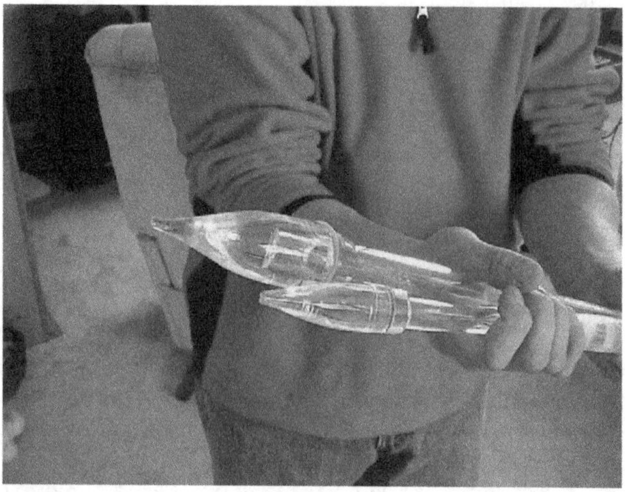

FIGURE 9.20 An optical preform, ready for the drawing tower.

The ceramic is then removed, leaving behind the fused silica that will become the fiber.

There are a number of other techniques for creating the preforms that are used to create fiber, but these are the principal techniques in use today.

The next step is to convert the preform into optical fiber.

Drawing the Fiber

To make fiber from a preform, the preform is mounted in a furnace at the top of a tall building called a drawing tower (Fig. 9.21). The bottom of the preform is heated until it has the consistency of taffy, at which time the soft glass is drawn down to form a thin fiber. When it strikes the cooler air outside the furnace, the fiber solidifies. Needless to say, the process is carefully managed to ensure that the thickness of the fiber is precise; microscopes are used to verify the geometry of the fiber, as shown in Fig. 9.22.

Other stages in the manufacturing process include monitoring processes to check the integrity of the product, a coating process that applies a protective layer, and a take-up stage where the fiber is wound onto reels for later assembly into cables of various types.

Optical Fiber

There are dozens of different types of fiber. Some of them are holdovers from previous generations of optical technology that are still in use and represented the best efforts of technology available at the

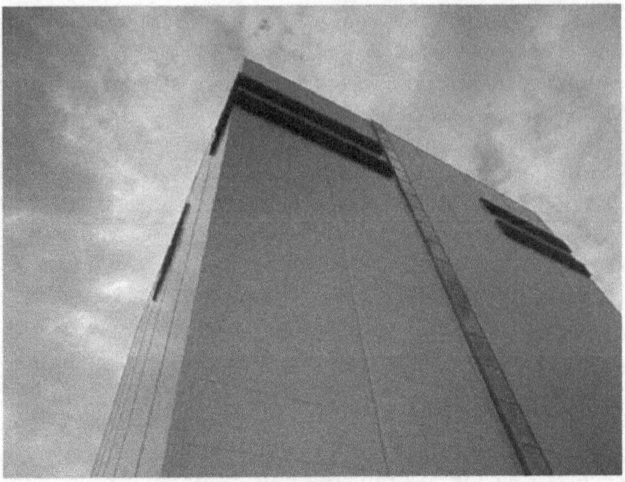

FIGURE 9.21 The drawing tower.

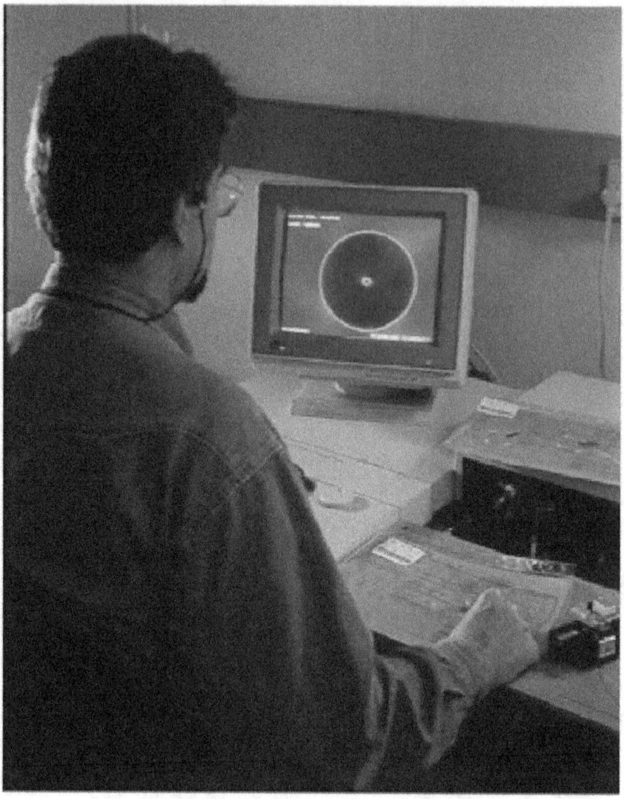

FIGURE 9.22 Technician monitoring the fiber manufacturing process.

time; others represent improvements on the general theme or specialized solutions to specific optical transmission challenges.

Generally speaking, there are two major types of fiber: multimode, which is the earliest form of optical fiber and is characterized by a large diameter central core, short distance capability, and low bandwidth; and single mode, which has a narrow core and is capable of greater distance and higher bandwidth. There are varieties of each that will be discussed in detail, later in the chapter.

To understand the reason for and philosophy behind the various forms of fiber, it is first necessary to understand the issues that confront transmission engineers who design optical networks.

Optical fiber has a number of advantages over copper: it is lightweight, has enormous bandwidth potential, has significantly higher tensile strength, can support many simultaneous channels, and is immune to electromagnetic interference. It does, however, suffer from several disruptive problems that cannot be discounted. The first of these is loss or attenuation, the inevitable weakening of the transmitted signal over distance that has a direct analog in the copper world. Attenuation is typically the result of two sub-properties, scattering and absorption, both of which have cumulative effects. The second is dispersion, which is the spreading of the transmitted signal and is analogous to noise.

Scattering

Scattering occurs because of impurities or irregularities in the physical makeup of the fiber itself. The best-known form of scattering is called *Rayleigh scattering*; it is caused by metal ions in the silica matrix and results in light rays being scattered in various directions, the result of which is an undesirable diffusion of the primary optical signal.

Rayleigh scattering occurs most commonly around wavelengths of 1000 nm and is responsible for as much as 90 percent of the total attenuation that occurs in modern optical systems. It occurs when the wavelengths of the light being transmitted are roughly the same size as the physical molecular structures within the silica matrix; thus, short wavelengths are affected by Rayleigh scattering effects far more than long wavelengths. In fact, it is because of Rayleigh scattering that the sky appears to be blue: the shorter (blue) wavelengths of light are scattered more than the longer wavelengths of light.

Absorption

Absorption results from three factors: hydroxyl (OH^-; water) ions in the silica; impurities in the silica; and incompletely diminished residue from the manufacturing process. These impurities tend to absorb the energy of the transmitted signal and convert it to heat, resulting in an overall weakening of the optical signal. Hydroxyl absorption occurs at 1.25 and 1.39 μm; at 1.7 μm, the silica itself starts to absorb energy because of the natural resonance of silicon dioxide.

The hydroxyl effect is an interesting one and is a phenomenon that most people today are intimately familiar with, although they don't realize it. When we heat food in a microwave oven, the reason it works is because of the hydroxyl ion. Microwave ovens emit energy at precisely the wavelength that is known to excite water, which is, of course, made up of hydroxyl and hydrogen ions. When the water in the food is excited by the energy being pumped into the oven's arming chamber, the hydroxyl ions get excited and start to vibrate, heating (and cooking) the food. This is also why satellite dishes sometimes stop working on rainy, snowy or cloudy days. If there is enough moisture in the air, the atmospheric hydroxyl ions absorb the energy emitted by the orbiting satellite and little of it reaches the receiving satellite dish. Pretty cool, eh?

Dispersion

As mentioned earlier, dispersion is the optical term for the spreading of the transmitted light pulse as it transits the fiber. It is a bandwidth-limiting phenomenon and comes in two forms: multimode dispersion, and chromatic dispersion. Chromatic dispersion is further subdivided into material dispersion and waveguide dispersion.

Multimode Dispersion

To understand multimode dispersion, it is first important to understand the concept of a mode. Figure 9.23 shows a fiber with a relatively wide core. Because of the width of the core, it allows light rays arriving from the source at a variety of angles (three in this case) to enter the fiber and be transmitted to the receiver. Because of the different paths that each ray, or mode, will take, they will arrive at the receiver at different times, resulting in a dispersed signal.

Now consider the system shown in Fig. 9.24. The core is much narrower, and only allows a single ray, or mode, to be sent down the fiber. This results in less end-to-end energy loss and avoids the dispersion problem that occurs in multimode installations.

Chromatic Dispersion

The speed at which an optical signal travels down a fiber is absolutely dependent upon its wavelength. If the signal comprises multiple wavelengths, then the different wavelengths will travel at different speeds, resulting in an overall spreading or smearing of the signal. As discussed earlier, chromatic dispersion comprises two subcategories: material dispersion and waveguide dispersion.

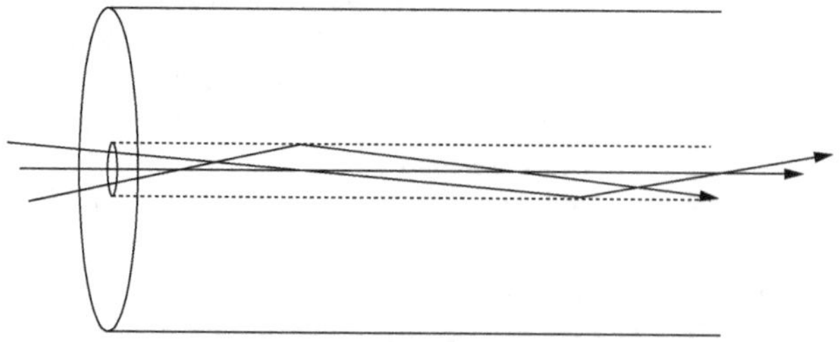

Figure 9.23 Multimode dispersion.

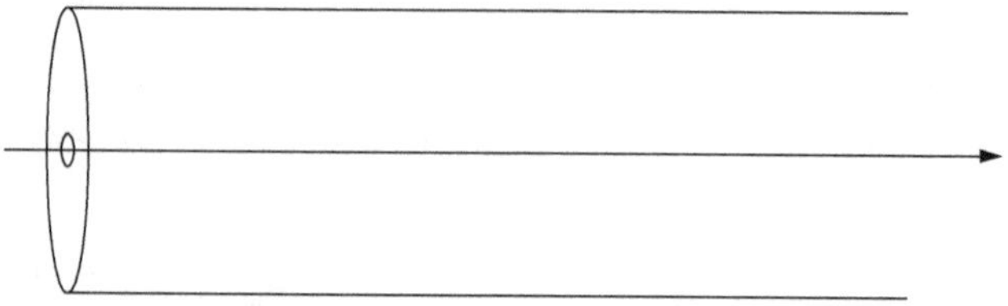

Figure 9.24 Single-mode dispersion.

Material Dispersion

Simply put, material dispersion occurs because different wavelengths of light travel at different speeds through an optical fiber. To minimize this particular dispersion phenomenon, two factors must be managed. The first of these is the number of wavelengths that make up the transmitted signal. An LED, for example, emits a rather broad range of wavelengths between 30 and 180 nm, whereas a laser emits a much narrower spectrum—typically less than 5 nm. Thus a laser's output is far less prone to be seriously affected by material dispersion than the signal from an LED.

The second factor that affects the degree of material dispersion is a characteristic called the center operating wavelength of the source signal. In the vicinity of 850 nm, red, longer wavelengths travel faster than their shorter blue counterparts, but at 1550 nm, the situation is the opposite: blue wavelengths travel faster. There is, of course, a point at which the two meet and share a common minimum dispersion level; it is in the range of 1310 nm, often referred to as the zero-dispersion wavelength. Clearly, this is an ideal place to transmit data signals, since dispersion effects are minimized here. As we will see later, however, other factors crop up that make this a less desirable transmission window than it appears.

Material dispersion is a particularly vexing problem in single-mode fibers.

Waveguide Dispersion

Because the core and the cladding of a fiber have slightly different indices of refraction, the light that travels in the core moves slightly slower than the light that escapes into and travels in the cladding. This results in a dispersion effect that can be corrected by transmitting at specific wavelengths where material and waveguide dispersion actually occur at minimums.

Putting It All Together

So what does all of this have to do with the high-speed transmission of voice, video, and data? A lot, as it turns out. Understanding where attenuation and dispersion problems occur helps optical design engineers determine the best wavelengths at which to transmit information, taking into account distance, type of fiber, and other factors that can potentially affect the integrity of the transmitted signal. Consider the graph shown in Fig. 9.25. It depicts the optical transmission domain, as well as the areas where problems arise. Attenuation (dB/km) is shown on the y-axis; wavelength (nm) is shown on the x-axis.

First of all, note that there are four transmission windows in the diagram. The first one is at approximately 850 nm, the second at 1310 nm, a third at 1550 nm, and a fourth at 1625 nm, the last two labeled 'C' and 'L' band, respectively. The 850-nm band was the first

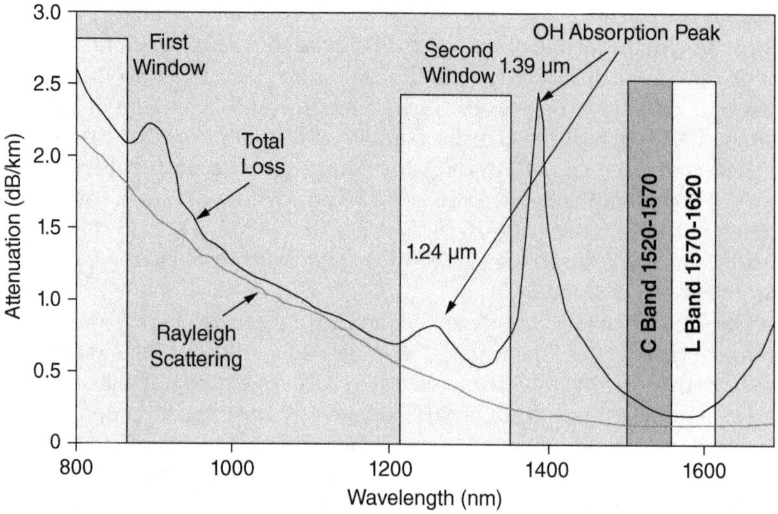

Figure 9.25 The optical transmission domain.

to be used because of its adherence to the wavelength at which the original LED technology operated. The second window at 1310 nm enjoys low dispersion; this is where dispersion effects are minimized. The so-called C-Band, at 1550 nm, has emerged as the ideal wavelength at which to operate long-haul systems and systems upon which *dense wavelength division multiplexing* (DWDM) has been deployed because (1) loss is minimized in this region, and (2) dispersion minimums can be shifted here.

Notice also that Rayleigh scattering is shown to occur at or around 1000 nm, while hydroxyl absorption by water occurs at 1240 and 1390 nm. Needless to say, network designers would be well-served to avoid transmitting at any of the points on the graph where Rayleigh scattering, high degrees of loss, or hydroxyl absorption have the greatest degree of impact. Notice also that dispersion, shown by the lower line, is at a minimum point in the second window, while loss, shown by the upper line, drops to a minimum point in the third window. In fact, dispersion is minimized in traditional single-mode fiber at 1310 nm, while loss is at minimums at 1550 nm. So the obvious question becomes this: Which one do you want to minimize—loss or dispersion?

Luckily, this choice no longer has to be made. Today, *dispersion-shifted fibers* (DSF) have become common. By modifying the manufacturing process, engineers can shift the point at which minimum dispersion occurs from 1310 to 1550 nm, causing it to coincide with the minimum loss point such that loss and dispersion occur at the same wavelength.

Unfortunately, while this fixed one problem, it created a new and potentially serious alternative problem. *Dense wavelength division multiplexing* has become a mainstay technology for multiplying the available bandwidth in optical systems. When DWDM is deployed over dispersion-shifted fiber, serious nonlinearities occur at the zero dispersion point which effectively destroy the DWDM signal. Think about it: DWDM relies on the ability to *channelize* the available bandwidth of the optical infrastructure and maintain some degree of separation between the channels. If dispersion is minimized in the 1559-nm window, then the channels will effectively overlay each other in DWDM systems. Specifically, a problem called four-wave mixing creates *sidebands* that interfere with the DWDM channels, destroying their integrity. In response, fiber manufacturers have created non-zero dispersion–shifted fiber (NZDSF) that lowers the dispersion point to near zero and making it occur just outside of the 1550-nm window. This eliminates the nonlinear four-wave mixing problem.

Fiber Nonlinearities

A classic business quote, imminently applicable to the optical networking world, observes "in its success lie the seeds of its own destruction." As the marketplace clamors for longer transmission distances with minimal amplification, more wavelengths per fiber, higher bit rates, and increased signal power, a rather ugly collection of transmission impairments, known as fiber nonlinearities, rises to challenge attempts to make them happen. These impairments go far beyond the simple concerns brought about by loss and dispersion; they represent a significant performance barrier.

The *special relationship* that exists between transmission power and the refractive index of the medium gives rise to four service-affecting optical nonlinearities: *self-phase modulation* (SPM), *cross-phase modulation* (XPM), *four-wave mixing* (FWM), and *intermodulation*.

Self-Phase Modulation

When self-phase modulation (SPM) occurs, chromatic dispersion kicks in to create something of a technological double-whammy. As the light pulse moves down the fiber, its leading edge increases the refractive index of the core, which causes a shift toward the longer wavelength, blue end of the spectrum. The trailing edge, on the other hand, decreases the refractive index of the core, causing a shift toward the shorter wavelength, red end of the spectrum. This causes an overall spreading or smearing of the transmitted signal, a phenomenon known as *chirp*. It occurs in fiber systems that transmit a single pulse down the fiber and is proportional to the amount of chromatic dispersion in the fiber: the more chromatic dispersion, the more SPM. It is counteracted with the use of large effective area fibers.

Cross-Phase Modulation

When multiple optical signals travel down the same fiber core, they both change the refractive index in direct proportion to their individual power levels. If the signals happen to cross, they will distort each other. While cross-phase modulation (XPM) is similar to SPM, there is one significant difference: while self-phase modulation is directly affected by chromatic dispersion, cross-phase modulation is only minimally affected by it. Large effective area fibers can reduce the impact of XPM.

Four-Wave Mixing

Four-wave mixing (FWM) is the most serious of the power/refractive index-induced nonlinearities today because it has a catastrophic effect on DWDM-enhanced systems. Because the refractive index of fiber is nonlinear, and because multiple optical signals travel down the fiber in DWDM systems, a phenomenon known as third-order distortion can occur that seriously affects multichannel transmission systems. Third-order distortion causes harmonics to be created in large numbers that have the annoying habit of occurring where the actual signals are, resulting in their obliteration.

Four-wave mixing is directly related to DWDM. In DWDM fiber systems, multiple simultaneous optical signals are transmitted across an optical span. They are separated on an ITU-blessed standard transmission grid by as much as 100 GHz (although most manufacturers today have reduced that to 50 GHz or better). This separation ensures that they do not interfere with each other.

Consider now the effect of dispersion-shifted fiber on DWDM systems. In DSF, signal transmission is moved to the 1550-nm band to ensure that dispersion and loss are both minimized within the same window. However, minimal dispersion has a rather severe unintended consequence when it occurs in concert with DWDM: because it reduces dispersion to near zero, it also prevents multichannel systems from existing because it does not allow proper channel spacing. Four-wave mixing, then, becomes a serious problem.

Several things can reduce the impact of FWM. As the dispersion in the fiber drops, the degree of four-wave mixing increases dramatically. In fact, it is worst at the zero-dispersion point. Thus, the intentional inclusion of a small amount of chromatic dispersion actually helps to reduce the effects of FWM. For this reason, fiber manufacturers sell non-zero dispersion shifted fiber, which moves the dispersion point to a point near the zero point, thus ensuring that a small amount of dispersion creeps in to protect against FWM problems.

Another factor that can minimize the impact of FWM is to widen the spacing between DWDM channels. This, of course, reduces the efficiency of the fiber by reducing the total number of available channels, and is therefore not a popular solution, particularly since the trend in the industry is to move toward narrower channel spacing as a way to increase the total number of available channels. Already several vendors have announced spacing as narrow as 5 GHz.

Finally, large effective area fibers tend to suffer less from the effects of FWM.

Intermodulation Effects

In the same way that cross-phase modulation results from interference between multiple simultaneous signals, intermodulation causes secondary frequencies to be created that are cross-products of the original signals being transmitted. Large effective area fibers can alleviate the symptoms of intermodulation.

Scattering Problems

Scattering within the silica matrix causes the second major impairment phenomenon. Two significant nonlinearities result: *stimulated Brillouin scattering* (SBS), and *stimulated Raman scattering* (SRS).

Stimulated Brillouin Scattering

Stimulated Brillouin scattering is a power-related phenomenon. As long as the power level of a transmitted optical signal remains below a certain threshold, usually on the order of 3 mW, stimulated Brillouin scattering is not a problem. The threshold is directly proportional to the fiber's effective area, and because dispersion-shifted fibers typically have smaller effective areas, they have lower thresholds. The threshold is also proportional to the width of the originating laser pulse: as the pulse gets wider, the threshold goes up. Thus, steps are often taken through a variety of techniques to artificially broaden the laser pulse. This can raise the threshold significantly, to as high as 40 mW.

Stimulated Brillouin scattering is caused by the interaction of the optical signal moving down the fiber with the acoustic vibration of the silica matrix that makes up the fiber. As the silica matrix resonates, it causes some of the signal to be reflected back toward the source of the signal, resulting in noise, signal degradation and a reduction of overall bit rate in the system. As the power of the signal increases beyond the threshold, more of the signal is reflected, resulting in a multiplication of the initial problem.

It is interesting to note that there are actually two forms of Brillouin scattering. When (sorry, a little more physics) electric fields that oscillate in time within an optical fiber interact with the natural acoustic resonance of the fiber material itself, the result is a tendency to backscatter light as it passes through the material. This is called

Brillouin scattering. If, however, the electric fields are caused by the optical signal itself, the signal is seen to cause the phenomenon; this is called stimulated Brillouin scattering.

To summarize, because of backscattering, SBS reduces the amount of light that actually reaches the receiver and cause noise impairments. The problem increases quickly above the threshold, and has a more deleterious impact on longer wavelengths of light. One additional fact: in-line optical amplifiers such as erbium-doped fiber amplifiers (EDFAs) add to the problem significantly. If there are four optical amplifiers along an optical span, the threshold will drop by a factor of four.

Solutions to SBS include the use of wider-pulse lasers and larger effective area fibers.

Stimulated Raman Scattering

Stimulated Raman scattering is something of a power-based crosstalk problem. In SRS, high-power, short wavelength channels tend to bleed power into longer wavelength, lower-power channels. It occurs when a light pulse moving down the fiber interacts with the crystalline matrix of the silica, causing the light to (1) be back-scattered and (2) shift the wavelength of the pulse slightly. Whereas SBS is a backward-scattering phenomenon, SRS is a two-way phenomenon, causing both back-scattering and a wavelength shift. The result is crosstalk between adjacent channels.

The good news is that SRS occurs at a much higher power level, close to a watt. Furthermore, it can be effectively reduced through the use of large effective area fibers.

Optical Amplification

As long as we are on the subject of Raman scattering, we should introduce the concept of optical amplification. This may seem like a bit of a non sequitur, but it really isn't; true optical amplification actually uses a form of Raman scattering to amplify the transmitted signal!

Traditional Amplification and Regeneration Techniques

In a traditional metallic analog environment, transmitted signals tend to weaken over distance. To overcome this problem, amplifiers are placed in the circuit periodically to raise the power level of the signal. There is a problem with this technique, however: in addition to amplifying the signal, amplifiers also amplify whatever cumulative noise has been picked up by the signal during its trip across the network. Over time, it becomes difficult for a receiver to discriminate between the actual signal and the noise embedded in the signal. Extraordinarily complex recovery mechanisms are required to discriminate between optical wheat and noise chaff.

In digital systems, regenerators are used to not only amplify the signal, but to also remove any extraneous noise that has been picked up along the way. Thus, digital regeneration is a far more effective signal recovery methodology than simple amplification.

Even though signals propagate significantly farther in optical fiber than they do in copper facilities, they are still eventually attenuated to the point that they must be regenerated. In a traditional installation, the optical signal is received by a receiver circuit, converted to its electrical analog, regenerated, converted back to an optical signal, and transmitted onward over the next fiber segment. This optical-to-electrical-to-optical conversion process is costly, complex and time-consuming. However, it is proving to be far less necessary as an amplification technique than it used to be because of true optical amplification that has recently become commercially feasible. Please note that optical amplifiers do not regenerate signals; they merely amplify. Regenerators are still required, albeit far less frequently.

Optical amplifiers represent one of the technological leading edges of data networking. Instead of the O-E-O process, optical amplifiers receive the optical signal, amplify it as an optical signal, and then retransmit it as an optical signal—no electrical conversion is required. Like their electrical counterparts, however, they also amplify the noise; at some point signal regeneration is required.

Optical Amplifiers: How They Work

It was only a matter of time before all-optical amplifiers became a reality. It makes intuitively clear sense that a solution that eliminates the electrical portion of the O-E-O process would be a good one; optical amplification is that solution.

You will recall that SRS is a fiber nonlinearity that is characterized by high-energy channels pumping power into low-energy channels. What if that phenomenon could be harnessed as a way to amplify optical signals that have weakened over distance?

Optical amplifiers are actually rather simple devices that as a result tend to be extremely reliable. The optical amplifier comprises the following: an input fiber, carrying the weakened signal that is to be amplified; a pair of optical isolators; a coil of doped fiber; a pump laser; and the output fiber that now carries the amplified signal. A functional diagram of an optical amplifier is shown in Fig. 9.26.

The coil of doped fiber, shown in Fig. 9.27, lies at the heart of the optical amplifier's functionality. Doping is simply the process of embedding some kind of functional impurity in the silica matrix of the fiber when it is manufactured. In optical amplifiers, this impurity is more often than not an element called erbium. Its role will become clear in just a moment.

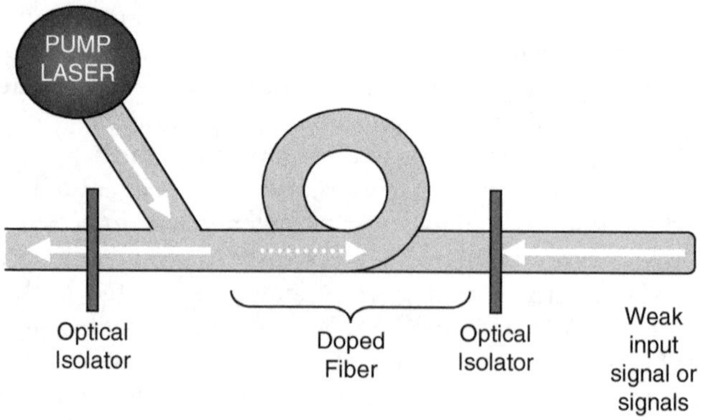

FIGURE 9.26 Schematic diagram of an optical amplifier.

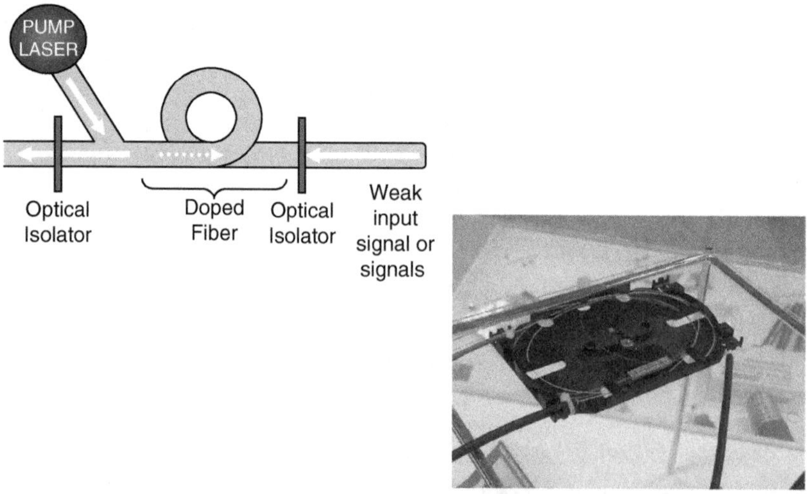

FIGURE 9.27 The coil of doped fiber inside an optical amplifier.

The pump laser shown in the upper left corner of Fig. 9.27 generates a light signal at a particular frequency, oftentimes 980 nm, in the opposite direction that the actual transmitted signal flows. As it turns out, erbium becomes atomically excited when it is struck by light at that wavelength. When an atom is excited by pumped energy, it jumps to a higher energy level (those of you who are recovering physicists will remember classroom discussions about orbital levels—$1S^1$, $1S^2$, $2S^1$, $2S^2$, $2P^6$, and so on.), then falls back down, during which it gives off a photon at a certain wavelength. When erbium is excited by light at 980 nm, it emits photons within the 1550-nm region—coincidentally the wavelength at which multichannel optical systems operate. So, when the weak, transmitted signal reaches the coil of erbium-doped fiber, the erbium atoms, now excited by the energy from the pump laser, bleed power into the weak signal at precisely the right wavelength, causing a generalized amplification of the transmitted signal. The optical isolators serve to prevent errant light from backscattering into the system, creating noise.

Erbium-doped fiber amplifiers (EDFAs) are highly proletariat in nature: they amplify anything, including the noise that the signal may have picked up. There will therefore still be a need at some point along the path of long-haul systems for regeneration, although far less frequently than in traditional copper systems. Most manufacturers of optical systems publish recommended span engineering specifications that help service providers and network designers take such concerns into account as they design each transmission facility.

Other Amplification Options

There are at least two other amplification techniques in addition to EDFAs that have recently come into favor. The first of these is called Raman amplification, which is similar to EDFA in the sense that it relies on Raman effects to do its task, but different for other rather substantial reasons. In Raman amplification, the signal beam travels down the fiber alongside a rather powerful pump beam, which excites atoms in the silica matrix that in turn emit photons that amplify the signal. The advantage of Raman amplification is that it requires no special doping: erbium is not necessary. Instead, the silica itself gives off the necessary amplification. In this case, the fiber itself becomes the amplifier!

Raman amplifiers require a significantly high-power pump beam (about a watt, although some systems have been able to reduce the required power to 750 mw or less) and even at high levels the power gain is relatively low. Their advantage, however, is that their induced gain is distributed across the entire optical span. Furthermore, it will operate within a relatively wide range of wavelengths, including 1310 and 1550 nm, currently the two most popular and effective transmission windows.

Semiconductor lasers have also been deployed as optical amplification devices in some installations. In semiconductor optical amplifiers, the weakened optical signal is pumped into the ingress edge of a semiconductor optical amplifier. The active layer of the semiconductor substrate amplifies the signal and regenerates it on the other side. The primary downside to these devices is their size: they are small, and their light-collecting ability is therefore somewhat limited. A typical single-mode fiber generates an intense spot of light that is roughly 10 μm in diameter; the point upon which that light impinges upon the semiconductor amplifier is less than a micron in diameter, meaning that a lot of the light is lost. Other problems also crop up including polarization issues, reflection, and variable gain. As a result, these devices are not in widespread use; EDFAs and Raman amplification techniques are far more common.

Optical Receivers

So far, we have discussed the sources of light, including LEDs and laser diodes; we have briefly described the various flavors of optical fiber and the problems they encounter as transmission media; now, we turn our attention to the devices that receive the transmitted signal.

The receive devices used in optical networks have a single responsibility: to capture the transmitted optical signal and convert it into an electrical signal that can then be processed by the end equipment. There may also be various stages of amplification to ensure that the signal is strong enough to be acted upon, and demodulation circuitry, which recreates the originally transmitted electronic signal.

Photodetector Types

While there are many different types of photosensitive devices, there are two used most commonly as photodetectors in modern networks: positive-intrinsic-negative (PIN) photodiodes, and avalanche photodiodes (APDs).

Positive-Intrinsic-Negative Photodiodes

Positive-intrinsic-negative (PIN) photodiodes are similar to the device described above in the general discussion of photosensitive semiconductors. Reverse biasing the junction region of the device prevents current flow until light at a specific wavelength strikes the substance, creating electron-hole pairs and allowing current to flow across the three-layer interface in proportion to the intensity of the incident light. While they are not the most sensitive devices available for the purpose of photodetection, they are perfectly adequate for the requirements of most optical systems. In cases where they are not considered sensitive enough for high-performance systems, they can be coupled with a preamplifier to increase the overall sensitivity.

Avalanche Photodiodes

Avalanche photodiodes (APD) work as optical signal amplifiers. They use a strong electric field to perform what is known as avalanche multiplication. In an APD, the electric field causes current accelerations such that the atoms in the semiconductor matrix get excited and create, in effect, an *avalanche* of current to occur. The good news is that the amplification effect can be as much as 30 to 100 times the original signal; the bad news is that the effect is not altogether linear and can create noise. APDs are sensitive to temperature and require a significant voltage to operate them—30 to 300 V depending on the device. However, they are popular for broadband systems and work well in the gigabit range.

We have now discussed transmitters, fiber media, and receivers. In the next section, we examine the fibers themselves, and how they have been carefully designed to serve as solutions for a wide variety of networking challenges and to forestall the impact of the nonlinearities described in this section.

Optical Fiber

As was mentioned briefly in a prior section, fiber has evolved over the years in a variety of ways to accommodate both the changing requirements of the customer community and the technological challenges that emerged as the demand for bandwidth climbed precipitously. These changes came in various forms of fiber that presented different behavior characteristics to the market.

Modes: An Analogy

The concept of *modes* is sometimes difficult to understand, so let me pass along an analogy that will help. Imagine a shopping mall that has a wide, open central area that all the shops open onto. An announcement comes over the PA system informing people that "The mall is now closed; please make your way to the exit." Shoppers begin to make their way to the doors, but some wander from store to store, window-shopping along the way, while others take a relatively straight route to the exit. The result is that some shoppers take longer than others to exit the mall because there are different modes.

Now consider a mall that has a single, very narrow corridor that is only as wide as a person's shoulders. Now when the announcement comes, everyone heads for the exit, but they must form a single file line and head out in an orderly fashion. If you understand the difference between these two examples, you understand single versus multimode fiber. The first example represents multimode; the second represents single mode.

Multimode Fiber

The first of these was multimode fiber, which arrived in a variety of different forms. Multimode fiber bears that name because it allows more than a single mode or ray of light to be carried through the fiber simultaneously because of the relatively wide core diameter that characterizes the fiber. And while the dispersion that potentially results from this phenomenon can be a problem, there are advantages to the use of multimode fiber. For one thing, it is far easier to couple the relatively wide and forgiving end of a multimode fiber to a light source than that of the much narrower single mode fiber. It is also significantly less expensive to manufacture (and purchase) and relies on LEDs and inexpensive receivers rather than the more expensive laser diodes and ultra-sensitive receiver devices. However, advancements in technology have caused the use of multimode fiber to fall out of favor; single-mode is far more commonly used today.

Multimode fiber is manufactured in two forms: step-index fiber, and graded-index fiber. We will examine each in turn.

Multimode Step-Index Fiber

In step-index fiber, the index of refraction of the core is slightly higher than the index of refraction of the cladding. Remember that the higher the refractive index, the slower the signal travels through the medium. Thus, in step-index fiber, any light that escapes into the cladding because it enters the core at too oblique an angle will actually travel slightly faster in the cladding (assuming it does not escape altogether) than it would if it traveled in the core. Of course, any rays that are reflected repeatedly as they traverse the core also take longer to reach the receiver, resulting in a dispersed signal that causes problems for the receiver at the other end. Clearly, this phenomenon is undesirable; for that reason, graded-index fiber was developed.

Multimode Graded-Index Fiber

Because of the dispersion that is inherent in the use of step-index fiber, optical engineers created graded index fiber as a way to overcome the signal degradation that occurred.

In graded-index fiber, the refractive index of the core actually decreases from the center of the fiber outward. In other words, the refractive index at the center of the core is higher than the refractive index at the edge of the core. The result of this rather clever design is that as light enters the core at multiple angles and travels from the center of the core outward, it is actually accelerated at the edge and slowed down near the center, causing most of the light to arrive at roughly the same time. Thus, graded-index fiber helps to overcome the dispersion problems associated with step-index multimode fiber. Light that enters this type of fiber does not travel in a straight line, but rather follows a parabolic path, with all rays arriving at the receiver at more or less the same time.

Graded-index fiber typically has a core diameter of 50 to 62.5 μm, with a cladding diameter of 125 μm. Some variations exist; there is at least one form of multimode graded-index with a core diameter of 85 μm, somewhat larger than those described above. Furthermore, the actual thickness of the cladding is important: if it is thinner than

20 μm, light begins to seep out, causing additional problems for signal propagation.

Graded-index fiber was commonly used in telecommunications applications until the late 1980s. Even though graded-index fiber is significantly better than step-index fiber, it is still multimode fiber and does not eliminate the problems inherent in being multimode. Thus was born the next generation of optical fiber: single-mode.

Single-Mode Fiber

There is an interesting mental conundrum that crops up with the introduction of single-mode fiber. The core of single-mode fiber is significantly narrower than the core of multimode fiber. Because it is narrower, it would seem that its ability to carry information would be reduced due to limited light-gathering ability. This, of course, is not the case. As its name implies, it allows a single mode or ray of light to propagate down the fiber core, thus eliminating the intermodal dispersion problems that plague multimode fibers. In reality, single-mode fiber is a stepped-index design, because the core's refractive index is slightly higher than that of the cladding. It has become the de facto standard for optical transmission systems, and takes on many forms depending on the specific application within which it will be used.

Most single-mode fiber has an extremely narrow core diameter on the order of 7 to 9 μm, and a cladding diameter of 125 μm. The advantage of this design is that it only allows a single mode to propagate; the downside, however, is the difficulty involved in working with it. The core must be coupled directly to the light source and the receiver in order to make the system as effective as possible; given that the core is approximately one-sixth the diameter of a human hair, the mechanical process through which this coupling takes place becomes Herculean.

Single-Mode Fiber Designs

The reader will recall that we spent a considerable amount of time discussing the many different forms of transmission impairments (nonlinearities) that challenge optical systems. Loss and dispersion are the key contributing factors in most cases, and do in fact cause serious problems in high-speed systems. The good news is that optical engineers have done yeoman's work creating a wide variety of single-mode fibers that address most of the nonlinearities.

Since its introduction in the early 1980s, single-mode fiber has undergone a series of evolutionary phases in concert with the changing demands of the bandwidth marketplace. The first variety of single-mode fiber to enter the market was called non-dispersion-shifted fiber (NDSF). Designed to operate in the 1310-nm second window, dispersion in these fibers was close to zero at that wavelength. As a result, it offered high bandwidth and low dispersion. Unfortunately, it was soon the victim of its own success. As demand for high-bandwidth transport grew, a third window was created at 1550 nm for single-mode fiber transmission. It provided attenuation levels that were less than half those measured at 1310 nm, but unfortunately was plagued with significant dispersion. Since the bulk of all installed fiber was NDSF, the only solution available to transmission designers was to narrow the line width of the lasers employed in these systems and to make them more powerful. Unfortunately, increasing the power and reducing the laser line width is expensive, so another solution emerged.

Dispersion-Shifted Fiber

One solution that emerged was dispersion-shifted fiber. With *dispersion-shifted fiber* (DSF), the minimum dispersion point is mechanically shifted from 1310 to 1550 nm by modifying the design of the actual fiber so that waveguide dispersion is increased. The reader will recall that waveguide dispersion is a form of chromatic dispersion that occurs because the light travels at different speeds in the core and cladding.

One technique for building DSF (sometimes called *zero dispersion-shifted fiber*) is to actually build a fiber of multiple layers. In this design, the core has the highest index of refraction and changes gradually from the center outward until it equals the refractive index of the outer cladding. The inner core is surrounded by an inner cladding layer, which is in turn surrounded by an outer core. This design works well for single wavelength systems, but experiences serious signal degradation when multiple wavelengths are transmitted, such as when used with DWDM systems. Four-wave mixing, described earlier, becomes a serious impediment to clean transmission in these systems. Given that multiple wavelength systems are fast becoming the norm today, the single wavelength limit is a showstopper. The result was a relatively simple and elegant set of solutions.

The second technique was to eliminate or at least substantially reduce the absorption peaks in the fiber performance graph so that the second and third transmission windows merge into a single larger window, thus allowing for the creation of the fourth window described earlier which operates between 1565 and 1625 nm—the so-called *L-band*.

Finally, the third solution came with the development of non-zero dispersion-shifted fiber (NZDSF). NZDSF shifts the minimum dispersion point so that it is close to the zero point, but not actually at it. This prevents the nonlinear problems that occur at the zero point to be avoided because it introduces a small amount of chromatic dispersion.

Why Does It Matter?

It is always good to go back and review why we care about such things as dispersion-shifting and absorption issues. Remember that the key to keeping the cost of network down is to reduce maintenance and the need to add hardware or additional fiber when bandwidth gets tight. DWDM, discussed in detail later, offers an elegant and relatively simple solution to the problem of the cost of bandwidth. However, its use is not without cost. Multi-wavelength systems will not operate effectively over dispersion-shifted fiber because of dramatic nonlinearities, so if DWDM is to be used, non-zero dispersion-shifted fiber must be deployed.

Optical Fundamentals Summary

In this section, we have examined the history of optical technology and the technology itself, focusing on the three key components within an optical network: the light emitter, the transport medium, and the receiver. We also discussed the various forms of transmission impairment that can occur in optical systems, and the steps that have been taken to overcome them.

The result of all this is that optical fiber, once heralded as a near-technological miracle because it only lost 99 percent of its signal strength when transmitted over an entire kilometer, has become the standard medium for transmission of high-bandwidth signals over great distances. Optical amplification now serves as an augmentation to traditional regenerated systems, allowing for the elimination of the optical-to-electrical conversion that must take place in copper systems. The result of all this is an extremely efficient transmission system that has the ability to play a role in virtually any network design in existence today.

Dense Wavelength Division Multiplexing

When high-speed transport systems such as SONET and SDH were first introduced, the bandwidth that they made possible was unheard of. The early systems that operated at OC-3/STM-1 levels (155.52 Mbps) provided volumes of bandwidth that were almost unimaginable. As the technology advanced to higher levels, the market followed a phenomenon known as *Say's law*, creating demand for the ever more available volumes of bandwidth. There were limits, however; today, OC-48/STM-16 (2.5 Gbps) is extremely popular, but OC-192/STM-64 (10 Gbps) represents the practical upper limit of SONET's and SDH's transmission capabilities given the limitations of existing time division multiplexing technology. The alternative is to simply multiply the channel count—and that's where WDM comes into play.

WDM is really nothing more than frequency division multiplexing, albeit at very high frequencies. The ITU has standardized a channel separation grid that centers around 193.1 THz, ranging from 191.1 to 196.5 THz. Channels on the grid are technically separated by 100 GHz, but many industry players today are using 50 GHz separation.

The majority of WDM systems operate in the C-band (third window, 1550 nm), which allows for close placement of channels and the reliance on EDFAs to improve signal strength. Older systems, which spaced the channels 200 GHz (1.6 nm) apart, were referred to simply as WDM systems; the newer systems are referred to as dense WDM systems because of their tighter channel spacing. Modern systems routinely pack 40 to 10 Gbps channels across a single fiber, for an aggregate bit rate of 400 Gbps.

How DWDM Works

As Fig. 9.28 illustrates, a WDM system consists of multiple input lasers, an ingress multiplexer, a transport fiber, an egress multiplexer, and of course, customer receiving devices. If the system has eight channels such as the one shown in the diagram, it has eight lasers and eight receivers. The channels are separated by 100 GHz to avoid fiber nonlinearities, or closer if the system supports the 50-GHz spacing. Each channel, sometimes referred to as a lambda (λ, the Greek letter and universal symbol used to represent wavelength), is individually modulated, and ideally the signal strengths of the channels should be close to one another. Generally speaking this is not a problem, because in DWDM systems the channels are closely spaced and therefore do not experience significant attenuation variation from channel to channel.

There is a significant maintenance issue that face operators of DWDM-equipped networks. Consider a 16-channel DWDM system. This system has 16 lasers, one for each channel, which means that the service provider must maintain 16 spare lasers in case of a laser failure. The latest effort underway is the deployment of tunable lasers, which allow the laser to be tuned to any output wavelength, thus reducing the volume of spares that must be maintained and by extension, the cost.

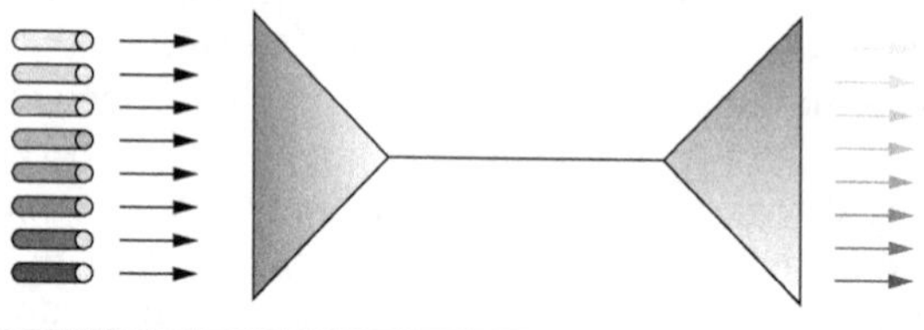

FIGURE 9.28 Wavelength division multiplexing (WDM).

So what do we find in a typical WDM system? A variety of components, including multiplexers, which combine multiple optical signals for transport across a single fiber; de-multiplexers, which disassemble the aggregate signal so that each signal component can be delivered to the appropriate optical receiver (PIN or APD); active or passive switches or routers, which direct each signal component in a variety of directions; filters, which serve to provide wavelength selection; and finally, optical add-drop multiplexers, which give the service provider the ability to pick up and drop off individual wavelength components at intermediate locations throughout the network. Together, these components make up the heart of the typical high-bandwidth optical network. And why is DWDM so important? Because of the cost differential that exists between a DWDM-enhanced network and a traditional network. To expand network capacity today by putting more fiber in the ground costs, on average, about $70K per mile. To add the same bandwidth using DWDM by changing out the endpoint electronics costs roughly one-sixth that amount. There is clearly a financial incentive to go with the WDM solution.

Optical Switching and Routing

DWDM facilitates the transport of massive volumes of data from a source to a destination. Once the data arrives at the destination, however, it must be terminated and redirected to its final destination on a lambda-by-lambda basis. This is done with switching and routing technologies.

The principal form of optical switching is really nothing more than a very sophisticated digital cross-connect system. In the early days of data networking, dedicated facilities were created by manually patching the end points of a circuit at a patch panel, thus creating a complete four-wire circuit. Beginning in the 1980s, digital cross-connect devices such as AT&T's digital access and cross-connect (DACS) became common, replacing the time-consuming, expensive, and error-prone manual process. The digital cross-connect is really a simple switch, designed to establish *long-term temporary* circuits quickly, accurately, and inexpensively.

Enter the world of optical networking. Traditional cross-connect systems worked fine in the optical domain, provided there was no problem going through the O-E-O conversion process. This, however, was one of the aspects of optical networking that network designers wanted to eradicate from their functional requirements. Thus was born the optical cross-connect switch.

The first of these to arrive on the scene was Lucent Technologies' Lambda router. Based on a switching technology called micro electrical mechanical system (MEMS), the Lambda router was the world's

Figure 9.29 A typical MEMS device. (*Image used courtesy of Lucent Technologies.*)

first all-optical cross-connect device. The product has since been discontinued, but the technology remains as a good illustration of optical switching.

MEMS relies on micro-mirrors, an array of which is shown in Fig. 9.29. The mirrors can be configured at various angles to ensure that an incoming lambda (wavelength) strikes one mirror, reflects off a fixed mirrored surface, strikes another movable mirror, and is then reflected out an egress fiber. The Lambda router and other devices like it offered switching speed, a relatively small footprint, bit rate and protocol transparency, non-blocking architecture, and highly developed database management. Fundamentally, these devices are very high-speed, high-capacity switches or cross-connect devices. They are not routers, because they do not perform layer three functions.

MEMS Alternatives

The problem with MEMS devices was their reliability. They were so small that the early implementations had a tendency to fail. Imagine a silicon chip the size of your little fingernail, on which are 256 tiny mirrors, each of which can be individually positioned by electronic command to a particular angle. At that scale, something's bound to fail. And they did: roughly 20 percent of the mirrors in the early devices stopped working after a short period of time. Clearly an alternative technology was required.

Several emerged. The first was a technique called piezoelectric beam steering, in which ceramic elements were used to guide the direction of the transmitted beam. Another was a technique that emerged from the world of inkjet printers; in this scheme, a tiny bubble injected by an inkjet nozzle refracted the inbound optical signal, changing its direction. A third technique was to use liquid crystal technology to create opaque *gates* that either permitted or blocked the transmission of light. Another technique used heat to change the refractive index of the material through which the light was being transmitted.

Optical networking, then, is a major component of the modern transport network. Together with switching technologies like frame relay and ATM, and physical layer standards like SONET, SDH, and copper-based transport technologies, the current network is able to handle the transport requirements of any traffic type while overlaying quality of service at the same time.

An Aside: A trip to the *Wave Venture*

While working in Singapore recently, a friend approached me and said, "I just finished reading your Optical Networking Crash Course book. And I noticed that you didn't have much to say in it about submarine cables." Yes, I replied, explaining that the book was really designed to explain the nature of terrestrial systems. I also added the observation that I had no experience in the submarine domain and didn't feel equipped to tackle that subject in detail in the book. "So you've never seen a cable laying ship?" he asked. "No," I replied. "Would you like to?" He countered. "Absolutely!" I said, probably a little more enthusiastically than I should have. He explained that a good friend of his was the president of Asia Global Crossing, which had a ship in port, and if I wanted he could probably set up a tour. So he went away to make a phone call, and when he returned he told me that it was all set up—we would tour the ship on Friday.

Friday morning arrived, and we took a taxi down to the industrial pier where we boarded an ancient, scabby workboat that chugged us out to the Wave Venture, anchored a few miles offshore (Fig. 9.30). Approaching the vessel, I was amazed at its size: a 142-m cable laying and maintenance ship, originally built as passenger ferry in 1982 and converted to a cable ship in 2000. She can stay at sea for as long as 40 days without resupplying food and equipment for her 62 crew members.

We boarded the ship via a starboard boarding ladder and gathered on the fantail, where Martin Swaffield, Master of the vessel, began the tour. We climbed up several flights of stairs to the driving

Figure 9.30 The *Wave Venture*, a cable laying ship in Singapore.

bridge where he explained the principal responsibilities of the ship and its crew. "Our primary job is cable maintenance. We locate cables, pull them up from the bottom, find the problem, repair it, and drop it back on the sea floor. We also lay cable: We can put down as much as 15 km a day, depending on the makeup of the bottom and obstacles we might have to go around, and we do it all remotely. As you'll see in a few minutes we have a robot that goes down and does all of the trenching and labor for us." I asked him what the biggest problem was that they faced.

"Unexploded ordinance," he responded. "We often pull up a cable that has been fouled in a fishing net, and in their attempts to free the drag net the fishing boat damages the cable. Unfortunately, a lot of unexploded ordinance has been dropped into the ocean over the years, sometimes intentionally, sometimes by accident. It's often the case that when we drag up the cable we find an unexploded mine or bomb in the net, dangling behind the stern of the ship. At that point we stop all operations, call whatever navy is closest and ask them to come deal with it."

From the driving bridge we strolled over to the positioning bridge, a separate control room that is used to hold the ship on station during cable operations. Naturally it faces toward the stern of the vessel so that Swaffield and his team can observe what is going on below. From the windows of the positioning bridge we looked down

Figure 9.31 The working deck of the *Wave Venture*, looking aft.

on the entire working area of the ship (Fig. 9.31). On the near right below us was the control room for the robotic cable layer; on the left, the robot itself, hanging from a giant crane. And far off in the distance was the stern of the ship where cables were retrieved for repair or dropped.

From the bridge we went back down to deck level where we walked into the covered area of the deck—looking for all the world like a giant garage (Fig. 9.32). On the left was a structure that resembled a restaurant walk-in refrigerator, which Martin confirmed was in fact a freezer where the repeaters are kept. "They cost a million dollars apiece; we keep them cold so that the electronics won't be shocked when we dump them overboard after installing them on cables."

On the right was a small room where fiber splicing is done (Fig. 9.33). A long, narrow window ran the length of the room, and cables to be spliced could be dragged through the window for repair.

This was also where I saw cross sections of submarine cable for the first time. The stuff is remarkable, and comes in a variety of forms depending on where it will be deployed. For deep water deployment, where the danger of a ship dropping an anchor on it or the likelihood of a fishing net getting snagged on the cable is relatively low, a fairly thin unarmored cable, shown in Fig. 9.34, is used. In shallower water,

Figure 9.32 Covered area of the working deck where the regenerators are stored as well as the cable repair facilities.

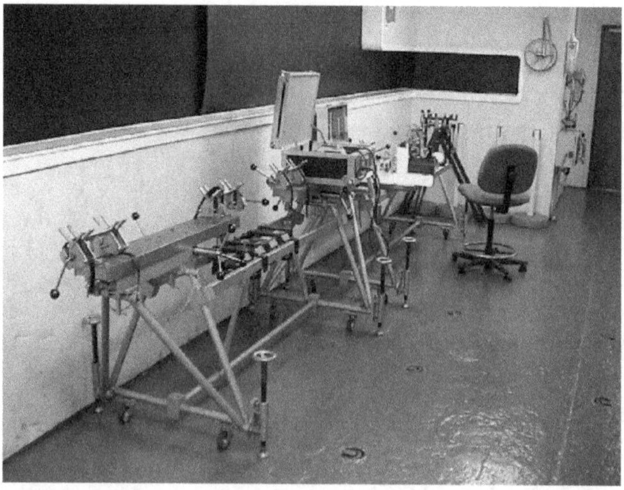

Figure 9.33 The fiber splicing room.

however, where crushing incidents are more common, a heavier, armored cable is deployed, shown in Fig. 9.35. Take a close look: the cable has four layers of armored strands, the largest of which are about the diameter of a pencil.

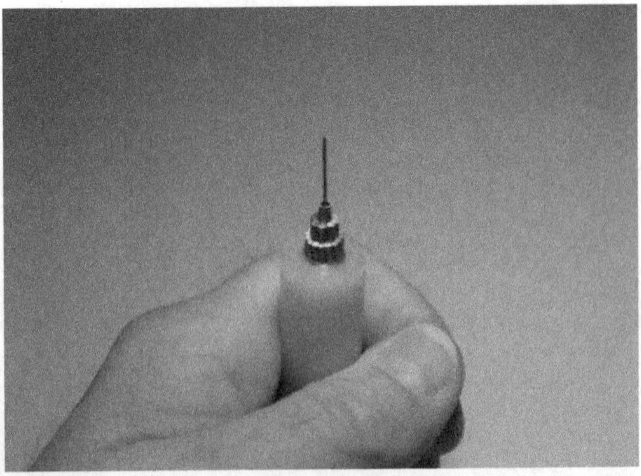

FIGURE 9.34 Unarmored submarine cable, used in deep water installations.

FIGURE 9.35 Heavily armored submarine cable, used in shallow water installations.

The robotic cable-laying device is a remarkable piece of machinery (Fig. 9.36). Controlled by an operator using a joystick like a video game, it is dropped on a cable to the bottom, where it either achieves neutral buoyancy or crawls across the bottom on tractor treads. It has thrusters to control its position, digging tools (including a giant circular rock saw that looks like something out of Roger Rabbit), and massive grappling arms with which it can gently pick up a fragile strand of unarmored cable or tear apart a rocky obstruction. It can operate in water as deep as 6,000 ft.

We then descended into the lower deck of the ship where the cable tanks are located (Fig. 9.37). These tanks—three in all, about

Figure 9.36 Remotely operated vehicle used for cable installation and repair in deep water.

Figure 9.37 Cable tanks filled with submarine cable, ready for installation on the sea floor.

FIGURE 9.38 Cable supports on the main deck.

50 ft in diameter and 11-ft deep, hold 1,200 miles of cable, and weigh about 1,700 tons. The cable is fed into the tanks from the main deck where it is coiled by hand, a process that takes days. Cable is pulled from the tanks as needed, and fed down the stern deck of the ship through the cable supports visible in Fig. 9.38.

No question about it, this is one of the coolest tours I've ever had the honor to take.

Chapter Summary

This concludes our long-winded coverage of transport solutions. Clearly they are remarkably diverse, but with that diversity and complexity comes an equally far-ranging degree of capability. We'll expand our understanding of that capability in our next chapter on the IP takeover.

Chapter Nine Questions

1. What is the difference between access and transport?

2. Explain the advantages and disadvantages of dedicated versus switched transport facilities.

3. Draw and label the switching hierarchy.

4. How do frame relay and ATM differ?

5. What is the difference between single mode and multimode fiber?

6. How does an EDFA work?

7. Even though optical fiber is immune from electromagnetic interference, it still suffers from transmission impairments. What are some of them?

CHAPTER 10

The IP Takeover

"On the Internet, nobody knows you're a dog."

—One dog to another while surfing the Web, in a cartoon by Peter Steiner

Most technology historians credit the invention of the Internet to a combination of the U.S. Department of Defense's Advanced Research Projects Agency (ARPA), the organization that created the ARPANET. They're also quick to credit Vinton Cerf and Robert Khan, the computer scientists who wrote TCP/IP, the fundamental protocol upon which the Internet operates. But there was someone else involved whose work predated all of them, and he rarely receives any credit.

In 2013, at a well-attended ceremony in London, Queen Elizabeth II awarded the 1 million pound Queen Elizabeth Prize for Engineering to five individuals: Tim Berbers-Lee considered to be the father of the World Wide Web; Vinton Cerf and Bob Khan, who wrote the now famous TCP/IP protocol that underpins the Internet's functionality; Marc Andreessen, who created Netscape, the first successful browser; and a little-known gentleman named Louis Pouzin whose contributions to the network we now depend on for just about everything are every bit as important as those of the other four who were honored that day.

Pouzin studied engineering at the École Polytechnique, after which he went to work for France's monopoly PTT. In the 1950s, he left the PTT and went to work for Bull, at the time France's answer to IBM, a move he undertook after reading an article in *le Monde* about IBM's intent to build computers that would automate vast amounts of the work done at the time by people, leaving people to do far more interesting things.

In the 1960s, he and his family left France and moved to Massachusetts where Pouzin wrote a program called RUNCOM, a timeshare program and the first example of shell-based command line environments ever created. His fascination with the world of computer science was just beginning.

Later that decade, Pouzin was asked to return to France to manage a research project designed to create the country's first national data network, a project called CYCLADES. Pouzin was familiar with the work undertaken by the U. S. Department of Defense's DARPA organization and its ongoing development of ARPANET (the precursor of the modern Internet), but after visiting a handful of American universities where the ARPANET research was being conducted, he concluded that the overall design of the network was bloated and inefficient. Most vexing to him was the fact that the network was connection-oriented, meaning that a connection had to be established on an end-to-end basis before data could be transmitted.

So Pouzin and his team set out to create something better, a model that required no call setup. Instead, in his design, packets would make their way independently across the network and be reassembled on the other end by the receiving machine. Lost packet? No problem—retransmission of individual packet was included in his design.

Vinton Cerf and Robert Khan, by this time tasked with overhauling the inefficient ARPANET, were paying close attention to Pouzin's work, which debuted in the early 1970s on a connection between Paris and Grenoble. Their redesign, which ultimately became the protocol suite known as the transmission control protocol (TCP) and the Internet protocol (IP) was based largely on the early successes of Mr. Pouzin and his team of researchers.

But as is often the case, no good deed goes unpunished, and Pouzin's work fell squarely in the gun sights of the legacy telephone company in France—not to mention every other European, state-controlled PTT. His connectionless network and the success it enjoyed flew in the face of their investments in connection-oriented switched networks. These were centralized, power-hungry, command-and-control-oriented, top-down, hierarchical organizations that ran centralized, power-hungry, command-and-control-oriented, top-down, hierarchical networks. The idea of a "spray-and-pray" architecture that decentralized power and made no delivery guarantees was anathema to them, and they revolted—loudly. Even though President Georges Pompidou was a staunch defender of Pouzin's work, after his death in 1974 that support disappeared and the vandals stormed the gates of Pouzin's kingdom. All funding for CYCLADES stopped, and Pouzin was sidelined. Meanwhile the PTT continued with its rollout of a connection-oriented network called TRANSPAC, the basis for France's famous Minitel network. It worked well until it fell behind and was unable to compete with, you guessed it, the Internet.

Perhaps the best vindication of Pouzin's work is Vinton's Cerf's observation during the London ceremony: "Recognition has come very, very late for Louis–unfairly so."

The Internet's Early Days

The physical Internet as we know it today began as a U. S. Department of Defense (DoD) project designed to interconnect four DoD research sites. In December of 1968, the government research agency known as the Advanced Research Projects Agency (ARPA) awarded a contract to consultancy Bolt, Beranek, and Newman (BBN) to design and build the packet network that would ultimately become the Internet. It had a proposed transmission speed of 50 kbps, and in September of 1969 the first node was installed at UCLA. Other nodes were installed on roughly a monthly basis at Stanford Research Institute (SRI), the University of California at Santa Barbara (UCSB), and the University of Utah. From those modest beginnings, the ARPANET grew to span the continental U. S. by 1971 and had connections to research facilities in Europe by 1973.

The original protocol selected for the ARPANET was called the network control protocol (NCP). It was designed to handle the emergent requirements of the low-volume architecture of the ARPANET network. As traffic grew, however, it proved to be inadequate to handle the load, and in 1974 a more robust protocol suite was implemented, based on the transmission control protocol (TCP), the ironclad protocol designed by Cerf and Khan for end-to-end network communications control. In 1978, a new design split the responsibilities for end-to-end versus node-to-node transmission among two protocols: the newly crafted Internet protocol (IP), designed to route packets from device-to-device, and TCP, designed to offer reliable, end-to-end communications. Since TCP and IP were originally envisioned as a single protocol they are now known as the TCP/IP protocol suite, a name that also incorporates a collection of protocols and applications that also handle routing, QoS, error control, and other functions.

One problem that occurred that the ARPANET planners didn't envision when they sited their nodes at college campuses was visibility. Naturally, they placed the switches in the raised floor facilities of the computer science department, and we know what is also found there: *undergraduus nerdus*, the dreaded computer science (or worse yet, engineering) student. In a matter of weeks the secret was out—ARPA's top-secret network was top-secret no longer. So, in 1983, the ARPANET was split into two networks. One half, still called ARPANET, continued to be used to interconnect research and academic sites; the other, called MILNET, was specifically used to carry military traffic and ultimately became part of Defense Data Network (DDN).

That year was also a good year for TCP/IP. It was included as part of the communications kernel for the University of California's UNIX implementation, known as 4.2BSD (Berkeley Software Distribution) UNIX.

ARPANET Growing Pains

Extension of the original ARPANET continued. In 1986, the National Science Foundation (NSF) built a backbone network to interconnect four NSF supercomputing centers and the National Center for

Atmospheric Research. This network, known as NSFNET, was originally intended to serve as a backbone for other networks, not as a standalone interconnection mechanism. Additionally, the NSF's appropriate use policy limited NSFNET to noncommercial traffic only. NSFNET continued to expand and eventually became what we know today as the Internet. And while the original NSFNET applications were multiprotocol implementations, TCP/IP was used for overall interconnectivity.

In 1994, a structure was put in place to reduce the NSF's overall role in the Internet. The new structure consists of three principal components. The first of these was a small number of network access points (NAPs) where Internet service providers (ISPs) would interconnect to the Internet backbone. The NSF originally funded four NAPs in Chicago (originally operated by Ameritech, now part of AT&T), New York (really Pensauken, NJ, operated by Sprint), San Francisco (operated by Pacific Bell, now part of AT&T), and Washington, D.C. (MAE-East, originally operated by MFS, later acquired by Verizon).

The second component was the Very High-Speed Backbone Network Service, a network that interconnected the NAPs and was originally operated by MCI. It was installed in 1995 and originally operated at OC-3 (155.52 Mbps), but was upgraded to OC-12 (622.08 Mbps) in 1997.

The third component was the routing arbiter, designed to ensure that appropriate routing protocols for the Internet were available and properly deployed.

ISPs were given five years of diminishing funding to become commercially self-sustaining. The funding ended in 1998 and starting at roughly the same time a significant number of additional NAPs were launched. As a matter of control and management, three tiers of ISP have been identified. Tier 1 refers to national ISPs that have a national presence and connect to at least three of the original four NAPs. Tier 2 refers to regional ISPs that primarily have a regional presence and connect to less than three of the original four NAPs. Finally, Tier 3 refers to local ISPs, or those that do not connect to a NAP but offer services via the connections of another ISP.

Managing the Internet

The Internet is really a network of networks interconnected by high-bandwidth circuits leased (for the most part) from various telephone companies. It is something of an erroneous conclusion to think of the Internet as a standalone network; in fact, it is made up of the same network components that make up corporate networks and public telephone networks. It's just a collection of leased lines that interconnect the routers that provide the sophisticated routing functions that make the Internet so powerful. As such, it is owned by no one and owned by everyone, but more importantly it is *managed* by no one and *managed* by everyone! There is no single authority that governs the Internet because the Internet is not a single monolithic entity but rather a collection of entities. Yet it runs well, perhaps better than other more centrally managed services. This, of course, was part of Louis Pouzin's thinking when he created his original CYCLADES network in France.

That said, there are certainly a number of organizations that provide oversight, guidance, and management for the Internet community. These organizations, described in some detail in the following sections, help to guide the developmental direction of the Internet with regard to such functions as communications standards, universal addressing systems, domain naming, protocol evolution, and so on.

Internet Activities Board

One of the longest standing organizations with Internet oversight authority is the Internet Activities Board (IAB), which governs administrative and technical activities on the Internet. It works closely with the Internet Society (ISOC), a non-governmental organization established in 1992 that coordinates such Internet functions as internetworking and applications development. ISOC is also chartered to provide oversight and communications for the IAB.

Internet Engineering Task Force

The Internet Engineering Task Force (IETF) is one of the best-known Internet organizations and is part of the IAB. The IETF establishes working groups responsible for technical activities involving the Internet such as writing technical specifications and protocols. Because of the organization's early, valuable commitment to the future of the Internet, the International Organization for Standardization (ISO) accredited the organization as an *official* international standards body at the end of 1994.

World Wide Web Consortium

On a similar front, the World Wide Web Consortium (W3C) has no officially recognized role but has taken a leading role in the development of standard protocols for the World Wide Web to promote its evolution and ensure its interoperability. Led by World Wide Web creator Tim Berners-Lee, the organization has more than 400 member organizations and is currently leading the technical development of the Web.

Other Organizations

Other smaller organizations play important roles as well. These include the Internet Engineering Steering Group (IESG), which provides strategic direction to the IETF and is part of the IAB; the Internet Research Task Force (IRTF), which engages in research that affects the evolution of the future Internet; the Internet Engineering Planning Group (IEPG), which coordinates worldwide Internet operations and helps ISPs achieve interoperability; and finally, the Forum of Incident Response and Security Teams, which coordinates Computer Emergency Response Teams (CERTs) in various countries.

Together these organizations ensure that the Internet operates as a single, coordinated *organism.* Remarkable, isn't it? The least managed network in the world (and arguably the largest) is also the most effectively managed!

Naming Conventions in the Internet

The Internet is based on a system of domains, that is, a hierarchy of names that uniquely identify the *operating regions* in the Internet. For example, IBM.com is a domain, as are Verizon.net, fcc.gov, and ShepardComm.com. The domains help to guide traffic from one user to another by following a hierarchical addressing scheme that includes a top-level domain, one or more sub-domains, and a host name. The postal service relies on a hierarchical addressing scheme, and it serves as a good analogy to understand how Internet addressing works. Consider the following address:

John Hardy
1237 Sunnyside Court
Matanuska, Alaska
USA

In reality the address is upside down because a package addressed to John must be read from the bottom-up if it is to be delivered properly. If we assume that the package was mailed from Zemlya Graham Bell[1] in Franz Josef Land. To begin the routing process the postal service will start at the bottom with USA, which narrows it down to a country. They will then go to Alaska, then on to Matanuska, then to the address in Matanuska, and finally to John. Messages routed through the Internet are handled in much the same way; routing arbiters deliver them first to a domain, then to a sub-domain, then on to

[1]There is an island in Franz Josef Land, near the Arctic Circle, called Zemlya (Russian for "land") Alexander Graham Bell. I would love to know why–if anyone knows, I'd appreciate an e-mail!

a hostname, where they are ultimately delivered to the proper account.

The assignment of Internet IP addresses was historically handled by the Internet assigned numbers authority (IANA), while domain names were assigned by the Internet network information center (InterNIC), which had overall responsibility for name dissemination while *regional NICs* handled non-U.S. domains. The InterNIC was also responsible for the management of the domain name system (DNS), the massive, distributed database that reconciles host names and IP addresses throughout the Internet.

The InterNIC and its overall role have gone through a series of significant changes in the last decade. In 1993 Network Solutions, Inc. (NSI) was given the responsibility of operating the InterNIC registry by the NSF and had exclusive authority for the assignment of such domains as .com, .org, .net, and .edu. NSI's contract expired in April 1998 but was extended several times because no alternate agency existed to perform the task. In October 1998, NSI became the sole administrator for those domains but a plan was created to allow users to register names in those domains with other companies. At roughly the same time responsibility for IP address assignments was migrated to a newly created organization called the American Registry for Internet Numbers (ARIN). Shortly thereafter, in March 2000, VeriSign acquired NSI.

The most recent addition to the domain management process is the Internet Corporation for Assigned Names and Numbers (ICANN). Established in late 1998, ICANN was appointed by the U.S. National Telecommunications and Information Administration (NTIA) to manage the DNS.

Of course, it makes sense that sooner or later the most common top-level domains, which include such well-known suffixes as .com, .gov, .net, .org, and .net, would be deemed inadequate. And sure enough, beginning in November 2000, the new top-level domains began to be released, all preceded by a "dot:"

- .aero, for the aviation industry
- .biz, for businesses
- .coop, for business cooperatives
- .info, for general use
- .museum, for (you guessed it)
- .name, for individuals
- .asia, for the Asia-Pac region
- .cat, for the Catalan language
- .int, for international organizations
- .jobs, for companies with jobs to advertise

- .mobi, for mobile-compatible Web sites
- .post, for postal administrations
- .pro, for professional or certified organizations
- .tel, for Internet communications services
- .travel, for travel agencies and tourism organizations
- .xxx, for adult entertainment and porn

It's interesting that as I was completing this chapter of the book, the Internet Corporation for Assigned Names and Numbers (ICANN) announced that they would soon release about 50 new top-level domains, some of which are listed below:

.academy, .bike, .build, .builders, .cab, .camera, .camp, .careers, .center, .clothing, .company, .computer, .construction, .contractors, .diamonds, .directory, .domains, .education, .email, .enterprises, .equipment, .estate, .gallery, .glass, .graphics, .guru, .holdings, .institute, .kitchen, .land, .lighting, .limo, .luxury, .management, .menu, .photography, .photos, .plumbing, .recipes, .repair, .shoes, .singles, .solutions, .support, .systems, .technology, .tips, .today, .training, .ventures, and .voyage

This list is expected to grow to several hundred new top-level domains over the course of the next few years.

Equally important, ICANN announced tentative plans to give up its role as the global overseer of Internet names and addresses, a move that has been rumored for years. Driven by a decision in the U.S. Department of Commerce, which has historically overseen the activities of ICANN, the move is most likely driven by international concerns over one country—and, more importantly, its government—having too much control and influence over the Internet.

The TCP/IP Protocol: What It Is and How It Works

TCP/IP has been around for a long time, and while we often tend to think of it as a single protocol that governs the Internet, it is in reality a fairly exhaustive collection of protocols that cover the functions of numerous layers of the protocol stack. And while OSI, discussed in an earlier chapter, has seven layers of protocol functionality, TCP/IP has only four, as shown in Fig. 10.1. We will explore each layer in moderate detail in the sections that follow. For purposes of comparison, however, OSI and TCP/IP compare functionally as shown in Table 10.1.

The Network Interface Layer

TCP/IP was created for the Internet with concept in mind that the Internet would not be a particularly well-behaved network. In other

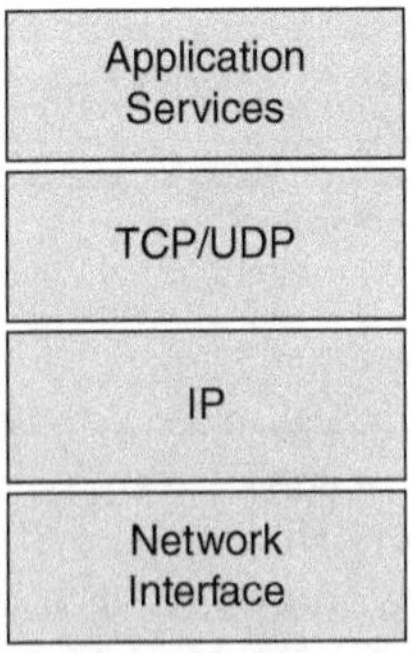

FIGURE 10.1 The four layers of the TCP/IP protocol stack.

TCP/IP Protocol Stack	OSI Reference Model
Network interface layer	Physical and data link layers (L1–2)
Internet layer	Network layer (L3)
Transport layer	Transport layer (L4)
Application services	Session, presentation, application layers (L5–7)

TABLE 10.1 A Functional Comparison of the OSI and TCP/IP Protocol Stacks

words, designers of the protocol made the assumption that the Internet would become precisely what it has become—a network of networks, using a plethora of unrelated and often conflicting protocols, transporting traffic with widely varying QoS requirements. The fundamental building block of the protocol, the IP packet, is designed to deal with all of these disparities, while TCP (and other related protocols, discussed later) take care of the QoS issues.

Two network interface protocols are particularly important to TCP/IP. The serial line Internet protocol (SLIP) and point-to-point protocol (PPP) are used to provide data link layer services in situations where no other data link protocol is present such as in leased line or older dial-up environments. Most TCP/IP software packages for desktop applications include these two protocols, even though dial-up is rapidly fading into near-oblivion in the presence of growing levels of broadband access. With SLIP or PPP, a remote computer can attach directly to a host and connect to the Internet using IP rather than being limited to an asynchronous connection.

The Point-to-Point Protocol

The point-to-point protocol, as its name implies, was created for the governance of point-to-point links. It has the ability to manage a variety of functions at the moment of connection and verification, including password verification, IP address resolution, compression

FIGURE 10.2 The PPP frame structure.

(where required), and encryption for privacy or security. It can also support multiple protocols over a single connection, an important capability for dial-up users that rely on IP or some other network layer protocol for routing and congestion control. It also supports inverse multiplexing and dynamic bandwidth allocation via the multilink-PPP protocol (ML-PPP), commonly used in ISDN environments where bandwidth supersets are required over the connection.

The PPP frame (Fig. 10.2) is similar to a typical high-level data link control (HDLC) frame, with delimiting flags, an address field, a protocol identification field, information and pad fields, and a frame check sequence for error control.

The Internet Layer

The Internet protocol is the heart and soul of the TCP/IP protocol suite and Internet itself—and perhaps the most talked about protocol in history. IP provides a connectionless service across the network which is sometimes referred to as an unreliable service because the network does not guarantee delivery or packet sequencing. IP packets typically contain an entire message or a piece (fragment) of a message that can be as large as 65,535 bytes in length. The protocol does not provide a flow control mechanism.

IP packets, like all packets, have a header that contains routing and content information (Fig. 10.3). The bits in the packet are numbered

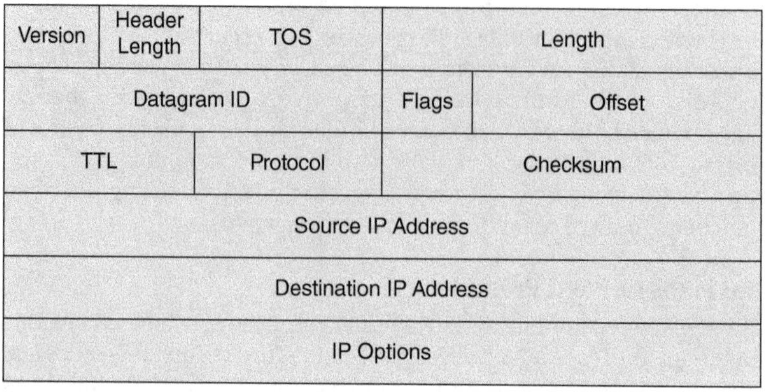

FIGURE 10.3 IP packet header.

from left-to-right starting at 0, and each row represents a single 32-bit word. An IP header must contain a minimum of five words.

IP Header Fields

The IP header contains approximately 15 unique fields. The version field identifies the version of IP that is being used to encode the packet (IPv4 vs. IP v6, for example). The Internet header length (IHL) field identifies the length of the header in 32 bit words. The maximum value of this field is 15, which means that the IP header has a maximum length of 60 octets.

The type of service (TOS) field: gives the transmitting system the ability to request different classes of service for the packets that it transmits into the network. The TOS field is not typically supported in IPv4, but can be used to specify a service priority (0–7) or route optimization.

The total length field indicates the length (in octets) of the entire packet, including both the header and the data within the packet. The maximum size of an IP packet is 64 KB (ok, 65,535 bytes).

When a packet is broken into smaller chunks (a process called fragmentation) during transmission, the identification field is used by the transmitting host to ensure that all of the fragments from a single message can be re-associated at the receiving end to ensure message reassembly.

The *flags* also play a role in fragmentation and reassembly. The first bit is referred to as the *more fragments* (MF) bit and is used to indicate to the receiving host that the last fragment of a packet has been received so that the receiver can reassemble the packet. The second bit is the *don't fragment* (DF) bit, which prevents packet fragmentation (for delay sensitive applications, for example). The third bit is unused and is always set to 0.

The fragment offset field indicates the relative position of this particular fragment in the original packet that was broken up for transmission. The first packet of a fragmented message will carry an offset value of 0, while subsequent fragments will indicate the offset in multiples of 8 bytes. Makes sense, no?

I love this field. A message is fragmented into a stream of packets and the packets are all sent on their merry way to the destination system. Somewhere along the way packet number 11 decide to take a detour and ends up in a routing loop on the far side of the world, trying in vain to reach its destination. To prevent this packet from living forever on the Internet, IP includes a time-to-live field (TTL). This configurable field has a value between 0 and 255 and indicates the maximum number of hops that this packet is allowed to make before it is discarded by the network. Every time the packet enters a router, the router decrements the TTL value by one; when it reaches zero, the packet is discarded and the receiving device will ultimately invoke error control to ask for a resend.

The protocol field indicates the nature of the higher layer protocol that is carried within the packet. Encoded options include values for ICMP (1), TCP (6), UDP (17), or OSPF (89).

The header checksum field is similar to a frame check sequence in HDLC, and is used to ensure that the received IP header is free of errors. Keep in mind that IP is a connectionless protocol; this error check does not check the packet; it only checks the header.

When transmitting packets, it is always a good idea to have a source address and a destination address. You can figure out what they are for.

Understanding IP Addresses

As Fig. 10.4 shows, IP addresses are 32 bits long. They are typically written as a sequence of four numbers that represent the decimal value of each of the address bytes. These numbers are separated by periods (dots in telecom parlance), and the notation is referred to as dotted decimal notation. A typical address might be 168.152.20.10. These numbers are hierarchical; the hierarchy is described below.

IP addresses are divided into two subfields. The network identifier (NET_ID) subfield identifies the sub-network that is connected to the Internet. The NET_ID is most commonly used for routing traffic between networks. On the other hand, the host identifier (HOST_ID) subfield identifies the address of a system (host) within a sub-network.

IP Address Classes

IP defines distinct address classes that are used to discriminate between different sized networks. Classes A, B, and C are used for host addressing, the only difference between them the length of the NET_ID subfield. A Class A address, for example, has an 8-bit NET _ID field and a 24-bit HOST_ID field. They are used for identification of very large networks and can identify as many as 16,777,214 hosts in each network. To date, only about 90 Class A addresses have been assigned.

Class B addresses have 16-bit NET_ID and 16-bit HOST_ID fields. They are used to address medium-sized networks and can identify as many as 65,534 hosts within each network.

192.168.1.1

FIGURE 10.4 A typical dotted-decimal IP address.

Class C addresses, which are far and away the most common, have a 24-bit NET_ID field and an 8-bit HOST_ID field. These addresses are used for smaller networks and can identify no more than 254 devices within any given network. There are 2,097,152 possible Class C NET_IDs which are commonly assigned to corporations that have fewer than 250 employees (or devices!).

There are two additional address types. Class D addresses are used for IP multicasting, such as transmitting a television signal to multiple recipients, while Class E addresses are reserved for experimental use.

Some addresses are reserved for specific purposes. A HOST_ID of 0 is reserved to identify an entire sub-network. For example, the address 168.152.20.0 refers to a Class C address with a NET_ID of 168.152.20. A HOST_ID that consists of all ones (usually written "255" when referring to an all-ones byte, but also denoted as "−1") is reserved as a broadcast address and is used to transmit a message to all hosts on a particular network.

Subnet Masking

One of the most valuable but least understood tools in IP protocol management is called the subnet mask. Subnet masks are used to identify the portion of the address that specifies the network or the sub-network for routing purposes. They can also be used to divide a large address into sub-networks or to combine multiple smaller addresses to create a single large *domain*. In the case of an organization subdividing its network, the address space is apportioned to identify multiple logical networks. This is accomplished by further dividing the HOST_ID subfield into a sub-network identifier (SUBNET_ID) and a HOST_ID.

Adding to the Alphabet Soup: CIDR, DHCP, NAT, and PAT

As soon as the Internet became popular in the early 90s, concerns began to arise about the eventual exhaustion of available IP addresses. For example, consider what happens when a small corporation of 11 employees purchases a Class C address. They now control more than 250 addresses, of which they may only be using 25. Clearly this is a waste of a scarce resource.

One technique that has been accepted for *address space conservation* is called classless interdomain routing (CIDR). CIDR effectively limits the number of addresses assigned to a given organization, making the process of address assignment far more granular—and therefore efficient. Furthermore, CIDR has had a secondary yet equally important impact: it has dramatically reduced the size of the Internet routing tables because of the pre-allocation techniques used for address space management.

Other important protocols include network address translation (NAT), which translates a private IP address that is being used to access the Web into a public IP address from an available pool of addresses, thus further conserving address space; port address translation (PAT) and network address port translation (NAPT), which allow multiple systems to share a single IP address by using different *port numbers.* Port numbers are used by transport layer protocols to identify specific higher layer applications.

Addressing in IP: The Domain Name System

Traditional IP addresses are 32 bits long, and while not all that complicated, most Internet users don't bother to memorize the dotted decimal addresses of their systems. Instead, they use natural language host names. Most hosts, then, must maintain a comparative *table* of both numeric IP addresses and natural language names. From a host perspective, however, the names are worthless; they must use the numeric identifiers for routing purposes.

Because the Internet continues to grow at a rapid clip, a system was needed to manage the growing list of new Internet domains. That system is called the domain name system (DNS). It is a distributed database that stores host names and IP address information for all of the recognized domains found on the Internet. For every domain there is an authoritative name server that contains all DNS-related information about that domain, and every domain has at least one secondary name server that also contains the information. A total of 13 root servers, in turn, maintain a list of all of the authoritative name servers.

How does the DNS actually work? When a system needs a another system's IP address based upon its host name, the inquiring system issues a DNS request to a local name server. Depending on the contents of its local database, the local name server may be able to respond to the request. If not, it forwards the request to a root server. The root server, in turn, consults its own database and determines the most likely name server for the request and forwards it appropriately.

Early Address Resolution Schemes

When the Internet first came banging into the public psyche, most users were ultimately connected to the Internet via an ethernet LAN. LANs use a local device address known as a medium access control (MAC) address, which is 48 bits long and non-hierarchical, which means that they cannot be used in IP networks for routing.

To get around this disparity, and to create a technique for relating MAC addresses to IP addresses, the address resolution protocol (ARP) was created. ARP allows a host to determine a receiver's MAC address when it only knows the device's IP address. The process is simple: the host transmits an ARP request packet that contains the MAC broadcast address. The ARP request advertises the destination IP address and asks for the associated MAC address. Since every station on the LAN hears the broadcast, the station that recognizes its own IP address responds with an ARP message that contains its MAC address.

As ARP became popular, other address management protocols came into play. Reverse ARP (RARP) gives a diskless workstation (a dumb terminal, for all intents and purposes) the ability to determine its own IP address, knowing only its own MAC address. Inverse ARP (InARP) maps, are used in frame relay installations to map IP addresses to frame relay virtual circuit identifiers. ATMARP and ATMInARP are used in ATM networks to map IP addresses to ATM virtual path/ channel identifiers. And finally, LAN emulation ARP (LEARP) maps a recipient's ATM address to its LAN emulation (LE) address, which is typically a MAC address.

Routing in IP Networks

There are three routing protocols that are most commonly used in IP networks: The routing information protocol (RIP), open shortest path first (OSPF), and the border gateway protocol (BGP).

OSPF and RIP are used primarily for intra-domain routing—within a company's dedicated network, for example. They are sometimes referred to as interior gateways protocols. RIP uses hop count as the measure of a particular network path's cost; in RIP, it is limited to 16 hops.

When RIP broadcasts information to other routers about the current state of the network, it broadcasts its entire routing table, resulting in a flood of what may be unnecessary traffic. As the Internet has grown, RIP has become relatively inefficient because it does not scale as well as it should in a large network. As a consequence, the open shortest path first (OSPF) protocol was introduced. OSPF is known as a link state protocol. It converges (spreads important information across the network) faster than RIP, requires less overall bandwidth, and scales well in larger networks. OSPF-based routers only broadcast changes in status rather than the entire routing table.

The BGP is referred to as an exterior gateway protocol because it is used for routing traffic between Internet domains. Like RIP, BGP is a distance vector protocol, but unlike other distance vector protocols BGP stores the actual route to the destination.

IP Version 6

Because of the largely unanticipated growth of the Internet since 1993 or so (when the general public became aware of its existence), it was roundly concluded that IP version 4 was inadequate for the emerging and burgeoning needs of Internet applications. In 1995, IP version 6 (IPv6) was introduced, designed to deal with the shortcomings of version 4. Changes included increased IP address space from 32 to 128 bits, improved support for differentiable quality-of-service requirements, and improvements with regard to security, confidentiality, and content integrity. IPv6 continues to be studied, and while it is in use its pace of adoption is relatively slow. Its arrival is inevitable, however; the question is when.

IP Version 6 (or IPv6 as it is commonly known) replaces IPv4, the 32-bit addressing scheme that we all know and love. At this point you may find yourself asking what happened to version 5. It's a good question, and I'll answer it now to avoid the suspense. As IPv4 was approaching the point in its implementation when concern was beginning to grow about its ability to provide enough IP addresses for all of the various unanticipated devices that suddenly needed them (smart meters, mobile devices, light switches, cars, coffee pots, electrical outlets, and the like), a new protocol called the Internet stream protocol emerged and was seriously considered as a viable replacement for IPv4 for about 7 minutes. It was quickly dismissed due to a number of technical deficiencies, and IPv6 rose to take its place. So IPv5 never stood a chance.

IPv4 stood the test of time and served the Internet community well. As the network became more sophisticated and more densely populated with both devices and people, however, it began to show its age. IPv4 uses a 32-bit addressing scheme, which means that the protocol can yield a maximum of 2^{32} possible addresses, which are just under 4.5 billion unique addresses. That worked fine until the world discovered the Internet and it became obvious that 4.3 billion just wasn't going to meet the need. Consider this: According to McKinsey, Cisco, and a number of other reliable sources, there are about 400 uniquely addressable devices connected to the Internet at any point in time in any given square kilometer of a typical urban area. By the end of 2015 or thereabouts, that number is expected to grow to *12,800 devices* in the same geographical area. That's one square kilometer in one city. Times how many cities? And how many kilometers do they cover? Clearly, the number of addressable devices will rapidly exceed the number of unique, available addresses.

The second problem that emerged was the inability of the large backbone routers that make up the Internet to handle very large routing tables. Third, a better, more effective security protocol was needed, given the degree of commercial traffic that was being moved across the Internet. And finally, IPv4 did very little to support discriminatory QoS.

IPv6 addresses (no pun intended) most of these issues rather well. First of all, instead of a 32-bit address it relies on a 128-bit address, which means that it can support 3.4 10^{38} unique addresses, which is enough.

IPv6 offers a number of advantages over its predecessor, including support for a greater number of multicast addresses; a much simpler header format, which radically increases packet processing speed; the elimination of address translation protocols like NAT, which are required in large networks; support for larger packet payloads, which simplifies processing efficiency and improves network throughout; supports a wide array of routing protocols such as BGP, OSPF, RIP, and so on; and uses a hierarchical network architecture to improve the efficiency of those same routing protocols over the wide area.

Of course, there are some challenges associated with bringing IPv6 into the IPv4 world, not the least of which is the fact that 32-bit and 128-bit addresses aren't mutually compatible—nor are the devices they're used in. As a result there are several techniques that can be used to span the gap between the two during the process of universal v4 to v6 conversion. One is a technique called dual stacking, in which routers run both protocols and support both; the second is VPN tunneling; and the third to use a network address translation (NAT) box. None of these is ideal, but they serve as adequate interim solutions until a single universal address scheme is adopted—a process which requires years of work.

IPv6 Header

The IPv6 header differs from the IPv4 header somewhat significantly, so I'll discuss the differences here. Please refer to Fig. 10.5.

The *version field* is a 4-bit field that does precisely what its name implies: It tells the receiving device that the packet that follows is either a version 4- or version 6-formatted packet.

The *traffic class field* conveys a sense of the priority or importance of the data contained in the packet. This field replaces the function of the Type of Service (ToS) field in IPv4.

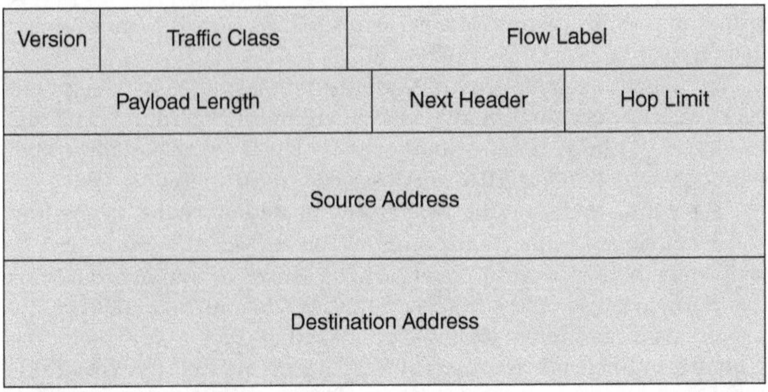

FIGURE 10.5 IPv6 header structure.

The *flow label field* is used to identify a stream of packets that originate from or are going to a common device, thus reducing end-to-end processing delay.

The *payload length field* indicates the size of the transported payload; it can be used to verify whether the entire payload has been received.

The *next header field* does two things. First, it indicates the nature of the protocol carried within the packet; second, it facilitates the use of extension headers that identify options associated with the packet.

The *hop limit field* replaces the time to live field in IPv4, and is used to prevent broadcast storms.

Transport Layer Protocols

We turn our attention now to layer four, the transport layer. Two key protocols are found at this layer: the transmission control protocol (TCP) and the user datagram protocol (UDP). TCP is an ironclad, absolutely guaranteed service delivery protocol with all of the attendant overhead you would expect. UDP, on the other hand, is a more *lightweight* protocol, used for delay sensitive applications like VoIP. Its overhead component is relatively light.

In TCP and UDP messages, higher layer applications are identified by port identifiers. The port identifier and IP address together form a socket, and the end-to-end communication between two or more systems is identified by a four-part complex address: the source port, the source address, the destination port, and the destination address. Commonly used port numbers are shown in Table 10.2

The Transmission Control Protocol

TCP provides a connection-oriented communication service across the network. It stipulates rule sets for message formats, establishing virtual circuit establishment and termination, data sequencing, flow control, and error correction. Most applications designed to operate within the TCP/IP protocol suite use TCP's reliable, guaranteed delivery services.

In TCP, the transmitted data entity is referred to as a segment because TCP does not operate in message mode: it simply transmits blocks of data from a sender and receiver. The fields that make up the segment, shown in Fig. 10.6, are described in the following text:

The source port and the destination port identify the originating and terminating connection points of the end-to-end connection as well as the higher-layer application. The sequence number identifies this particular segment's first byte in the byte stream, and since the sequence number refers to a byte count rather than to a segment, the sequence numbers in sequential TCP segments are not, ironically, numbered sequentially.

Port #	Protocol	Service	Port #	Protocol	Service
7	TCP	echo	80	TCP	http
9	TCP	discard	110	TCP	pop3
19	TCP	chargen	119	TCP	nntp
20	TCP	ftp-control	123	UDP	ntp
21	TCP	ftp-data	137	UDP	netbios-ns
23	TCP	telnet	138	UDP	netbios-dgm
25	TCP	smtp	139	TCP	netbios-ssn
37	UDP	time	143	TCP	imap
43	TCP	whois	161	UDP	snmp
53	TCP/UDP	dns	162	UDP	snmp-trap
67	UDP	bootps	179	TCP	bgp
68	UDP	bootpc	443	TCP	https
69	UDP	tftp	520	UDP	rip
79	TCP	finger	33434	UDP	traceroute

TABLE 10.2 Commonly Used Port Numbers

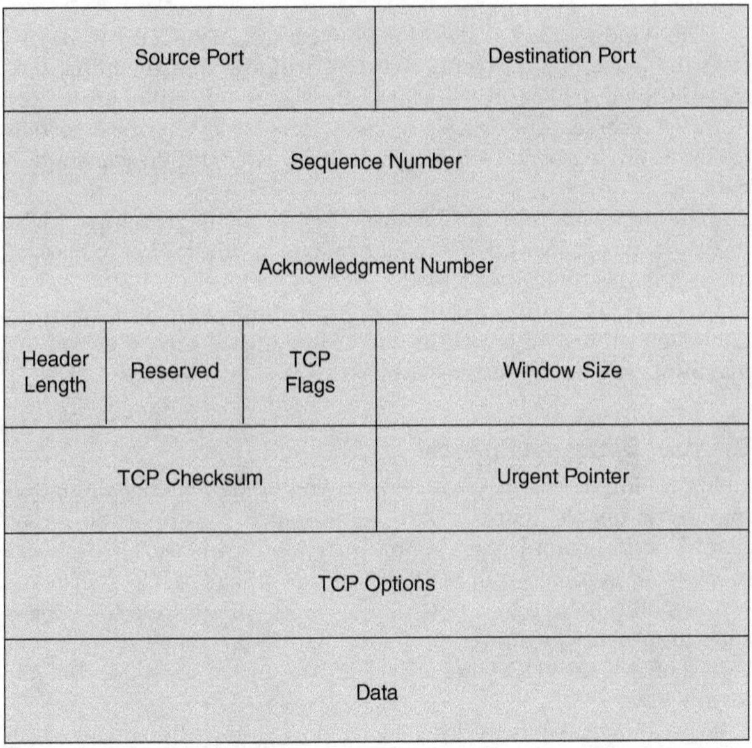

FIGURE 10.6 TCP structure.

The acknowledgment number is used by the sender to acknowledge to the transmitter that it has received the transmitted data. In practice, the field identifies the sequence number of the next byte that it expects from the receiver. The data offset field identifies the first byte in this particular segment; in effect it indicates the segment header length.

TCP relies on a collection of control flags that do in fact control certain characteristics of the virtual connection. They include an urgent pointer field significant (URG) which indicates that the current segment contains high-priority data and that the urgent pointer field value is valid; an acknowledgment field significant (ACK), which indicates that the value contained in the acknowledgment number field is valid; a push function (PSH) flag, which is used by the transmitting application to force TCP to transmit data immediately that it currently has buffered, without waiting for the buffer to fill; a reset connection (RST) flag, which is used to immediately terminate an end-to-end TCP connection; a synchronize sequence numbers (SYN) flag, which is used to establish a connection, and to indicate that the segments carry the proper initial sequence number; and finally, a finish (FIN) Flag, which is set to request a normal termination of a TCP connection in whatever direction the segment is traveling.

The window field is used for flow control management. It contains the value of the permitted receive window size, the number of transmitted bytes that the sender of the segment is willing to accept from the receiver. The checksum field offers bit-level error detection for the entire segment, including both the header and the transmitted data.

The urgent pointer field is used for the management of high-priority traffic as identified by a higher layer application. If so marked the segment is typically allowed to bypass normal TCP buffering.

The last field is the options field. At the time of the initial connection establishment this field is used to negotiate such functions as maximum segment size and selective acknowledgement (SACK).

The User Datagram Protocol

UDP provides connectionless service. And while *connectionless* often implies unreliable, that is a bit of a misnomer. For applications that require nothing more than a simple query and response, UDP is ideal because it involves minimal protocol overhead. UDP's primary responsibility is to add a port number to the IP address to create a socket for the application.

The fields of a UDP message (Fig. 10.7) are described in the following text:

The source port identifies the UDP port used by the sender of the datagram, while the destination port identifies the port used by the

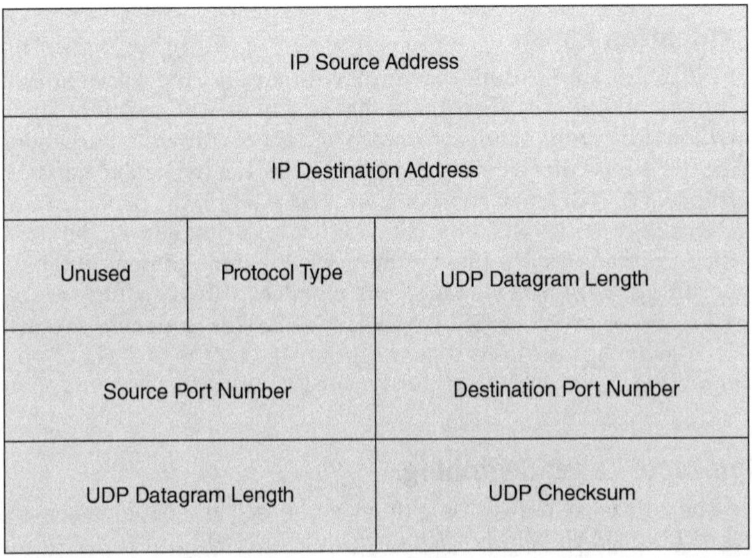

FIGURE 10.7 UDP structure.

datagram receiver. The length field indicates the overall length of the UDP datagram.

The checksum provides the same primitive bit error detection of the header and transported data as we saw with TCP.

The Internet Control Message Protocol

The Internet control message protocol (ICMP) is used as a diagnostic tool in IP networks to notify a sender that something unusual happened during transmission. It offers a healthy repertoire of messages, including destination unreachable, which indicates that delivery is not possible because the destination host cannot be reached; echo and echo reply, used to check whether hosts are available; parameter problem, used to indicate that a router encountered a header problem; redirect, which is used to make the sending system aware that packets should be forwarded to another address; source quench, used to indicate that a router is experiencing congestion and is about to begin discarding datagrams; TTL exceeded, which indicates that a received datagram has been discarded because the time-to-live field (sounds like a soap opera for geeks, doesn't it?) reached 0; and finally, timestamp and timestamp reply, which are similar to echo messages except that they timestamp the message, giving systems the ability to measure how long is required for remote systems to buffer and process datagrams.

The Application Layer

The TCP/IP application layer protocols support the actual applications and utilities that make the Internet—well, useful. They include the border gateway protocol (BGP), the domain name service (DNS), the file transfer protocol (FTP), the hypertext transfer protocol (HTTP), open shortest path first (OSPF), the packet internetwork groper (PING; how can you *not* love that name), the post office protocol (POP), the simple mail transfer protocol (SMTP), the simple network management protocol (SNMP), the secure sockets layer protocol (SSL), and TELNET. This is a small sample of the many applications that are supported by the TCP/IP application layer.

Multiprotocol Label Switching

The next phase of network evolution is IP-based and dependent on a protocol called *multiprotocol label switching* (MPLS).

The advantages that the legacy telephone network offers to voice and data users are well known. It is very well managed; it is predictable in terms of delay and other behaviors; and it offers good, "five-nines" performance. The downside of the legacy network is that it is static, inflexible, and not particularly resilient when failures occur.

On the other side of the fence, let's examine its contender and would-be replacement, IP. IP is extremely dynamic, very flexible, and is able to survive network failures with aplomb—no interruption of service. However, it is also unmanaged; completely unpredictable in terms of delay and other service-affecting behaviors; and is known for *best-effort* service. Interesting, isn't it? They are 180° out of phase with each other.

What if we could create a network that offers the best of both: A network that is extremely dynamic, very flexible, and able to survive network failures with no interruption of service, and at the same time is well managed, predictable in terms of delay and other behaviors, and able to offer "five-nines" performance. Sound like a dream? Meet MPLS. MPLS combines the best of both worlds to create a network that brings together the best features of the two networks.

When establishing connections over an IP network, it is critical to manage traffic queues to ensure the proper treatment of packets that come from delay-sensitive services such as voice and video. To do this, packets must be differentiable, that is, identifiable so that they can be classified properly. Routers, in turn, must be able to respond properly to delay-sensitive traffic by implementing queue management processes. This requires that routers establish both normal and high-priority queues, and handle the traffic found in high priority routing queues faster than the arrival rate of the traffic.

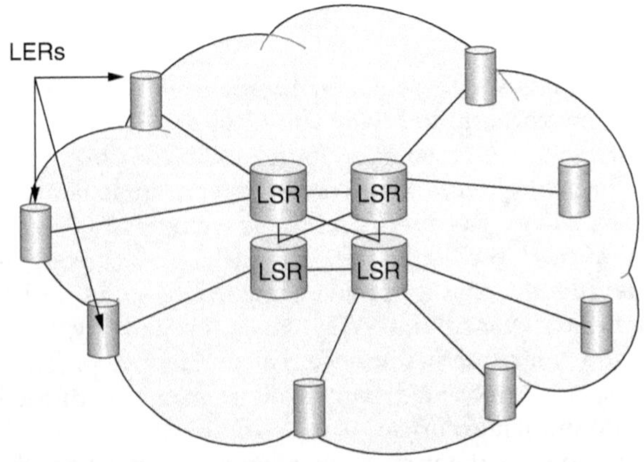

FIGURE 10.8 Architecture of an MPLS network.

MPLS delivers QoS by establishing virtual circuits known as label switched paths (LSPs) which are in turn built around traffic-specific QoS requirements. An MPLS network, such as that shown in Fig. 10.8, comprises label switch routers (LSRs) at the core of the network and label edge routers (LERs) at the edge. It is the responsibility of the LERs to set QoS requirements and pass them on to the LSRs responsible for ensuring that the required levels of QoS are achieved. Thus a router can establish LSPs with explicit QoS capabilities and route packets to those LSPs as required, guaranteeing the delay that a particular flow encounters on an end-to-end basis. It's interesting to note that some industry analysts have compared MPLS LSPs to the trunks established in the voice environment.

MPLS uses a two-part process for traffic differentiation and routing. First, it divides the packets into forwarding equivalence classes (FECs) based on their quality of service requirements, then maps the FECs to their next hop point. This process is performed at the point of ingress at the edge of the network. Each FEC is given a fixed-length *label* that accompanies each packet from hop to hop; at each router, the FEC label is examined and used to route the packet to the next hop point, where it is assigned a new label.

MPLS is a *shim* protocol that works closely with IP to help it deliver on QoS guarantees. Its implementation allows for the eventual dismissal of such architectures as frame relay and ATM as a required layer in the multimedia network protocol stack. Already, forward-thinking carriers are building MPLS networks that connect to IP backbones; TELUS, Canada's most advanced carrier, has built an all-IP backbone that runs MPLS and which allows them to deliver a broad array of services with five-nines of carrier-grade reliability.

Final Thoughts

IP is here, and it is slowly replacing legacy networking technologies in the local, metro, core, and wide area environments. Together with MPLS it offers the perfect solution for reliability, flexibility, adaptability, and high-quality service delivery, all over a single network.

But wait, as they say, there's more. As we noted in Chap. 7 when I took you through the hairball diagram, IP makes convergence possible. Today the average telephone company operates a plethora of disparate networks including ATM, voice, frame relay, ISDN, VoIP, wireless, and their Internet service. All of these require dedicated spares inventories, dedicated management systems, dedicated billing systems, and human expertise.

Imagine now what happens when we build a network that looks like the network in Fig. 10.9—a single network fabric that is covered with diverse network interfaces that allow it to converge all of that complexity and cost into a single low-cost, high-performance network. Now, instead of running seven networks, we run one, manage one, pay for one, maintain one, and understand one. And that one network talks to any other network through its diverse interfaces. Did the cost structure just decrease by a factor of seven? Of course not. Did it decrease by a third? Or by half? No doubt.

As we'll see in Chap. 11, IP also plays a major role in the accelerating evolution of IT. Pay attention—this is where it *really* gets interesting.

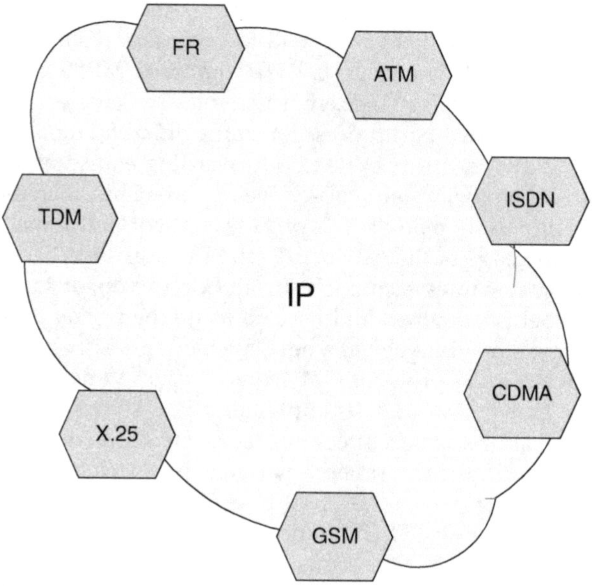

Figure 10.9 A universal, protocol-agnostic network, made possible by IP.

Chapter Ten Questions

1. Draw and label the TCP/IP protocol stack.

2. How do TCP and UDP differ? How are they each used, and why?

3. What is the difference between a label-switched router and a label edge router?

4. Why is MPLS such an important protocol in the modern data networking marketplace?

CHAPTER 11

The IT Mandate

> *We are drowning in information, while starving for wisdom. The world henceforth will be run by synthesizers, people able to put together the right information at the right time, think critically about it, and make important choices wisely.*
>
> —E. O. Wilson

Over the course of the last ten chapters, we've spent a lot of time talking about technology and how it works. We've also talked about the fact that this industry that does so many important things for so many other industries is going through wrenching changes that are at best, disruptive.

So far we've spent the majority of out time talking about the media and telecom industries, two of the three legs of our TMT stool. In this chapter, we're going to shift our focus and talk about the information technology (IT) sector, an equally important domain in its own right.

You'll recall that we made the point earlier in the book that telecom provides the ability to access and transport desirable content, the content industry creates that desirable content, and the IT or technology sector has the responsibility to store it, track it, analyze its consumption, and monetize it. As a result the three sectors are mutually interdependent and together represent one of the most important service ecosystems in the world today.

But there is trouble in paradise, and it is non-trivial. The trouble, however, has nothing to do with technology per se. It has everything to do with the TMT industry's ability to adapt to the changing marketplace, to understand the shift that we discussed in the beginning of the book, and the players' abilities to adapt to the changes that are underway. Phil Evans and Thomas Wurster said it best in *Blown to Bits:*

> A greater vulnerability than legacy assets is a legacy mindset. It may be easy to grasp this point intellectually, but it is profoundly difficult to practice. Managers must put aside the presuppositions of the old competitive world and compete according to totally new rules of engagement. They must make decisions at a different speed, long before the numbers are in place and the plans formalized. They must acquire

> totally new technical and entrepreneurial skills, quite different from what made their organizations (and them personally) so successful. They must manage for maximal opportunity, not minimum risk. They must devolve decision-making, install different reward structures, and perhaps even devise different ownership structures. They have little choice. If they don't deconstruct their own businesses, somebody else will do it to them.[1]

Change in the Wind

Before we jump into our discussion of IT, let's take a few minutes to discuss the nature of these changes because in many cases the IT part of our techno-triumvirate is well-positioned to deal with them. The challenges tend to fall into two categories: human and technology. Human challenges are things like personnel management, skill maintenance, poor communication, silo-style organizational management, and recognition and reward. Technology challenges are things like security, mobility, data access, bandwidth inadequacies, and physical space and resource allocation. I have often observed that human challenges are more often than not resolved with technology solutions, while technology challenges are typically resolved by people-based solutions. For example, the problems associated with inadequate or poor personnel management, a lack of skill maintenance, poor communication, and silo-style organizational management can be resolved through the judicious application of communications technologies, online skills inventories, and cross-organization communications strategies. On the other hand, problems associated with security, mobility, data access, bandwidth inadequacies, and physical space and resource allocation can be dealt with most effectively through people-oriented measures. Security experts, for example, will tell you that 80 percent of security breaches have nothing to do with technology; they occur because the weakest link in the security chain is the people involved in its execution. Space allocation can be dealt with through work-at-home programs, while the cost of mobile access can be handled very effectively by simply letting employees bring their own devices to work—a phenomenon we'll discuss later in the chapter called bring your own device (BYOD).

But these changes present radical shifts for legacy leadership, shifts that they are ill-equipped to deal with. The changes worry them, and for good reason, because they represent a threat to the status quo—a threat captured nicely in the quote, given before.

So what is it that worries executives in the technology space? Well, it's a collection of forces; let's examine them individually.

---

[1] Evans, Philip and Thomas Wurster. *Blown to Bits: How the New Economics of Information Transforms Strategy.* Boston; Harvard Business School Press, 2000. Page 66.

Loss of Control

Let's face it—the market is in charge. Instead of service providers of all kinds dictating to the market what they can have (and that they'll like it!), the market now has choices and is quick to exercise them. This is where the mantra we chanted earlier really becomes important: *If you want to gain influence over a market, you must give up control of that market.* But this runs counter to the prevailing legacy corporate culture of command and control.

Reinvention

Companies must accept the fact that business-as-usual may not be the correct course of action to satisfy the shifting demands of the future. Jack Welch, former chief executive of GE, was always quick to say, "destroy your business—change or die. When the pace of change *outside* the business exceeds the pace of change *inside* the business, the end is near."

Decline in Revenue and/or Margin

There was a time not all that long ago when a 1.544 Mbps T1 facility that crossed a metro area and connected two business locations earned the telephone company $40,000 per month. Today, a business can get orders of magnitude more bandwidth for a tiny fraction of that price. Deployed access lines, the measure of telephony revenue, are in free fall; the average telephone company loses about 11 percent of their deployed lines every year, a frightening prospect for an industry that once viewed voice as the cash cow of the entire industry. And what about software? The other day I was looking at a sophisticated photo manipulation application for my iPad at the Apple Apps store. What I was *really* doing was mulling over the price—9 dollars. Nine dollars?!? Really?!?!? For an application as sophisticated as the one I was considering? How quickly we forget that it wasn't all that long ago that we paid hundreds if not thousands of dollars for software. The rules are changing, and they aren't pretty—unless, of course, you're the consumer.

Infrastructure Evolution

A single local switch (a Lucent Technologies 5ESS, Nortel DMS100, Siemens EWSD, or Ericsson AXE) costs about $16 million in loaded cost and nominally supports 100,000 subscribers. That means that to support the population of—let's say, New York City (8 million people and change), service providers must install 80 switches—at a cost of about $130 million. That's just one city, albeit the largest in the country. Meanwhile a comparable IP router configured to handle a similar number of VoIP calls costs a fraction of that yet when deployed in an

IP/MPLS environment can offer service that is at least as good as that offered by legacy TDM switch fabrics. But these companies are locked into their in-place infrastructure because of accounting requirements: These devices represent such large investments that they are amortized over decades, and no corporation will willingly throw away investment dollars.

Burgeoning Competition

When competition was linear and direct, life was good. IBM competed against Unisys, Verizon competed with AT&T, Warner Music competed with Vivendi, CBS, NBC, and ABC all competed against each other, Apple competed against Microsoft. Now IBM must compete with the likes of Cisco and manufacturers of blade servers like Hitachi; Verizon and AT&T are plagued by the unanticipated competition from wireless players and VoIP providers; entertainment networks now face a growing challenge from YouTube, HULU, and other sources; and Apple and Microsoft must now look over their shoulder on a regular basis to keep an eye on the goings-on at Google. This nonlinear, indirect model of competitive behavior is disturbing to those involved, but it represents a form of Nirvana for the consumers of the products and services that the industry segments create. It also drives the players to improve their game, to step it up in response to similar actions from their competitive peers. Consider the diagram in Fig. 11.1. This is the classic cycle of technology innovation. Insightful companies begin with an innovation and through successful strategy and marketing manage to attract early adopters, who carry the innovation into the market as a new product and prove its value to the mass market. Before long, mass consumption kicks in and soon the

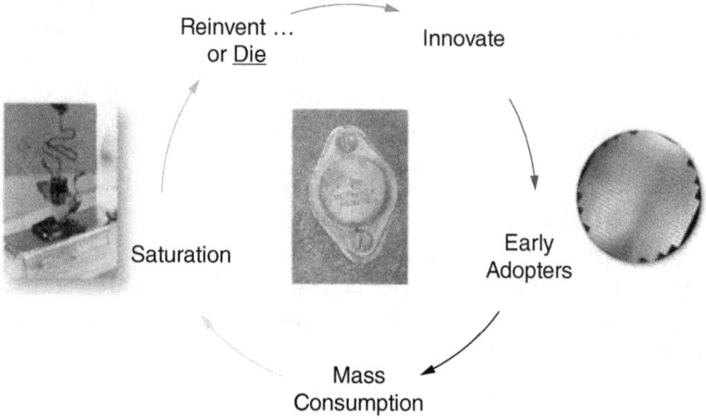

FIGURE 11.1 The cycle of technology innovation.

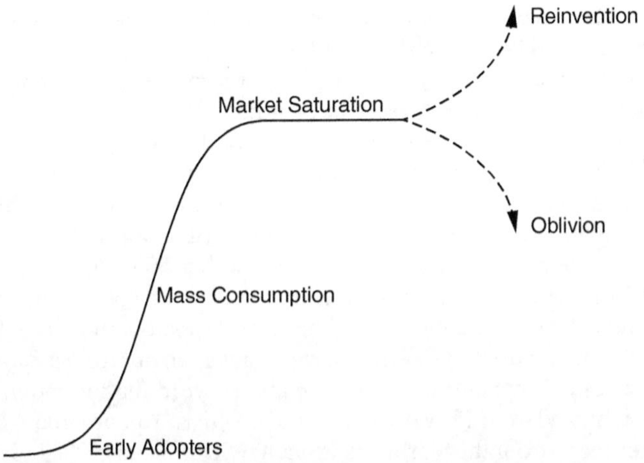

Figure 11.2 The decision point: re-innovate, or descend into oblivion?

market is saturated—not a bad problem to have. At that point, however, companies must heed the advice we cited earlier from Jack Welch: "reinvent ... or die." If they reinvent, then the cycle renews; if not, they begin a long, slow descent into oblivion—and death, as shown in Fig. 11.2.

Incidentally, Fig. 11.1 also illustrates a *real* reinvention cycle. The photograph on the left that looks like a fishing lure or somebody's bad idea of modern art is actually the world's first transistor—clearly not ready for primetime. It was created at Bell Labs by a group of researchers that included John Bardeen, William Shockley, and Walter Brattain and is on display in the lobby of Bell Labs Headquarters in Murray Hill, New Jersey. The device in the middle is a power transistor that represents a follow-on generation of the device, and on the extreme right is the current state-of-the-semiconductor art: a wafer made up of thousands of transistors that can be sliced and diced into discrete devices and installed in chipsets.

Customer Loyalty and Wallet Share

If you are like most people in the world who are coffee drinkers, you seek out Starbucks to get your buzz on. You'll happily stand in line for a 1/2 hour to get your usual medium half-caf no-foam non-fat naked bold grande extra shot of white chocolate and 4 pumps peppermint vanilla soy latte coffee (coffee? Really?), when 7 ft down the mall or concourse there's a perfectly good little coffee place selling perfectly good coffee for a fraction of the price—and there's no line. Why do we do this to ourselves?

The answer is rather interesting, and it's something we talked about briefly in the first chapter of the book.

Think about the last time you walked into a Starbucks and paid five dollars for that urn of coffee in your hand that requires two hands to lift. Actually, you *didn't* pay five dollars for the coffee; the coffee only cost $1.50. But you paid 50 cents for the cool jazz that's playing on the stereo system. You paid 50 cents for the ambiance, the low light, the comfortable, overstuffed chairs. You paid 50 cents for the fact that no one is hassling you to leave. You paid another 50 cents for the free Wi-Fi. You paid 50 cents for the fact that they write your name on the cup, and call you by name when your order is ready. And 50 cents for the fact that you don't order from a clerk or server, but from a *Barista,* a word co-opted by Starbucks from the Italian word for bartender. And did you buy a large? Please. That would be crass. You ordered a *Venti,* another word from the Starbucks lexicon, which means "20" in Italian and refers to the size of the drink in ounces. Cha-ching—50 cents.

I hope you understand that I'm not criticizing Starbucks—on the contrary, I'm complementing them on their marketing genius. Starbucks doesn't sell coffee—far from it. Starbucks sells an experience, one element of which happens to be a decent cup of coffee. And that's the secret. The key to success today is not to offer the best quality of service, product, or solution; that's what poker players refer to as table stakes. They key is to offer the best quality experience that can be had when purchasing or shopping for a product that in most cases is at least somewhat commoditized. Let's face it, coffee doesn't get much closer to being a commodity product, does it? Remember, the definition of a commodity is a product that can only be differentiated on the basis of price. If you're the cheapest, you win the game. Starbucks manages to be pricey, and *still* wins the game. So why does this work? Why do some companies manage to stand out head and shoulders above the competition, even when their products are for all intents and purposes identical?

Noted author and TED presenter Simon Sinek says that people buy from you not because of what you do but because of why you do it. The example he uses is Apple. Apple makes fine computers and other devices, but they don't make computing things any better than a Windows machine. Yet people's desire to buy from Apple often approaches a level of religious fervor, and part of the reason is that Apple creates a superior buying and use experience. Sinek's contention is that most companies start with the product story: "We make great computers; they're well designed, easy to use, and beautiful. You should buy one." Apple, on the other hand, saves the product pitch until last, beginning instead with their own deeply rooted philosophy: "We believe that everything we do should challenge the status quo. As a result our products are well designed, easy to use, and beautiful. We just happen to make great computers. Want to buy one?" The difference is simple, yet utterly profound, and has enormous relevance for the topic of this particular chapter—and, in all honesty, those that preceded it as well. IT is a commodity product. I can buy cloud computing services from Amazon, IBM, HP, Google, Apple, and more than a hundred others. But let's face it—storage is storage, a hosted app is a hosted app, computing cycles are computing cycles. For IT providers to truly succeed in today's hypercompetitive market, they must first understand the changes that the market expects from them and in some cases is foisting upon them, and must also understand that the key to their success—indeed, their very survival—is to offer a quality of experience that far exceeds that of the competition.

Beware the Distractions of the Everyday

In the face of this need to change it is crucial that technology leaders heed the warning of the late Peter Argyris, that intelligent people tend to espouse theories of action that have little to do with their actual behavior. Even though they accept the fact that change is needed they are often distracted by the mundane exigencies of the day. For example, telco executives might be distracted by a small uptick in their core voice and messaging products, while the potential for massive growth in other areas languishes with inadequate attention. Or, IT executives will spend millions to acquire a company that resembles them in many ways as a way to grow their footprint or eliminate a competitor, but they won't spend $100K on a startup acquisition because it's foreign to them. This creates a double jeopardy situation: On the one hand, the amount of money is too much for a middle manager to spend on a flyer, but isn't enough to capture the attention of a senior level executive. As a result the opportunity lies idle and disappears.

Another example is a scenario that I refer to as the *subliminal demotion* problem. Many technology executives began their careers in the technology trenches—working in the central office, running a system in a data center, and the like. As a consequence, they know the technology well, are comfortable with it, and *like* it. Many times I have seen a scenario develop where an otherwise highly skilled and capable technology executive allow him or herself to gravitate away from the strategic responsibilities of the job and instead refocus on the tactical. In other words, they begin to concern themselves with such issues as system uptime and security and application availability stats, rather than such strategic issues as long-term planning. As a consequence, *nobody* does the strategic work and the organization suffers.

So what do we do about this? The first thing is to understand the environment we're operating in. That includes the customer, the evolving technology base, the moves being made by the competition, and technological trends that are shaping the TMT industry as a whole. If we understand the trends then we can develop an understanding of the impact they will have on our customers and ourselves and craft effective responses to them. So in this next section we're going to talk about trends, but before I begin let me state one caveat that should inform your thinking about them. I find that people tend to view trends somewhat incorrectly, almost as if they were a moment in time. For example, we're going to spend some time in this chapter about a phenomenon called *bring your own device.* In spite of what you may read, BYOD is not a trend. The trend is the growing tendency for employees to bring their personal devices to work and to use them for work-related functions. I'm not trying to be picky here; I want to be very, very clear—as you'll see, this is important, particularly when you're having a technology conversation with a client.

So with that in mind let's look at some of the tectonic changes that are affecting the IT organization.

The Changing Face of IT

We begin our discussion of IT transformation by looking at what is perhaps the most important part of the transformation, what I refer to as the migration from legacy or traditional IT to *IT 2.0* for lack of a better name. Please refer to Fig. 11.3.

We're going to look at IT through seven different lenses: infrastructure; output; focus; the vision shift that is underway; who the decision-maker is; the value that IT brings; and the overall evolutionary

Traditional IT		**IT 2.0**
Technology Drives IT	**Evolution**	Business Drives IT
Operational; Transactional; Silo	**Value**	Transformational; Analytical; Collaborative
IT Manager/Director/VP/CIO/CTO	**Decision-Maker**	Marketing/Sales/Customer Experience/CEO/CIO/CTO/R&D
Cost Center Chief Information Officer Chief Technology Officer	**Vision Shift**	Profit Center Chief Analytics Officer Chief Trends Officer
Data Management; Internal (Needs of Business); QoS; Efficiency	**Focus**	Data Analysis; External (Needs of Market); QoE; Effectiveness
Technology Services	**Output**	Business Services
Dedicated, Inflexible	**Infrastructure**	Shared, Virtual

FIGURE 11.3 The evolution of IT.

direction of IT as a functional entity. We'll start at the bottom and work our way to the top of the diagram.

Infrastructure

The early days of IT were characterized by size: enormous mainframes like the ones shown in Fig. 11.4 that occupied vast rooms, took hours to boot and cost tens of millions of dollars. The infrastructure was dedicated and inflexible; if IT management wanted to add computing capacity, the first thing they had to do was buy real estate adjacent to the data center because adding capacity meant adding a new mainframe, and that required a massive expansion of the data center. Machines tended to be application-centric; there was typically a mainframe for operations, a mainframe for billing, a mainframe for maintenance, another for the employee benefits processes, yet another for centralized accounting and customer care.

Today, that's no longer the case. Modern data center resources are shared among all concurrent processes using a virtual allocation methodology that distributes resources on-demand, making for a much more efficient model of resource allocation. And instead of mainframe computers, most data centers are based on a blade server model such as the one shown in Fig. 11.5. As a result, resources can be allocated on-demand in a very efficient and cost-effective manner.

This is accomplished through a technology philosophy called *virtualization*. The word "virtual" should make you think of the Wizard of Oz's now-famous quote: "Pay no attention to that man behind the curtain." Virtualization is a process whereby computing resources can be made to appear different than their physical configuration would indicate. Figure 11.6 illustrates the process of virtualization.

FIGURE 11.4 A room filled with mainframe computers.

Figure 11.5 A blade server–a far cry from the days of room-filling mainframes!

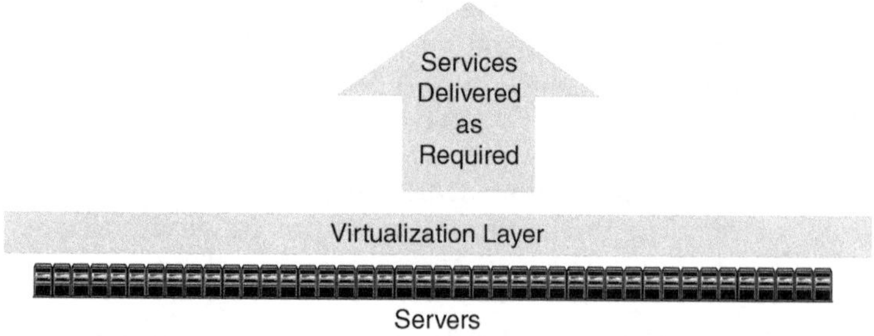

Figure 11.6 The process of virtualization. Pay no attention to that man behind the curtain!

Here we have an array of blade servers (they're called blades because they are really nothing more than circuit cards with edge connectors that are inserted into a chassis that is rack mounted and wired to the backplane of the data center so that all of the devices can communicate with each other) in a data center. Sitting functionally on top of the servers is a virtualization software layer, which is the Wizard of Oz part of the equation.

Above the virtualization layer we find the users of the data center resources. If one of those users (which can be a person or a process) asks the data center to allocate resources to a process they are running, they are actually "talking" to the virtualization layer. The virtualization layer assesses their requirements and allocates enough server capacity to handle the job. It might be half of a server's capacity; it might be multiple server blades. The point is that the user has no need to know what's going on behind the scenes. As far as they know they have been allocated what they require to get the job done.

This is what companies like VMware do. VM is an acronym for virtual machine. The VMware software sits between the massive resources in the data center and the users of those resources and acts as an allocation supervisor, doling out computing resources on demand and as required. It operates as if the data center is one very large chunk of computing and storage capacity. When a user asks for resources, it assesses their requirements and slices off an appropriate size piece for them to use. As a result the precious and very expensive resources in the IT cloud are never wasted, not the case in the days of legacy IT. Virtualization software can make hundreds of small servers look like a single mainframe, or can make a mainframe look like hundreds of smaller servers. That's the beauty of the model, and is clearly a major shift for legacy IT.

Output

There was a time when the output of IT was ... IT. The information systems organization of yore ran extraordinarily complex systems and applications that generated technical information. With a certain amount of effort those services could be made to serve the business, but the effort was often over-the-top. IT created services in its own image, and the business just had to deal with that.

I remember very well my own days in the telephone company data center. Every night at midnight or so we would send out a system-wide broadcast notifying the user community that the online systems would be coming down in 10 minutes. We would then disable the network going into the systems and go into what we called batch mode, which was the process that ran every night during which all of the databases were backed up and the various reports were run that served the needs of the user community.

One of the most important reports we ran was called the PDSO30 (for those of you who are fans of the movie *Office Space*, think "TPS Report"). The job took four hours to run on a massive, dedicated mainframe, and the output of the job every night was 127 boxes of fanfold paper, covered with hexadecimal numbers. My operators carefully stacked the boxes of paper in the hallway outside the secure area, and at 6 a.m. every morning, a mysterious service came with a pallet jack and took the boxes away. Where they went, I never knew.

Anyway, this report was crucial. One night, the unthinkable happened: A hard drive crashed, which meant that all of the data on that drive was lost. So we went into recovery mode, which meant that a new drive hard to be installed (incidentally, these were the days when a hard drive was roughly the size of a washer-dryer and held 80 MB) and all of the files had to be reloaded from magnetic tape, a process that took hours. It also meant that because we didn't have ready access to the transaction from the day before we wouldn't be able to run any of the jobs that ran every night during the graveyard shift, including PDSO30. So I called my director, woke him up and warned him that he would probably be getting calls during the night and into the following online day about the lack of PDSO30.

It took seven hours but we finally got everything restored. I called my boss and gave him the good news, and asked him how many escalation calls he received during the night and early morning. "None," he replied. "Not a single one." I expressed perplexity; obviously, I reasoned, PDSO30 is crucial—after all, it generates 127 boxes of paper every night. "Stop running it," my boss instructed me. Incredulous, I responded, "stop running it? We can't do that—it's PDSO30!" He smiled and said, "stop running it. If somebody complains, we'll put it back on the schedule."

We never ran it again.

The point of the story is that legacy IT organizations tend to be self-governing fiefdoms. Today, however, their focus has shifted away from technology services toward the delivery of business services that are focused on the needs of the business units that IT serves. This is where things really start to change: IT no longer exists for the sake of IT, but rather for the sake of the entities that depend on it.

Focus

Legacy IT has always had a tactical focus, concerned about the management of the data they shepherd. They are traditionally inwardly focused on the needs of the business, on the need to deliver QoS, on the efficiency of everything they do.

IT 2.0 has a very different perspective and direction. Instead of being tactical, the new IT is being asked to be more strategic in its focus. Instead of concerning themselves with data management, they spend more time focusing on data analysis and what it yields, on the external needs of the market, on quality of experience (QoE) rather than quality of service (QoS). They're more concerned with effectiveness (doing the right things) than efficiency (doing things right).

Vision Shift

Traditionally, IT has always been perceived as a necessary evil in the business, a very expensive albatross that drains the balance sheet and is quietly resented by all. The visionary direction of the organization has always been created by (no surprise) the chief technology officer (CTO) and the chief information officer (CIO). But how the mighty have fallen. Today, it's dawning on many organizations that IT, while costly, doesn't have to be the technological equivalent of owning a boat (a hole in the water that you pour money into); with a little bit of tweaking it can become a profit center, especially if the IT resources can be virtualized so that there is no waste! And because of the need to identify a market for IT services, a number of new executive titles have emerged (like we don't already have enough chief executive whatevers): the chief trends officer, responsible for tracking the long-term implications of trends that affect the business; the chief analytics officer, responsible for understanding the implications of the output of analysis of vast stores of data; and the chief people officer, responsible to ensure that the right kind of people are assigned to the IT tasks that must be performed and *not* part of human resources. For example, as Big Data (discussed shortly) becomes an important analytics tool, one of the human capabilities that must be present to understand the output of Big Data is industrial psychology.

Finally, Gartner studies show that somewhere between five and six percent of all enterprises today have appointed a chief digital officer, responsible for embedding digital capabilities into everything the company does.

Decision-Maker

As we noted earlier, legacy IT organizations tended to be self-governing entities, and as such the principal decision-makers were benevolent shills for IT—the IT manager or director, the VP of technology services or the CTO or CIO. Today, however, because of the newfound recognition of the importance of IT as a facilitator of superior QoE, the budget for IT has left IT in many cases and now sits squarely and irretrievably in the business units. The chief marketing officer, the chief operations officer, the chief financial officer, the chief customer advocate, are now responsible for IT's destiny. This is perhaps the single biggest and most wrenching change that the IT organization is going through—in fact, many industry pundits now describe IT as "the new procurement organization." In other words, the business unit heads say, "In order to satisfy the needs of our most important customers, we have to be able to anticipate everything they do and be there with it before anyone else is. The only way I can do that is through customer analytics. So IT, I don't care how you do it, or what vendor you select, but here's your budget—go make it happen."

Value

There's no question whatsoever that IT has always brought value to the enterprise. But in recent years, the nature of that value has morphed, evolving from an operationally focused model centered on the day-to-day transactions of the business and operating under a silo mentality and structure, to an organization that must now focus on transforming the business, discretely analyzing the forces that affect it, and creating a collaborative, knowledge-driven organization.

This focus on the need to share knowledge within the enterprise is one of the most powerful ways in which IT can justify its existence in the hyper-competitive knowledge-centric world. In *Givers Take All: The Hidden Dimension of Corporate Culture,* McKinsey cites a study they undertook of professional services organizations to determine why some of them rose above the others in terms of success and customer relevance. The results stunned them.

They identified what they thought was a comprehensive list of factors that drive a unit's effectiveness—only to discover, after parsing the data, that the most important factor wasn't on their list. The critical factor wasn't having stable team membership and the right number of people. It wasn't having a vision that is clear, challenging, and meaningful. Nor was it well-defined roles and responsibilities; appropriate rewards, recognition, and resources; or strong leadership.

Rather, the single strongest predictor of group effectiveness was the amount of help that analysts gave to each other. In the highest-performing teams, analysts invested extensive time and energy in coaching, teaching, and consulting with their colleagues.

No organization in the modern enterprise is better positioned to create and drive a knowledge-sharing culture than IT.

Evolution

Finally, we reach the top-most level of our IT evolution hierarchy, where we note that instead of technology driving the direction, focus, and spend of IT, that responsibility now falls to the business itself–which is where it should be, after all.

Tracking the Trends and the Shaping Forces

There are literally hundreds of current trends and shaping forces at work in the technology world today. Many of them are vertical industry-specific and won't enter into our discussion here; others are extremely transient and therefore not worth discussing at this level.

Others, however, are powerful enough, universally relevant enough, and important enough to deserve coverage in the pages of this book. We're going to talk about trends that will be around for some time to come and that will have lasting impact on the TMT industries and the companies that depend on them.

For the purposes of this discussion we're going to differentiate between *trends* and *shaping forces.* Please refer to Fig. 11.7.

Trends versus Shaping Forces: What's the Difference?

To my way of thinking, trends are long-term, technology-centric tendencies that markets have to behave in a particular way. In Fig. 11.7

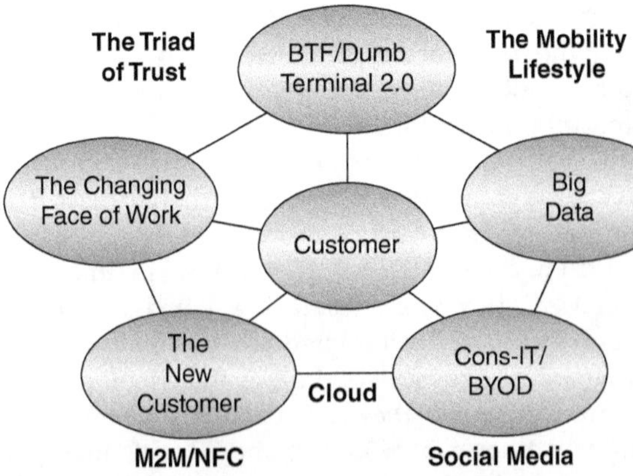

FIGURE 11.7 Trends and shaping forces.

the trends make up the five points of the star: beyond the firewall (BTF) and Dumb Terminal 2.0; Big Data; consumerization of IT and BYOD; the new customer; and the changing face of work. These are trends, which means that while they will have long-term impact and will be with us for some time to come, they are indeed transient.

Shaping forces, on the other hand, are those inexorable, unavoidable, deliberate, tectonic-in-scope changes that irretrievably change the industry landscape forever. In Fig. 11.7 they surround the star: the mobility lifestyle, social media, cloud, machine-to-machine communications and near-field communications, and a view on security that I call the triad of trust. We begin with the shaping forces.

The Mobility Lifestyle

For the purposes of this conversation it is important to point out a very important observation: Mobility is not a technology—it's a lifestyle choice. There's no question that mobility is rapidly replacing fixed line access, especially among younger people, but this is happening, not because of a technological advantage but because of a lifestyle advantage: mobile technology supports a mobile lifestyle.

Consider these numbers from Cisco's visual networking index. Year-over-year, mobile connection speeds tend to double. Think about it: LTE today offers connection speeds that rival those of Wi-Fi! But let's put that number into perspective: The top one percent of all mobile data subscribers in the world generates 20 percent of the carried traffic. Those are smartphone and tablet users for the most part, but about those smartphones: They represent about 15 percent of all the handsets in use around the world today, yet they generate about 80 percent of the total traffic, because they're smart, meaning they have data capability. But here's an interesting observation for you.

We all know that smartphones generate huge flows of data, but a typical tablet in the hands of a typical user generates *15 times the traffic* generated by a smartphone, and a laptop generates 22 times the traffic of a smartphone. Imagine!

One of the benefits that comes from studying the trends and shaping forces is the ability to narrow the gap between *what we believe* and *what we know*—particularly important in a technology world beset by so much myth and legend. Here's a fascinating observation from Deloitte's *Media Democracy Report* (2013). Imagine the following: you, a Baby Boomer, have been invited to participate in a very exciting short-term (six weeks) study of the use of technology by different generations. You arrive for the first meeting and enter the room, where you find 19 other baby boomers and 20 millennials, all sitting around a table and enjoying the free coffee, Red Bull, and cookies provided by the host organization.

"Welcome to our first meeting!" says the host. "You've been asked to take part in a study of the use of different media access devices over the course of the next six weeks, after which we'll ask you to come back and debrief with a series of questions. As you leave today, you will be given two gifts, which are yours to keep: a computer, which can be a PC or Mac, laptop or desktop, and a top-of-the-line tablet, either an iPad or a Samsung Galaxy. All we ask is that you pay attention to the ways in which you use these devices over the next month-and-a-half so that you can give us detailed feedback when the time comes. Any questions? No? Ok, thanks again for coming; please pick up your devices from the folks in the lobby on the way out."

Six weeks later you return for the debrief where you are cordially greeted by your host. As he hands out cookies he says, "I have some bad news for you. Unfortunately our funding has been cut, and as a result you may only keep one or the other of the devices you were given—either the computer or the tablet. We're very sorry, but at least you get to keep one. Now, on to the debrief."

Here's my question to you, reader: which of the devices did the millennials keep? Which did the baby boomers keep? If you answered that the boomers kept the computer and the millennials the tablet, you'd be among the vast majority of people who answered the question. You would also be wrong. Huh?

Here's why—and why it is so important to think about the trends *in toto* and not just on face value. In fact, the millennials overwhelmingly wanted to keep the computer, while the boomers left with tablets. Why? Because millennials tend to be *content creators*, which is very difficult to do on a tablet. Boomers, however, tend to be *content consumers*—they want to lie on the couch and watch Netflix programs on their tablets. Interesting, isn't it?

Mobility as a lifestyle choice is permanently changing the face of the market in every way you can imagine. It's a powerful shaping force.

Social Media

In 2009, Coca-Cola conducted an experiment. Their goal? To determine whether social media could be used to effectively disseminate a viral message to a large population. Their tactic? A marketing campaign built around the idea of personal happiness.

Here's how it played out. Coke conducted its experiment[2] on the campus of St. Johns University in Queens, New York, during final exam week. Nobody was happy; no one was eating properly. They modified a Coke machine in the cafeteria to dispense not only Coke, but happiness (watch the video in the footnote—or just go to YouTube and search for "Coca-Cola Happiness Machine") in the form of toys, gifts, pizza, and sub sandwiches—and then they sat back and watched. The video of what happened next went viral (as I sit here writing this, I see that it has had just under 6 million views), but it's the mechanism that got it there that really matters. A white paper written by the Stanford Graduate School of Business[3] tells the story in detail, but Coke created the video, posted it to YouTube without fanfare, then posted a *single* message on one person's Facebook page that said, "This is kind of cool—check it out." The entire campaign cost Coke $50,000 including modification of the Coke dispensing machine, removal of the wall behind the machine, and so on. Within 10 days they had a million views; in two months that number climbed to two million.

Coke based its strategy on its examination of millennial behavior and the insights they gleaned in the process. Two facts stood out. First, if a millennial wants to research a particular topic in order to know more about it, Google isn't their first stop, their friends are. They trust their friends significantly more than they trust online resources. So they first ask their friends for insight, then they use Google and other resources to fill in the gaps.

Second, they learned that if a millennial learns something interesting that they think ought to be shared, they first post in on a friend's Facebook page, not their own. They allow their friend(s) to determine whether it's worthy of sharing.

With those two factoids in mind, Coke created what has now become one of the great social media experiments of all time.

What Social Media Is—and Isn't

When we think of social media, what comes to mind are thoughts of Facebook, Twitter, and Google+. What *doesn't* typically come to mind are LinkedIn, Amazon, the Apple Apps Store or the Home Depot or B&H Photo Web sites. Or the FitBit site that you visit to see how many

---

[2] Watch the YouTube video here: http://www.youtube.com/watch?v=lqT_dPApj9U.
[3] Dispensing Happiness: How Coke Harnesses Video To Spread Happiness. Stanford Graduate School of Business; Case M-335, June 9, 2010.

calories you've burned, steps you've taken or to check in with your friends to see how *their* exercise and weight loss plans are going. Those are all examples of social media, and they are profoundly changing the way companies market their products, understand their customers, and position themselves competitively. Indeed, a recent study shows that 80 percent of business people believe that companies should regularly and routinely review social media sites to see what customers are saying about them, and 70 percent of consumers want to be able to use social media to find technical support and billing expertise in real-time. Consider, for example, Best Buy's Blue Shirt Nation is a homegrown social media environment that is used by 30,000 or so employees to develop innovative ideas, provide feedback to management and accelerate the delivery of customer service. And their TWELPFORCE application has revolutionized customer service in many ways and served as a model for other businesses. TWELP is a contraction of TWITTER HELP; TWELPFORCE is Best Buy's Twitter-based customer support service. If you can ask your question in 140 characters or less (the limit of a Twitter tweet), they guarantee you'll have a response within 1 minute. So given the choice of calling an 800 number and listening to an earnest voice say, "Thank you for calling Best Buy. Your call is important to us. Please stay on the line" followed by interminably long minutes of Barry Manilow music, sending a tweet with an expected response in under a minute, which will you pick? We'll talk more about this phenomenon when we discuss the new customer later in the chapter.

A couple of years ago I attended Mobile World Congress in Barcelona, a massive convention dedicated to global mobility that is held in the downtown convention center every year. My hotel was two blocks from the event; it took me five minutes to walk there—except for the third day I was there. As you know, Spain's economy didn't weather the global economic crisis well and one of the results of that was that the government raised university tuition pretty dramatically. In the face of a slow economy this was unacceptable to students, and they decided to make their feelings known. Their logic went something like this: "This week, one of the largest conventions in the world is taking place in Barcelona. 70,000 attendees will pass though the main gate every day, and the media will be there in force. We should show them how we feel, don't you think?"

On Wednesday morning at 9 a.m., one of the students sent the following tweet (translated from Spanish) from his Blackberry:

> @MWC: Govt decision to raise tuition is unfair. Let's show our displeasure. Meet in Plaza España at noon near main gate of MWC. Bring friends.

44,000 people showed up. Figure 11.8 is the view from my hotel room; that day it took me over an hour to navigate my way through

FIGURE 11.8 A protest in Barcelona during Mobile World Congress. Two hours after a Tweet was sent, 44,000 people showed up.

the throngs of people who showed up. So just out of curiosity, how long does it take *you* to motivate 44,000 people to do *anything?*

This is the extraordinary power of social media. Here's another example. In Cairo, in Tahrir Square, protesters scrawled a message on a wall during the Arab Spring protests that struck me profoundly in its simplicity. The message is written in French, but its impact could not be clearer. The graffiti says, "Thank you Facebook, Thank you Twitter, for our freedom."

Clearly, Facebook and Twitter didn't grant freedom to the protesters in Cairo. What they *did* do, however, was provide a window into a world they had never seen before, a window that showed them in real-time the global support they had for what they were attempting to accomplish.

One of the concerns that is often voiced by the leadership of large corporations is that social media (especially Facebook) poses a security threat and a massive waste of time and should therefore be prohibited from use while at work. Those beliefs are absurd for the following reasons. First, employees are going to use it whether you want them to or not, on their personal devices—which you can't prohibit them from bringing to work (nor should you, as we'll see in a few more pages). Second, social media is no more of a security threat than e-mail and should be treated with a similar level of circumspection. And as for the issue of wasting time, is it any more of a time-waster than standing outside multiple times a day to smoke? Or walking down to the cafeteria with friends for coffee several times a day? Or taking part in the proverbial water cooler conversations that are part of every workplace? The truth is that in most cases, the people who protest the use of social media in the workplace are typically those who don't use it themselves and therefore don't understand it. These are the same people who lambasted attempts to have employees work at home, claiming that they would do far less work—in the face of every study ever done that demonstrates conclusively that employees who telecommute are 30 percent more productive, across the board, than those who work in the office.

Here's the other thing about social media that must be considered. *It is the mechanism by which your younger employees communicate.* It is not optional, and if you want to attract and retain younger employees (and trust me—you do, as you'll discover toward the end of this chapter), you have to develop a workplace that is friendly to the way they live, work, and play. And that includes availability of social media.

Should it be scrutinized for abuses? Absolutely, every company should put into place a fair use policy for all employees that governs the use of social media in the workplace. But an employee who is spending too much time socializing, whether interpersonally or via social media, and not spending enough time working, isn't a social media problem—it's a management problem.

Cloud

The next shaping force that we'll consider is the growing use of cloud technologies today. Let's begin with a definition:

> *Cloud computing* enables ubiquitous, convenient, on-demand network access to a shared pool of configurable computing resources (e.g., networks, servers, storage, applications, and services) that can be rapidly provisioned with minimal management effort or service provider interaction.
>
> —National Institute of Standards and Technology (NIST)

Now let's say that again, this time in English. I begin with a question: If you're not in the business of running a data center, why are you running a data center? This question lies at the heart of the debate over ubiquitous cloud computing and is one reason that IT organizations are feeling the pressure to recreate their relevance in the enterprise.

Simply put, here's what cloud services make possible: the cost-effective delivery of computing, application, and storage resources on-demand; much more efficient availability of both human and IT resources according to the perceived needs-of-the-business; security in the fact that online information assets are always secure and up-to-date; the ability to perform data mining and analytics as part of the organization's mission to deliver a superior customer experience; and finally, cloud makes it possible to redeploy expensive and often scarce IT resources as a way to support business-critical operations.

In *Competing in a Digital World*,[4] McKinsey makes the case that thanks to Cloud computing, IT has reached a fork in the road and is now going in two very different directions.

IT will undoubtedly play a more significant role as software becomes a larger part of the company and the product. This will require structural changes for both leadership and operations. CIOs will increasingly have a greater voice in strategic discussions. Operationally, there will be more cross-fertilization among specialties, with IT employees placed directly in the line of business to help push through desired product and process changes.

Some executives are beginning to experiment with a new management model consisting of two categories: *factory IT* and *enabling IT*. Factory IT encompasses the bulk of an organization's IT activities. It applies lessons from the production floor on scale, standardization, and simplification in order to drive efficiency, optimize delivery, and lower unit costs. Enabling IT focuses on helping organizations respond more effectively to changing business needs and gain a competitive advantage by spurring innovation and growth. Enabling IT also requires technology leaders to think more expansively about their role and how the systems they manage affect the business as a whole. They need to keep in mind that software assets are ubiquitous and span every part of their organization—and in some cases its business partners and customers—which influences the organization in not-so-obvious ways.

What McKinsey calls "factory IT" are those IT functions that affect efficiency, whereas "enabling IT" affects the effectiveness of the organization. Clearly, factory IT is not mission critical—important, but not so important that it can't be done at least as effectively by an external cloud provider. Enabling IT is a different story: those are IT functions that create competitive advantage and drive organizational strategy, and should be done in-house.

---

[4] *Competing in a Digital World*. McKinsey and Company, February 2013.

Characteristics of Cloud

So just how new is cloud? Sorry to disappoint you, but it isn't all that new. It's been around since 1968 or so when the first mainframe computers arrived on the scene with applications on-board that could support hundreds if not thousands of simultaneous users, all sharing access to a common, expensive, and often scarce resource. In fact, one of my colleagues believes that the first example of a cloud solution was the Library of Alexandria, built in the third century B.C. I kind of like that.

Cloud computing can be defined using four basic characteristics—network access, pooled, scalable resources, usage-based pricing, and on-demand provisioning.

Network Access

The degree to which cloud-based resources can be accessed over traditional, network resources available to businesses is one measure of cloud's utility. This includes traditional Internet, VPN, and even secure wireless options, as well as gigabit ethernet and more traditional access modalities such as T1 and E1. These readily available access methods are a boon for the service provider.

Pooled, Scalable Resources

A well-designed cloud deployment will facilitate sharing of computing and hosting resources a large number of both public and private users. Users must have the ability to scale their usage as required as a way to control cost and take advantage of *utility computing;* the implication, of course, is that the resources they consume are virtual and therefore easily scalable to handle both usage spikes and slow periods.

Usage-Based Pricing

As we mentioned earlier one of the advantages of a cloud solution is its ability to deliver on the promise of utility computing. Not only can customers scale up and down depending on their real-time needs, they can also take advantage of tiered pricing plans and the fact that most cloud offerings have no setup costs or fixed terms of service.

On-Demand Provisioning

On-demand provisioning relies on a simple-to-use Web interface that gives the customer access to the cloud so that they can self-provision as required. The beauty of this model for the service provider is that the customer does all of the work to provision the service they require; the customer pays for the right to do all of the work; and the customer reports a higher level of customer satisfaction. Clearly, Tom Sawyer was onto something when he offered to let Ben Rogers paint his fence for a small sum of money!

As we saw earlier in the book, cloud and IP work hand-in-glove and are part of the ongoing migration to unified communications and managed services. Unified communications refers to the functional integration of a variety of communications modalities including voice (both TDM and VoIP), instant messaging, chat, videoconferencing, and interactive whiteboards, as well as the various elements of unified messaging—e-mail, SMS, fax, and voice mail. The force behind unified communications is *presence,* the ability of the network to recognize that a person is online and available. It also means (and this is where the *unified* part of the name comes into play) that a person can leave a voice mail message, for example, and the recipient can receive it as a media attachment to an e-mail or as a text transcription of the original message.

Remember our discussion about factory versus enabling IT earlier in the chapter? Unified communications relies on the services of a *managed services provider* to handle many of the functions that can best be done by an external cloud provider—otherwise known as a managed service or hosting provider. Managed service providers handle such functions as system backup, data recovery, security, network monitoring, maintenance and management, online storage, software version management, supply chain management, and a host of telecommunications services such as voice, e-mail, contact center provisioning, videoconferencing, and other communications options.

Cloud Growth—Real or Imagined?

Trust me—it's real. By the end of 2015, cloud traffic is expected to grow by a factor of 12, from 130 exabytes (that's 130,000,000,000,000,000,000 bytes, or 130×10^{18}) to 1.6 zettabytes (1,600,000,000,000,000,000,000, or 1.6×10^{21}). That is a *lot* of zeroes, and it's the equivalent of 22 trillion hours of streaming music, five trillion hours of Web conferencing or 1.6 trillion hours of HD video. And while those numbers don't necessarily apply to the business world, these do: According to Gartner,[5] by 2015, 43 percent of all enterprise organizations will have the majority of their IT in the cloud compared to only three percent back in 2011. They also observe that most CIOs expect that the deployment of cloud solutions will between 35 and 50 percent of infrastructure and operational resources.

Manifestations of Cloud

In today's world, cloud comes in three major flavors. The first is called infrastructure-as-a-service (IaaS). Infrastructure-as-a-service provides the customer with access to a virtual server, meaning that they can control cloud-based resources as if they were manipulating

[5] Gartner, *Reimagining IT*: The 2011 CIO Agenda, 2011.

their own dedicated server, when in fact they're using a shared resource. What the user has the ability to do, however, is start and stop processes running on their server, and access and configure both the servers and whatever storage resources are associated with them. In this model, the customer pays for what they use, which is why IaaS is sometimes called *utility computing,* since it's consumed and paid for the same way that gas, water, and power are consumed and paid for. Amazon Web Services, or Elastic Cloud, is an example of this option.

Another example is software-as-a-service (SaaS), examples of which are Gmail, Facebook, Google Apps, Microsoft Office 365, Salesforce.com and AOL, just to name a few. In a SaaS scenario, the vendor supplies cloud-based applications and a little client that runs on the user's device, providing access to the hosted app from any device they like, as long as they have a broadband network connection.

The last example of cloud services, and perhaps the most sophisticated of all, is called platform-as-a-service (PaaS). Platform-as-a-service is not for everybody; it is used by sophisticated IT organizations that want to integrate a variety of processes into a single complex application. In PaaS environments the cloud provider makes available a collection of tools and libraries that can be used by the developer to create a sophisticated environment for the user. Examples are Force.com and the Google App Engine.

Cloud Use Cases

Let's take a quick look at typical cloud use cases, as shown in Fig. 11.9.

If we examine cloud applications from two different perspectives—the level of performance necessary to successfully execute the application and the extent to which the cloud-based resources must be available to the application for successful implementation, we find that our use cases fall into four categories. Let's look at them, beginning with the least critical of the four.

Low Performance, Low Availability

Services that require relatively low performance and aren't affected if the resource occasionally encounters availability problems include archive and backup functions as well as SaaS. None of these applications are particularly time-sensitive; if the backup application encounters an issue during its run, it will wait a few seconds and try again.

Low Performance, High Availability

These applications include unified communications, hosted collaboration, access to user content, machine-to-machine communications applications, and SaaS. Notice also that at the bottom of the box we have the letters *NAS*. This is an acronym for *network-attached storage* that, along with the *storage area network* (SAN) we'll discuss shortly.

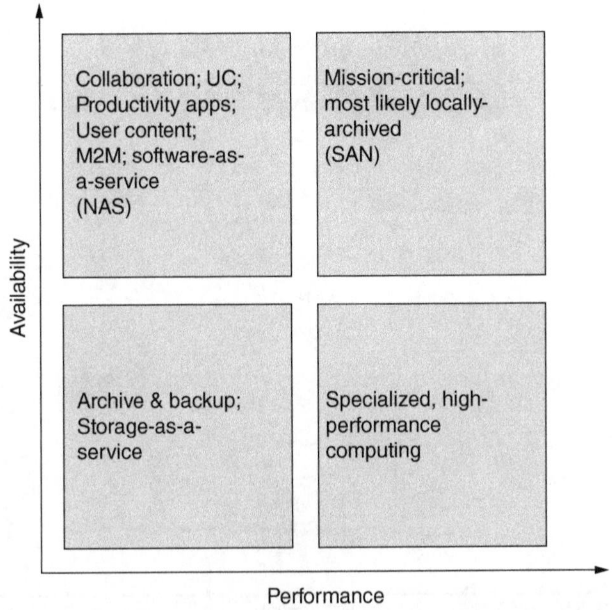

FIGURE 11.9 Typical cloud use cases.

Low Availability, High Performance

These tend to be somewhat specialized applications that tend to be used sparingly; but when they are used, they require massive amounts of cloud resource. They include analysis of polling data, interpretation of oil field drilling data, flight simulation, and so on.

High Availability, High Performance

At the top of the demand curve we find mission-critical applications that absolutely must run, no matter what. They might include, for example, customer billing records, network monitoring and management, or medical records. These would most likely fall into McKinsey's *Enabling IT* category. Note also that we have the designation of SAN in this box, which, along with NAS, we'll explain now.

SAN versus NAS—What's the Difference?

Storage area networks and network-attached storage accomplish much the same things but do so in very different ways. Let's begin with network-attached storage (NAS). NAS is a less expensive alternative to SAN; in this scenario storage devices (hard drives, for example) are connected to the server. They are used to store files, and when their capacity is depleted we simply add another. Access to these devices is typically via common access schemes such as IP and ethernet.

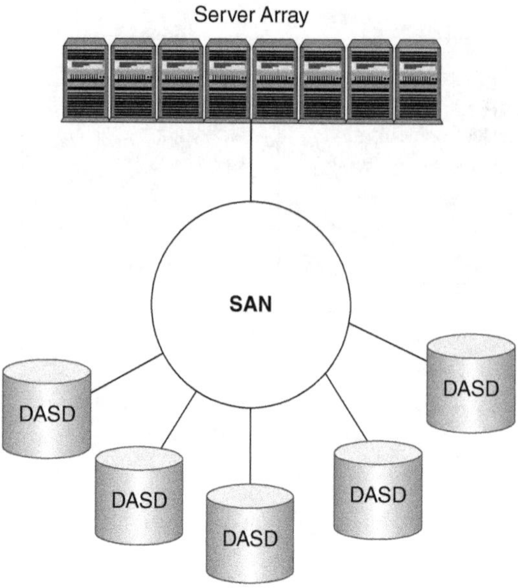

Figure 11.10 A storage area network (SAN).

A SAN, on the other hand, is much more sophisticated, as you can see in Fig. 11.10.

In a SAN environment, storage devices (here labeled DASD, for direct access storage device) are attached to a shared, high-speed network and can be accessed by any device that needs capacity. This model is block-based and is seen by the servers as a single large storage resource. As such it is more expensive than network-attached storage, but then again, it's used for mission-critical applications. Access is typically via fibre channel, although some manufacturers are now switching over to gigabit ethernet and IP. The result? There is a convergence underway of SAN and NAS.

How Cloud Enables Business

Here's what we know about businesses today in terms of what they're facing and what cloud technologies might be able to do to help. First and foremost, the unpredictable growth of enterprise data combined with the (correct) perception of its competitive and strategic value will drive the use of third-party cloud providers for one simple reason: internal IT organizations are not adequately capable of predicting hardware requirements for computing and storage, required levels of network bandwidth, or classes of transport service. Furthermore, seamless interoperability between enterprise functionality and cloud providers is not an option; user and cloud providers alike will demand standards. And this, by the way, is not just an enterprise or small business phenomenon; federal, provincial, and local governments are also in the mix and are accelerating their own migration to cloud services as they increasingly realize the cost savings it brings. And the result of all this? The role of the in-house data center is called into question, not because it is suddenly irrelevant or unnecessary in the face of third-party cloud providers, but because of the need to focus exclusively on mission-critical operations, leaving utility functions to the third-party provider.

So where does this lead? As cloud solutions become an integral part of the business toolbox, a number of interesting things happen. First, the in-house IT organization evolves its focus from a technology orientation to a service orientation. As this happens, the timeframes for service delivery begin to shorten with the shift from QoS as a priority to QoE as the *only* priority. Suddenly the organization finds itself in a position where it can respond much more efficiently and effectively to customer requests for service, one manifestation of which is improved organizational agility.

Furthermore, as organizational sophistication increases and differentiation between IaaS, PaaS, and SaaS become a reality in terms of implementation, a number of other benefits accrue as well. First, IT staff can be reallocated within the organization as required to improve effectiveness and reduce cost. Second, as businesses seek out greater cloud service sophistication to handle their increasingly complex IT workloads, platform-as-a-service providers such as RightScale, Engine Yard, and Heroku will respond. Solution providers will emerge to serve as trusted brokers to help businesses migrate to hosted IT environments. At the same time, PaaS providers will help the business world accelerate software development and perform much faster IT prototyping, thanks to the ability to shift their thinking from factory IT to enabling IT. Meanwhile, IaaS providers such as Amazon Web Services, Rackspace, and Verizon's Terremark will continue to push the limits of cloud infrastructure; much will be learned over the next few years form their efforts.

The numbers don't lie. Studies of personnel demand indicate that demand for cloud-ready IT personnel will continue to grow, increasing by about 26 percent per year for the foreseeable future. According to IDC, revenues from innovative cloud practices could reach $1.1 trillion per year by 2016 and create 8.8 million cloud-related jobs.

So why is cloud important today? Because companies are coming to realize that running a data center in-house may not be the best use of their financial and intellectual resources. Third party cloud providers are mature, sophisticated, and capable—and definitely worth considering. Should every business outsource everything to a cloud provider? Probably not, but there's no reason why a large percentage of the data center work that is done in-house today can't be migrated, freeing up expensive and capable resources for other things. It also enables productivity improvements for those who *use* cloud services, making possible what my colleague Joe Candido and I like to call the "any time, any place, any device" phenomenon. With cloud connectivity, location, access device, and the connection itself become irrelevant—they all work, as long as the connection is broadband. This is important: none of these cloud services work without dependable broadband.

A Final Word: Colocation

One option that is often discussed during cloud conversations is colocation, or CoLo, as it's often called. There are both advantages and disadvantages to a CoLo play; let me discuss them here before we leave cloud behind.

First, let me define what it means. The word first came into the technology lexicon during the heady telecom bubble years at the end of the twentieth century. (I love being able to say that.) In an attempt to advance the broadband options available to customers and promote competition, regulatory decisions were made that allowed for a new family of service providers called competitive local exchange carriers (CLECs). They were formed to compete with the legacy carriers, now known as incumbent local exchange carriers (ILECs). The ILECs referred to the CLECs as *bypassers* for reasons we'll explain shortly.

The Telecommunications Act of 1996 (TA96) lies at the center of all this controversy; it was written to foster competition at the local loop level. Unfortunately, by most players' estimations, it did precisely the opposite.

The introduction of new technologies always conveys a temporary advantage to the first mover. Innovation involves significant risk; the temporary first mover advantage allows first movers to extract profits as a reward for taking that initial risk, thus allowing them to recover the cost of innovation. As technology advances, the first mover position usually erodes, so the advantage is fleeting. The relationship, however, fosters a zeal for ongoing entrepreneurial development.

Under the regulatory rules imposed by TA96 (and adopted by a number of other countries, by the way), incumbent providers were required to open their networks through a process called element unbundling, and were required to lease those unbundled resources—including new technologies—at a wholesale price to their competitors. In the minds of regulators this would create a competitive marketplace. In fact, the opposite happened. The incumbents, faced with the very real need to invest heavily in new technologies such as DSL at the time to meet the growing broadband needs of their customers, now faced a very different challenge: A legally mandated requirement to make those new investments available for lease to their competitors so that the competitors could resell them to the same customers. As you might imagine, the incumbents saw this as a disincentive to invest, and as a result they slowed down their cash infusions into the network, and innovation began to take place at what came to lovingly be known as "telco time." Under the wholesale unbundling requirements of TA96, the rewards for investment in innovative technology were socialized, while the risks undertaken by the incumbents were privatized. Why should they invest and take substantial economic risk, they argued, when they were required by law to immediately share the rewards with their competitors?

Ultimately, the truth came down to this: success in the access marketplace, translated as sustainable profits, relied on network ownership—period. In other words, cable companies were (and still are) well positioned to compete against telcos because they have a physical network that enters the customer's home or business. Wireless players and satellite players do as well. This is called *facilities-based competition,* and it works. Because of the behavior of the marketplace and loyal consumers, most CLECs failed to attract a viable customer base and quickly learned that this model could not succeed. The failure of the CLECs was one major contributor to the collapse of the telecom bubble in 2001 that sent the tech industry into a tailspin and vaporized 7 trillion dollars in market value almost overnight.

But back to our colocation story. When CLECs decided to compete against the ILECs over their own networks, they were allocated space in the incumbents' central offices so that they would have a place to install their equipment. Typically this was a cage made of chain link fencing with a locking door; inside the cage were equipment racks, light, and power. Think of the CoLo cage as the CLECs embassy on the foreign soil of the ILEC—ILEC personnel weren't allowed to enter unless asked; it was considered sovereign CLEC territory.

So what does this have to do with cloud services? Well, one option for companies that currently run their own data centers is to physically relocate them to a third-party provider's IT facility, thus freeing up expensive corporate real estate. The company's equipment is now in a cage in a remote but secure facility, and they can redeploy the space that the IT equipment originally occupied. But let's be clear: Colocation is a real estate play, pure and simple, and in most cases is not a particularly good business model for the following reasons. In the final analysis, it's identical to running an in-house data center, with the following disadvantages: The data center is now remote, which means that employees must travel to and from headquarters to work with the equipment; their ability to respond quickly is now diminished because of the remoteness of the center; control over the environment is somewhat restricted; and finally, no one at the colocation facility is intimate with or vested in the operation or security of your equipment.

There's no question that a hosting solution costs more than colocation, but the level of support is much higher. Think hard before making a final decision. Your needs may be adequately met by a CoLo play, but make sure before you make a final decision.

Machine-to-Machine and Near Field Communications

We've already mentioned machine-to-machine (M2M) and near field communications (NFC) earlier in the book so we won't repeat ourselves here, but let's add a bit of richness to our previous conversation. M2M is defined as a collection of technologies that allow wireless and wired devices to communicate with each other and exchange information. The model typically relies on a sensor or meter of some kind to capture an environmental event such as light level, motion, temperature, number of products in inventory, and so on. These sensors can be passive devices such as the RFID tag that you find randomly placed within the pages of expensive books at the bookstore, or active devices such as traffic monitors or medical devices placed on a patient's body. The fitness bracelets that many people wear today that collect exercise and activity data and upload it via their smartphone to a resident app and then on to a social media site for comparison with other users is another example of an active sensor.

Today one of the most widely used sensors is the personal smartphone. Equipped with digital compasses, GPS, cellular triangulation capabilities, and accelerometers, smartphones use location-based services and other "opt-in" techniques to share personal data in real-time. The information is then transmitted through a network application that converts it into intelligence. Thanks to the proliferation of low-cost semiconductors, IP networking, connectivity, and a willingness on the part of the user to share their information in return for some form of in-kind value, M2M has become nearly ubiquitous. Typical applications for M2M include fleet tracking, customer purchase and behavior analysis, in-building temperature control, medical device and patient monitoring, inventory tracking and control, and traffic and accident monitoring.

Near Field Communications

One of my favorite TV commercials of all-time is a smartphone ad. In it, dad is preparing to leave on a business trip. As he gets into the taxi in front of his house his kids run over with their mom's smartphone, telling him that they created a video for him to watch on the plane. They bump their phone against his, and magically the video is now on his device.

Mom now comes over and kisses him goodbye, saying, "I made a video for you too—but you probably don't want to watch it on the plane." The look on his face is priceless.

The process of *bumping* the phones together demonstrates NFC. You don't actually have to bump the devices; when they are placed in close proximity to each other (that's the near field part) the two devices are capable of exchanging files.

The application that holds the most promise for NFC is contactless payment systems to facilitate mobile payment and mobile banking capabilities. Google has operated a version of this technology for quite some time now in the form of Google Wallet. Customers store their credit card and retail loyalty program information in their virtual Google Wallet on their smartphone; by simply waving the phone close to a paired payment device, the information is securely captured and transacted.

NFC is also finding a home for itself in the world of social media, allowing users to exchange files, contacts, photos, videos, or gaming information. It is also being examined as a possible facilitator of digital identification.

Let's explore this idea for a moment because I believe that it is profoundly important. In the fall of 2013, Apple launched the iPhone 5s, which included an optional on-screen thumbprint reader that can be used to secure the phone. We all know that fingerprints are universally unique—no two people anywhere on the planet have the same combination of whorls, loops, lines, arches, or deltas. Today the fingerprint reader is used to secure a mobile device. But what else could it be used for? How about voting? Today we take our right to vote for granted in the developed western world. We show up at the polling place, cast our vote, and leave without a second thought. In many other parts of the world, however, voting is beset by fraud, corruption, and violence at the polling places, the result of which is that many people don't show up to take part in the democratic process for fear of violence or reprisal. But what if they could vote from their living rooms, using a secure and confidential mobile connection, with the transaction guaranteed by their digital fingerprint? Is it possible that Apple may have found a way to democratize the democratic process?

The Triad of Trust

Sales Person: So, what do you think of security?

Customer: I'm for it.

Thus endeth the conversation.

Having a conversation about network and IT security—indeed, merely *thinking* about the topic—can be something of a dead-end exercise if it isn't done properly and with the appropriate breadth of scope. So whenever I undertake a security discussion, I don't talk about security—I talk about what I like to call the triad of trust, illustrated in Fig. 11.11.

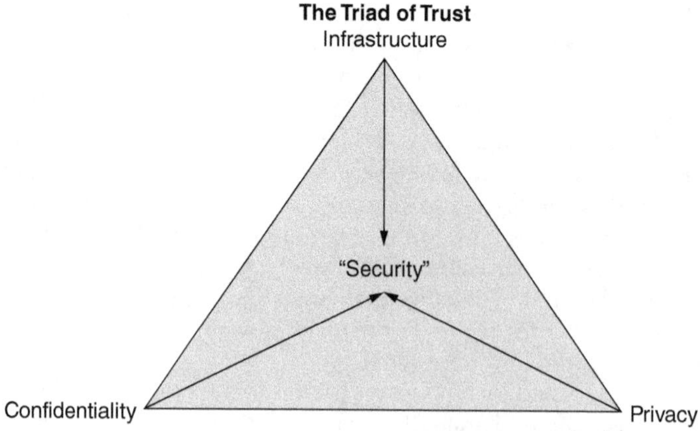

FIGURE 11.11 The triad of trust.

As the diagram shows, security comprises three elements: infrastructure, confidentiality, and privacy. Infrastructure includes exactly what the name implies: Do you have an in-place, good quality firewall? Are your virus definitions up-to-date? Are you running the various forms of intrusion detection software as well as malware and adware filters? Are your people properly trained to identify threats and respond to them properly? Do you have a security best practices policy in place that all employees are required to be familiar with and sign off on? Again, straightforward, typical infrastructure protection issues.

Confidentiality and privacy are different—and different. In spite of the fact that the two terms are often used interchangeably, I see them as very different concerns. Confidentiality is an agreement, whether legal or tacit, executed via a contract or a handshake, between two entities and based on mutual trust. For example, if I use a particular online service like Amazon, Best Buy, or a favorite department store, and they have always treated me well and demonstrated their commitment to protect my personal information, I may agree to keep a credit card on file with them for future purchases. This is a confidentiality agreement: I trust them to protect my personal information.

Privacy is something very different. Privacy, to my way of thinking, is a decision I undertake to exchange something of value for something of equal or greater value. Every single person reading this book has done this at one time or another. How many times have you agreed to provide a company with your e-mail address in exchange for the right to "download that free white paper"? We've all done it, knowing that there will undoubtedly be a flurry of commercial messaging that follows as a result. But I know that up front and make a deliberate and conscious decision that it will be worth the minor inconvenience to have access to the information contained in the paper.

On to the Trends

That concludes our discussion of the shaping forces that are having such a profound impact on the TMT industry. Now we turn our attention to the trends, beginning with a phenomenon called beyond the firewall.

Beyond the Firewall

There was a time not all that long ago when all of the information a person needed to do their job was safely housed on enterprise databases behind the corporate firewall. Rarely was there a need to venture beyond the walls of the digital keep into the unknown. Today, however it's a very different story. Enterprise employees now require access to data resources that lie both behind and beyond the firewall, including public research databases, Google, social networks, Internet-based applications, Salesforce.com, and in many cases their customers' networks. Furthermore, they often have to gain access from remote and unsecured locations. The result? The firewall is merely a suggestion, given that employees must operate on both sides of it if they are to do their jobs effectively. This, by the way, is one of the reasons that hosted security delivered by cloud providers is one of the most sought-after third-party services. By moving the firewall into the cloud, and *forcing* all communications to pass through the cloud—and therefore the firewall—on the way into the corporate data stores, a much higher level of data security can be provided.

This is non-trivial information. In 2013, more than a third of the IT budget in most corporations was spent on cloud services, and nearly half of all enterprises had the majority of their IT services hosted with a cloud provider. By the end of 2016, it is estimated that the global cloud market spend will be a very respectable $200 billion annually.

The Roaming Worker

Consider the network diagram in Fig. 11.12. At the bottom is a LAN that lies behind the corporate firewall and provides secure connectivity to the database on the right end of the LAN segment for employees A and C. Employee B, however, is working remotely, in this case connected to the Internet on the right side of the cloud. This employee also needs access to corporate data resources. Similarly, all three employees also need access to the external resources shown, including connectivity with other vendors and perhaps even customer networks. By moving the firewall function into the cloud, all problems are solved—including the need to eliminate the tyranny of geography.

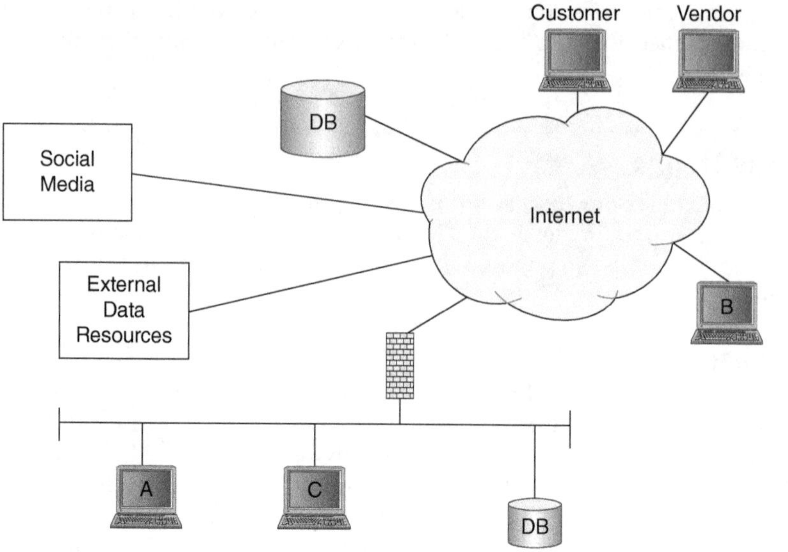

Figure 11.12 The role of the cloud in the modern network.

Earlier in the book we mentioned Frances Cairncross' "Death of Distance" article; now we add the death of geography to the mix.

Dumb Terminal 2.0

Dumb Terminal 2.0 refers to the fact that the role of the mobile device is changing in the face of growing cloud penetration and application availability. Mobile traffic volumes (in lockstep with a growing move toward the mobile lifestyle) continues to grow beyond all expectations, and demand for application and multimedia-ready mobile devices continues at a torrid pace. The challenge is that mobile devices are expected to be diversely capable, display-focused, rich with available memory, easy to use, while at the same time expected to work well in both consumer and enterprise settings. But there's a challenge. Applications are becoming increasingly diverse and resource (and bandwidth) dependent, while at the same time growing increasingly personal: Each user creates a unique relationship (and experience) with their device based on its capability and the application itself. The applications, in turn, create massive data flows that must be stored, analyzed, and acted upon—and which cannot be stored on the device itself, because of the inherent storage limitations of a small device.

But dumb terminals don't have to be mobile devices—a Smart TV, for example, qualifies. Is it a computer that does TV or a TV that has computing capability? This convergence of television content and Internet access, made possible by broadband access and IP networking, gives the subscriber the ability to watch not only broadcast and on-demand television programming, but also home videos, photos, and computer applications.

By now you have undoubtedly realized that the trends and shaping forces are interlinked in an inextricable fashion. It's almost impossible to talk about one without talking about one or more of the others. For example, as more data resources move into the cloud, the role of the mobile device changes, and we discover that the experience is less about storage and more about broadband access and having high-quality display. Ultimately the cloud adds to the experience—as long as the access is adequate for the task at hand. This is one of the reasons that the TMT industry is so powerful: it is internally connected in multiple ways, each element with every other.

An Aside: HTML5 and Why It Matters

Those of you with iPads or iPhones have no doubt had the lovely experience of being denied access to a YouTube video because it was formatted in Flash—and Apple doesn't allow Flash to execute on its portable devices. But there is a reason, and it's probably a good one. Apple is ferocious about delivering the best possible customer experience they can—period. Nothing is more important to them. So when they introduced the iPhone and iPad, they concluded that Adobe (the makers of Flash) was unable to deliver a version of Flash that would meet the exacting needs of Apple's guaranteed experience. They concluded that Flash used too much memory, was a power hog, and had some bugs that were irreconcilable. So Apple chose to deny the execution of Flash-based files on its devices because of the risk of a poor customer experience. Some argue that Steve Jobs went too far with this decision; others support it wholeheartedly. We won't resolve it here except to say that an alternative has emerged that makes the whole argument moot, and that's the arrival of HTML5.

HTML5

As you know, the hypertext markup language (HTML) is the data representation methodology used to create content for Web sites. HTML5 is the next generation of HTML, now adapted to handle the avalanche of rich and diverse multimedia that is now generated and consumed by mobile devices. It is designed to converge all of the pre-existing versions of HTML and is therefore the ideal solution for cross-platform mobile applications. Ideally, HTML5 will run in the cloud and enable vast numbers of HTML5-ready applications, all accessible from a range of devices and display types. As a result, the devices evolve from being specialized to being multipurpose, with HTML5 standardizing the manner in which all content delivered to a viewing device of any kind is encoded. Ultimately the applications, the content, the support infrastructure, and the management functions all reside in the cloud and are encoded via HTML5, eliminating issues associated with version control or data encoding and representation. As a result the corporation can focus instead on delivering a superior customer experience, simplifying access to critical online resources, and catalyzing improved levels of collaboration and co-creation in the enterprise. This concept is central to the success of Dumb Terminal 2.0 implementations; network-based capabilities now adapt to the user and their work style, not the other way around. The power of Dumb Terminal 2.0 lies in the combination of a powerful and capable device (laptop, tablet, book reader, mobile device); ubiquitous broadband connectivity; and a robust cloud environment that translates user activity into monetizable value and a unique customer experience. It also means that user authentication is central to the argument; as we noted earlier, cloud-based security accomplishes this handily.

Most enterprise organizations face the same or similarly disruptive issues including stale or declining relationships with key clients; a sense of declining relevance; a growing need to be more timely and responsive to client requests for product or service-related actions; growing customer demand for alternative engagement mechanisms, including longer reach hours (more on that later); the ever-present need to reduce operating expenses without negatively affecting customer service; a need to develop unique or custom-appearing products and services for customers; and finally, the nagging fear that competitive differentiation capabilities are eroding. But as companies increasingly turn to cloud-based options for their enterprise IT functions, the promise of Dumb Terminal 2.0 shines brightly. By migrating so much of the device's functions into the network, the device becomes the experience delivery mechanism—and provides an opportunity to harvest, analyze, and act upon customer behavior data.

Remember when the telephone company mantra was "reach out and touch someone"? Dumb Terminal 2.0 takes that concept to a whole new level by enabling the ultimate multi-screen experience. Regardless of whether the content is delivered via a TV, PC, tablet, laptop, mobile device, eBook reader, or something else entirely, whoever controls the device and the content on it controls the customer experience. As Apple has said so many times, that's the secret of success. There is *nothing* more important.

Big Data

Let's see. Amazon pays close attention to what you're reading, what format you read it in, how often you buy books, what genres you appear to like, who you buy gifts for, and how much you're willing to spend. They then make cheery recommendations to you about books, gifts, and other products. And you *like* it.

Target Corporation exhaustively analyzes massive amounts of seemingly unrelated information about market segments and then, based on the results of their analyses, changes the product information mix they send to different customers. And customers scream foul. They *hate* it.

Why? What's the difference? In both cases a retailer analyzes customer-buying habits and makes recommendations on the basis of that analysis. In one case, the results are received warmly; in the other, they are rejected with screams of "Invasion of privacy!" echoing off the computer screen.

I find this to be rather interesting. A recent study conducted by Target showed that, given certain market circumstances, they could predict with unerring accuracy whether a woman was pregnant when (1) certain marketplace factors aligned and (2) she purchased a particular brand of hand crème. This came to a head when the father of a teenage girl stormed into a Target store and demanded to see the manager. "How dare you send product literature for pregnant women to my 17 year-old daughter!" he screamed. Turns out she was indeed pregnant; she just hadn't gotten around to telling her parents yet.

So here's my question: Do people get incensed over this kind of analysis and what it reveals, or over the fact that companies are now capable of actually doing it to such a fine level of granularity? I suspect the latter. My guess is that they let Amazon's analyses slide because they're innocuous: we're talking about books, after all, and they're recommending them to you. How nice. But when a retailer demonstrates that they may know or, more importantly, are *able* to know more about your family than you do, people get twitchy.

This is where the touch point between humans and technology gets really, really interesting. Technology is readily accepted when it serves people—meaning *me*. As soon as it begins to serve others, people start to think a bit differently about it. And when it provides insight to others about ME, then they get downright obstreperous.

So what should a retailer do? One option is to make the public use of such data opt-in only. In other words, the consumer must give permission to the retailer to use collected data about them to make recommendations. The consumer should not be allowed to control behavioral information; that's important ammunition for the competitive weapons that all businesses wield. But to reassure the consumer that the information will not be used for nefarious purposes goes a long way. Of course, let's define nefarious. What are the limits? Who gets to define them? How can information actually be used to further a business' ability to most effectively serve its customers? Clearly there will be legislation or regulation emerging around this, but for now, it's a very important, very contentious and utterly fascinating conversation. Stay tuned—we haven't heard the last of this one.

What Is Big Data?

Wikipedia's definition of Big Data is a good one:

> *"Data sets whose size is beyond the ability of commonly used software tools to capture, manage, and process the data within a tolerable elapsed time. Big data sizes are a constantly moving target currently ranging from a few dozen terabytes to many petabytes of data in a single data set."*

So what does this really mean? Big Data refers to a large and usually complex collection of data that is so large and so complex that it can't be adequately analyzed using traditional database management tools or traditional data processing applications. These are collections of data that can single out discrete business trends, anticipate criminal behavior before it happens, predict weather to an eerie degree of accuracy, identify the early stages of an epidemic, and a host of other outcomes. This is big business: the Big Data industry is currently worth about $200 billion, and all of the major players, among them Oracle, IBM, Microsoft, SAP, EMC, HP and others, have spent more than $15 billion to acquire software firms that specialize in data management and analytics.

Let me give you some examples of Big Data in action. UPS is one of the best users of Big Data analytics in the world today. For years they have studied the methods they rely on to deliver packages including driver habits, selected routes, and vehicle usage. After analyzing massive volumes of data they concluded that they could increase efficiency, lower their overall cost, and improve safety by doing one simple thing: eliminating all left-hand turns. That's right. Unless there is simply no other way to get there, you will *never* see a UPS vehicle turn left. Ever.

Did it work? Since 2004 when the study was undertaken, UPS has eliminated millions of miles of unnecessary travel, saved 10 million gallons of fuel, and reduced their CO_2 emissions by 100,000 metric tons—the equivalent of removing 5,300 cars from the road for an entire year.

Here's another example. Researchers at Kaggle, a pattern recognition startup, performed an exhaustive analysis of car purchase histories. One of the things they learned is that if you want to buy a used car and want to be assured that it will be in good shape, buy an orange one. Weird colors tend to be a means of self-expression and if the previous owner saw the car as an extension of her or himself, chances are they took better-than-usual care of it.

Here's another interesting Big Data fact. The Global Virus Forecasting Initiative[6] is charged with predicting and coordinating relief efforts for major epidemics. They rely on Big Data analytics to generate insights

---

[6]Check out the work performed by the Global Viral Forecasting Initiative at their Web site, http://globalviral.org.

that help them act more decisively and accurately. In a business where minutes count, they analyze everything from blog entries to cell phone propagation patterns to news reports from a variety of sources to create predictive analytics that indicate precursors to pandemic viral infections.

Here's another example that is fascinating in its simplicity. A small African mobile company learned that it could predict impending attacks in Congo by monitoring the sale of prepaid calling cards, which skyrocket prior to an attack. But it isn't because people are making calls about what is about to happen; it's because the cards are denominated in U. S. dollars and the local population wants to have something stable and valuable that they can use as currency when the chaos begins.

My personal belief is that Big Data is the new crude oil, and analytics is the new refinery. Industries are lining up to monetize the promise of this new technology family and in the same way we derive dozens of products from crude oil, we'll see similarly diverse derivatives from Big Data. But remember: No network, no Big Data. It's as simple as that.

Structured versus Unstructured Data

Big Data falls into two categories: structured and unstructured. Structured Big Data includes anything that can be inserted into a spreadsheet or a SQL database for analysis purposes, which means that structured data is numerical.

Unstructured data, on the other hand, has no discernible structure and is therefore much more difficult to analyze. Unstructured data, which tends to be largely text-based and is believed to make up the vast majority of all Big Data stores, includes such things as photographic metadata, text documents, e-mail messages, Web page contents, Tweets, health records, arrest records, and so on. There is major growth in both, but it is the unstructured form of Big Data that holds the greatest potential value—if satisfactory analytical tools can be developed to *mine* the value contained within it. Think about it: modern data analytics have the ability to discern relationships among data elements as diverse as all of those listed above and more importantly to make predictions that have genuine business value.

Implications of Big Data

According to research firm IDC, 1.8 zettabytes (1.8 trillion gigabytes) of data were replicated in 2011, nine times the data volume produced in 2006. And the revenues associated with it are equally significant: IDC's research indicates that Big Data will enjoy a compound annual growth rate (CAGR) of 40 percent for the next few years, seven times the growth rate of the TMT industry at large. So the numbers are large, but that doesn't necessarily lead to value. In fact, most Fortune 500 companies will fail to gain the right kind of value from their Big Data efforts for the simple reason that there is a lot more to Big Data than collection and storage. It must be distributed to the right people for analysis and examination, and must be acted upon rapidly if its value is to have impact. Big Data, like most types of data, is typically fragmented, which makes analysis and correlation efforts difficult. Furthermore, it has a half-life—its value decays rapidly, which means that time is of the essence when it comes to deriving value from Big Data analytics. It is crucial, therefore, when undertaking Big Data efforts, to think about the following questions before beginning the analytics process:

- What questions are we trying to answer?
- What are our analyzable data assets?
- Is there a data priority involved?
- What does success look like based on the outcome of the first question?

Without answers to these kinds of incisive questions, the 30 petabytes of data transmitted by AT&T every day, the 24 petabytes carried by Google every day, the 90 trillion e-mail messages sent every year, and the two million transactions generated by Wal-Mart every hour are nothing more than a vast sea of meaningless ones and zeroes.

The *Real* Value of Big Data

All data, whether it be corporate, medical, geophysical, market-dependent, social media, or retail purchase histories is, by its very nature structured in silos of potential knowledge. In the same way that physicists talk about potential versus kinetic energy, here we talk about potential versus kinetic *knowledge*. By aggregating data across many different silos, an intelligent understanding of causality emerges across all of the business units, leading to far more informed and comprehensive decisions. What Big Data really does is accelerate the enterprise ability to understand and adapt to the market faster than the competition. That's competitive advantage; the ability to understand, anticipate, and adapt to market shifts defines the market leader.

Refining Big Data

When oil companies pump crude oil into a refinery, the process of cracking it into various distillates yields gasoline, home heating oil, diesel fuel, kerosene-based jet fuel, heavy lubricating oils, liquefied natural gas, asphalt, road oil, a plethora of petrochemicals, and many other products. According to the American Petroleum Institute, those refined substances are used to manufacture antihistamines, antiseptics, artificial hearts, aspirin, bandages, cameras, candles, clothing, computers, crayons, credit cards, dentures, deodorant, diapers, DVDs, dyes, fertilizers, food preservatives, footballs, garbage bags, glue, golf balls, hair dryers, heart valve replacements, house paint, ink, insecticides, lipstick, medical and surgical equipment, nylon, pacemakers, perfumes, safety glass, shampoo, shaving cream, toothpaste, and vitamin capsules to name but a few of the diverse products that emerge from the catalytic cracking processes of the modern refinery.

In the same way that refineries covert crude oil into many diverse products, so too do sophisticated analytics break Big Data, the crude of the IT industry, into a vast number of sophisticated sub-products. I believe that Big Data is today's crude oil, and that analytics is the new refinery. I also believe that we are on the verge of seeing a new oil rush—but this one will be a fight for customer insight rather than oil leases.

Consumerization of IT Bring Your Own Device

Imagine living and working in a world where technology decisions are driven by influences that lie outside IT and where devices connect seamlessly to the corporate network, regardless of their brand or operating system. In this world, the user dictates their requirements to the IT organization, and IT complies, and employees use their personal devices for access to corporate resources.

Consumerization of IT is all about collaboration, co-creation, improved customer relationships, knowledge-sharing, and shared innovation. It is closely linked to the BYOD phenomenon, which we'll explore in the next section.

Bring Your Own Device

Many organizations already allow their employees to supplement their corporate device with smartphones, tablets, laptops, and home PCs to achieve optimal flexibility, mobility, and productivity. The research bears this out. According to Forrester, 37 percent of enterprises already allow employees to connect personal smartphones to corporate networks; 34 percent allow personal tablets to connect to the corporate network, and out of 350 large enterprises that were surveyed, 83 percent allow employees to use personal devices to access their e-mail. A far smaller percentage of them, however, allow those same employees access to sales force automation or supply chain management applications.

Reasons cited for resistance to BYOD fall into two main categories. The first is a belief that BYOD will cause IT costs to skyrocket. And while there are reasons to support this theory, the facts don't support it. Today, enterprise IT organizations argue, they can dictate which devices people must use—phone, laptop, PC, and the like. That way they can specify a particular device architecture and configuration and a specific operating system, not to mention a limited number of supported applications. But as soon as employees bring their own device to use at work, IT must support it when there's a problem. And it isn't just a mobile phone, they argue: It's also tablets, laptops, desktop machines, and so on.

The other argument that corporate IT uses to resist a BYOD strategy is security. "As soon as we let people bring their own devices to work and use them for work-related communications, we open a massive security hole that we can't control. What happens if they lose their device with corporate data on it?

Let me address both of these arguments, neither of which holds a drop of water. With regard to skyrocketing costs, as Tony Soprano would say, Fuggedaboutit. People are going to use whatever device allows them to do their job most effectively, IT policy be damned. A recent article that crossed my desk pointed out that over 70 percent of enterprise employees already use non-sanctioned applications on their phones to support their ability to do their jobs well.

Here's the other piece of the cost question, stated as an equation. This is complicated, so spend some time with it:

BYOD = BYOIT

When I have a problem with my iPhone, I don't call Apple for support—I call my daughter, my son, or my friend Joe. If they aren't around or can't help, I go to Google, where I enter the error code and stand back, lest I be buried under an avalanche of solutions to my problem.

The point is that when employees bring their own devices to work, they also bring their own tech support. Every company that has decided to permit a BYOD strategy has been pleasantly surprised to find that their IT costs did not become infinitely more complex or costly.

With regard to the issue of security, again, this is a non-starter. Let's begin with the most obvious response to the question about losing the personal device with corporate data resident on it. How is that any different than losing a corporate-issued device? Both have data on them; both can be hacked. Ownership has nothing to do with it, particularly if the company has put into place a BYOD strategy that adheres to best practices for data security. Furthermore, most corporations today use VPN tunneling to allow a remote device to gain access to the corporate database, usually via the cloud. As long as the tunnel is in place, data flows between the host and the mobile device. As soon as the session terminates or times out, the data, which was never resident on the device to begin with, vanishes from the screen. No data? No threat.

What's interesting about this is that it can't be stopped: in fact, nearly 100 percent of companies from the same survey reported that roughly 30 percent of their employees were already using personal computing devices for work-related activities, and a growing percentage of them support the use of personal devices in the workplace. Let's face it: BYOD is not an "if" question; it's a "when" question.

The New Customer

Make no mistake about it: the market is in charge. As a provider of services, the best you can hope for is to understand them better than the competition, give them what they want, anticipate their wants, needs, and demands, and be as multimodal as possible in the manner in which you engage with them. Many years ago, Tom Peters used a quote in *In Search of Excellence* that I have never forgotten:

> Probably the most important management fundamental that is being ignored today is staying close to the customer to satisfy his needs and anticipate his wants. In too many companies, the customer has become a bloody nuisance whose unpredictable behavior damages carefully made strategic plans, whose activities mess up computer operations, and who stubbornly insists that purchased products should work.

The truth is, customers want to be actively involved in the relationship they have and the service they receive from a service provider. For a traditional, legacy organization, that translates into *giving up control*. There is no other choice, of course; the enterprise must be willing and able to connect with customers in different ways as part of a multimodal customer service experience. That means blowing up the traditional script used in the contact center, allowing the customer to engage with you using e-mail, texting, SMS, Facebook Twitter, Google+, even a personal telephone call, in addition to the contact center. The customer makes the rules, now more than ever before.

We do know this: The marketing budget dedicated to customer loyalty and customer retention will double by the end of 2015, and the Universal Customer History Record is the number one most important customer service trend in the domain of the contact center today. Peer evaluation sites like Yelp, Angie's List, Zagat, and others have exploded in the past four years, largely because customers are demanding a quality experience that surpasses previous expectations.

Catering to the Generations: Watching the Market Grow Up

In our discussion about the new customer it would be irresponsible to ignore the generational nature of changing market demographics, particularly the impact of one generation in particular: the millennials.

From work done by William Strauss and Neil Howe, and later by Morley Winograd and Mike Hais (see the citations in the bibliography), we know that generational change is a repeating cycle of four, 20-year cycles, each cycle being a generation. A generation is defined as a group of people who share a common place in history and therefore develop a common set of beliefs and values that tend to be common across the entire generation. It is often said that generations create history, and history creates generations—true enough. And while there are always exceptions to the rules of behavior, the law of

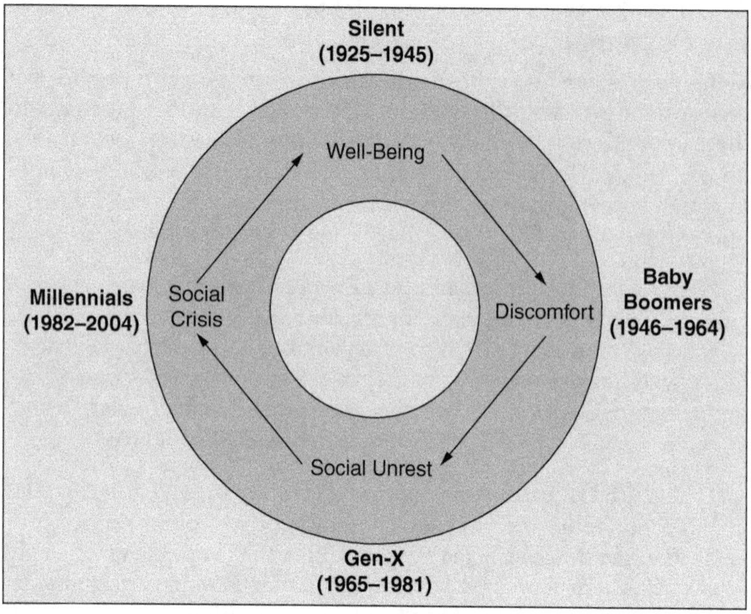

FIGURE 11.13 The cycle of generations.

large numbers tells us that over a sufficiently large population, we will be able to identify behavioral characteristics that are statistically common—and relevant.

Please refer to Fig. 11.13 for the discussion that follows. We're going to discuss each of the last four generations so that you have a sense of how they relate to one another. This is very, very important: understand generational theory and you understand human behavior, market dynamics, politics, and the like.

Every generation is born for about 20 years, after which the next generation arrives. There is always a certain amount of overlap between the generations; people born during this overlap period are called *cuspers* because they are born on the cusp of a generational shift. These people tend to make very good leaders and managers because they have one foot firmly planted in each of two generations and have the ability to view the world through both generational prisms.

The Cycle of Life

Societies as a general rule pass through four behavioral phases. Powerful forces drive these phases including the evolution of social values, political power shifts, balances in demographics and social makeup, and economic upturns or downturns. The phases blend one into the next like the seasons of life, with each season lasting about 20 years, together making up an 80-year (or so) cycle. The first cycle, by convention, is a time of social strength, of growth and optimism during which business and government institutions grow stronger and the individual's influence weakens as individuals place more trust in the institution—banks, health care institutions, large corporations. During this period the individual worker becomes relatively anonymous and places his or her trust in the institution to guide society along. Think back to the 1950s and 1960s when seemingly everybody worked for a large oil company, or a bank, or IBM, or AT&T (remember that there was a time when AT&T had a staggering one million employees). This is the period when comedian Bob Newhart released his "Organization Man" album that made light of Corporate America. During this period a social norm emerges and flourishes as the strength of the previous norm declines. The most recent occurrence of this evolutionary phase was during the Truman, Eisenhower, and Kennedy presidencies in the United States. During that time the United States became powerful and confident but also became highly conformist. A societal "Borg" formed as everyone marched to the dance of the large corporation. This was the "leave it to Beaver" era when the corporate uniform was the white button-down shirt, dark blue suit, dark tie (maybe a stripe or two if you were feeling particularly zippy that day), briefcase, and hat. At the same time, however, a feeling of spiritual emptiness gripped the country that prompted follow-on generations such as the baby boomers to seek greater meaning in their lives. We'll delve into that later in the chapter.

The second phase of the generational cycle is a period of realization that bubbles just below the surface for some time. During this period the individual, who has become anonymous with the vesting of their social power in the institutions, begins to question the social fabric that led to this anonymity. Increasingly disturbed by the sense of being increasingly powerless, a strong probing of social mores ensues. This period becomes loud and passionate, and, prompted by a sense of spiritual emptiness that has begun to grow, attacks on the existing social order take place as new values arise and begin to take effect. Think about the rise of such movements as EST and LifeSpring, both targeted at the self-affirmation process of the 1960s and 1970s. Most recently this phenomenon was seen during the Carter and Reagan presidencies, a time that saw such manifestations as the sometimes-violent unrest of the 1960s. A certain *moral courage* came into play as the values of the past were openly rejected and a sense of personal liberation and a glimmering of individual power emerged.

In phase three, society is in wholesale revolt against the institution. The individual rises in importance, rejecting the anonymity of the corporate workplace. A search for greater meaning in life strengthens as the institution weakens. During this period, individuals become much stronger and more influential elements of the social fabric, institutions weaken as they decline in the public trust, and the values of the inbound generational *regime* take root and diminish the incumbent values. As the quest for meaning peaks, a sense of moral restlessness grips the country. The controversial presidencies of the elder George Bush, Bill Clinton, and Bush the Younger fall into this category.

Finally, phase four arrives amidst a strong sense of change. This is a period of secular crisis in society, a period when individuals go in search of relevance in their lives. This final phase is typically characterized by crisis and social emergency, a time of strong social upheaval that continues as the new social infrastructure takes effect. George Bush's second term experienced this with the devastation of 9/11 and the ensuing crisis in the Middle East.

These four evolutionary periods have been cycling for as long as historians have been studying cultural change, and as might be expected sociologists give each generation a name. Most recently the four archetypical generations, in the same order as the four phases described above, are the Hero Generation (born 1901–1924), the Silent Generation (born 1925–1945), the baby boomers (born 1946–1964), and Generation-X (born 1965–1981). The most recent arrival, the millennials, were born between 1982 and 2004. This is critical because their generational cycle, their time in the spotlight, has ended; the oldest of them are now mid-level managers; the youngest of them are pre-teens. The oldest of them are having their own children now and the cycle refreshes yet again.

Note that we have just named five generations, yet in the last few paragraphs we described four archetypes. In fact, the millennials are the beginning of the next cycle and are identical in every way to the hero generation, the generation that rebuilt following World War II, the generation that Tom Brokaw called "the greatest generation" in his book of the same name. This is a generation of fixers, a generation determined to right the wrongs left behind by the prior two generations—at least that's how they see their role. And they are extraordinarily good at it.

Today we find ourselves at the end of an inner-driven, introspective, baby boomer-centric era. The narcissistic baby boomers, with their emphasis on accountability, wealth accumulation, and community values, left their mark on succeeding generations. Following the boomers are the Gen-Xers, an alienated group of kids who felt abandoned by their dual-income parents. And behind them come the millennials, sometimes erroneously called Gen-Y (wrong because Y implies that they are an extension of the Xers—something they decidedly are *not*).

Baby Boomers

Each of these most recent generational groupings has definable characteristics that follow them throughout their lives. Baby boomers are ideological to a fault, highly judgmental, unwaveringly focused on values, and inordinately narcissistic. They are the most egocentric generation to come along in a very long time (enough about you, what about me?), a characteristic that often manifests itself in the workplace. Baby boomers feel an inordinately strong need to be right and will often argue their own point of view unceasingly in meetings to ensure that they are heard. When it comes to work they are driven; remember that these are the children of the people who grew up enduring the impact of the Great Depression, people who saw their accumulated wealth disappear on one, single, fateful day. As a result they inculcated in their boomer children the belief that wealth is fleeting and should therefore be amassed and diversified. Baby boomers are the dual income generation, a generation for whom work is life and for whom the line between the two is increasingly ephemeral. In fact, boomers tend to define themselves by their work. If you want to see evidence of this, introduce yourself to a boomer, and during the handshake watch what happens. The first question out of their mouth will be, "so, what do you do?" not, "do you have a family," "do you have a dog," or "what are your hobbies." Work for boomers is something they *are*, not something they *do*.

Their silent generation parents also instilled in them a strong set of values that, good or bad, stayed with them for the rest of their lives and permeated everything they did. If you want to see an example of this relationship between the two generations, you need go no farther than Netflix. Find an old episode of the 1950s TV series *Leave it to Beaver*—it doesn't matter which one, they all have the same story—and watch. Here's what you'll see. Every day, dad (Ward Cleaver) gets up, puts on his suit, tie, and hat, picks up his briefcase, and goers off to work *at the company*. We never know *what* company, because it doesn't matter. His wife, June, gets up every day and puts on her dress, high heels, pearls, and apron so that she can clean house and make dinner.

The two boys, Theodore (the Beaver) and his big brother Wally, get up and head off to school but somewhere between school and home—*every episode*—they get into some kind of trouble. The story, then, is about how they can hide it from their parents, which of course, they can't. Why? Because of social networking. That's right—the mom network. All the moms talked with one another on the phone, and by the time the boys got there, the network had kicked in, and the neighborhood equivalent of the Oracle at Delphi was on the front porch, arms crossed, tapping her foot, looking sternly at the boys. Ordering them up to their room, she added tension to the scene with this line: *Wait 'til your father gets home.*

Father gets home and hears the tale from his wife. He goes upstairs, sits on the edge of the bed, and listens to their tale of woe, occasionally shaking his head and nodding sagely. At the end of the story he says two things, *every episode.* To the boys, he says, "I hope you've learned something from this." To his wife, he says, "Boys will be boys."

Boomers were indulged as children, but were also drive by the deeply instilled values that their parents passed on to them. Interestingly, boomers also have a love-hate relationship with authority of all kinds (think 1960s, Berkeley, Kent State, the Blue Meanies of Chicago). These are the people who smoked so much dope in the 60s that they don't even *remember* the 60s, yet if they catch their own children smoking it they ground them for the rest of their lives. Baby boomers are perfectionists by nature, somewhat spiritual, and often community oriented. They are fairly optimistic and involved in life, concerning themselves with youth (their own as much as that of their children), health, and wellness. They are also somewhat schizophrenic when it comes to relationships; they have the highest divorce rate in history, often because work supersedes all else. The all-important work-life balance is often lacking in their lives.

Generation X

Generation Xers, on the other hand, are strikingly different from their baby boomer predecessors. They are skeptical and cynical about life, as you might expect. These are the children of the baby boomers after all. Because both of their parents worked, their perception is that they grew up in the modern-day equivalent of a Charles Dickens novel, raising themselves and living on their own. As one Gen-Xer said to me (partially) in jest, "I lived on the street and had to kill small animals to eat." These are the latchkey kids who came home from school to an empty house and had to fend for themselves, because both parents worked. As you would imagine they became self-reliant, action-oriented, and highly self-accountable. They are also far more balanced in terms of the division of time between their work and personal lives. A typical interaction at work between a baby boomer boss and a Gen-X employee might go something like this: "Look, boss, I understand that to get paid, I have to work hard for you—*for 8 hours.* After that, I go home. And as for this misplaced, pathological idea you have that I'm going to carry a pager and a second cell phone so that you can call me after hours? That's just not going to happen." Imagine what happens when a driven baby boomer manages a group of Gen-Xers. I refer to Xers as the "Jerry Maguire generation." Think about it: "Show me the money. You want something extra from me? Then I expect something in return that goes above and beyond. Working longer hours so that I can tell people that I worked longer hours is not a reward. Time with my family, that's a reward." The most common verbal response to exercised authority from a Gen-Xer is—preceded by an exaggerated sigh and roll of their eyes— "Whatever..."

Millennials

Now we turn our attention to the millennials. The millennials are as different from Gen-X as Xers are from boomers. Millennials tend to be confident, team-oriented, and remarkably, refreshingly conventional. Unlike the Gen-Xers, who eschewed everything that would identify them with their parents, it's ok for millennials to be smart. And *also* unlike the Gen-Xers, the millennials actually *like* their parents.

This is a very large generation, larger, actually than the baby boomers, who up until the arrival of the millennials were the largest generation in human history. Part of this is because they are the children of two generations—the Xers, who wanted to have children, and the late-stage (younger) boomers, who did as well.

Millennials are an optimistic, practical, high-achieving generation that demonstrates the lowest levels of teen pregnancy, drug abuse, alcohol abuse, and violent crime that we have seen since we started measuring them. They believe that they have the potential to be a great generation. The good news is that they're right. The question to ask, however, is why do they believe it? The answer isn't surprising: *Because they've been told their entire lives how great they are.* This is a generation that was indulged as children. They weren't allowed to fail; everybody got a trophy just for showing up at the ball field. The problem is that this behavior bleeds over into the workplace, and for legacy-thinking managers, it makes no sense and has no place. But we don't have the right to blame them for the behavior they exhibit; we created them, so it's our job to learn how to deal with it and to help them blend their learned behavior with the way the real world works. Here are a few anecdotes, all true, that will help you understand the millennial mind.

The Gamer

I ran an executive leadership program at a major university a couple of years ago, and when we got to the part of the presentation where generational theory entered the conversation, a woman in the class raised her hand. Her husband, who was also in the class and worked for the same company, groaned and dropped his head into his hands, mumbling, "I know what story she's going to tell." With that she laughed and suggested that her husband tell the story. He did.

"As you know, I'm the executive vice-president of operations for the firm (a very large service provider in Canada) and my wife is the executive assistant to the CEO," he began. "We've both been with the company for more than 25 years. A few weeks ago my phone rang, and it was my wife, calling from the CEO's office. 'Do you have an employee named John LaChance?' she asked me? Confirming that I did and that he was a new-hire with two weeks with the company under his belt, I asked why she was asking about my newest employee. 'Because he's here in the CEO's office,' she explained. At this point I could see my career burning down around me, and I responded, 'What the hell is he doing in the CEO's office?!?!?' Laughing, she responded, 'It's kind of interesting. He walked in and introduced himself to me, told me he was a new hire and who his boss was, and said, 'I don't know what a CEO does but I think if I did it might help me do my job better. Can I ask him?' Well, the CEO was standing in the door of his office at the time going through the mail and overheard the conversation, and said, 'Sure, come on in.'

"What happened next?!?" I asked her. 'I don't know,' she replied. 'They just left for lunch.'

At this point the husband looked at me at the front of the classroom and said, "son of a bitch! I've been with this company for almost 30 years and *I've* never had lunch with the CEO!" I looked right back at him and responded, "yeah, but you've also never asked, have you?"

Millennials aren't afraid to die—they'll happily try things to see what might happen. If they die, it's okay—they always get another life!

The Barrister

Another participant in a different executive leadership program told me this story. "My husband and I have a 16 year-old daughter. On Friday and Saturday nights she has an 11:30 PM curfew. Last Friday night she came home an hour late but didn't call, so my husband and I waited up for her. When she got home she walked in the door and said, 'I know, I know…' We sent her to her room and told her that we'd talk about it in the morning.

The next morning our daughter came downstairs with her Mac PowerBook under her arm. Without saying a word she opened it and turned it around so that we could see the screen. On it was a four-slide PowerPoint presentation (shown in Fig. 11.14). Slide one, statement of the problem. Slide two, short-term implications. Slide three, long-term implications. Slide four, 'So here are the three punishment options from which you may choose.'"

When I asked her what their response was, she said, "What do you think? We felt like we were on Let's Make a Deal. We'll take door number three."

Millennials are a generation of attorneys. They will negotiate with anyone, for anything. Adults call it arguing; they call it getting what they want.

The Adventurer

While walking with a friend of mine in his neighborhood in Los Angeles a few months ago we encountered one of his neighbors, a young woman with her three-year-old child. We chatted for a while, but as we stood there I couldn't help but notice that the child was wearing a bike

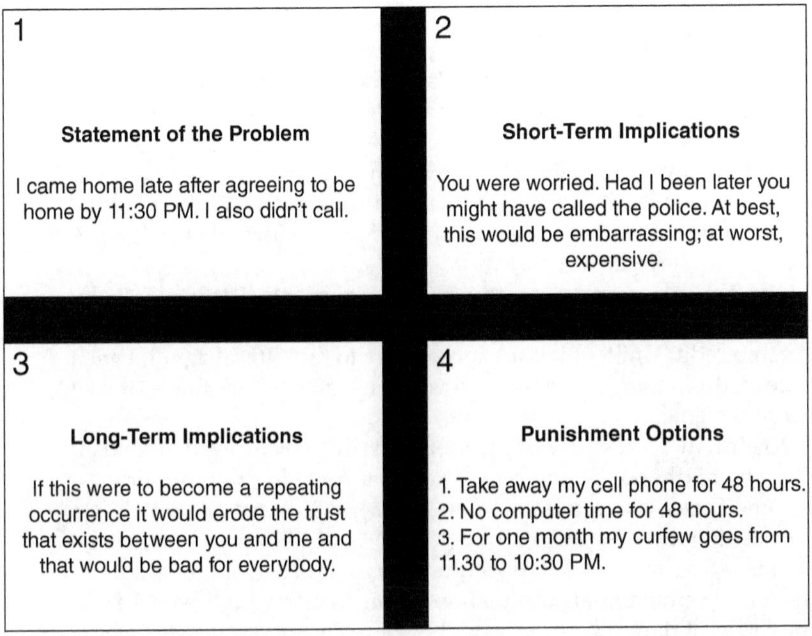

FIGURE 11.14 Millennials make rather good negotiators, as this PowerPoint deck illustrates.

helmet, elbow pads, and kneepads. *And, she was in a stroller*. Not on a dirt bike, not on her way to a roller derby competition. I couldn't help but think, "God forbid that this child should learn about gravity by falling down and scraping her knee." Discretion got the best of me and I kept my mouth shut, but the stories about overprotected millennial children are true. I have to point back at the parents, though, not at the kids. In fact, two new words have been added to the generational lexicon because of them: *boomerang kids* and *helicopter parents*.

Boomerang kids are those kids who finish university and come right back home to live, typically because the economy is poor and there are no jobs to be had in their field. Interestingly when they arrive they find that their parents have preserved their rooms in precisely the state they were in when they left four years ago, a shrine to their remarkably special kids.

Helicopter parents refers to the parents who simply can't let go when their kids grow up and fledge. To this point, I am personally aware of three different universities, one in Canada, two in the United States, which are building dormitories specifically for parents—not because they're enrolled as students but because they have to be physically close to their kids to make sure that they survive the rigors of university life. Really?!? Yes, I'm afraid it's true.

The Bright Side of Millennials

It's easy to poke fun at millennials as they struggle to adapt to the modern workplace, but let's get serious for a moment and recognize a few things that should change the way you think about them.

First, they are the single largest human generation that has ever lived on this planet. Let me say that in a different way: They are the single largest human generation of employees, customers, and competitors we have ever seen.

Second, they are extraordinarily team-oriented. Whereas baby boomers are wont to say, "It's alright for the team to win as long as I get the credit;" millennials are more likely to say, "It's alright if I get a little credit as long as the team wins." They are goal-oriented—it's all about the goal.

Third, they are extremely team-centric. It's not an accident that all of their childhood TV and movie heroes are groups, not individuals: Sesame Street, Fraggle Rock, Ghostbusters, Teenage Mutant Ninja Turtles, Power Rangers, Barney—and Friends. Even recent cinema events reflect this bias. Harry Potter is based on a group of wizards-in-training (millennials all) that as a team will save the world from the forces of darkness, represented by Severus Snape (clearly Gen-X) and guided to a certain extent by the all-seeing Albus Dumbledore and Minerva McGonagall (boomers to the core).

Television content reflects the generational bias as well. Today, Gen-X is in its prime earning years as late-stage middle managers on their way to executive positions. As a result, primetime advertising targets them, as does programming. Remember, this is a generation that is also referred to as the latchkey kids—they essentially raised themselves, which means that they were put into a position where they had to fend for themselves because no on else did. And what about television content that targets them? Well, think about it. Virtually 100 percent of primetime programming today has the same theme: A strong competitive environment where everybody dies…except one. This is the universal theme of so-called "reality TV."

As millennials enter their peak earning period over the course of the next few years, watch how the profile of primetime television content changes from where it is today to content that is more team-oriented and that has a strong social message. It's already beginning to happen; look at the new programs that are appearing on the scene in which the heroes are freshly minted police and firefighter rookies, learning the ropes from the older generation but doing things their own way while at the same time being somewhat deferential to those who came before.

Fourth, millennials want to be connected to the world 24 hours-per-day, seven days-per week, 365 days-per-year. That's all they ask. Peer connectivity is central to their lives, which is why it's not an accident that social media in all its various manifestations came along when it did—and why, if you have a policy that prohibits its use in the workplace, you need to strike it down *now.* Not tomorrow; *now.*

Fifth, millennials have no loyalty to brand or company. They are loyal to two things: their soul, meaning the things that are deeply, personally important to them, and their 3,500 closest friends on Facebook.

Millennials are on a quest for meaning. They're looking for relevance, for a way to make a difference in their lives, and that is true for their career as much as it is for their personal lives. If their job gives them a sense of relevance, they will stay in it for a very long time. Place them in a position, however, that has them doing the same meaningless task repeatedly with no sense of social value, and they will leave without warning. It may not happen today because of the questionable economy and sparse availability of jobs, but it will.

An Example: TELUS

Some companies have cracked the code on this and internalized it as part of their culture. TELUS is a large service provider headquartered in western Canada. But unlike other companies, they have learned to adapt to the different generations that make up their workforce and put into place cultural elements that change the way they do things. For example, TELUS hires millennials every year and puts them into a work program within which they rotate through every organization in the company over a two-year period, learning the inner workings of the entire corporation. Their responsibility is three-fold: (1) learn the job they are in; (2) make recommendations for improvement; and (3) determine where their individual skills and preferences would best serve TELUS and the communities it serves.

TELUS also has a corporate value that states, "we give where we live." It captures the soul of the company and is attractive to everyone, especially millennials. The company donates money, expertise, and people to the communities where they and their customers live and where they offer services. Their charitable contributions are legendary—and real.

But their focus on generational acumen goes well beyond charitable giving. They rely heavily on social media to stay in two-way contact with customers. For example, if a customer has a service problem with TELUS, they are encouraged to post it publicly on TELUS' Facebook page. TELUS responds immediately—and publicly—with a proposed solution. Do they do everything right? Of course not. But they have made a commitment as a company to embrace the inevitable change that the millennials bring to the workplace and to seek out the value that comes with that change.

Motivating Millennials

Whenever I lecture about this subject there is always a handful of people who come up to me laughing, telling me that I just described their kids. And it's true: the law of large numbers says that as long as the statistical sample is sufficiently large, it is valid to make general assumptions about the behavior of social groups. So here are some additional facts. Because millennials play by the rules and are morally-driven, they naturally expect others to do so as well. Baby boomers, take note: *If you make a commitment to a millennial, and then fail to meet the commitment, you will hear about it for the rest of your natural life.*

So what is the best way to motivate millennials in the workplace? First, pay attention to the characteristics listed earlier and to the degree possible structure the workplace around them. Recognize millennials' high level of required social interaction. Use experiential learning and team assignments wherever and whenever possible. Give them freedom with regard to where, when, and how they do their jobs. Put work in a nice place, like their homes, and encourage telecommuting. As morally driven as this generation is, an employer will not be disappointed with the results. Note that the self-policing millennials do not tolerate delays in themselves and others—they are often seen to be unrealistically impatient—and will therefore deliver on time. They also will not tolerate *being managed.* Just because they don't do the work the way you would do it is not a reason to assume that the work will be done poorly.

Next, make the work they are assigned meaningful. Nothing will turn off a millennial faster than work that has no perceived value. Remember, they are looking for meaning, so give it to them. At the same time, they like variety, so give millennials a chance to learn continuously and reward their learning with diverse, ever-changing jobs.

Remember, millennials look at work differently than generations that came before them. They're not necessarily looking for a *career*—they're looking for *meaningful work.* If the work is meaningful and challenging, they may become long-term employees.

Finally, give them plenty of continuous feedback. When assigning work, state the desired outcome as clearly as you possibly can, then step out of the way and let them run with it. It *will* get done, and will most likely exceed expectations—*provided the work is meaningful and challenging.* I can't state this strongly enough.

Conclusion

Because the millennials are a functional repeat of the famously capable greatest generation, they are ideally suited to inherit the chaos of the first half of the twenty-first century. They will rebuild and strengthen the institutions that stabilize the country, will create a longed-for sense of community and belonging, and will restore order and purpose, leading the country out of the crisis that plagues it today. Fear not: We're in good hands.

Networks and content are only as valuable as the people using them, and the value is directly proportional to the degree to which they bring value to those users. For millennials, technology is a lifestyle choice, not a visible set of tools that they occasionally use. They expect to be fully connected, all the time, and want to have seamless access to content from a broad variety of devices, both mobile and fixed, without hassle. The network to them is immaterial because in the minds of millennial customers, *access* means *access to my stuff*, not to the network. The network must adapt to ensure that it delivers the right content, to the right device, in the hands of the right person, in the right form, with the right features, at the right time, for the right price—period and end of discussion. In other words, the network must learn to adapt to the requirements of the user, instead of the user having to modify their behavior because of the physical limitations of the network. And because millennials are not only employees but also highly influential customers and competitors, it is critical to start thinking now about how the network and the resources that it makes available *to the largest generation in human history* must evolve to accommodate their needs.

Think also about the fact that all of these trends that we have discussed so far reflect brilliantly the needs, wants and desires of the millennial generation. The benefits of Dumb Terminal 2.0, Big Data, BYOD, mobility, social media, and machine-to-machine communications, all lend themselves to the millennial lifestyle.

Any questions? Good. Let's move on to our last trend: the changing face of work.

The Changing Face of Work

We all remember when telecommuting became a reality; the bosses all said that it would become a colossal waste of time because people wouldn't really work when they were at home, and that it would become a monstrous management challenge. That, of course, proved to be wrong; as we noted earlier, studies show that telecommuting employees are 30 percent more productive at home than they are at the office.

Today, things have changed. Employees aren't asking to telecommute; their employers are *asking* them to, for a variety of very good reasons. It helps to reduce real estate cost, including parking, heat, lighting, food, and the like; employees are safer, since they don't spend as much time on the road commuting between home and work; the company's overall carbon footprint goes down because of the reduced number of commuters; and finally employees are not only more productive but also more content.

Of course, for this model to work, employees must be connected, and that means broadband access, security, capable end-point devices, application availability, and perhaps unified communications.

Final Thoughts

In this chapter, we have explored a collection of technology-centric shaping forces and trends that universally affect every company and individual and that for the most part fall into the IT realm. Each of them is interesting in its own right, but what really matters is the extent to which they are interlinked. I encourage you to study Fig. 11.7 and look for the linkages between the trends and shaping forces. The linkages are far more important than the trends themselves, because they lead to a better understanding of the impact that TMT technologies have on the companies that depend on them.

Chapter Eleven Questions

1. Pick any one of the trends or shaping forces in Fig. 11.7 and come up with a functional linkage between it and every other item in the diagram.

2. Create a list of five things that you will do differently at work to harness the power of the millennials who work there.

3. How would you influence baby boomers differently than millennials or Gen-Xers?

4. How will television and cinema content change in the next five years?

Closing Thoughts

Many years ago, while still living in California and working for Pacific Bell, I began my writing career by submitting feature articles to local magazines in the Bay Area. I always gravitated toward offbeat subjects, because they made my stories interesting—and desirable.

One day I sat down to write a feature piece, the subject of which has long since been lost in the mists of time. But as I began I realized that I didn't really know what a feature article was, even though I'd been writing them for several years. (For the record, a feature story is a "human interest" piece that is not tied to a particular news event.) Feature stories usually discuss concepts or ideas that are specific to a particular market and are often quite detailed. For example, a story about the history of a landmark building that is being renovated, or a few days spent on a cable-laying ship in Singapore, or a piece on the behind-the-scenes goings-on of a particular town qualify as feature stories.

Anyway, I grabbed the dictionary off the shelf (this was years before the Web and digital dictionaries were still a dream) and turned to the entry on *feature story*. And there I saw it: directly across the gutter from feature was the word *feces*. Never being one to ignore an intriguing pathway, I followed it. The dictionary entry for feces led me off on two side trails ("see also..." is my favorite part of any definition), which were equally intriguing. The first was *scatology;* the second, *coprology*. So off I went, down the linguistic rabbit hole. Scatology, it turns out, is the study of animal droppings, a field important to wildlife biologists because from scat can be determined all kinds of things about a particular species including what they eat, how often they eat it, how healthy the prey was, whether the animal that dropped the scat has any parasites, how widely the animal ranges, and all sorts of other things that are actually quite interesting, in a weird sort of way. So I called my alma mater's biology department and was soon on the phone with a genuine scatologist, a wildlife biologist who specializes in the analysis of wildlife ... droppings, which are known as *scat* to those who care enough to name such things. We ultimately met and I spent a day with him in the field, and I'm delighted to tell you that it was one of the most interesting days I have ever had.

The other trail that feces took me down was *coprolite.* A coprolite is, and I'm not making this up, a fossilized dinosaur dropping. A paleo-scat, as it were. I have one on my desk, a gift from a friend with a good sense of humor. Once again I got on the phone and called Berkeley, the paleontology department this time, where I soon found myself talking to a coprologist—yes, there is such a person. How do you explain *that* at a dinner party? Anyway, he agreed to meet with me and once again I had one of those rare and wonderful days, learning just how fascinating the stuff is (was?) that came out of the north end of a south-bound dinosaur. He showed me how they slice the things on a very fine diamond saw and then examine them under a high-power microscope to identify the contents, just as the scatologist did with owl pellets and coyote scat.

Had I not allowed myself to fall prey to serendipity (definition: "A "happy accident" or "pleasant surprise;" specifically, the accident of finding something good or useful while not specifically searching for it), I never would have found those remarkable people, and never would have written what turned out to be one of most popular articles in that particular magazine.

Another time, this time more recently, my wife and I were out walking with our dogs in the field near our house. At one point near the end of the walk I turned around to check on the dogs and saw one of our dachshunds rolling around on his back the way all dogs do when they find something disgustingly smelly. This was no exception: He had found the carcass of some recently dead animal, too far gone to identify but not so far gone that it didn't smell absolutely horrible. I dragged him home with my wife following about 30 ft behind me and gave him the bath of baths to eliminate the smell. Anyway, once he smelled like a dog again I felt that old curiosity coming on so

I went downstairs to my office and began to search Google for the source of that horrible smell that's always present in dead things.

I found it. The smell actually comes from two chemicals, both of which are so perfectly named that whoever named them clearly had a good time doing so. The first of them is called *cadaverine*; the second, *putrescine*. Can you think of better names for this stuff? Interestingly, putrescine is used industrially to make a form of nylon.

So what's the point of this wandering tale? Storytellers are always looking for sources, and I am, first and foremost, a storyteller. The question I get more often than any other is, "where do all of your stories come from?" The answer, of course, is multifaceted, but in many cases I find stories because I go looking for them but leave my mind open to the power of serendipity.

These closing thoughts make the case for the need to knowledge-share within an organization, whether a work group or an entire enterprise.

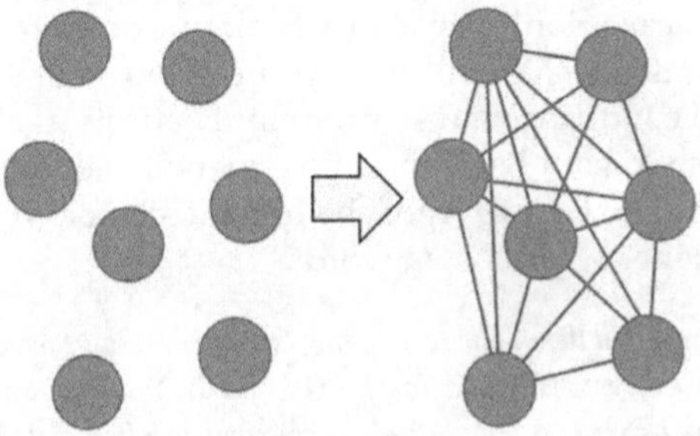

Imagine that each of the circles in the following figure is a person in a company who possesses some amount of knowledge that is relatively unique. The left side of the diagram shows seven knowledge silos that have no ability to share with each other. The right side, however, is fully connected—a mesh network, as it is called. Every node (person) has access to the knowledge of every other node (person). So the real value of a connected enterprise that has adopted a knowledge-sharing culture doesn't lie in the individual knowledge silos, but rather in the connections that exist *between* the knowledge silos. This is the phenomenon that Silicon Valley pioneer Bob Metcalfe, the inventor of ethernet, was describing when he coined the concept of Metcalfe's law: *The power of a fully connected network increases as a function of the square of the number of nodes in the network.* In other words, when new nodes (knowledge bases) are added to the network (enterprise), the value of the entire knowledge assembly doesn't go up incrementally—it goes up exponentially. Think of the value of the World Wide Web, or of such Web-based resources as Wikipedia. Their value doesn't lie in any one element of the network, but rather in the power of the connected whole. So what happens when your enterprise, or that of your customer, adopts a knowledge-sharing philosophy based simply on the idea that every person in the organization accepts the responsibility to share what they know when they come to know it, and to accept the fact that they don't have the right to decide whether a knowledge nugget is important to someone else. That's *their* responsibility. The only requirement placed on each employee's shoulders is to share. How hard is that?

This book is a collection of stories that together represent a journey through one of the most fascinating and important industries on the planet today—in fact, beyond the planet. Voyager 2 is NASA's longest-lived galactic probe; launched in 1977 and now nearly 10 billion miles from Earth, the spacecraft continues to send transmissions to NASA. At that distance, traveling at the speed of light, a message takes just under 17 hours to reach Earth. That's telecommunications with an attitude, folks.

You've just spent quite a bit of time going through this book, and I thank you for that commitment—and hope you got some value from it. But please don't stop here: Seek out other resources, both literary and human, that will help you expand your knowledge of this remarkable industry that so profoundly affects us all.

Alexander Graham Bell did some remarkable things in his life as an inventor, but he never forgot the human side of what he was all about, as evidenced by his own words:

> *Don't keep forever on the public road, going only where others have gone. Leave the beaten track occasionally and dive into the woods. You will be certain to find something you have never seen before. Of course it will be a little thing, but do not ignore it. Follow it up, explore around it; one discovery will lead to another, and before you know it you will have something worth thinking about to occupy your mind. All really big discoveries are the result of thought.*

Thanks again for taking this journey with me. Stay in touch.

Appendix

Photo Credits

All historical photographs are used with the kind courtesy of the Lucent Technology historical archives. My thanks to Lucent (now Alcatel-Lucent) for their permission to reproduce the images.

All other images are from the author's collection.

Bibliography

"The Internet's Fifth Man." *The Economist*, Nov. 30, 2013.

Dispensing Happiness: How Coke Harnesses Video to Spread Happiness. Stanford Graduate School of Business, June 9, 2010.

"Surfing a Digital Wave, or Drowning? The Future of Corporate IT." *The Economist*, Dec. 7, 2013.

Adams, Douglas. *A Hitchhiker's Guide to the Galaxy.* Pan MacMillan Adult, 2002.

Anderson, Chris. The Long Tail. *Wired*, Oct. 2004.

Anderson, Chris. *The Long Tail: Why the Future of Business is Selling Less of More.* Hyperion , New York, NY. 2006.

Cairncross, Francis. *The Death of Distance: How the Communications Revolution is Changing Our Lives.* HBR Press, March 2001.

Cisco Visual Networking Index, 2012. Available at: *www.cisco.com/c/en/.../visual-networking-index-vni/.*

Cisco Global Cloud Index, 2012. Available at: *http://www.cisco.com/c/en/us/solutions/service-provider/global-cloud-index-gci/index.html.*

Deloitte Media Democracy Report. Available at: *http://www.deloitte.com/assets/Dcom-UnitedKingdom/Local%20Assets/Documents/Industries/TMT/uk-tmt-state-of-the-media-democracy-report-2012.pdf.*

Engineering and Operations in the Bell System: Rey, R. F. *Engineering and Operations in the Bell System, Second Edition.* AT&T Bell; Murray Hill, New Jersey, 1983.

Frost and Sullivan. *World Unified Communications Markets: Business Models Evolve as Technologies Mature.* Feb. 2010.

Frost and Sullivan. *Worldwide Enterprise Telephony Platform and Endpoint Markets.* June 2010.

Grant, Adam. "Givers Take All: The Hidden Dimension of Corporate Culture." *McKinsey Quarterly*, April 2013.

Hais, Michael, and M. Winograd. *Headwaters of the Arab Spring.*

Hais, Michael, and M. Winograd. *Millennial Majority.*

Hais, Michael, and M. Winograd. *Millennial Makeover: MySpace, YouTube, and the Future of American Politics.*

Hais, Michael, and M. Winograd. *Millennial Momentum: How a New Generation is Remaking America.*

Herzfeld and Andy. *Revolution in the Valley: The Insanely Great Story of How the Mac Was Made.*

Isaacson, Walter. *Steve Jobs.*

Isenberg, David. *The Rise of the Stupid Network. http://www.isen.com/stupid.html.*

Kao, K. C. and G. A. Hockham. "Dielectric-Fibre Surface Waveguides for Optical Frequencies." *Proc. IEE* 113 (7): 1151–58, 1966.
Peters, Tom. *In Search of Excellence.*
Philip, E. and T. Wurster. *Blow to Bits: How the New Economics of Information Transforms Strategy.* Harvard Business School Press, Boston, 2000.
Sarrazin, Hugo, and J. Sikes. *Competing in a Digital World.* McKinsey and Company, Feb. 2013.
Shannon, Claude. *A Mathematical Theory of Communication.* University of Illinois Press, 1949.
Sinek, Simon. *Start With Why: How Great Leaders Inspire Everyone to Take Action.* 2009.
Sinek, Simon. *TED talk,* posted to: YouTube, August 2011. *http://www.youtube.com/watch?v=d2SEPoQEgqA.*
Standage, Tom. *The Victorian Internet.*
Strauss, William, and N. Howe. *The Fourth Turning.*

Web Resources and Links

"Apps Rocket Toward $25 billion in Sales." *Wall Street Journal.* Accessed Mar. 4, 2013. *http://online.wsj.com/news/articles/SB10001424127887323293704578334401534217878.*
"Budget-Friendly Alternatives to Cable TV." *Kiplinger.* Accessed Oct. 2011. *http://www.kiplinger.com/article/spending/T057-C000-S002-budget-friendly-alternatives-to-cable-tv.html.*
"Latest Telecommuting Statistics from Global Workplace Analytics.com." Accessed Nov. 4, 2013. *http://www.globalworkplaceanalytics.com/telecommuting-statistics.*
"Various Reports from the National Cable and Telecommunications Association." Accessed Aug. 3, 2013. *https://www.ncta.com/.*
"Various Mobile Statistics from Mobile Thinking." Accessed Oct. 12, 2013. *http://mobithinking.com/mobile-marketing-tools/latest-mobile-stats.*

Steve's Learning Resources

One of the most common questions that I get in my workshops is, "What do you read to stay on top of all the goings-on in the world?" It's a good question, and one that I'll address in this post.

As many of you know, I work in the world of technology, at the intersection between telecom, IT, and media. I spend my time studying the ebb and flow of these three sectors and, more importantly, the impact that they have on various verticals including healthcare, transportation, education, government, manufacturing, and a number of others. My job as a consulting analyst is to distill the vast amount of data generated by the interaction among different verticals and the technologies they encounter to create insight. So it might surprise you to know that in my quest to stay on top of the happenings in telecom, IT, and media I don't read any trade journals—as in none. Zip. Nada. Huh?

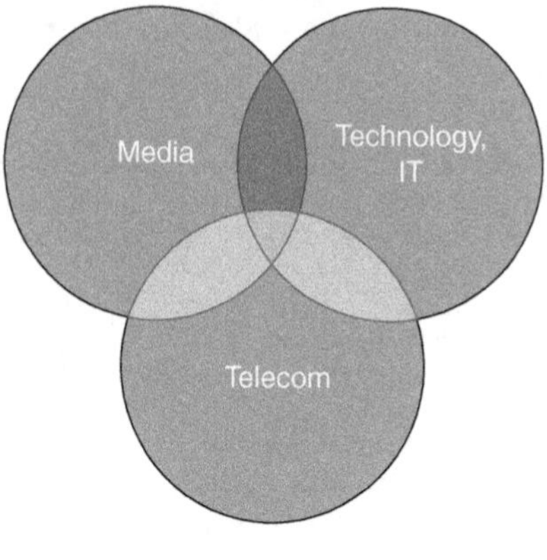

Here's the reason. I know the technologies in these sectors very well because I've worked in this world for more than 30 years, often in very technical roles. I've written more than 50 books about the inner workings of these technologies so I'm comfortable with them. Is my knowledge 100 percent current? Nope. But it's current enough that I know what I don't know and also know when I have to retreat to the standards or to my fellow authors to refresh my knowledge of one technology or another.

So where do I cast my attention? What sources do I seek out? What do I read? Who do I talk to? These are good questions so I'll answer them here as best I can. This is a bit of a moving target, and some of these questions I'm still trying to formulate answers to. So bear with me and we'll see what I come up with.

What I Read

I'll begin with magazines and journals (this doesn't include the photography magazines I receive every month, even though they're instrumental in my personal storytelling craft). As you'll see they're pretty diverse.

Magazines and Journals

- *The McKinsey Quarterly.* This business journal is fantastic for insight into the high-level leadership trends that are imposing their will on the industries they serve. The *Quarterly* is available in digital format for tablet readers and is excellent.
- *The Economist.* Perhaps the best weekly news magazine (although they refer to themselves as a newspaper) being published today. *The Economist* provides a truly global view of the goings-on around the world, dividing itself into sections that address each region of the world as well as technology and other specific areas of interest. Every quarter they also publish a *Technology Quarterly*, in which they address the hot tech topics that the world is seeing. The *Economist* is available in a digital format as well.
- *National Geographic.* Perhaps the best monthly collection of stories in print, *National Geographic's* supremely talented writers and photographers are on the bleeding edge of storytelling every month, creating compelling content that delivers on-the-mark every month. I strongly recommend the digital edition simply because in addition to the articles, subscribers also get video content and interactive maps that dramatically enhance the richness of the written content.
- *Business Week. Business Week* has a decidedly American bent to it, but that's ok: Even though the articles are largely about business issues in the United States, that's not completely true. Articles in *Business Week* are increasingly global in scope and provide insights that are similar to those provided by *The Economist* in terms of broad perspective. Easy to read, often funny, *Business Week* is also available in a digital edition.
- *Foreign Policy Magazine.* Before you start questioning me about a magazine that focuses on foreign policy, hear me out. Mark Twain once wrote that "Travel is fatal to bigotry, prejudice, and narrow-mindedness." And while you may not be able to travel as widely as you would like, magazines like *Foreign Policy* give you the next best thing. Rich with far-ranging articles about globalization, politics, foreign policy, and many other topics that are rich with storytelling lore. The history of the organization is fascinating and will give you an insight into the value that this remarkable magazine brings. Like the others, it is also available in a digital edition.
- *The Atlantic Monthly.* This wonderful publication advertises itself as the publication that reports on "news and analysis on politics, business, culture, technology, national, international, and life." They do

this, and more. Few magazines are as richly wide-ranging as this one. Also available in a digital edition for tablet readers.

- *Wall Street Journal.* What can I say? The truth is that I don't read the *Wall Street Journal* every day—far from it. I do, however, read the news flashes I receive in e-mail as a subscriber to the WSJ's digital edition. Those flashes often send me digging into the full-blown version of the paper in search of a more detailed article. For example, a couple of weeks ago I received notification of an article that had just appeared in the Journal called "Why Big Data Matters." This happened to be an area of focus for me at that point in time, and the article was rich with exactly the content I needed. Again, digital edition available.
- *Local Newspaper.* The local newspaper of whatever town or city I happen to be in. I can't emphasize this enough. Relevance is a critical element of effective storytelling, and one of the best ways to ensure relevance is to imbue your stories with local color. Nothing is more irritating to an audience in another country than a presenter who insists on using references from their own country. If you're working in Canada, take the time to learn something about Canada! Draw maps of Canada! Use Canadian cities as references! Don't talk about Starbucks, talk about Tim's!
- *Deloitte Publications.* A remarkable variety of them are available at the company's Web site. Deloitte's Technology, Media, and Telecommunications (TMT) practice publishes a range of documents which are free and available to anyone who wants to download them.

Authors

It's hard to provide a concise list of books that I recommend because I am a voracious reader. I average about 120 books a year that cover a broad swath of things that keep me informed about the international marketplace that I'm so interested in. But I also read authors who are first and foremost good storytellers. I won't recommend specific books here (maybe in a later post—let me know if you have specific requests), but I *will* recommend specific authors. Here are my favorites (I recommend *anything* they've ever written):

- *Rick Bragg.* One of the world's best contemporary storytellers. For a taste of good writing, read *All Over but the Shoutin.*
- *Thomas Friedman.* The author of *The Lexus and the Olive Tree* and *The World is Flat* this New York Times journalist and bestselling author is great at providing a view of the world and its complex issues through a storyteller's view.
- *Robert D. Kaplan.* Equal parts historian, political scientist, travel writer, and 100 percent storyteller, Kaplan's books offer gripping views of the complex world we inhabit. Start with *To The Ends of the Earth.*
- *Fareed Zakaria.* The host of CNN's GPS program has written several books, the best of which is called *The Post-American World.* This is not a book about the decline of America, but rather about the rise of everybody else. A truly well-presented case with far-reaching implications.
- *John McPhee.* McPhee asks the questions that nobody else thinks about in his many, many books. This guy knows how to tell a story better than most. I recommend *Uncommon Carriers* as a good place to get started on his books.
- *William Least Heat-Moon.* I credit Bill with being the man who made me realize that I wanted to be a writer. His seminal book, *Blue Highways,* came out in the early 1980s and he has written a number of well-received travel essay collections and stories since then. Start with *Blue Highways,*

but be prepared: halfway through the book you will find yourself packing the car for a four-month road trip.

- *Tom Peters*. Equal parts curmudgeon, visionary, and consultant, Peters enchanted the world with *In Search of Excellence* in the early 1980s and continues to churn out thought-provoking material. Start anywhere, but be sure to read *Excellence*. It's a bit dated but still as good today as it was when it first appeared.

Podcasts

I spend a lot of time in airplanes, rental cars, and hotel rooms, so I also listen to Podcasts that keep me informed about specific things. They include

- *The Classic Tales* (great way to "read" classic literature without actually reading the stories).
- *This American Life* (from National Public Radio). Sometimes bizarre, always entertaining and thought-provoking stories about life in the United States.
- *This Week in Technology* (TWIT). The host and crew discuss a wide range of entertaining technology stories every week. Very, very good.
- *National Geographic Video Shorts*. Available through iTunes University and at the National Geographic site, these are lectures presented by the Geographic's best people. Fascinating and absolutely rich with stories.
- *TED*. Founded by Richard Saul Wurman (author of *Information Anxiety*) in 1984, TED (abbreviation for technology, entertainment, and design) brings together the best and the brightest in a series of conferences that take place several times a year around the world. Available at the TED Web site or as an iTunes video Podcast, I consume TED videos the same way I consume daily vitamins—at least one every day. One of the best sources of inspired thinking I know.

Books

Finally, books. This is particularly difficult because I read a *lot*. I'm not going to recommend any particular titles because we'd be here until the end of time. So if you are looking for books on a particular topic, please reach out to me and I'll come back with one or more titles on the subject.

There are others, of course, and I suspect I've left out some of the best ones. If you are interested in a particular topic or medium, please send me a message at *Steve@ShepardComm.com* and I'll come back to you with suggestions.

Common Industry Acronyms

AAL	ATM adaptation layer
AARP	AppleTalk address resolution protocol
ABM	Asynchronous balanced mode
ABR	Available bit rate
AC	Alternating current
ACD	Automatic call distribution
ACELP	Algebraic code-excited linear prediction
ACF	Advanced communication function

ACK	Acknowledgment
ACM	Address complete message
ACSE	Association control service element
ACTLU	Activate logical unit
ACTPU	Activate physical unit
ADCCP	Advanced data communications control procedures
ADM	Add/drop multiplexer
ADPCM	Adaptive differential pulse code modulation
ADSL	Asymmetric digital subscriber line
AFI	Application family identifier (RFID)
AFI	Authority and format identifier
AI	Application identifier
AIN	Advanced intelligent network
AIS	Alarm indication signal
ALU	Arithmetic logic unit
AM	Administrative module (Lucent 5ESS)
AM	Amplitude modulation
AMI	Alternate mark inversion
AMP	Administrative module processor
AMPS	Advanced mobile phone system
ANI	Automatic number identification (SS7)
ANSI	American National Standards Institute
APD	Avalanche photodiode
API	Application programming interface
APPC	Advanced program-to-program communication
APPN	Advanced peer-to-peer networking
APS	Automatic protection switching
ARE	All routes explorer (source route bridging)
ARM	Asynchronous response mode
ARP	Address resolution protocol (IETF)
ARPA	Advanced Research Projects Agency
ARPANET	Advanced Research Projects Agency Network
ARPU	Average revenue per user
ARQ	Automatic repeat request
ASCII	American standard code for information interchange
ASI	Alternate space inversion
ASIC	Application-specific integrated circuit
ASK	Amplitude shift keying
ASN	Abstract syntax notation
ASP	Application service provider
AT&T	American Telephone and Telegraph
ATDM	Asynchronous time division multiplexing
ATM	Asynchronous transfer mode
ATM	Automatic teller machine
ATMF	ATM forum
ATQA	Answer to request A (RFID)
ATQB	answer to request B (RFID)
ATS	Answer to select (RFID)
ATTRIB	Attribute (RFID)
AU	Administrative unit (SDH)
AUG	Administrative unit group (SDH)
AWG	American wire gauge
B2B	Business-to-business

B2C	Business-to-consumer
B8ZS	Binary 8 zero substitution
BANCS	Bell administrative network communications system
BBN	Bolt, Beranak, and Newman
BBS	Bulletin board service
Bc	Committed burst size
BCC	Blocked calls cleared
BCC	Block check character
BCD	Blocked calls delayed
BCDIC	Binary coded decimal interchange code
Be	Excess burst size
BECN	Backward explicit congestion notification
BER	Bit error rate
BERT	Bit error rate test
BGP	Border gateway protocol (IETF)
BIB	Backward indicator bit (SS7)
B-ICI	Broadband intercarrier interface
BIOS	Basic input/output system
BIP	Bit interleaved parity
B-ISDN	Broadband integrated services digital network
BISYNC	Binary synchronous communications protocol
BITNET	Because it's time network
BITS	Building integrated timing supply
BLSR	Bidirectional line switched ring
BOC	Bell Operating Company
BPRZ	Bipolar return to zero
bps	Bits per second
BRI	Basic rate interface
BRITE	Basic rate interface transmission equipment
BSC	Binary synchronous communications
BSN	Backward sequence number (SS7)
BSRF	Bell System reference frequency
BTAM	Basic telecommunications access method
BUS	Broadcast unknown server
B-VoIP	Broadband VoIP
BYOD	Bring your own device
C/R	Command/response
CAD	Computer-aided design
CAE	Computer-aided engineering
CAGR	Compound annual growth rate
CAM	Computer-aided manufacturing
CAP	Carrierless amplitude/phase modulation
CAP	Competitive access provider
CAPEX	Capital expense
CARICOM	Caribbean community and common market
CASE	Common application service element
CASE	Computer-aided software engineering
CASPIAN	Consumers against privacy invasion and numbering (RFID)
CAT	Computer-aided tomography
CATIA	Computer-assisted three-dimensional interactive application
CATV	Community antenna television
CBEMA	Computer and Business Equipment Manufacturers Association
CBR	Constant bit rate

CBT	Computer-based training
CC	Cluster controller
CCIR	International radio consultative committee
CCIS	Common channel interoffice signaling
CCITT	International telegraph and telephone Consultative Committee
CCS	Common channel signaling
CCS	Hundred call seconds per hour
CD	Collision detection
CD	Compact disc
CDC	Control Data Corporation
CDMA	Code division multiple access
CDPD	Cellular digital packet data
CD-ROM	Compact disc-read only memory
CDVT	Cell delay variation tolerance
CEI	Comparably efficient interconnection
CEPT	Conference of European Postal and Telecommunications Administrations
CERN	European Council for Nuclear Research
CERT	Computer Emergency Response Team
CES	Circuit emulation service
CEV	Controlled environmental vault
CGI	Common gateway interface (Internet)
CHAP	Challenge handshake authentication protocol
CHL	Chain home low RADAR
CICS	Customer information control system
CICS/VS	Customer information control system/virtual storage
CID	Card identifier (RFID)
CIDR	Classless interdomain routing (IETF)
CIF	Cells in frames
CIO	Chief information officer
CIR	Committed information rate
CISC	Complex instruction set computer
CIX	Commercial internet exchange
CLASS	Custom local area signaling services (Bellcore)
CLEC	Competitive local exchange carrier
CLLM	Consolidated link layer management
CLNP	Connectionless network protocol
CLNS	Connectionless network service
CLP	Cell loss priority
CM	Communications module (Lucent 5ESS)
CMIP	Common management information protocol
CMISE	Common management information service element
CMOL	CMIP over LLC
CMOS	Complementary metal oxide semiconductor
CMOT	CMIP over TCP/IP
CMP	Communications module processor
CNE	Certified NetWare engineer
CNM	Customer network management
CNR	Carrier-to-noise ratio
CO	Central office
CoCOM	Coordinating Committee on Multilateral Export Controls
CODEC	Coder-decoder
COMC	Communications controller

CONS	Connection-oriented network service
CORBA	Common object request brokered architecture
COS	Class of service (APPN)
COS	Corporation for open systems
CPE	Customer premises equipment
CPU	Central processing unit
CRC	Cyclic redundancy check
CRM	Customer relationship management
CRT	Cathode ray tube
CRV	Call reference value
CS	Convergence sublayer
CSA	Carrier serving area
CSMA	Carrier sense multiple access
CSMA/CA	Carrier sense multiple access with collision avoidance
CSMA/CD	Carrier sense multiple access with collision detection
CSU	Channel service unit
CTI	Computer telephony integration
CTIA	Cellular telecommunications industry association
CTO	Chief technology officer
CTS	Clear to send
CU	Control unit
CVSD	Continuously variable slope delta modulation
CWDM	Coarse wavelength division multiplexing
D/A	Digital-to-analog
DA	Destination address
DAC	Dual attachment concentrator (FDDI)
DACS	Digital access and cross-connect system
DARPA	Defense advanced research projects agency
DAS	Dual attachment station (FDDI)
DAS	Direct attached storage
DASD	Direct access storage device
DB	Decibel
DBS	Direct broadcast satellite
DC	Direct current
DCC	Data communications channel (SONET)
DCE	Data circuit-terminating equipment
DCN	Data communications network
DCS	Digital cross-connect system
DCT	Discrete cosine transform
DDCMP	Digital data communications management protocol (DNA)
DDD	Direct distance dialing
DDP	Datagram delivery protocol
DDS	DATAPHONE digital service (sometimes digital data service)
DDS	Digital data service
DE	Discard eligibility (LAPF)
DECT	Digital European cordless telephone
DES	Data encryption standard (NIST)
DID	Direct inward dialing
DIP	Dual inline package
DLC	Digital loop carrier
DLCI	Data link connection identifier
DLE	Data link escape
DLSw	Data link switching

DM	Delta modulation
DM	Disconnected mode
DM	Data mining
DMA	Direct memory access (computers)
DMAC	Direct memory access control
DME	Distributed management environment
DMS	Digital multiplex switch
DMT	Discrete multitone
DNA	Digital network architecture
DNIC	Data network identification code (X.121)
DNIS	Dialed number identification service
DNS	Domain name service
DNS	Domain name system (IETF)
DOD	Direct outward dialing
DOD	Department of Defense
DOJ	Department of Justice
DOV	Data over voice
DPSK	Differential phase shift keying
DQDB	Distributed queue dual bus
DR	Data rate send (RFID)
DRAM	Dynamic random access memory
DS	Data rate send (RFID)
DSAP	Destination service access point
DSF	Dispersion-shifted fiber
DSI	Digital speech interpolation
DSL	Digital subscriber line
DSLAM	Digital subscriber line access multiplexer
DSP	Digital signal processing
DSR	Data set ready
DSS	Digital satellite system
DSS	Digital subscriber signaling system
DSSS	Direct sequence spread spectrum
DSU	Data service unit
DTE	Data terminal equipment
DTMF	Dual tone multifrequency
DTR	Data terminal ready
DVRN	Dense virtual routed networking (crescent)
DWDM	Dense wavelength division multiplexing
DXI	Data exchange interface
E/O	Electrical-to-optical
EAN	European Article Numbering System
EBCDIC	Extended Binary Coded Decimal Interchange Code
EBITDA	Earnings before interest, tax, depreciation, and amortization
ECMA	European computer manufacturer association
ECN	Explicit congestion notification
ECSA	Exchange Carriers Standards Association
EDFA	Erbium-doped fiber amplifier
EDI	Electronic data interchange
EDIBANX	EDI Bank Alliance Network Exchange
EDIFACT	Electronic Data Interchange for Administration, Commerce, and Trade (ANSI)
EFCI	Explicit forward congestion indicator
EFTA	European Free Trade Association

EGP	Exterior gateway protocol (IETF)
EIA	Electronics Industry Association
EIGRP	Enhanced interior gateway routing protocol
EIR	Excess information rate
EMBARC	Electronic mail broadcast to a roaming computer
EMI	Electromagnetic interference
EMS	Element management system
EN	End node
ENIAC	Electronic numerical integrator and computer
EO	End office
EOC	Embedded operations channel (SONET)
EOT	End of transmission (BISYNC)
EPC	Electronic product code
EPROM	Erasable programmable read-only memory
EPS	Earnings per share
ERP	Enterprise resource planning
ESCON	Enterprise system connection (IBM)
ESF	Extended superframe format
ESOP	Employee stock ownership plan
ESP	Enhanced service provider
ESS	Electronic switching system
ETSI	European Telecommunications Standards Institute
ETX	End of Text (BISYNC)
EVA	Economic value added
EWOS	European Workshop for Open Systems
FACTR	Fujitsu Access and Transport System
FAQ	Frequently asked questions
FASB	Financial Accounting Standards Board
FAT	File allocation table
FCF	Free cash flow
FCS	Frame check sequence
FDA	Food and Drug Administration
FDD	Frequency division duplex
FDDI	Fiber distributed data interface
FDM	Frequency division multiplexing
FDMA	Frequency division multiple access
FDX	Full-duplex
FEBE	Far-end block error (SONET)
FEC	Forward error correction
FEC	Forward equivalence class
FECN	Forward explicit congestion notification
FEP	Front-end processor
FERF	Far-end receive failure (SONET)
FET	Field effect transistor
FHSS	Frequency hopping spread spectrum
FIB	Forward indicator bit (SS7)
FIFO	First in first out
FITL	Fiber in the loop
FLAG	Fiber ling across the globe
FM	Frequency modulation
FOIRL	Fiber optic inter-repeater link
FPGA	Field programmable gate array
FR	Frame relay

FRAD	Frame relay access device
FRBS	Frame relay bearer service
FSDI	Frame size device integer (RFID)
FSK	Frequency shift keying
FSN	Forward sequence number (SS7)
FTAM	File transfer, access, and management
FTP	File transfer protocol (IETF)
FTTC	Fiber to the curb
FTTH	Fiber to the home
FUNI	Frame user-to-network interface
FWI	Frame waiting integer (RFID)
FWM	Four wave mixing
GAAP	Generally accepted accounting principles
GATT	General agreement on tariffs and trade
GbE	Gigabit ethernet
Gbps	Gigabits per second (billion bits per second)
GDMO	Guidelines for the Development of Managed Objects
GDP	Gross domestic product
GEOS	Geosynchronous Earth orbit satellites
GFC	Generic flow control (ATM)
GFI	General format identifier (X.25)
GFP	Generic framing procedure
GFP-F	Generic framing procedure-frame-based
GFP-X	Generic framing procedure-transparent
GMPLS	Generalized MPLS
GOSIP	Government open systems interconnection profile
GPS	Global positioning system
GRIN	Graded index (fiber)
GSM	Global system for mobile Communications
GTIN	Global trade item number
GUI	Graphical user interface
HDB3	High density, bipolar 3 (E-Carrier)
HDLC	High-level data link control
HDSL	High-bit-rate digital subscriber line
HDTV	High-definition television
HDX	Half-duplex
HEC	Header error control (ATM)
HFC	Hybrid fiber/coax
HFS	Hierarchical file storage
HIPAA	Health Insurance Portability and Accountability Act
HLR	Home location register
HPPI	High-performance parallel interface
HSDPA	High-speed downline packet access
HSPA	High-speed packet access
HSUPA	High-speed upline packet access
HSSI	High-speed serial interface (ANSI)
HTML	Hypertext markup language
HTTP	Hypertext transfer protocol (IETF)
HTU	HDSL transmission unit
I	Intrapictures
IaaS	Infrastructure-as-a-service
IAB	Internet Architecture Board (formerly Internet Activities Board)
IACS	Integrated access and cross-connect system

IAD	Integrated access device
IAM	Initial address message (SS7)
IANA	Internet address naming authority
ICMP	Internet control message protocol (IETF)
ICT	Information and communications technology
IDP	Internet datagram protocol
IEC	Interexchange carrier (also IXC)
IEC	International Electrotechnical Commission
IEEE	Institute of Electrical and Electronics Engineers
IETF	Internet Engineering Task Force
IFRB	International Frequency Registration Board
IGP	Interior gateway protocol (IETF)
IGRP	Interior gateway routing protocol
ILEC	Incumbent local exchange carrier
IM	Instant messenger (AOL)
IML	Initial microcode load
IMP	Interface message processor (ARPANET)
IMS	Information management system
InARP	Inverse address resolution protocol (IETF)
InATMARP	Inverse ATMARP
INMARSAT	International Maritime Satellite Organization
INP	Internet nodal processor
InterNIC	Internet network information center
IP	Intellectual property
IP	Internet protocol (IETF)
IPO	Initial product offer
IPX	Internetwork packet exchange (NetWare)
IRU	Indefeasible rights of use
IS	Information systems
ISDN	Integrated services digital network
ISO	International Organization for Standardization
ISO	Information Systems Organization
ISOC	Internet Society
ISP	Internet service provider
ISUP	ISDN User Part (SS7)
IT	Information technology
ITU	International Telecommunication Union
ITU-R	International Telecommunication Union-Radio Communication Sector
IVD	Inside vapor deposition
IVR	Interactive voice response
IXC	Interexchange carrier
JAN	Japanese article numbering system
JEPI	Joint electronic paynets initiative
JES	Job entry system
JIT	Just in time
JPEG	Joint Photographic Experts Group
JTC	Joint technical committee
KB	Kilobytes
Kbps	Kilobits per second (thousand bits per second)
KLTN	Potassium lithium tantalate niobate
KM	Knowledge management
LAN	Local area network

LANE	LAN emulation
LAP	Link access procedure (X.25)
LAPB	Link access procedure balanced (X.25)
LAPD	Link access procedure for the D-channel
LAPF	Link access procedure to frame mode bearer services
LAPF-Core	Core aspects of the link access procedure to frame mode bearer services
LAPM	Link access procedure for modems
LAPX	Link access procedure half-duplex
LASER	Light amplification by the stimulated emission of radiation
LATA	Local access and transport area
LCD	Liquid crystal display
LCGN	Logical channel group number
LCM	Line concentrator module
LCN	Local communications network
LD	Laser diode
LDAP	Lightweight directory access protocol (X.500)
LEAF®	Large effective area fiber® (Corning product)
LEC	Local exchange carrier
LED	Light emitting diode
LENS	Lightwave efficient network solution (centerpoint)
LEOS	Low earth orbit satellites
LER	Label edge router
LI	Length indicator
LIDB	Line information database
LIFO	Last in first out
LIS	Logical IP subnet
LLC	Logical link control
LMDS	Local multipoint distribution system
LMI	Local management interface
LMOS	Loop maintenance operations system
LORAN	Long-range radio navigation
LPC	Linear predictive coding
LPP	Lightweight presentation protocol
LRC	Longitudinal redundancy check (BISYNC)
LS	Link state
LSI	Large-scale integration
LSP	Label switched path
LSR	Label switched router
LTE	Long-term evolution (4th-generation wireless)
LU	Line unit
LU	Logical unit (SNA)
MAC	Media access control
MAN	Metropolitan area network
MAP	Manufacturing automation protocol
MAU	Medium attachment unit (ethernet)
MAU	Multistation access unit (token ring)
MB	Megabytes
MBA™	Metro Business Access™ (ocular)
Mbps	Megabits per second (million bits per second)
MD	Message Digest (MD2, MD4, MD5) (IETF)
MDF	Main distribution frame
MDU	Multi-dwelling unit

MEMS	Micro electrical mechanical system
MF	Multifrequency
MFJ	Modified final judgment
MHS	Message handling system (X.400)
MIB	Management information base
MIC	Medium interface connector (FDDI)
MIME	Multipurpose Internet mail extensions (IETF)
MIPS	Millions of instructions per second
MIS	Management information systems
MITI	Ministry of International Trade and Industry (Japan)
MITS	Micro instrumentation and telemetry systems
ML-PPP	Multilink point-to-point protocol
MMDS	Multichannel, multipoint distribution system
MMF	Multimode fiber
MNP	Microcom networking protocol
MON	Metropolitan optical network
MoU	Memorandum of understanding
MP	Multilink PPP
MPEG	Motion Picture Experts Group
MPLS	Multiprotocol label switching
MPOA	Multiprotocol over ATM
MPλS	Multiprotocol lambda switching
MRI	Magnetic resonance imaging
MSB	Most significant bit
MSC	Mobile switching center
MSO	Mobile switching office
MSPP	Multi-service provisioning platform
MSVC	Meta-signaling virtual channel
MTA	Major trading area
MTBF	Mean time between failure
MTP	Message transfer part (SS7)
MTSO	Mobile Telephone Switching Office
MTTR	Mean time to repair
MTU	Maximum transmission unit
MTU	Multi-tenant unit
MVNO	Mobile virtual network operator
MVS	Multiple virtual storage
NAD	Node address (RFID)
NAFTA	North American free trade agreement
NAK	Negative acknowledgment (BISYNC, DDCMP)
NAP	Network access point (Internet)
NARUC	National Association of Regulatory Utility Commissioners
NAS	Network attached storage
NASA	National Aeronautics and Space Administration
NASDAQ	National Association of Securities Dealers Automated quotations
NATA	North American Telecommunications Association
NATO	North Atlantic Treaty Organization
NAU	Network accessible unit
NCP	Network control program
NCSA	National Center for Supercomputer Applications
NCTA	National Cable Television Association
NDIS	Network driver interface specifications
NDSF	Non-dispersion-shifted fiber

NetBEUI	NetBIOS extended user interface
NetBIOS	Network basic input/output system
NFS	Network file system (Sun)
NIC	Network interface card
NII	National information infrastructure
NIST	National Institute of Standards and Technology (formerly NBS)
NIU	Network interface unit
NLPID	Network layer protocol identifier
NLSP	NetWare link services protocol
NM	Network module
Nm	Nanometer
NMC	Network management center
NMS	Network management system
NMT	Nordic mobile telephone
NMVT	Network management vector transport protocol
NNI	Network node interface
NNI	Network-to-network interface
NOC	Network operations center
NOCC	Network operations control center
NOPAT	Net operating profit after tax
NOS	Network operating system
NPA	Numbering plan area
NREN	National Research and Education Network
NRZ	Non-return to zero
NRZI	Non-return to zero inverted
NSA	National Security Agency
NSAP	Network service access point
NSAPA	Network service access point address
NSF	National Science Foundation
NTSC	National Television Systems Committee
NTT	Nippon Telephone and Telegraph
NVB	Number of valid bits (RFID)
NVOD	Near video on demand
NZDSF	Non-zero dispersion-shifted fiber
OADM	Optical add-drop multiplexer
OAM	Operations, administration, and maintenance
OAM&P	Operations, administration, maintenance, and provisioning
OAN	Optical area network
OBS	Optical burst switching
OC	Optical carrier
OEM	Original equipment manufacturer
O-E-O	Optical-electrical-optical
OLS	Optical line system (Lucent)
OMAP	Operations, maintenance, and administration part (SS7)
ONA	Open network architecture
ONS	Object name service
ONU	Optical network unit
OOF	Out of frame
OPEX	Operating expense
OS	Operating system
OSF	Open Software Foundation
OSI	Open systems interconnection (ISO, ITU-T)
OSI RM	Open systems interconnection reference model

OSPF	Open shortest path first (IETF)
OSS	Operation support systems
OTDM	Optical time division multiplexing
OTDR	Optical time-domain reflectometer
OUI	Organizationally unique identifier (SNAP)
OVD	Outside vapor deposition
OXC	Optical cross-connect
P/F	Poll/final (HDLC)
PaaS	Platform-as-a-service
PAD	Packet assembler/disassembler (X.25)
PAL	Phase alternate line
PAM	Pulse amplitude modulation
PANS	Pretty amazing new stuff
PBX	Private branch exchange
PCB	Protocol control byte (RFID)
PCI	Peripheral component interface
PCM	Pulse code modulation
PCMCIA	Personal Computer Memory Card International Association
PCN	Personal communications network
PCS	Personal communications services
PDA	Personal digital assistant
PDH	Plesiochronous digital hierarchy
PDU	Protocol data unit
PIN	Positive-intrinsic-negative
PING	Packet Internet groper (TCP/IP)
PKC	Public key cryptography
PLCP	Physical layer convergence protocol
PLP	Packet layer protocol (X.25)
PM	Phase modulation
PMD	Physical medium dependent (FDDI)
PML	Physical markup language
PNNI	Private network node interface (ATM)
PON	Passive optical networking
POP	Point of presence
POSIT	Profiles for open systems interworking technologies
POSIX	Portable operating system interface for UNIX
POTS	Plain old telephone service
PPM	Pulse position modulation
PPP	Point-to-point protocol (IETF)
PPS	Protocol parameter selection (RFID)
PRC	Primary reference clock
PRI	Primary rate interface
PROFS	Professional office system
PROM	Programmable read-only memory
PSDN	Packet switched data network
PSK	Phase shift keying (RFID)
PSK	Phase shift keying
PSPDN	Packet switched public data network
PSTN	Public switched telephone network
PTI	Payload type identifier (ATM)
PTT	Post, telephone, and telegraph
PU	Physical unit (SNA)
PUC	Public utility commission

PUPI	Pseudo-unique PICC identifier
PVC	Permanent virtual circuit
QAM	Quadrature amplitude modulation
Q-bit	Qualified data bit (X.25)
QLLC	Qualified logical link control (SNA)
QoE	Quality of experience
QoS	Quality of service
QPSK	Quadrature phase shift keying
QPSX	Queued packet synchronous exchange
R&D	Research & development
RADAR	Radio detection and ranging
RADSL	Rate adaptive digital subscriber line
RAID	Redundant array of inexpensive disks
RAM	Random access memory
RARP	Reverse address resolution protocol (IETF)
RAS	Remote access server
RATS	Request for answer to select (RFID)
RBOC	Regional Bell Operating Company
READ_DATA	Read data from transponder (RFID)
REQA	Request A (RFID)
REQB	Request B (RFID)
REQUEST_SNR	Request serial number (RFID)
RF	Radio frequency
RFC	Request for comments (IETF)
RFH	Remote frame handler (ISDN)
RFI	Radio frequency interference
RFID	Radio frequency identification
RFP	Request for proposal
RFQ	Request for quote
RFx	Request for X, where 'X' can be proposal, quote, information, comment, etc.
RHC	Regional Holding Company
RHK	Ryan, Hankin, and Kent (Consultancy)
RIP	Routing information protocol (IETF)
RISC	Reduced instruction set computer
RJE	Remote job entry
RNR	Receive not ready (HDLC)
ROA	Return on assets
ROE	Return on equity
ROI	Return on investment
ROM	Read-only memory
RO-RO	Roll-on roll-off
ROSE	Remote operation service element
RPC	Remote procedure call
RPR	Resilient packet ring
RR	Receive ready (HDLC)
RSA	Rivest, Shamir, and Aleman
RTS	Request to send (EIA-232-E)
S/DMS	SONET/digital multiplex system
S/N	Signal-to-noise ratio
SAA	Systems application architecture (IBM)
SAAL	Signaling ATM adaptation layer (ATM)
SaaS	Software-as-a-service

SaaS	Storage-as-a-service
SABM	Set asynchronous balanced mode (HDLC)
SABME	Set asynchronous balanced mode extended (HDLC)
SAC	Single attachment concentrator (FDDI)
SAK	Select acknowledge (RFID)
SAN	Storage area network
SAP	Service access point (generic)
SAPI	Service access point identifier (LAPD)
SAR	Segmentation and reassembly (ATM)
SAS	Single attachment station (FDDI)
SASE	Specific applications service element (subset of CASE, application layer)
SATAN	System administrator tool for analyzing networks
SBS	Stimulated Brillouin scattering
SCCP	Signaling connection control point (SS7)
SCM	Supply chain management
SCP	Service control point (SS7)
SCREAM™	Scalable control of a rearrangeable extensible array of mirrors (Calient)
SCSI	Small computer systems interface
SCTE	Serial clock transmit external (EIA-232-E)
SCUBA	Self-contained underwater breathing apparatus
SDH	Synchronous digital hierarchy (ITU-T)
SDLC	Synchronous data link control (IBM)
SDS	Scientific data systems
SEC	Securities and exchange commission
SECAM	Sequential color with memory
SELECT	Select transponder (RFID)
SELECT_ACKNOWLEDGE	Acknowledge selection (RFID)
SELECT_SNR	Select serial number (RFID)
SF	Superframe format (T-1)
SFGI	Startup frame guard integer (RFID)
SGML	Standard generalized markup language
SGMP	Simple gateway management protocol (IETF)
SHDSL	Symmetric HDSL
S-HTTP	Secure HTTP (IETF)
SIF	Signaling information field
SIG	Special interest group
SIO	Service information octet
SIP	Serial interface protocol
SIP	Session initiation protocol
SIR	Sustained information rate (SMDS)
SLA	Service level agreement
SLIP	Serial line interface protocol (IETF)
SM	Switching module
SMAP	System management application part
SMB	Small-to-medium business
SMDS	Switched multimegabit data service
SMF	Single mode fiber
SMP	Simple management protocol
SMP	Switching module processor
SMR	Specialized mobile radio

SMS	Standard management system (SS7)
SMTP	Simple mail transfer protocol (IETF)
SNA	Systems network architecture (IBM)
SNAP	Subnetwork access protocol
SNI	Subscriber network interface (SMDS)
SNMP	Simple network management protocol (IETF)
SNP	Sequence number protection
SNR	Serial number (RFID)
SOHO	Small-office, home-office
SONET	Synchronous optical network
SPAG	Standards Promotion and Application Group
SPARC	Scalable performance architecture
SPE	Synchronous payload envelope (SONET)
SPID	Service profile identifier (ISDN)
SPM	Self-phase modulation
SPOC	Single point of contact
SPX	Sequenced packet exchange (NetWare)
SQL	Structured query language
SRB	Source route bridging
SRP	Spatial reuse protocol
SRS	Stimulated Raman scattering
SRT	Source routing transparent
SS7	Signaling System 7
SSCC	Serial shipping container code
SSL	Secure socket layer (IETF)
SSP	Service switching point (SS7)
SSR	Secondary surveillance RADAR
SST	Spread spectrum transmission
STDM	Statistical time division multiplexing
STM	Synchronous transfer mode
STM	Synchronous transport module (SDH)
STP	Signal transfer point (SS7)
STP	Shielded twisted pair
STS	Synchronous transport signal (SONET)
STX	Start of text (BISYNC)
SVC	Signaling virtual channel (ATM)
SVC	Switched virtual circuit
SXS	Step-by-step switching
SYN	Synchronization
SYNTRAN	Synchronous transmission
TA	Terminal adapter (ISDN)
TA96	Telecommunications Act of 1996
TAG	Technical Advisory Group
TASI	Time assigned speech interpolation
TAXI	Transparent asynchronous transmitter/receiver interface (physical layer)
TCAP	Transaction Capabilities Application Part (SS7)
TCM	Time compression multiplexing
TCM	Trellis coding modulation
TCP	Transmission control protocol (IETF)
TDD	Time division duplexing
TDM	Time division multiplexing
TDMA	Time division multiple access

TDR	Time domain reflectometer
TE1	Terminal equipment type 1 (ISDN capable)
TE2	Terminal equipment type 2 (non-ISDN capable)
TEI	Terminal endpoint identifier (LAPD)
TELRIC	Total element long-run incremental cost
TIA	Telecommunications Industry Association
TIRIS	TI RF Identification Systems (Texas Instruments)
TIRKS	Trunk integrated record keeping system
TL1	Transaction language 1
TLAN	Transparent LAN
TM	Terminal multiplexer
TMN	Telecommunications management network
TMS	Time-multiplexed switch
TMT	Telecommunications, media, and technology
TOH	Transport overhead (SONET)
TOP	Technical and office protocol
TOS	Type of service (IP)
TP	Twisted pair
TR	Token ring
TRA	Traffic routing administration
TSI	Time slot interchange
TSLRIC	Total service long-run incremental cost
TSO	Terminating screening office
TSO	Time-sharing option (IBM)
TSR	Terminate and stay resident
TSS	Telecommunication standardization sector (ITU-T)
TST	Time-space-time switching
TSTS	Time-space-time-space switching
TTL	Time to live
TU	Tributary unit (SDH)
TUG	Tributary unit group (SDH)
TUP	Telephone user part (SS7)
UA	Unnumbered acknowledgment (HDLC)
UART	Universal asynchronous receiver transmitter
UBR	Unspecified bit rate (ATM)
UCC	Uniform Code Council
UDI	Unrestricted digital information (ISDN)
UDP	User datagram protocol (IETF)
UHF	Ultra high frequency
UI	Unnumbered information (HDLC)
UNI	User-to-network interface (ATM, FR)
UNIT™	Unified Network Interface Technology™ (Ocular)
UNMA	Unified network management architecture
UNSELECT	Unselect transponder (RFID)
UPC	Universal product code
UPS	Uninterruptable power supply
UPSR	Unidirectional path switched ring
UPT	Universal personal telecommunications
URL	Uniform resource locator
USART	Universal synchronous asynchronous receiver transmitter
USB	Universal serial bus
UTC	Coordinated universal time
UTP	Unshielded twisted pair (physical layer)

UUCP	UNIX-UNIX copy
VAN	Value-added network
VAX	Virtual address extension (DEC)
vBNS	Very high-speed backbone network service
VBR	Variable bit rate (ATM)
VBR-NRT	Variable bit rate-non-real-time (ATM)
VBR-RT	Variable bit rate-real-time (ATM)
VC	Venture capital
VC	Virtual channel (ATM)
VC	Virtual circuit (PSN)
VC	Virtual container (SDH)
VCC	Virtual channel connection (ATM)
VCI	Virtual channel identifier (ATM)
VCI	Virtual channel identifier (ATM)
VCSEL	Vertical cavity surface emitting laser
VDSL	Very high-speed digital subscriber line
VDSL	Very high bit rate digital subscriber line
VERONICA	Very easy rodent-oriented netwide index to computerized archives (Internet)
VGA	Variable graphics array
VHF	Very high frequency
VHS	Video home system
VID	VLAN ID
VIN	Vehicle identification number
VINES	Virtual networking system (Banyan)
VIP	VINES Internet protocol
VLAN	Virtual LAN
VLF	Very low frequency
VLR	Visitor location register (wireless)
VLSI	Very large scale integration
VM	Virtual machine (IBM)
VM	Virtual memory
VMS	Virtual memory system (DEC)
VOD	Video-on-demand
VP	Virtual path
VPC	Virtual path connection
VPI	Virtual path identifier
VPN	Virtual private network
VR	Virtual reality
VSAT	Very small aperture terminal
VSB	Vestigial sideband
VSELP	Vector-sum excited linear prediction
VT	Virtual tributary
VTAM	Virtual telecommunications access method (SNA)
VTOA	Voice and telephony over ATM
VTP	Virtual terminal protocol (ISO)
WACK	Wait acknowledgment (BISYNC)
WACS	Wireless access communications system
WAIS	Wide area information server (IETF)
WAN	Wide area network
WAP	Wireless application protocol (wrong approach to portability)
WARC	World administrative radio conference
WATS	Wide area telecommunications service

WDM	Wavelength division multiplexing
WIN	Wireless in-building network
WISP	Wireless ISP
WTO	World Trade Organization
WWW	World Wide Web (IETF)
WYSIWYG	What you see is what you get
xDSL	x-type digital subscriber line
XID	Exchange identification (HDLC)
XML	Extensible markup language
XNS	Xerox network systems
XPM	Cross-phase modulation
ZBTSI	Zero-byte time slot interchange
ZCS	Zero-code suppression

Dr. Steven Shepard

Dr. Steven Shepard is the founder of the Shepard Communications Group in Williston, Vermont, co-founder of the Executive Crash Course Company, and founder of Shepard Images. A professional author, photographer, and educator with more than 30 years of experience in the technology industry, he has written books and articles on a wide variety of topics. His books include

- *Commotion in the Ocean: A Technical Commercial Diving Manual.* National Association of Underwater Instructors, 1979.
- *A Matter of Last Resort: The Story of Byron Hot Springs.* Contra Costa County Historical Society, 1987.
- *Managing Cross-Cultural Transition: A Handbook for Corporations, Employees, and Their Families.* Aletheia Publications, New York, 1997.
- *Telecommunications Convergence: How to Profit from the Convergence of Technologies, Services, and Companies.* McGraw-Hill, New York, 2000.
- *A Spanish-English Telecommunications Dictionary.* Shepard Communications Group, Williston, Vermont, 2001.
- *An Optical Networking Crash Course.* McGraw-Hill, New York, 2001.
- *SONET and SDH Demystified.* McGraw-Hill, New York, 2001.
- *Telecom Crash Course.* McGraw-Hill, New York, 2001.
- *Telecommunications Convergence,* 2d Ed. McGraw-Hill, New York, 2002.
- *Videoconferencing Demystified.* McGraw-Hill, New York, 2002.
- *Metro Networking Demystified.* McGraw-Hill, New York, 2002.
- *RFID Demystified.* McGraw-Hill, New York, 2004.
- *Telecom Crash Course,* 2d Ed. McGraw-Hill, New York, 2005.
- *VoIP Crash Course.* McGraw-Hill, New York, 2005.
- *IMS Crash Course.* McGraw-Hill, New York, 2006.
- *WiMAX Crash Course.* McGraw-Hill, New York, 2006.
- *How to do Everything with VoIP.* McGraw-Hill, New York, 2006.
- *Managing Supply Chain Technology.* Aspatore Publishing, 2006.
- *Using VoIP to Empower Your Business.* Business Week, 2007.
- *The Telecom Economy: Charting a Path in Uncertain Times.* McGraw-Hill, New York, 2008.
- *Road Scholar: How to Start and Operate an Independent Consultancy.* Executive Crash Course Press, 2009.

- *World View: Images from a Life of Travel.* See Life Productions, Incorporated.
- *Reverse Engineering the Future: A Prescription for Change Leadership.* Executive Crash Course Press, 2009.
- *A Year in Southridge Meadow.* See Life Productions, Incorporated, 2010.
- *Giving Up Control: Strategies for Success in the User-Generated Economy.* Executive Crash Course Press, 2010.
- *The Deliberate Photographer.* Shepard Communications Group, 2012.
- *The Deliberate Storyteller.* Shepard Communications Group, 2012.
- *Passage to Burma.* A photojournalistic collaboration with Scotty Stulberg, 2012.
- *Whatever Happened to Mister Duncan: Mapping the Memory of Childhood.* Shepard Communications Group, 2013.
- *Telecom Crash Course: A Primer for the Telecom, Media, and Technology Sectors.* McGraw-Hill, New York, 2014.

For examples of Steve's video presentations, including his recent TED talk, please visit *http://www.youtube.com/user/TheShepardComm.*

Dr. Shepard received his undergraduate degree in Spanish and Romance Philology from the University of California at Berkeley (1976), his Masters Degree in International Business from St. Mary's College (1985), and his PhD at the Da Vinci Institute in Rivonia, South Africa (2009). He spent 11 years with Pacific Bell in San Francisco in a variety of capacities followed by 10 years with Hill Associates in Colchester, Vermont before forming the Shepard Communications Group in early 2000. He is a senior fellow of the Da Vinci Institute of South Africa; adjunct professor in the University of Southern California's Marshall School of Business; a founding director of the African Telecoms Institute; former chairman of the Vermont Telecommunications Authority, the organization tasked with managing the distribution of broadband stimulus money in the State of Vermont; member of the board of advisors of the Telecommunications Industry Association (TIA); and an Emeritus member of the Board of Trustees of Champlain College. He is also the resident director of the University of Southern California's Executive Leadership and Advanced Management Programs, and adjunct professor at Thunderbird University, the University of Vermont, Wharton University, Emory University, The Ivey School, Champlain College, and St. Michael's College.

Dr. Shepard specializes in international issues in technology with an emphasis on the social implications of technological change, technology infrastructure development, strategy creation, technical marketing and strategic technical sales; the deployment of social media as an organizational collaboration tool; the development of multilingual educational programs; and, through the Executive Crash Course Company, the effective use of multimedia. He has written and directed more than 40 videos and films and written technical presentations on a broad range of topics for companies and organizations worldwide. He has written and photographed in more than 90 countries, serving clients across many different industries including telecommunications, IT, media, advertising, healthcare, and government, to name a few. He is fluent in Spanish and routinely publishes and delivers presentations in that language. Global clients include major telecommunications manufacturers, service providers, software development firms, multinational corporations, universities, professional services firms, NGOs, advertising firms, venture capital firms, and regulatory bodies.

He lives in Vermont with his wife Sabine, who has put up with him for more than 30 years.

Index

J

K

L

M

N

Q

R

U

V

W

CPSIA information can be obtained
at www.ICGtesting.com
Printed in the USA
BVHW040424090519
547751BV00007B/87/P

9 780071 832663